5th Global Conference on Polymer and Composite Materials (PCM 2018)

IOP Conference Series: Materials Science and Engineering
Volume 369

Kitakyushu, Japan
10 – 13 April 2018

ISBN: 978-1-5108-6484-9
ISSN: 1757-8981

Printed from e-media with permission by:

Curran Associates, Inc.
57 Morehouse Lane
Red Hook, NY 12571

CURRAN ASSOCIATES INC. proceedings.com

Some format issues inherent in the e-media version may also appear in this print version.

Printed by Curran Associates, Inc. (2018)

For permission requests, please contact the Institute of Physics at the address below.

Institute of Physics
Dirac House, Temple Back
Bristol BS1 6BE UK

Phone: 44 1 17 929 7481
Fax: 44 1 17 920 0979

techtracking@iop.org

Additional copies of this publication are available from:

Curran Associates, Inc.
57 Morehouse Lane
Red Hook, NY 12571 USA
Phone: 845-758-0400
Fax: 845-758-2633
Email: curran@proceedings.com
Web: www.proceedings.com

Table of contents

Volume 369

2018 5th Global Conference on Polymer and Composite Materials (PCM 2018)

10–13 April 2018, Kitakyushu City, Japan

Accepted papers received: 21 May 2018
Published online: 12 June 2018

IOP Conf. Series: Materials Science and Engineering **369** (2018) 011001 doi:10.1088/1757-899X/369/1/011001

PCM 2018 Preface

The *5th Global Conference on Polymer and Composite Materials (PCM 2018)* was held from April 10th -13th, 2018, in Kitakyushu City, Kokura region, Japan. The conference, with 187 participants, had a technical program consisting of 3 international keynote speakers, 120 oral presentations, and a poster session with 53 works. The conference had parallel sessions with dedicated selected subjects related to Polymers and Composite Materials.

The present issue of IOP Conference Series: Materials Science and Engineering (MSE) records the proceedings of PCM 2018 and contains 46 specially selected manuscripts submitted to the conference. The electronic submission and handling of manuscripts via the conference website, including the selection of reviewers and evaluation of manuscripts, were identical to the procedures applied to manuscripts submitted as regular contributions for international per-reviewed publications.

The organization of this conference and the preparation of proceedings volume would have been impossible without the tremendous efforts and dedication of many individuals, especially from Ms. Yajun (Yoko) Ye, who oversaw the organization of the conference and the program, and a large team of reviewers with their timely submission of quality reports. We express our sincere thanks to all authors and presenters for their contributions. We also thank very much to the sponsors from local companies, universities and academic association close to Kitakyushu, they are Kitakyushu Convention Bureau, Kyusyu Institute of Technology, Fukuoka University, University of the Ryukyus, Oita University and The Japan Society of Materials Science (Kyushu Branch), Hinode Ltd., Yaskawa Electric Corporation, Shiromizu Iron Works Co. Ltd., Japan Casting & Forging Corp., Toyota Motor Kyushu, Inc., Shin-Nippon Nondestructive Inspection Co. Ltd., Kumagaigumi Co. Ltd., Takagi Co. Ltd., Kitakyushu Convention & Visitors Association.

The 6th Global Conference on Polymer and Composite Materials (PCM 2019) will be held during May 22nd-25th, 2019, in Bangkok, Thailand. Located in the Chao Phraya River delta in Thailand's central plains, the capital of Thailand is a large city known for its heavily decorated architecture and vibrant street life. Bangkok is home of many opulent temples, like the Grand Palace, the sacred Wat Phra Kaew, and Wat Pho with an enormous reclining Buddha. Thailand is the second-largest economy in Southeast Asia, and recognized for adopting sophisticated, cutting-edge technologies for use in key export industries like electronics, automobile manufacturing, polymers, and food production. Furthermore, Thai food is famous around all the world due to a vast menu of appetizing dishes, being Bangkok cuisine notorious by its Teochew (Chinese) and Portuguese influences. All participants are welcome to the renewed and vibrant PCM 2019! More information will be available at the conference website: http://www.cpcmconf.org.

Prof. Dr. Esteban Broitman
General Chairman PCM2018
SKF Research & Technology Development, Netherlands, May 2018

Peer review statement

All papers published in this volume of *IOP Conference Series: Materials Science and Engineering* have been peer reviewed through processes administered by the proceedings Editors. Reviews were conducted by expert referees to the professional and scientific standards expected of a proceedings journal published by IOP Publishing.

IOP Conf. Series: Materials Science and Engineering **369** (2018) 012001 doi:10.1088/1757-899X/369/1/012001

Production of al-si alloy feedstocks using the solvent hot mixing method

J Q Ni[1], K Q Han[1,2] and M H Yu[1,2]

[1]State Key Laboratory for Modification of Chemical Fibers and Polymer Materials, College of Materials Science and Engineering, Donghua University, Shanghai 201620, P R China

E-mail: hankeqing@dhu.edu.cn/ yumuhuo@dhu.edu.cn

Abstract. Powder injection molding is a promising low-cost technique for net shape processing of metal and ceramic components. This study aimed to investigate a new method for preparing aluminium (Al) – silicon (Si) alloy feedstock using the solvent hot mixing process. For this purpose, micron-sized Al-Si (20 wt. %) alloy powder was mixed with a binder consisting of 55 wt. % carnauba wax, 45 wt. % high-density polyethylene, and 3 wt. % stearic acid in a hot xylene bath. The scanning electron microscopy technique, thermogravimetric analysis, density measurement and torque measurements were used to verify the homogeneity of the feedstock. Moreover, the feedstock was chosen to perform the molding, debinding cycle and sintering. An Al-Si (20 wt. %) alloy part was successfully produced using this new method.

1. Introduction

Aluminum (Al) and its alloys are widely used in traffic, industrial production, electronics, communication, aeronautics, and military fields because of its superior properties such as low density, excellent specific mechanical properties, good resistance to stress corrosion, and high thermal and electrical conductivity [1-4]. Hypereutectic Al-silicon (Si) alloys and its composites have attracted much attention in electronic packaging applications due to their low density, high wear resistance, low thermal expansion coefficient and excellent thermal conductivity [5,6]. The requirement for packing materials is increasing, as electronic packaging has become an emerging technology featuring smaller size, higher integration, and complex geometries. Efforts are ongoing to fabricate hypereutectic Al-Si alloy and its composites using different technologies. Sumitomo Electric Industries, Japan, has developed an Al (60 wt.%) and Si (40 wt.%) composite for electronic packaging using the traditional powder metallurgy technology [7]. Hogg investigated the microstructure of a spray Al-Si (30 wt.%) alloy used in electronic packaging applications [8]. Zhang produced a 70 vol.% SiCp/Al-Si (12 wt.%) composite for electronic packaging using pressure infiltration method [9]. Besides, Liang Tian produced an Al-calcium (Ca) composite used in national high-voltage system using the powder metallurgy and severe deformation method. This method can produce an extremely smooth interface, which is free of cracks and pores [10,11].

The powder injection moulding (PIM) is a near net shape technology combining the advances of conventional powder technology and plastic injection molding to mass fabricate small, complex, precision metal or ceramic parts [12-14]. This potential and attractive technology can be used for fabricating miniature and complex metal and ceramic packing materials. We would investigate the process of Al-Si (20 wt.%) alloy parts by PIM technology, and this study is the first part: preparing the feedstocks. In general, the PIM process consists of four main steps: feedstock preparation, molding, debinding, and sintering [12,15]. In the feedstock preparation process, conventionally, the binder and the powder material are mixed thoroughly with a mechanical mixer at a temperature above the higher melting component of the binders [14-16]. It could realize satisfactory mixing performance, and obtain homogeneous feedstock. However, it needs complex mixer equipment and a high mixing temperature.

In this study, the Al-Si alloy feedstock was prepared using solvent hot mixing with a xylene solvent bath to simplify the mixing process of feedstock and reduce the mixing temperature. The proposed binder system was dissolved in a 110℃ xylene bath and mixed using an electric stirring bar. Then, xylene was removed through distillation to obtain the feedstock. The scanning electron microscopy (SEM) technique, thermogravimetric (TG) analysis, density measurement, and torque measurements were used to verify the homogeneous of feedstock. Moreover, debinding and sintering were also conducted to show the feasibility of feedstock produced by solvent hot mixing.

2. Experiments

2.1. Materials

In this study, commercial air atomised Al-Si (20 wt.%) alloy powder having a mean particle size of 6.67 μm was used. The shape and morphology of the alloy powder are shown using the SEM micrograph in figure 1. The granule size distribution measured using a laser granulometry (BT-9300H) is shown in figure 2. Small amounts of pure magnesium (Mg) and tin (Sn) powders were used as sintering aids. Mg could react with the oxide, inducing an oxide film fracture and breakup [17-19]. Sn was a beneficial additive for the enhanced liquid-phase sintering of aluminium and its alloy. A trace amount of Sn could improve the wetting characteristics and behaviour of aluminium [20,21]. A multi-component binder system comprising 45 wt.% high-density polyethylene (HDPE), 52 wt. % carnauba wax (CW), and 3 wt.% stearic acid (SA) was used (table 1). HDPE was a minor binder component, which retains the form of feedstocks after solvent debinding and prior to sintering. CW was a major binder component, which could reduce the viscosity of the molding feedstock and promote flowability. SA served as a surfactant, which improved the interaction between the power and the binder.

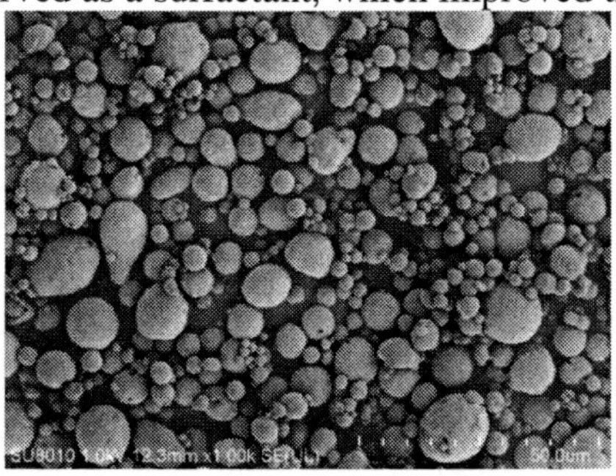

Figure 1. SEM micrograph of original Al-Si alloy powder.

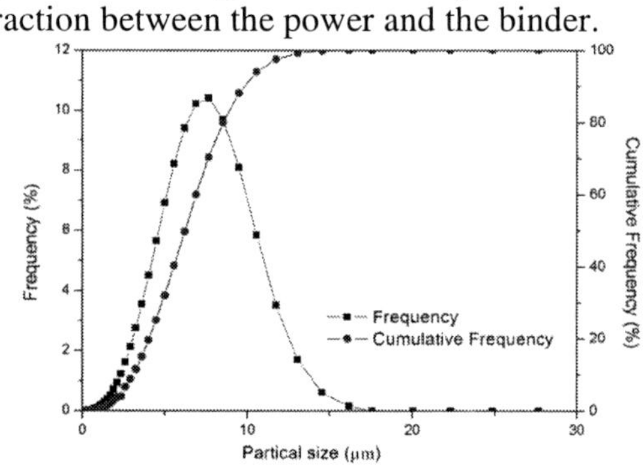

Figure 2. Particle size distributions of Al-Si alloy powder.

Table 1. Characteristics of binder components used in this study.

Binder constituent	Supplier	Density(g/cm^3)	Melting point (℃)
High density polyethylene (HDPE)	Exxon Mobil	0.956	142
carnauba wax (CW)	LUSEN Chemical	0.990-0.999	82-86
stearic acid(SA)	JINDA shuangpeng oil company	0.9408	56-69.6

2.2. Methods
In total, 85 wt. % of powder loading (82 wt.% Al-Si alloy, 2 wt.% Sn, and 1 wt.% Mg), and the binder is the rest 15 wt.%. The feedstock was prepared in a flask containing an appropriate amount of xylene at 110℃. At this temperature, the respective components of the binder could dissolve in xylene. At first, all of the binder components were added into the xylene solvent, and the solution was stirred at 300 rpm at 110℃ using a stainless steel stirring bar. Until all binder components completely dissolved, the metal powder was gradually added into the solution and continually stirred for an additional 2 hours to homogeneous the mixture. Finally, the xylene was distilled off to obtain the feedstock.

The micromorphology of the feedstock using a SEM (Hitachi SU-8010), torque measurement usingg a rheometer (ARES—RFS), the TG analysis using a thermal analyzer (TGA, Q5000IR), and the density measurements using a densitometer (Quantachrome UPY-20T) were performed to study the homogeneity of the feedstock.

The feedstocks were injection molded into tensile test bars using a LX-MIM120 injection molder. Solvent debinding was performed in xylene at 60℃ for 48 h. Thermal debinding and sintering were conducted in a thermal schedule as shown in table 2 in a glass tube furnace. The heating process was designed on the basis of TG analysis results of the feedstock. The furnace was evacuated to less than 5.0×10^{-2} torr before commencing heating. Then, nitrogen was backfilled at 100℃ and the nitrogen gas flow was maintained at 0.5 L/min for debinding and sintering. Some sacrificial Mg powders were placed beside the samples to scavenge oxygen.

3. Results and discussion

3.1. Homogeneity analysis
The homogeneity of the feedstock influenced the injection mouing and sintering process[16,22]. Therefore, the homogeneity of the feedstock was verified by observing its micromorphology, torque value, density and TG analysis.

The SEM micrograph of the feedstock in figure 3 illustrates that the powder and binder were uniformly mixed in this feedstock without any obvious bubble and hole.

Table 2. Thermal debinding and sintering schedule.

Stage	Heating rate (°C /min)	Debinding/ sintering temperature (°C)	Hold time (min)
1	1	ambient temperature to 200	0
2	0.5	200 to 500	1
3	0.8	500 to 550	180
4	furnace cooling	ambient temperature	0

Supati [23] observed variation in the torque value of the feedstock during dynamic rheology measurement to predict its homogeneity. If the torque of the feedstock had a steady value, uniform mixing was considered to occur [23]. As shown in figure 4, the torque dropped quickly at the beginning of the test, which, according to Ghanbari [22] was due to the loss of feedstock structure, resulting in the reduction of viscosity. Then, the torque remained stable. The stability of the torque meant the stability of the viscosity, and the homogenous of the feedstock.

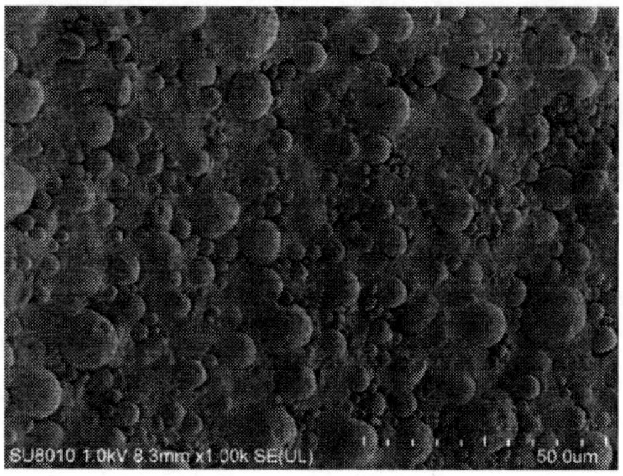

Figure 3. SEM micrograph of feedstock produced by solvent hot mixing.

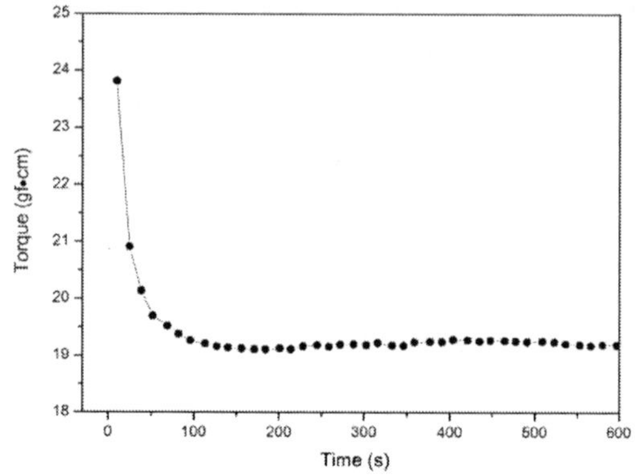

Figure 4. Torque rheological diagram of the feedstock at 160℃.

Figure 5 shows a comparison of the measured densities of the feedstock calculated using a nitrogen pycnometer. The standard variation of the pycnometer density measured in five random samples was extremely small, indicating a thorough homogeneity of the prepared feedstocks. Figure 6 shows the thermograms of the feedstocks measured in five random samples. The five curves displayed the same behavior and nearly overlapped. This phenomenon verified the excellent homogeneity of the feedstock again.

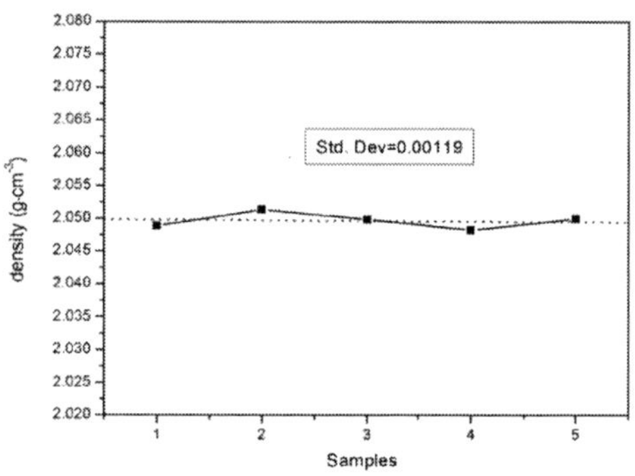

Figure 5. Densities of randomly selected portions of the feedstock.

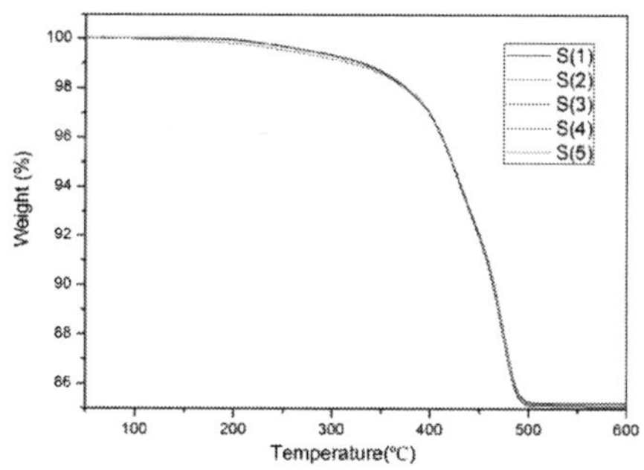

Figure 6. Tg analysis of randomly selected portions of the feedstock. The heating rate was 10℃/min.

3.2. Debinding and sintering

The subsequent injection molding and sintering progress was conducted to investigate the feasibility of the feedstocks during the process. The mold injection sample called "green sample" is shown in figure 7(a), which illustrates that the green sample had no porosity and cracks on the surface. The mold part had remarkable properties.

IOP Conf. Series: Materials Science and Engineering **369** (2018) 012001 doi:10.1088/1757-899X/369/1/012001

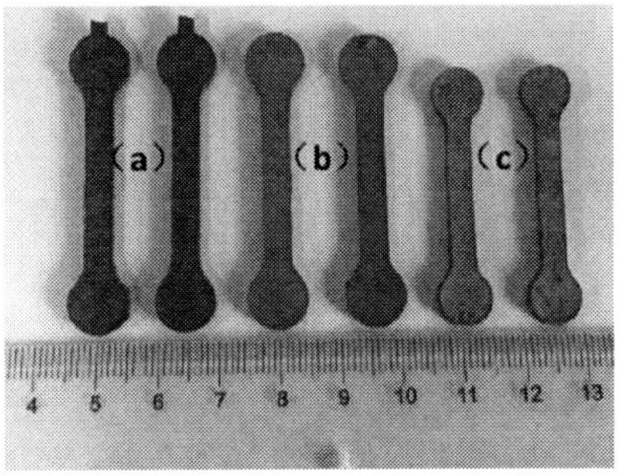

Figure 7. (a) Green sample, (b) solvent debinding sample, and (c) sintered sample.

The subsequent progress was binder removal. Traditionally, a combination of solvent and thermal debinding process was used for this[14,24]. Solvent debinding was conducted by immersing the parts into a bath with xylene at 60℃ for 48 h; in general 8.20 wt.% of the feedstock could be extracted, which is approximately 99.40 wt.% of the soluble ingredients of binder. After solvent debinding, the green samples were free of defects and chosen for sintering. The solvent debinding sample is shown in figure 7(b).

Thermal debinding and sintering were conducted in a single heating sequence. Knowledge of the thermogram for the solvent debinding feedstock allowed one to predict the thermal decomposition behavior of the binder in the feedstock. Figure 8 shows the weight loss when the solvent debinding feedstock was heated at 10℃/min in nitrogen. Binder decomposition started at 300℃, and the dissociation speed was low, between 300℃ and 400℃. When the temperature was more than 400℃, the degradation rate obviously increased followed by a rigorous weight loss to about 92.6 wt.% at about 490℃ and the binder was completely decomposed. This indicated that about 8.2 wt.% binder was extracted after solvent debinding. The result was consistent with the thermograms of "green samples".

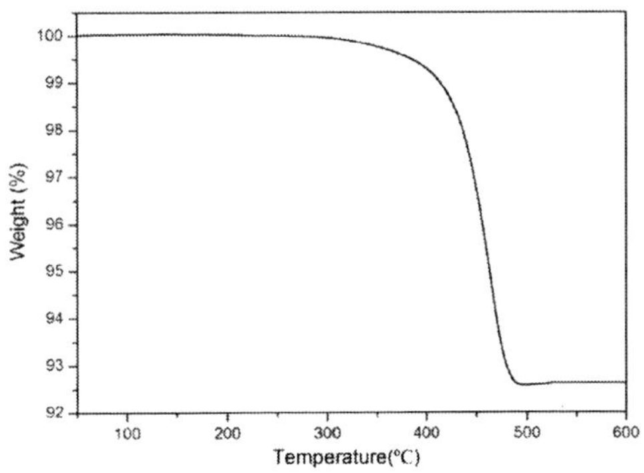

Figure 8. TG analysis of the solvent debinding feedstock.

Based on the thermogram of the solvent debinding feedstock and taking into account that high heating rates introduced cracks and blisters into the samples, the debingding and sintering thermal cycle was established, shown in table 2. Figure 7(c) shows a sintered part. The sintered part maintained the shape well and had a uniform shrinkage.

4. Conclusion

A homogeneous Al-Si (20 wt.%) alloy + Sn (2 wt.%) + Mg (1 wt.%) feedstock with HDPE, CW, and SA as the binders was manufactured using the solvent hot mixing method. The SEM, TG analysis, density measurement, and torque measurements verified the high homogeneity of the feedstock. An Al-Si (20 wt.%) alloy + Sn (2 wt.%) + Mg (1 wt.%) part was successfully produced by a process similar to PIM. The part exhibited satisfactory geometric stability and was free of distortion and macrocracks.

References

[1] Akhlaghi F and Zare-Bidaki 2009 *Wear* **266** 37-45
[2] Gökçe A, Fındık F and Kurt A O 2011 *Mater. Charact.* **62** 730-5
[3] Rahimian M, Ehsani N, Parvin N, Baharvandi H R 2009 *J. Mater. Process. Technol.* **209** 5387-93
[4] Tatar C and Özdemir N 2010 *Physica B: Condensed Matter* **405** 896-9
[5] Wang F, Xiong B, Zhang Y, Zhu B, Liu H and Wei Y 2008 *Mater. Charact.* **59** 1455-7
[6] Zhang Q, Wu G, Jiang L and Chen G 2003 *Mater Chem & Physics* **82** 780-5
[7] Chen Y and Chung D 1994 *J Mater Sci* **29** 6069-75
[8] Hogg S, Lambourne A, Ogilvy A and Grant P 2006 *Scripta Mater.* **55** 111-4
[9] Zhang Q, Xiu Z Y, Song M H and Wu G H 2005 *Mater. Sci. Forum, Trans Tech Publ* **88** 1-4
[10] Liang T, Russell A, Riedemann T, Mueller S and Anderson I 2017 *Mater. Sci. & Eng. A* **690** 348-54
[11] Tian L, Anderson I, Riedemann T, Russell A and Kim H 2013 *Elect. Power Sys. Res.* **105** 105-14
[12] Saidin H and Azuddin M 2014 *Appl. Mechan. & Mater.* **465-466** 1250-4
[13] Tay B Y, Loh N H, Tor S B, Ng F L, Fu G and Lu X H 2009 *Powder Technol.* **188** 179-82
[14] Thomas-Vielma P, Cervera A, Levenfeld B and Várez A 2008 *J. Eur. Ceram. Soc.* **28** 763-71
[15] Abdoos H, Khorsand H and Yousefi A A 2014 *Iranian Polymer J.* **23** 745-55
[16] Tan L K and Ma J 2004 *The Ninth Intersociety Conference on Thermal and Thermomechanical Phenomena in Electronic Systems* **451** 451-4
[17] Kimura A, Shibata M, Kondoh K, Takeda Y, Katayama M, Kanie T and Takada H 1997 *Appl. Phys. Lett.* **70** 3615-7
[18] Macaskill I A, Donaldson I W and Bishop D P 2013 *Powder Metall.* **49** 314-22
[19] Pieczonka T, Schubert T, Baunack S and Kieback B 2008 *Materials Science & Engineering A* **478** 251-6
[20] Liu Z Y, Sercombe T B and Schaffer G B 2008 *Powder Metall.* **51** 78-83
[21] Schaffer G B, Sercombe T B and Lumley R N 2001 *Mater. Chem. & Physics* **67** 85-91
[22] Ghanbari A, Alizadeh M, Ghasemi E, Rad R Y and Ghaffari S 2015 *Sci. & Eng. of Comp. Mater.* **22** 549-54
[23] Supati R, Loh N H, Khor K A and Tor S B 2000 *Mater. Lett.* **46** 109-14
[24] Liu Z Y, Kent D and Schaffer G B 2009 *Mater. Sci. & Eng. A s* **513–514** 352-6

IOP Conf. Series: Materials Science and Engineering **369** (2018) 012002 doi:10.1088/1757-899X/369/1/012002

Tailorable drug capacity of dexamethasone-loaded conducting polymer matrix

K Krukiewicz[1,2,3]

[1]Department of Physical Chemistry and Technology of Polymers, Faculty of Chemistry, Silesian University of Technology, M. Strzody 9, 44-100 Gliwice, Poland
[2]Centre for Research in Medical Devices, National University of Ireland, Galway, Newcastle Road, Galway, Ireland

E-mail: katarzyna.krukiewicz@polsl.pl

Abstract. The unique properties of conducting polymers, which are in the same time biocompatible and electrically responsive materials, make them perfect candidates for controlled drug release systems. In this study, the electrically-triggered controlled release system based on dexamethasone-loaded poly (3, 4-ethylenedioxypyrrole) (PEDOP) matrix is described. It is shown that the electropolymerization conditions can facilitate or suppress the formation of PEDOP/Dex matrix, as well as they can have the effect on its electrochemical performance. The release experiments performed in three different modes show that the drug capacity of PEDOP matrix increases with the increase in Dex concentration in the step of matrix synthesis, and higher Dex concentrations make it easier to control the amount of Dex released in an electrically-triggered mode. These results confirm the importance of the careful optimization of immobilization conditions to maximize drug capacity of matrix and maintain its electrochemical properties.

1. Introduction

Apart from their unquestionable success in organic electronics, conducting polymers have been recently discovered as robust and efficient materials for biomedical engineering [1,2]. The major advantage of this type of materials lies in the fact that together with proven biocompatibility [3]. They are able to change their physicochemical properties as the result of electrical stimulus [4]. Those unique properties make conducting polymer perfect candidates for the electrically-triggered controlled drug release systems. This has been already confirmed by numerous literature reports describing the immobilization and release of anti-inflammatory drugs [5], antibiotics [6] and anti-cancer agents [7,8].

Dexamethasone (Dex) is a highly potent glucocorticoid drug with a strong anti-inflammatory and immunosuppressant effect, known from its efficiency against cerebral edema. Therefore, it is frequently used to minimize the negative effects of the surgical procedures occurring within the brain tissue, including implantation of neural electrodes [9]. To prevent the negative effects resulting from the exposition of the whole body to high dosages of dexamethasone, local delivery of drug at the implant / tissue interface appears as a promising option. As summarized by Aqrawe *et al* [10], this concept can be easily implemented by the application of drug-loaded conducting polymer matrix as a bioactive coating for the neural electrodes, providing the simultaneous decrease in the risk of inflammatory response and enhancement in the electrical communication between neural tissue and electrode.

Several conducting polymers, including polypyrrole and polyterthiophene, have been already applied as the carriers for dexamethasone. Polypyrrole, one of the most extensively researched drug-carrier among conducting polymers, was found to possess low drug capacity in the range from 6 μg/cm^2 [11] to 16 μg/cm^2 [12]. Stevenson *et al* [13] showed that the amount of dexamethasone released from polyterthiophene exceeded the therapeutic levels, reaching up to 80 μg/cm^2. Due to its chemical structure derived from polypyrrole and the regularity of structure (polymerization in the 2- and 5-position) resembling polyterthiophene, poly (3, 4-ethylenedioxypyrrole) (PEDOP), is supposed to possess superior physicochemical properties and higher drug capacity than both aforementioned compounds [14]. This is why PEDOP is believed to be a promising candidate in a role of a controlled drug delivery matrix, especially for the negatively charged dexamethasone.

Recently, it has been shown that PEDOP can serve as an efficient drug-carrier for ibuprofen [15]. Ibuprofen was, however, treated more like a model drug than a real therapeutic agent. In this paper, the more potent drug, dexamethasone, is chosen in order to ensure than the released amounts of drug will have the desired therapeutic effects. It is presented how through the careful choice of the electropolymerization conditions, mainly the concentration of drug in electrolytic solution, the drug capacity and the release modes of conducting polymer matrices can be tailored. Dexamethasone was immobilized through the *in situ* method basing on cyclic voltammetry. The obtained PEDOP/Dex matrices were analyzed in terms of their electrochemical performance (charge storage capacity), surface morphology, surface chemistry and drug capacity, showing the high variability and tunability of these properties. Release studies were performed for three different modes, i.e. Dex was spontaneously eluted, released potentiodynamically and potentiostatically.

2. Materials and methods

2.1. Materials

3, 4-ethylenedioxypyrrole, EDOP (2% w/v in THF, Sigma-Aldrich), Dexamethasone 21-phosphate disodium salt, Dex (≥ 98%, Sigma-Aldrich), potassium chloride (BioReagent, ≥ 99.0%, Sigma Aldrich), sodium chloride (BioReagent, ≥ 99.0%, Sigma Aldrich), disodium hydrogen phosphate (BioReagent, ≥ 99.0%, Sigma Aldrich), potassium dihydrogen phosphate (99.0%, Sigma Aldrich) were used as received. Grade 1 (R > 10 MΩ·cm^{-1}) deionized water was employed as solvent.

2.2. Drug immobilization

CH Instruments 620a Electrochemical Workstation was used to perform electrochemical immobilization procedure in a three-electrode set-up consisting of a Pt foil working electrode (1 cm^2, 0.5 mm thickness, 99.99% trace metals basis), Ag/AgCl reference electrode and a Pt foil counter electrode. The process of drug immobilization was based on electrochemical polymerization of monomer, EDOP (5 mM) in phosphate buffer saline solution (PBS, 0.15 M NaCl, 0.0025 M KCl, 0.01 M Na_2HPO_4, 0.002 M KH_2PO_4, pH = 7.4) containing dexamethasone 21-phosphate disodium salt (Dex) (0 mM – 100 mM). The polymerization was performed in the potential range from -0.3 V to 0.8 V (vs. Ag/AgCl) within 25 CV cycles at a scan rate of 0.1 V/s. This potential range has been found to be non-destructive for dexamethasone.

2.3. Matrix characterization

The voltammetric experiments were performed using CHI620a in the same three-electrode electrochemical system as described earlier. Cyclic voltammetric (CV) scans were carried out in PBS solution, in the potential range from -0.3 V to 0.8 V (vs. Ag/AgCl) at a scan rate of 0.1 V/s during 3 CV cycles. Cyclic voltammograms were used to determine charge storage capacity (CSC), calculated as the electric charge integrated under corresponding CV curve during one CV cycle [16]:

$$CSC = \int_{t1}^{t2} I(t)dt$$

where t_1 is the beginning of CV cycle, t_2 is the end of CV cycle, I is the current.

The chemical structure of PEDOP and PEDOP/Dex was characterized using IR Perkin Elmer Spectrum Two spectrometer. IR spectra of Dex, PEDOP and PEDOP/Dex were acquired in the range between 1200 and 1750 cm^{-1} using Diamond UATR accessory. The surface morphology of samples was studied using a Phenom Pro-X scanning electron microscope operating at 10 kV.

2.4. Drug release

Dex release from PEDOP matrix was carried out in PBS solution. Drug concentration was determined through time-resolved UV/Vis spectroscopy (HP 8453 Spectrophotometer) with the use of a calibration curve plotted for absorbance at λ_{max} = 240 nm versus drug concentration ($y = 8.43 \cdot x$, R^2 = 0.991). Open circuit conditions were used for the spontaneous release process, while electrically-triggered drug release was performed by applying 5 CV cycles between -0.5 V and 0.3 V (vs. Ag/AgCl) at a scan rate of 0.1 V/s (potentiodynamic release) and a reduction potential of -0.5 V (vs. Ag/AgCl) for 600 s (potentiostatic release). The range of potentials has been determined basing on the optimization procedure for PEDOP [15], which showed the highest drug release occurring at -0.5 V.

3. Results and discussion

3.1. Electrochemical performance

As it has been shown by Svirskis *et al* [17], the *in situ* method of drug immobilization, in which conducting polymer matrix is electrochemically polymerized from the solution containing drug molecules, is often applied in the formation of drug-loaded conducting materials. If the drug-to-be-immobilized is of an ionic nature, preferably anionic, it can serve as a counter ion for the positively charged polymeric chain and be reversibly immobilized *via* ionic interactions. The application of cyclic voltammetry as an electropolymerization technique allows to observe the formation of drug-loaded polymer layer, as it is presented in figure 1(a), where the gradually increasing currents confirm the formation of a conductive deposit.

The common problem associated with the application of conducting polymers as drug delivery systems is the optimization of drug content to maximize drug capacity of matrix and maintain its electrochemical properties. While drug capacity can be determined in the release experiments, the use of cyclic voltammetry as the electropolymerization technique enables to assess the electrochemical parameters, such as charge storage capacity (CSC), directly from CV curves. Such set of voltammograms recorded for different Dex concentrations is presented in figure 1(b). It is evident that at the same number of potential scans, the final level of the resulting currents is lower in case of PEDOP/Dex (100 mM) than for pristine PEDOP or PEDOP/Dex (5 mM). The same can be observed when CV curves are integrated and CSC is calculated (figure 1(c)). The 30% decrease in CSC is observed when the concentration of Dex in electropolymerization solution is increased from 5 mM to 100 mM. This decrease in electrochemical performance confirms that Dex concentration of 100 mM can be considered as the upper threshold concentration above which electrochemical properties of conducting polymer matrix cannot be maintained. What is interesting, the enhancing effect of small quantities of drug (5 mM) on the electrochemical performance could be seen. This effect is supposed to originate from the increase in chain separation of the growing chain in a similar way to other bulky dopants, such as for ibuprofen-doped PEDOP [15].

The FT-IR spectra of PEDOP/Dex (figure 1(d)) performed prior to the controlled release studies proved that Dex had been incorporated into PEDOP matrix. Characteristic for Dex peaks at 1715 cm^{-1}, 1449 cm^{-1}, 1240 cm^{-1}, as well as the strong carbonyl signal at 1667 cm^{-1} [12] were observed in the spectra of PEDOP/Dex. The intensity of signals was found to increase with the increasing concentration of Dex, showing the highest content of this drug on the surface of PEDOP/Dex (100 mM). SEM images of PEDOP and PEDOP/Dex (figure 2) evidence the discrepancy between the surface morphologies of samples, confirming the formation of a thin and non-continuous conducting polymer deposit in case of PEDOP/Dex (100 mM) matrix.

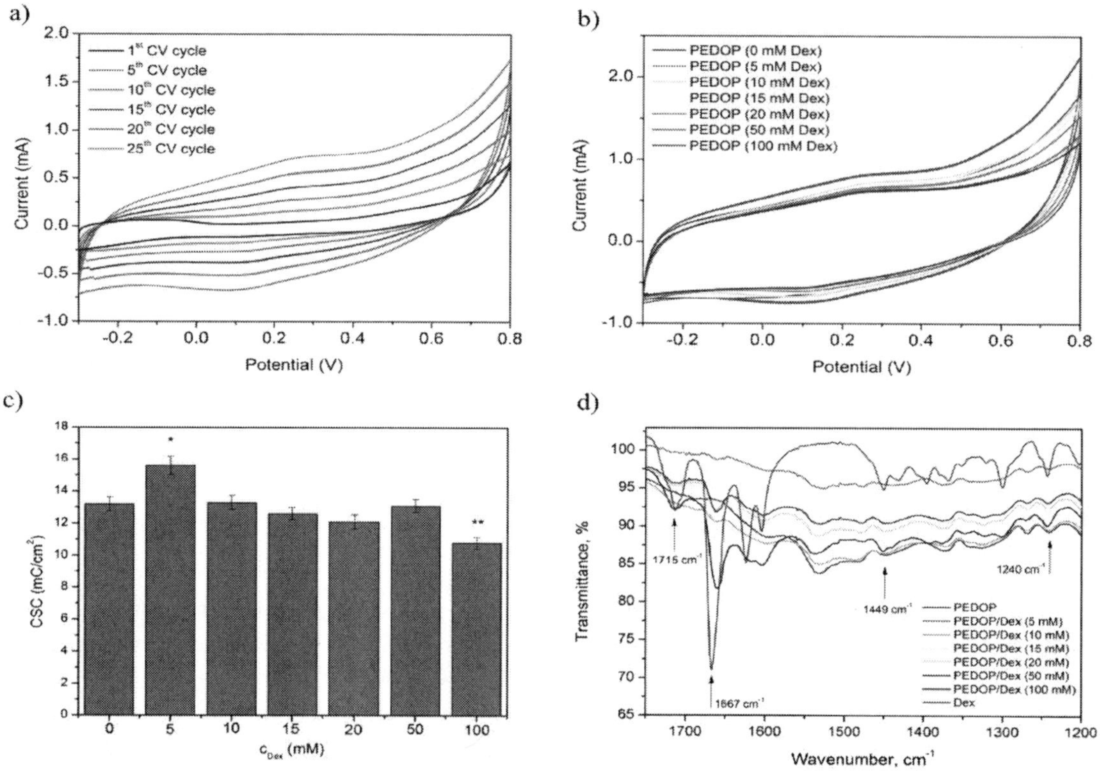

Figure 1. Typical CVs of EDOP polymerization in the presence of Dex (10 mM) (a), CVs of 25th cycle of polymerization for different Dex concentrations (b), the comparison of CSC (c) and FTIR spectra of PEDOP and PEDOP/Dex matrices (d).

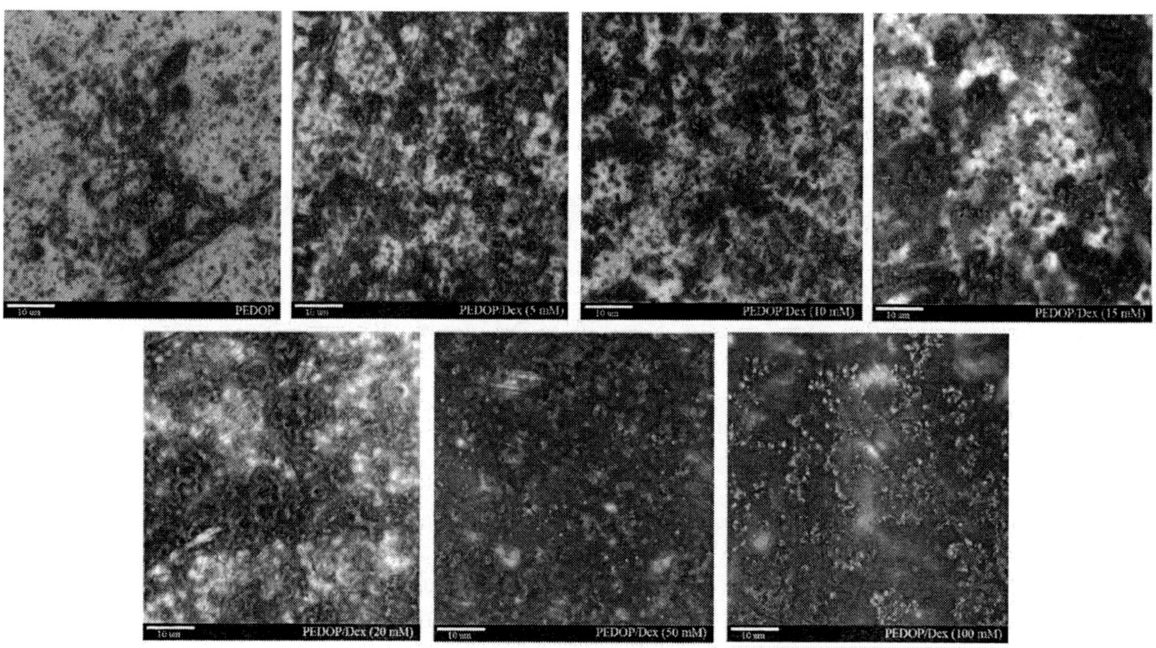

Figure 2. SEM images of PEDOP and different PEDOP/Dex matrices.

3.2. Electrically-triggered drug release

Since conducting polymers are electrically responsive materials, the electrical trigger in the form of potential or current stimulus is able to change the oxidation state of matrix causing drug release. Among few types of electrical stimulation techniques, two modes are considered as the most effective, i.e. potentiodynamic mode realized *via* cyclic voltammetry and potentiostatic mode realized *via* chronoamperometry. Cyclic linear potential sweep in potentiodynamic mode (figure 3(a)) allows the matrix to release drug when it is reduced and to uptake the anions from the environment, mainly chlorides and phosphates, when it is oxidized. In this way the neutral character of matrix is maintained and the process of release is not supposed to have any deteriorating effect on the integrity of matrix. The amount of released drug is controlled through the changes in potential range and number of CV cycles.

Potentiostatic process, on the other hand, consists of the constant application of a negative potential (figure 3(b)) resulting in the reduction of polymer matrix. Drug release is controlled by the magnitude of potential applied and time of process. Although longer release times and more negative potentials can increase the release rate, they pose a serious threat to the matrix itself, because constantly reduced matrix is prone to destruction. In this study, potentiodynamic method is used as a preferential release technique, and the potentiostatic method is applied to verify the presence of traces of drugs that have not been released in the potentiodynamic mode.

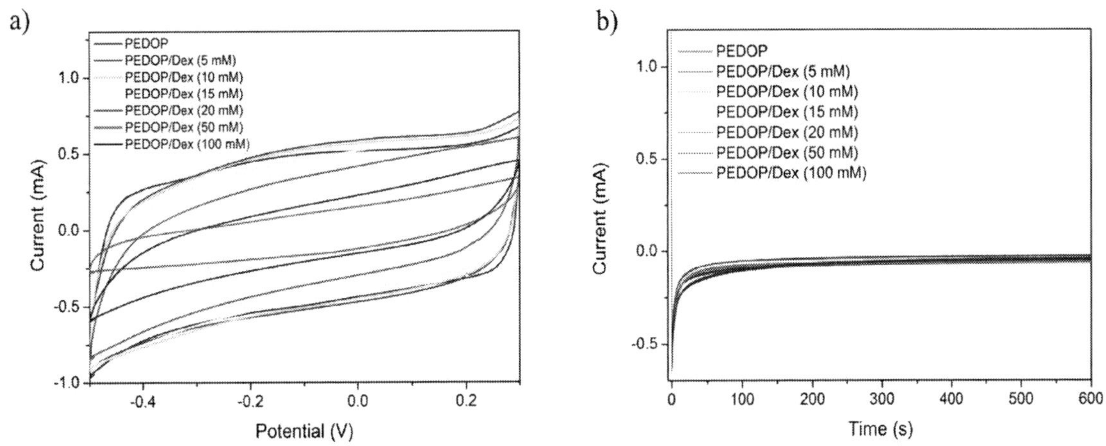

Figure 3. Typical curves of potentiodynamic (5th CV cycle) (a) and potentiostatic (b) release.

3.3. Drug capacity of PEDOP/Dex

The absorbance peak at 240 nm (figure 4(a)) was used to monitor the release of dexamethasone form PEDOP/Dex matrices in different release modes. The achievable concentrations of Dex released via spontaneous and two electrically-triggered modes as a function of Dex concentration in electropolymerization solution (figure 4(b)) show the substantial increase in drug capacity of PEDOP/Dex with increasing amounts of Dex. The cumulative release, being the sum of Dex concentrations released via all three modes, is increasing from 23 μM up to 116 μM when the concentration of Dex in electrolytic solution is increased from 5 mM to 100 mM. It shows that by the careful choice of electropolymerization conditions, mainly Dex concentration, it is possible to produce PEDOP/Dex matrices with a defined drug capacity tailored to the specific needs. What is more, the amount of drug released under all release modes exceeds the therapeutic levels of Dex, which is effective at the concentrations of 0.2 – 0.68 μmol [17]. In this work, up to 116 μM was released from PEDOP/Dex(100 mM), much higher than compared with previously reported polypyrrole/Dex (10 μmol) [12] and polyterthiophene/Dex (50 μmol) [13].

Apart from the differences in total drug capacity, the variation in drug concentration can have the effect on the efficiency of different release modes, both spontaneous and electrically triggered. It can be observed that while the amounts of drug released from PEDOP/Dex (5 mM) *via* three different modes are almost identical, there is a significant discrepancy in achievable concentrations among the release modes for PEDOP/Dex (50 mM). In the latter, the potentiodynamic release is evidently dominating over both spontaneous and potentiostatic release, showing the superiority of PEDOP/Dex (50 mM) as a controlled release system. The substantial increase in the amount of Dex released *via* spontaneous mode between PEDOP/Dex (50 mM) and PEDOP/Dex (100 mM) confirms that the high Dex concentrations have the negative effect on the matrix integrity and favors spontaneous drug elution rather than electrically-triggered release.

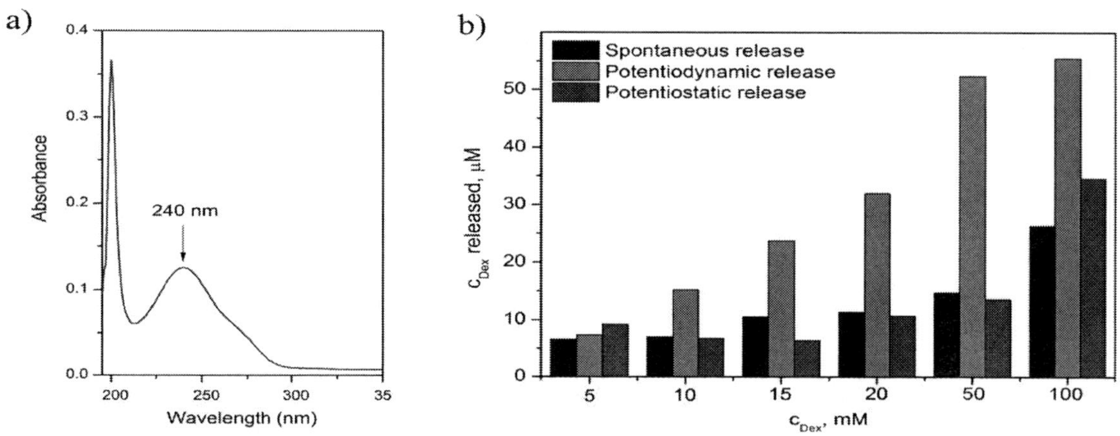

Figure 4. UV-Vis spectrum of Dex with a chemical structure of Dex as the inset (a) and the amounts of Dex released through various modes as a function of Dex concentration in polymerization solution (b).

4. Conclusions

In this study, the electrically-triggered controlled release system based on dexamethasone-loaded PEDOP matrix is described. It is shown that the electropolymerization conditions, mainly the concentration of Dex, can facilitate (at 5 mM) or suppress (at 100 mM) the formation of PEDOP/Dex matrix, as well as have the effect on its electrochemical performance in terms of charge storage capacity. Drug release experiments show that the drug capacity of PEDOP matrix increases with the increase in Dex concentration in the step of matrix synthesis, and higher Dex concentrations make it easier to control the amount of Dex released in an electrically-triggered mode. The results confirm the importance of the careful optimization of immobilization conditions to maximize drug capacity of matrix and maintain its electrochemical properties.

Acknowledgments

The author is grateful to the National Science Centre in Poland for financing the research in the framework of PRELUDIUM (2012/07/N/ST5/01878) and SONATA (2016/23/D/ST5/01306), as well as to the Silesian University of Technology for the Rector's grant in the area of scientific and development research (04/040/RGJ17/0054).

References

[1] Guimard N K, Gomez N and Schmidt C E 2007 *Prog. Polym. Sci.* **32** 876-921
[2] Ravichandran R, Sundarrajan S, Vanugopal J R, Mukherjee S and Ramakrishna S 2010 *J. R. Soc. Interface* **7** S559-79
[3] Kaur G, Adhikari R, Cass P, Bown M and Gunatillake P 2015 *RSC Advances* **5** 37553-67

[4] Balint R, Cassidy N J and Cartmell S H 2014 *Acta Biomater.* **10** 2341-53
[5] Krukiewicz K, Stokfisz A and Zak J K 2015 *Mater. Sci. Eng. C* **54** 176-81
[6] Esrafilzadeh D, Razal J M, Moulton S E, Stewart E M and Wallace G G 2013 *J. Control. Release* **169** 313-20
[7] Alizadeh N and Shamaeli E 2014 *Electrochim. Acta* **130** 488-96
[8] Krukiewicz K, Cichy M, Ruszkowski P, Turczyn R, Jarosz T, Zak J K, Lapkowski M and Bednarczyk-Cwynar B 2017 *Mater. Sci. Eng. C* **73** 611-5
[9] Kozai T D Y, Jaquins-Gerstl A S, Vazquez A L, Michael A C and Cui X T 2016 *Biomater.* **87** 157-69
[10] Aqrawe Z, Montgomery J, Travas-Sejdic J and Svirskis D 2017 *Trends Biotechnol.* **35** 93-5
[11] Moulton S E, Imisides M D, Shepherd R L and Wallace G G 2008 *J. Mater. Chem.* **18** 3608-13
[12] Wadhwa R, Lagenaur C F and Cui X T 2006 *J. Control. Rel.* **110** 531-41
[13] Stevenson G, Moulton S E, Innis P C and Wallace G G 2010 *Synth. Met.* **160** 1107-14
[14] Krukiewicz K, Jarosz T, Herman A P, Turczyn R, Boncel S and Zak J K 2016 *Synth. Met.* **217** 231-6
[15] Krukiewicz K, Zawisza P, Herman A P, Turczyn R, Boncel S and Zak J K 2016 *Bioelectrochem.* **108** 13-20
[16] Krukiewicz K and Zak J K 2014 *J. Mater. Sci.* **49** 5738-45
[17] Svirskis D, Travas-Sejdic J, Rodgers A and Garg S 2010 *J. Control. Release* **146** 6-15
[18] Shain W, Spataro L, Dilgen J, Haverstick K, Retterer S, Isaacson M, Saltzman M and Turner J N 2003 *IEEE T. Neura. Sys. Req.* **11** 186-8

IOP Conf. Series: Materials Science and Engineering **369** (2018) 012003 doi:10.1088/1757-899X/369/1/012003

Intensity of singular stress fields of an embedded fibre under pull-out force

D Chen[1], G W Zhang[1], R Takaki[1], A Inoue[1], K Tsuboi[1], Y Sano[1] and N-A Noda[1,2]

[1]Department of Mechanical and Control Engineering, Kyushu Institute of Technology Sensui-Cho 1-1 Tobata-Ku, Kitakyushu-Shi, Fukuoka, Japan

E-mail: noda@mech.kyutech.jp

Abstract. Previous experimental studies of fibre pull-out test show two dangerous points on the interface. Failure usually occurs at the bonded end of the fibre (Point A) or at the entry point on the surface of the matrix (Point E). Both points have different singular stress fields which causes crack initiation, crack propagation, and final failure. In this paper, intensity of singular stress fields (ISSF) at the fibre bonded end A and ISSF at the intersection point E of the fibre and the surface are discussed. The analysis method focuses on calculating the finite element method (FEM) stress radio by using a reference model and an unknown model. In the unknown model and the reference model, same FEM mesh pattern is applied. To analyse the ISSF at A, the body force method solution is used as the reference model. To analyse the ISSF at E, the reciprocal work contour integral method (RWCIM) solution is used as the reference model. Then, the two ISSFs are compared and discussed by varying the fibre embedded length l_{in}. When l_{in} is shorter, the singular stress at A is larger than the singular stress at E. When l_{in} is longer, the singular stress at E is larger than the ISSF at A.

1. Introduction

Fibre reinforced composites are very common nowadays in construction, marine industry, aerospace technology, transportation sector, consumer products and industrial equipment [1-4]. In fibre reinforced composites, fibres are mainly used to enhance load carrying capacity. Indeed, composites are produced when two or more materials or phases are used together to give a combination of properties that cannot be achieved otherwise [4]. The mechanical properties of fibre-reinforced plastics are governed, to a great extent, by the adhesion of the matrix to the fibre surface, i.e. the interface bond strength [5]. Many different alternative test set-ups and experimental techniques have been developed in recent years to gain more insight into the basic mechanisms dominating the properties of the fibre/matrix interface. One of the most popular is the pull-out test in which a single fibre or bar is pulled out off the surrounding matrix and the corresponding load versus displacement relation, $P(\omega)$, is recorded [6]. However, the inhomogeneous stress distribution on the interface with stress concentrations at the fibre entry point and at the fibre bonded end cannot be indicated by $P(\omega)$. The resistance $P(\omega)$of the debonding and pull-out processes depends on the bond properties of the fibre/matrix interface and the interfacial bond area. Commonly, the failure is assumed to occur when the maximum load is reached. However, pull-out experiments revealed that the interface crack forms long before the maximum load is reached [7]. Besides, a typical force-displacement curve from the pull-out test was shown. The 'kink' in the figure indicates crack initiation [8].

S Zhandarov *et al* found the $P(\omega)$ curve of pull-out and micro-bond tests is similar, i.e. crack propagation may starts from the fibre entry point E [8-10]. However, K-H Tsai *et al* observed that crack initiate at the fibre bonded end A during the fibre pull-out test [11,12]. In a rod pull out test that very similar to fibre pull-out test, C. Atkinson, et al. observed crack initiation sometimes occur at point A and sometimes occur at Point E [13].

2-D analytical model for the embedded fibre under pull-out force is shown in figure 1. As mentioned above, stress singularities arise at the fibre entry point E as well as at the fibre bonded end A in the theory of elasticity under idealized geometrical conditions, i.e. at sharp transitions between the two phases, even though no interface crack has formed [7]. The order of singularity arising at Point A and at Point E is not constant, but varies with the angle between the two surfaces as well with the elastic constants of fibre and matrix. The determination of stress intensity factors (i.e. ISSF) remains reserved to the analysis of an interface crack, even though the situation is more complex compared with the homogeneous case [7]. There are two different types of specific singular stress fields at point A as well as at point E. However, little research on these two singular stress fields is found. Therefore, attention is paid to discuss the ISSF by varying the fibre bonded length l_{in}.

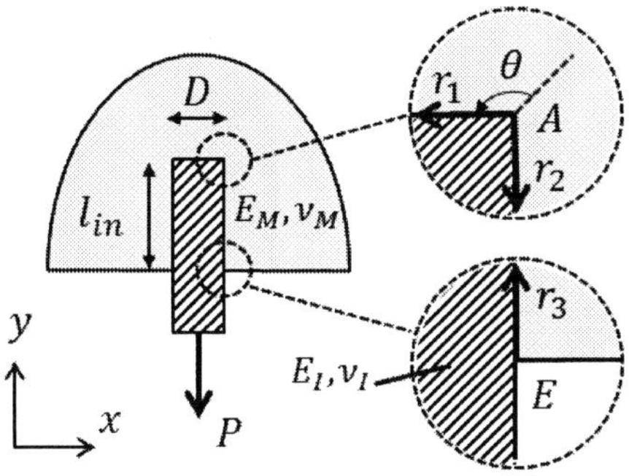

Figure 1. Reinforcing fibre under pull-out force.

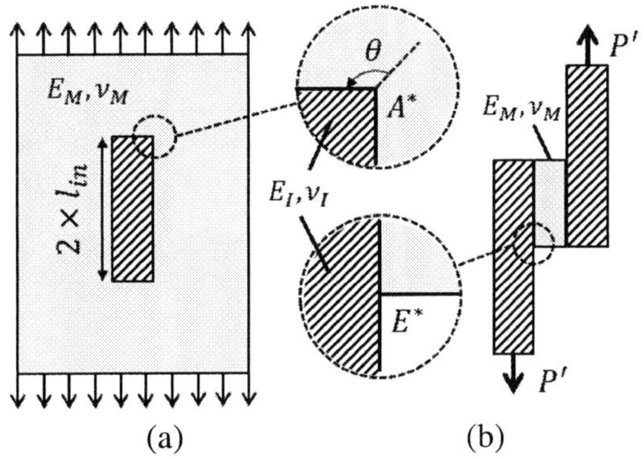

Figure 2. (a) Rectangular inclusion in infinite plate and (b) Single lap joint.

As is shown in figure 2(a), a rectangular inclusion is embedded in an infinite matrix under tension. Since the stress concentration zone is very small compared with the fibre diameter, the singular stress field at point A in Figure 1 is similar to that at point A^* in figure 2. Therefore, configuration shown in figure 2(a) can be used as the reference problem of the fibre bond end. The order of singularity arising at point A^* was discussed by D H Chen *et al* [14-16] Singular stress field at point E in figure 1 and at point E^* in figure 2 is similar. As is shown in figure 2(b), Point E^* is the corner of the interface of a single lap joint, at which ISSF was discussed by Miyazaki *et al* [17,18]. The singular stress field in this case can be examined by the interaction line integral method (RWCIM method) [19].

2. The two specific singular stress fields

As the order of singularity arising at point A and at point E is different, the singular stress field at point A and at point E has to be calculated separately.

2.1. Singular stress around corner A

On the one hand, stress distribution near corner A can be expressed as follows [14-16,20]:

$$\sigma_{ij,k} = \frac{K^A_{I,\lambda_1^A}}{r^{1-\lambda_1^A}} f^I_{ij,k}(\theta) \pm \frac{K^A_{II,\lambda_2^A}}{r^{1-\lambda_2^A}} f^{II}_{ij,k}(\theta) \qquad (ij = r, \theta, r\theta;\ k = M, I) \qquad (1)$$

In order to calculate the normal stress $\sigma_y^A(r_1)$ on the interface r_1, $ij = \theta$ and $\theta = 3\pi/4$ are used as shown in figure 1. Similarly, normal stress $\sigma_x^A(r_2)$ on the interface r_1 can be calculated from equation

(1) when $\theta = -3\pi/4$. Here, $f^{I}_{ij,k}(\theta)$ and $f^{II}_{ij,k}(\theta)$ can be calculated from equation (2) [14-16]. K^{A}_{I,λ_1^A} and K^{A}_{II,λ_2^A} are ISSFs, which are used for mode I and Mode II failure seperately. $K^{A}_{\sigma,\lambda_1^A} = K^{A}_{I,\lambda_1^A} \cdot f^{I}_{ij,k}(\theta)$ and $K^{A}_{\sigma,\lambda_2^A} = K^{A}_{II,\lambda_2^A} \cdot f^{II}_{ij,k}(\theta)$.

For matrix M $(-3\pi/4 \leq \theta \leq 3\pi/4)$ [15,16]

$$\begin{aligned} f^{I}_{\theta,M}(\theta) &= \frac{\lambda_1^A}{\sqrt{2\pi(\alpha-\beta)}}\{[-\lambda_1^A(\alpha-\beta)\cos(\lambda_1^A\pi/2) + (1-\beta)\sin(\lambda_1^A\pi)] \times \cos[(\lambda_1^A+1)\theta]\} \\ &\quad +(\lambda_1^A+1)(\alpha-\beta)\sin(\lambda_1^A\pi/2) \times \cos[(\lambda_1^A-1)\theta] \\ f^{II}_{\theta,M}(\theta) &= \frac{\lambda_2^A}{\sqrt{2\pi(\alpha-\beta)}}\{[-\lambda_2^A(\alpha-\beta)\cos(\lambda_2^A\pi/2) - (1-\beta)\sin(\lambda_2^A\pi)] \times \sin[(\lambda_2^A+1)\theta]\} \\ &\quad +(\lambda_2^A+1)(\alpha-\beta)\sin(\lambda_2^A\pi/2) \times \sin[(\lambda_2^A-1)\theta] \end{aligned} \tag{2}$$

Here, α, βdenote Dundurs bimaterial parameters, and shear modulus G_I and G_M can be transformed from Young's Modulus E_I, E_M and Poisson's Ratio v_I and v_M.

$$\alpha = \frac{G_M(\kappa_I+1)-G_I(\kappa_M+1)}{G_M(\kappa_I+1)+G_I(\kappa_M+1)} \qquad \beta = \frac{G_M(\kappa_I-1)-G_I(\kappa_M-1)}{G_M(\kappa_I+1)+G_I(\kappa_M+1)}$$
$$\kappa_i = \begin{cases} (3-v_i)/(1+v_i) & (Plain\ stress) \\ (3-4v_i) & (Plain\ strain) \end{cases} \qquad (i = M, I) \tag{3}$$

Singular indexes λ_1^A, λ_2^A around the corner A can be calculated from the following equations (4) [15]. Here, the singular indexes λ_1^A, λ_2^A have real values in the range $(0 < \mathrm{Re}(\lambda_i^A) < 1)$ if $\beta(\alpha-\beta) > 0$, which are equivalent to the order of singularity. Here, $\gamma = \pi/2$.

$$\begin{aligned} D_1(\alpha,\beta,\gamma,\lambda) &= (\alpha-\beta)^2\lambda^2[1-\cos(2\gamma)] + 2\lambda(\alpha-\beta)\sin(\gamma)\{\sin(\lambda\gamma) + \sin[\lambda(2\pi-\gamma)]\} \\ &\quad +2\lambda(\alpha-\beta)\beta \cdot \sin(\gamma)\{\sin[\lambda(2\pi-\gamma)] - \sin(\lambda\gamma)\} \\ &\quad +(1-\alpha^2) - (1-\beta^2)\cos(2\lambda\pi) + (\alpha^2-\beta^2)\cos[2\lambda(\gamma-\pi)] = 0 \\ D_2(\alpha,\beta,\gamma,\lambda) &= (\alpha-\beta)^2\lambda^2(1-\cos 2\gamma) - 2\lambda(\alpha-\beta)\sin(\gamma)\{\sin(\lambda\gamma) + \sin[\lambda(2\pi-\gamma)]\} \\ &\quad -2\lambda(\alpha-\beta)\beta \cdot \sin(\gamma)\{\sin[\lambda(2\pi-\gamma)] - \sin(\lambda\gamma)\} \\ &\quad +(1-\alpha^2) - (1-\beta^2)\cos(2\lambda\pi) + (\alpha^2-\beta^2)\cos[2\lambda(\gamma-\pi)] = 0 \end{aligned} \tag{4}$$

2.2. Singular stress around corner E

On the other hand, singular stress distribution around the corner E along the interface r_3, as is shown in figure 1, can be expressed as follows [17,18]:

$$\sigma_x^E(r) = \frac{K^{E}_{\sigma,\lambda_1^E}}{r^{1-\lambda_1^E}} + \frac{K^{E}_{\sigma,\lambda_2^E}}{r^{1-\lambda_2^E}} \tag{5}$$

The value of singular index λ_1^E and λ_2^E can be determined from equation (6) [20-22]. Detail calculation of $K^{E}_{\sigma,\lambda_1^E}$ and $K^{E}_{\sigma,\lambda_2^E}$in reference problem can be found in [17,18].

$$\begin{aligned} &4\sin^2(\pi\lambda)\left\{\sin^2\left(\frac{\pi\lambda}{2}\right) - \lambda^2\right\}\beta^2 + 4\lambda^2\sin^2(\pi\lambda)\alpha\beta + \left\{\sin^2\left(\frac{\pi\lambda}{2}\right) - \lambda^2\right\}\alpha^2 - 4\lambda^2\sin^2(\pi\lambda)\beta \\ &-2\left\{\lambda^2\cos(2\pi\lambda) + \sin^2\left(\frac{\pi\lambda}{2}\right)\cos(\pi\lambda) + \frac{1}{2}\sin^2(\pi\lambda)\right\}\alpha + \sin^2\left(\frac{3\pi\lambda}{2}\right) - \lambda^2 = 0 \end{aligned} \tag{6}$$

2.3. Singular index of different material combinations

Some results of singular index from equation (4) and equation (6) are shown in table 1. It can be verified that the singularity arising at point A and at point E is different.

Table 1. Singular index λ value for some material combination.

	λ_1^A	λ_2^A	λ_1^E	λ_2^E
$\alpha = -0.90$ $\beta = -0.40$	0.752493	0.748645	0.800102	0.997323
$\alpha = -0.90$ $\beta = -0.20$	0.796059	0.699861	0.657967	0.999111
$\alpha = -0.80$ $\beta = -0.40$	0.772894	0.847628	0.794890	0.988598
$\alpha = -0.80$ $\beta = -0.20$	0.809649	0.791916	0.651598	0.996246
$\alpha = -0.9071$ $\beta = -0.2016$	0.763235	0.621844	0.659165	0.999232

3. Method of analysis

Finite Element Method (FEM) was widely used for many engineering applications [23-28]. In this analysis software MSC Marc was used. Stress distribution along the interface (r_1, r_2, r_3) were calculated by applying the same mesh pattern to the pull-out model and reference model. Thus stress ratio$\left[\sigma_{I,FEM}^{A}(r)/\sigma_{I,FEM}^{A*}(r)\right]$ can be calculated between the pull-out model and the reference model. Similar method was used in [17,18].

As is shown in equation (7), $\sigma_{I,FEM}^{A}(r)$ is calculated from the stress distribution $\sigma_y^A(r_1)$ along the interface r_1 and $\sigma_x^A(r_2)$ along the interface r_2 by using the pull-out model (figure 1). $\sigma_{I,FEM}^{A*}(r)$ is calculated from the stress distribution $\sigma_y^{A*}(r_1)$ along the interface r_1 and $\sigma_x^{A*}(r_2)$ along the interface r_2 by using the reference model (figure 2). Material properties are shown in table 2. As mentioned above, Fibre in figure 1 and Inclusion in figure 2(b) are set to be same material properties.

$$2\sigma_I^A(r) = \sigma_y^A(r_1) + \sigma_x^A(r_2), (r = r_1 = r_2); \quad 2\sigma_{II}^A(r) = \sigma_y^A(r_1) - \sigma_x^A(r_2), (r = r_1 = r_2). \quad (7)$$

Table 2. Material properties.

	Matrix Epon 828	Fibre (Inclusion) E-Glass
Young's Modulus	$E_M = 3.3$ GPa	$E_I = 75$ GPa
Poisson's Ratio	0.35	0.17

Table 3. FEM Stress Ratio of Symmetrical type.

Smallest mesh size $e_{min} = 3.8 \times 10^{-8}$			Smallest mesh size $e_{min} = 1.27 \times 10^{-8}$		
$\frac{r}{e_{min}}$	$\sigma_{I,FEM}^{A}(r)$ [MPa]	$\frac{\sigma_{I,FEM}^{A}(r)}{\sigma_{I,FEM}^{A*}(r)}$	$\frac{r}{e_{min}}$	$\sigma_{I,FEM}^{A}(r)$ [MPa]	$\frac{\sigma_{I,FEM}^{A}(r)}{\sigma_{I,FEM}^{A*}(r)}$
0.0	13.20	1.136	0.0	16.53	1.136
0.5	10.88	1.135	0.5	13.62	1.136
1.0	8.383	1.135	1.0	10.49	1.135
1.5	7.741	1.134	1.5	9.681	1.135
2.0	7.603	1.134	2.0	9.507	1.135

In order to verify that the stress ratio between reference model and pull-out model is independent of mesh size, different mesh size e_{min} is used. As shown in table 3, for different e_{min} the value of the symmetrical type stress $\sigma^{A}_{I,FEM}(r)$ is different, while the ratio of stress distribution $[\sigma^{A}_{I,FEM}(r)/\sigma^{A*}_{I,FEM}(r)]$ is constant. Same situation happen to $\sigma^{A}_{II,FEM}(r)$ and $[\sigma^{A}_{II,FEM}(r)/\sigma^{A*}_{II,FEM}(r)]$, which is known as Skew-symmetrical type.

The normalized stress intensity factor F_I^* and F_{II}^* can be acquired on the basis of body force method [14]. Here F_I^* and F_{II}^* of the reference problem can be expressed as follows, in which $l_2 = D/2$.

$$F_I^* = K^*_{I,\lambda_1^A}/(\sigma^\infty \sqrt{\pi} l_2^{1-\lambda_1^A}) \qquad F_{II}^* = K^*_{II,\lambda_2^A}/(\sigma^\infty \sqrt{\pi} l_2^{1-\lambda_2^A}) \tag{8}$$

The normalized stress intensity factor of the fibre pull-out problem can be defined similarly as follows:

$$F_I = K^A_{I,\lambda_1^A}/[(P/2l_2)\sqrt{\pi} l_2^{1-\lambda_1^A}] \qquad F_{II} = K^A_{II,\lambda_2^A}/[(P/2l_2)\sqrt{\pi} l_2^{1-\lambda_2^A}] \tag{9}$$

As is shown in equations (10), by using the FEM stress ratio mentioned above and F_I^* acquired by body force method, F_I for the fibre pull-out problem can be calculated. So is F_{II}.

$$\frac{F_I}{F_I^*} = \frac{\sigma^{I}_{\theta,FEM}(r)}{\sigma^{I*}_{\theta,FEM}(r)} \qquad \frac{F_{II}}{F_{II}^*} = \frac{\sigma^{II}_{\theta,FEM}(r)}{\sigma^{II*}_{\theta,FEM}(r)} \tag{10}$$

4. Numerical results and discussion

In short fibre reinforced composite, the aspect ratio $2l_{in}/D$ of short fibres is about 30 [29]. In this study, we assume the diameter of the fibre, i.e. the width of the inclusion, is D = 21 μm in the 2-D model. The total fibre length should be l = 630 μm . Therefore, the embedded length of the fibre l_{in}, from 30 μm to 1000 μm, was considered in the material combination shown in table 2.

4.1. ISSF at point A of different l_{in}

The ISSF values K^A_{σ,λ_1^A} and K^A_{σ,λ_2^A} correspond to λ_1^A and λ_2^A, respectively. It is found that, K^A_{σ,λ_1^A} and K^A_{σ,λ_2^A} decrease as the embedded length l_{in} increases.

For example when l_{in} = 150μm, $K^A_{\sigma,\lambda_1^A} = 0.02334 MPa \cdot m^{1-0.7632}$, whereas when l_{in} = 300μm, $K^A_{\sigma,\lambda_1^A} = 0.01494\ MPa \cdot m^{1-0.7632}$. The value of K^A_{σ,λ_1^A} for l_{in} = 300μm is 36.0% less than that of l_{in} = 150μm. Besides, when l_{in} = 150μm, $K^A_{\sigma,\lambda_2^A} = 0.01519\ MPa \cdot m^{1-0.6218}$, whereas when l_{in} = 300μm, $K^A_{\sigma,\lambda_2^A} = 0.01017\ MPa \cdot m^{1-0.6218}$. The value of K^A_{σ,λ_2^A} for l_{in} = 300μm is 33.0% less than that of l_{in} = 150μm. That is, the ISSF of the $1/2$ embedded case is about 2/3 of that of 1/4 embedded case.

4.2. ISSF at point E

The ISSF values K^E_{σ,λ_1^E} and K^E_{σ,λ_2^E} correspond to λ_1^E and λ_2^E, respectively. It is found that, K^E_{σ,λ_1^A} and K^E_{σ,λ_2^A} decrease as the embedded length l_{in} increases.

For example, when about 1/4 of the length of the fibre is embedded, l_{in} = 150μm, when about half of the fibre is embedded l_{in} = 300μm. When l_{in} = 150μm, $K^E_{\sigma,\lambda_1^E} = 0.03689 MPa \cdot m^{1-0.6592}$, whereas when l_{in} = 300 μm, $K^E_{\sigma,\lambda_1^E} = 0.03219\ MPa \cdot m^{1-0.6592}$. The value for l_{in} = 300μm is 12.7% less than that of l_{in} = 150 μm. Besides, when l_{in} = 150μm, $K^E_{\sigma,\lambda_2^E} = -0.04929\ MPa \cdot m^{1-0.9992}$, whereas when l_{in} = 300 μm, $K^E_{\sigma,\lambda_2^E} = -0.05350\ MPa \cdot m^{1-0.9992}$. The value for l_{in} = 300μm is 8.54% less than that of l_{in} = 150 μm. That is, the ISSF of the 1/2 embedded case is about 10% less than that of 1/4 embedded case.

4.3. Compare of ISSF at point A and at point E

As the singular indexes (λ_1^A, λ_2^A), (λ_1^E, λ_2^E) are different at point A and at point E, it is not possible to compare ISSF directly. In order to compare the magnitude of the ISSF at point A and at Point E. Before crack initiation the radial tensile stress greatly exceed the shear stresses [7,30]. Therefore, $\sigma_y^A(r_1)$ and $\sigma_x^E(r_3)$ is calculated from the ISSFs. Here, $\sigma_x^E(r_3)$ is the radial stress distribution along interface r_3 in figure 1. As is shown in figure 3 when l_{in} = 100μm, the peculiar stress at point A is larger than the specific stress at point E. Therefore, the crack initiation may occur from point A. However, figure 4 shows that when l_{in} = 1000μm, radial stress at point A is larger than the specific stress at point E, and crack initiation may occur from point E.

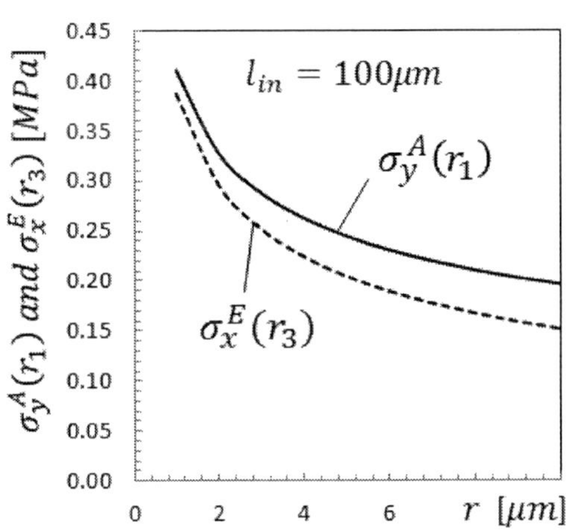

Figure 3. The dependence of $\sigma_y^A(r_1)$ and $\sigma_x^E(r_3)$ on r if $l_{in} = 100\mu m$

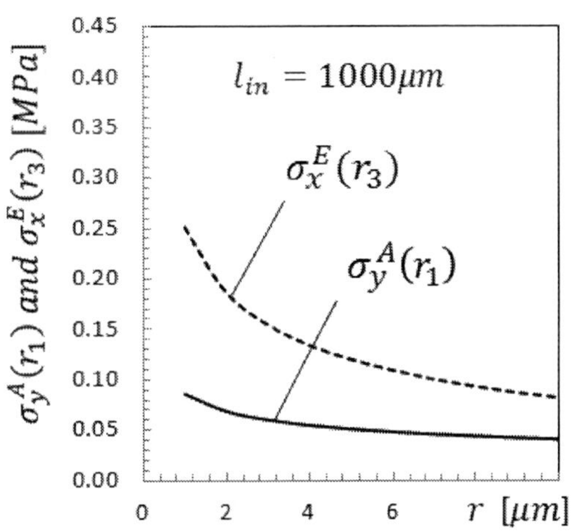

Figure 4. The dependence of $\sigma_y^A(r_1)$ and $\sigma_x^E(r_3)$ on r if $l_{in} = 1000\mu m$

5. Conclusions

In this study, the singular stress fields generated at the entry point and at the bond end of an embedded fibre under pull-out force are discussed. Stress distribution around the two areas was compared, and the following conclusions were obtained.

- The ISSF K^A_{σ,λ_1^A} and K^A_{σ,λ_2^A}at point A decreases as the embedded length l_{in} gets longer.
- The ISSF K^E_{σ,λ_1^E} *and* K^E_{σ,λ_2^E} at point E decreases as the embedded length l_{in} gets longer.
- For the selected material combination the singular stress at the entry point is larger than that at the bonded end if $l_{in} >$ 300μm. On the other hand, the crack of embedded fibres may occur from bonded end if $l_{in} <$ 300μm..

References

[1] Frikha M, Nouri H, Guessasma S, Roger F and Bradai C 2017 Interfacial behavior from pull-out tests of steel and aluminium fibres in unsaturated polyester matrix *J. Mater. Sci.* **52** 13829-40
[2] Mouritz A P, Gellert E, Burchill P and Challis K 2001 Review of advanced composite structures for naval ships and submarines *Composite Structures* **53** 21-42
[3] Nikhamkin M, Sazhenkov N and Samodurov D 2017 Fatigue fracture of fiber reinforced polymer honeycomb composite sandwich structures for gas turbine engines *J. Phys.: Conf. Ser.* **843** 012029
[4] Bagherpour S 2012 *Fibre Reinforced Polyester Composites Intech* doi:10.5772/48697
[5] Zhandarov S F and Pisanova E V 1997 The local bond strength and its determination by fragmentation and pull-out tests *Composites Science and Technology* **57** 957-64
[6] Banholzer B, Brameshuber W and Jung W 2006 Analytical evaluation of pull-out tests-the

inverse problem *Cement & Concrete Composites* **28** 564-71

[7] Marotzke C and Qiao L 1997 Interfacial crack propagation arising in single-fiber pull-out tests *Composites Science and Technology* **57** 887-97

[8] Pisanova E, Zhandarov S, Mäder E, Ahmad I and Young R J 2001 Three techniques of interfacial bond strength estimation from direct observation of crack initiation and propagation in polymer-fibre systems *Composites: Part A* **32** 435-43

[9] Zhandarov S and Mäder E 2005 Characterization of fiber/matrix interface strength: applicability of different tests, approaches and parameters *Composites Science and Technology* **65** 114-60

[10] Scheer R J and Nairn J A 1995 A comparison of several fracture mechanics methods for measuring interfacial toughness with microbond tests *The Journal of Adhesion* **53** 45-68

[11] Wang C 1997 Fracture mechanics of single-fibre pull-out test *Journal of Materials Science* **32** 483-90

[12] Tsai K-H and Kim K-S 1996 The micromechanics of fibre pull-out *Journal of the Mechanics and Physics of Solids* **44** 1147-77

[13] Atkinson C, Avila J, Betz E and and Smelser R E 1982 The rod pull out problem, theory and experiment *Journal of the Mechanics and Physics of Solids* **30** 97-120

[14] Noda N-A, Wang Q, Uemura Y and Kawashima Y 1998 Singular integral equation method in the analysis of interaction between rectangular inclusions *JSME International Journal Series A: Mechanics and Material Engineering* **41** 303-8

[15] Chen D H and Nisitani H 1993 Singular stress field near the corner of jointed dissimilar materials *Journal of Applied Mechanics, Transactions ASME* **60** 607-13

[16] Chen D H and Nisitani H 1992 Analysis of intensity of singular stress field at fibre end (2nd Report, Results of Calculation) *Transactions of the Japan Society of Mechanical Engineers, Series A* **58** 143-8

[17] Miyazaki T, Noda N-A, Li R, Uchikoba T and Sano Y 2013 Debonding criterion for single lap joints from the intensity of singular stress field *Journal of the JIEP* **16** 143-51

[18] Miyazaki T, Noda N-A, Uchikoba T, Li R and Sano Y 2014 Proposal of a convenient and accurate method for evaluation of debonding strength *Transaction of the JSAE* **45** 895-901

[19] Carpenter W C and Byers C 1987 A path independent integral for computing stress intensities for V-notched cracks in a bi-material *International Journal of Fracture* **35** 245-68

[20] Bogy D B 1968 Edge-bonded dissimilar orthogonal elastic wedges under normal and shear loading *Trans ASME J Appl. Mech.* **35** 460-6

[21] Bogy D B 1971 Two edge-bonded elastic wedges of different materials a nd wedge angles under surface tractions *Trans ASME J Appl. Mech.* **38** 377-86

[22] Noda N A, Shirao R, Li J and Sugimoto J-S 2007 Intensity of singular stress fields causing interfacial debonding at the end of a fiber under pullout force and transverse tension *International Journal of Solids and Structures* **44** 4472-91

[23] Noda N-A, Suryadi D, Kumasaki S, Sano Y and Takase Y 2015 Failure analysis for coming out of shaft from shrink-fitted ceramic sleeve *Engineering Failure Analysis* **57** 219-35

[24] Noda N-A, Miyazaki T, Li R, Uchikoba T and Sano Y 2015 Debonding strength evaluation in terms of the intensity of singular stress at the interface corner with and without fictitious crack *Int. J. Adhes. Adhes* **61** 46-64

[25] Noda N-A, Uchikoba T, Ueno M, Sano Y, Iida K, Wang Z and Wang G 2015 Convenient debonding strength evaluation for spray coating based on intensity of singular stress *ISIJ International* **55** 2624-30

[26] Wang Z, Noda N-A, Ueno M and Sano Y 2016 Optimum design of ceramic spray coating evaluated in terms of intensity of singular stress field *Steel Research International* **88** 1600353

[27] Miyazaki T, Noda N-A, Ren F, Wang Z, Sano Y and Iida K 2017 Analysis of intensity of singular stress field for bonded cylinder and bonded pipe in comparison with bonded plate *Int. J. Adhes. Adhes.* **77** 118-37

[28] Noda N-A, Chen X, Sano Y, Wahab M A, Maruyama H and Fujisawa R 2016 Effect of pitch difference between the bolt-nut connections upon the anti-loosening performance and fatigue life *Materials and Design* **96** 476-89
[29] Noda N-A, Takase Y and Iizuka T 2005 Generalized stress intensiy factors at the fibre end in fibre reinforced plastics *Transactions of the JSME* **71** 36-42
[30] Marotzke C 1993 Influence of the fiber length on the stress transfer from glass and carbon fibers into a thermoplastic matrix in the pull-out test *Comp. Interfaces* **1** 153-66

Intensity of singular stress field for three-dimensional butt joint to evaluate the adhesive strength

N-A Noda[1,3], K Tsuboi[1], R Takaki[1], F Ren[1], M R Aridi[1], Y Sano[1], Y Takase[1] and T Miyazaki[2]

[1] Department of Mechanical Engineering, Kyushu Institute of Technology Sensui-Cho 1-1 Tobata-Ku, Kitakyushu-Shi, Fukuoka, Japan
[2] Department of Mechanical Engineering, University of the Ryukyus Azasenbaru 1, Nishihara-Cho, Nakagami-Gun, Okinawa, Japan

E-mail: noda@mech.kyutech.ac.jp

Abstract. Adhesive joints are extensively used in various manufacturing processes in different industrial sectors because of its high fatigue resistance. Different materials properties cause the singular stress field, whose intensity is depending on the adhesive joint geometry. Our previous studies show that debonding strength can be expressed as a constant value of the critical intensity of singular stress field (ISSF) by using two-dimensional butt joint models. By considering real specimen geometry, in this paper, the ISSFs on the interface outer edges of three-dimensional butt joints are analysed by varying the adhesive thicknesses. A mesh-independent technique combined with three-dimensional finite element method (FEM) is shown to evaluate the ISSF. The ISSF distributions on the interface outer edges are analysed in comparison with the previous two-dimensional results. It is found that the critical ISSF considered 3D geometry is almost constant independent of the adhesive thickness.

1. Introduction

Nowadays composite materials become one of the most important structural materials in various industrial fields to replace previous conventional joining technologies. However, different materials properties cause singular stress at the end of the interface, which may lead to failure of bonding portions in structures. Several problems on debonding strength have been studied recently [1-3]. By using butt and single lap joints, the study showed that the tensile strength of the butt joints decreased as adhesive thickness increases while the shear strength did not affected by the adhesive thickness of single lap joints [4]. The same result that related to adhesive strength and adhesive thickness has been observed [5-7]. On the other hand, the study of the effects of surface roughness, adhesive thickness and combined stresses on the adhesive strength for various metals bonded with epoxy resin has been also discussed [8].

Moreover, the experimental studies also justified that the result will be affected by the residual strain of the adhesive layer [9-12]. Suzuki [13] experimentally observed the specimens where S35C JIS medium carbon steel plates are bonded by epoxy resin and it is found that the adhesive strength decreases with increasing adhesive thickness h and tends to be a constant when $h/W \geq 1$, where W is the width of specimen. In addition, Akisanya and Meng [14] discussed the experimental adhesive strength for the butt joints with rectangular cross section. To explain those experimental results, Noda *et al* [15-16] proposed a useful method to calculate the intensity of singular stress field (ISSF) at the adhesive dissimilar joint by focusing on the stresses at the edge calculated by finite element method(FEM). FEM can be used for many engineering applications conveniently [17-19].They found that the ISSF at the interface corner of two-dimensional butt joint decreases with decreasing the adhesive thickness. Also, Noda *et al* [20] obtained the critical ISSF for Suzuki's experimental specimen by using two-dimensional butt joint models, and found that the adhesive strength σ_c can be evaluated as a constant critical ISSF ($K_{\sigma c} = const$) for carbon steel/epoxy resin. In this paper, by considering real specimen 3D geometry, the ISSFs on the interface outer edges of 3D butt joints will be analysed by varying the adhesive thicknesses to confirm the validity of previous 2D modelling. The reference solution in figure 1(b) will be studied under arbitrary material combinations.

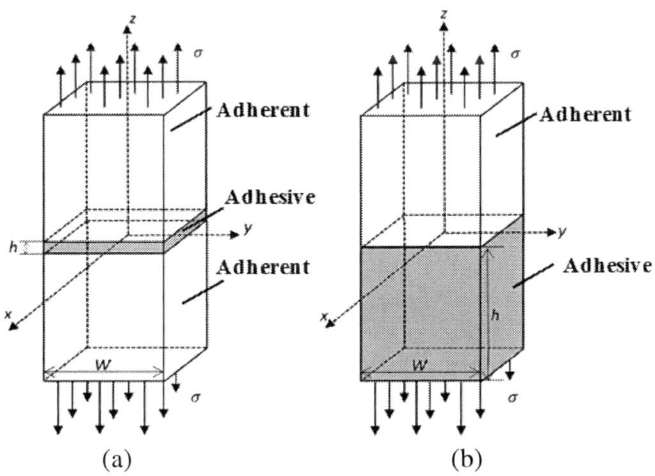

Figure 1. Three-dimensional butt joint when $h/W = 0.01\,(a)$ and $h/W = 1.0\,(b)$.

2. Three-dimensional mesh-independent technique to obtain ISSF

The most popular ISSF is known as the stress intensity factor for cracks. To obtain more general ISSF for evaluating interface strength, several analytical techniques were applied [21-28]. One of the most used numerical modelling techniques is the finite element method, which can be used for many engineering applications conveniently [29-35]. In order to obtain the ISSF, a mesh-independent technique was proposed for two-dimensional butt joints. The details are indicated in [15,16,20].

By considering the real specimen geometry, in this study, 3D butt joint as shown in figure 1(a) will be analysed. In order to apply the mesh-independent technique to 3D problem, a reference 3D solution is necessary. Therefore, 3D butt join for $h/W = 1.0$ as shown in figure 1(b) is analysed under arbitrary material combinations. Then, the adhesively bonded specimen used by Suzuki [13] is considered as comparison.

At this stage, the elastic properties for the materials are assumed as Young's modulus E=210 GPa and Poisson's ratio ν=0.3 for S35C, and E=3.14 GPa and ν=0.37 for epoxy resin [13]. Figure 2 shows that in the interior interface area $0 \le x, y < 0.45$ the accurate FEM stress is obtained as $|\sigma_z - 1| < 0.002$ independent of the mesh size. However, the FEM stress values near the outer of interface sides are inaccurate since they are different depending on the mesh size due to the existence of the singular stress field along the outer interface side. Focusing on the outer interface side from $y=0$ to $y=W/2$ (see table 1), the constant FEM stress decreases first as shown in figure 2 and then increases rapidly when y is close to the outer interface corner. This is because another stronger singular stress field exists around the outer interface corner. In this paper, the ISSF distribution along the outer interface side in the three-dimensional joint will be focused.

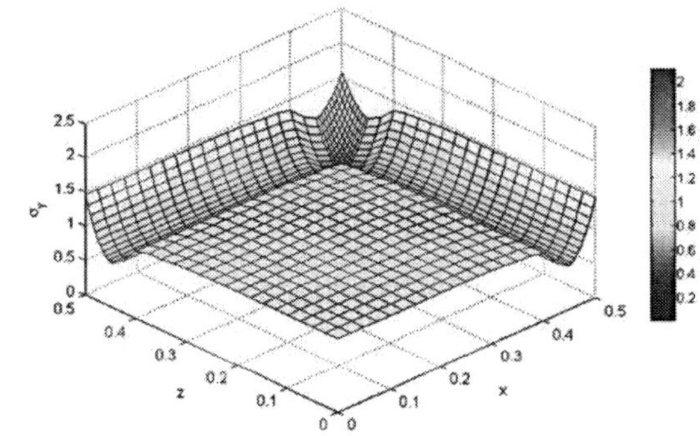

$|\sigma_z - 1| < 0.002$ on $0 \leq x, y < 0.45$
(a) e_{min}=0.05

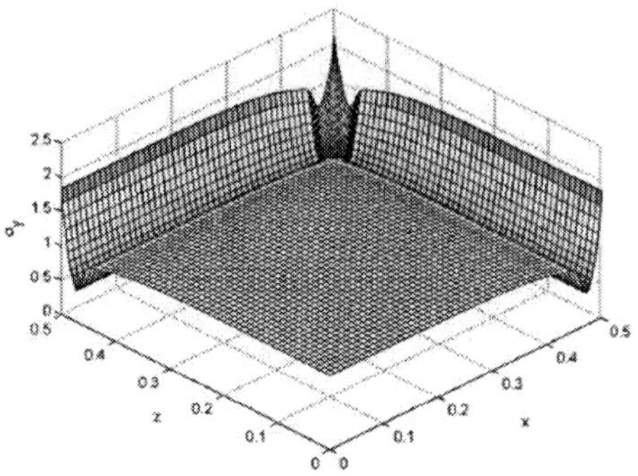

$|\sigma_z - 1| < 0.002$ on $0 \leq x, y < 0.45$
(b) e_{min}=0.025

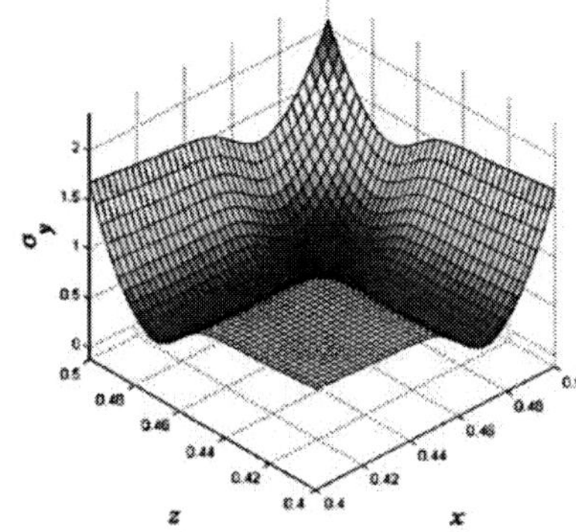

(c) Detail of x, y=0.4~0.5

Figure 2. FEM stress distribution σ_z of three-dimensional butt joint on $z = \pm h/2$ when $E_1 = 210\text{GPa}$, $\nu_1 = 0.3, E_2 = 3.14\text{GPa}$, $\nu_2 = 0.37$.

Table 1. FEM stress distributions for three-dimensional joint under tension obtained by different mesh sizes when E_1=210 GPa, ν_1=0.3, E_2=3.14 GPa and ν_2=0.37.

y	Minimum mesh size e_{min}=1/3200 around the edge			Minimum mesh size e_{min}=1/12800 around the edge		
	$\sigma^{3D}_{y,h/W=0.01}$	$\sigma^{3D}_{y,h/W\geq 1}$	$\frac{\sigma^{3D,FEM}_{y,h/W=0.01}}{\sigma^{3D,FEM}_{y,h/W\geq 1}}$	$\sigma^{3D}_{y,h/W=0.01}$	$\sigma^{3D}_{y,h/W\geq 1}$	$\frac{\sigma^{3D}_{y,h/W=0.01}}{\sigma^{3D}_{y,h/W\geq 1}}$
0.000	3.282	13.006	0.252	4.941	19.540	0.253
0.053	3.282	12.991	0.253	4.939	19.513	0.253
0.105	3.283	12.978	0.253	4.939	19.498	0.253
0.158	3.284	12.956	0.253	4.941	19.471	0.254
0.211	3.285	12.931	0.254	4.942	19.418	0.255
0.263	3.287	12.908	0.255	4.945	19.390	0.255
0.316	3.290	12.900	0.255	4.950	19.382	0.255
0.368	3.294	12.944	0.254	4.957	19.444	0.255
0.421	3.303	13.129	0.252	4.970	19.718	0.252
0.447	3.311	13.374	0.248	4.982	20.082	0.248
0.474	3.302	13.933	0.237	4.968	20.931	0.237
0.500	4.483	31.002	0.145	7.538	52.086	0.145

IOP Conf. Series: Materials Science and Engineering **369** (2018) 012004 doi:10.1088/1757-899X/369/1/012004

Table 1 shows the FEM stress distribution along the outer interface side in three-dimensional butt joint for $h/W=0.01$ and $h/W\geq 1$. It is seen that the FEM stress $\sigma^{3D}_{y,h/W=0.01}$ and $\sigma^{3D}_{y,h/W\geq 1}$ are inaccurate since they are different depending on the mesh size. However, the FEM stress ratio $\sigma^{3D}_{y,h/W=0.01}/\sigma^{3D}_{y,h/W\geq 1}$ is very accurate since they are the same independent of the mesh size. Therefore, for example, by focusing on the middle point of the side, the ISSF ratio can be expressed as the FEM stress ratio as shown in equation (1) [15,16,20].

$$\frac{K_\sigma^{3D}}{K_\sigma^{REF}}=\frac{F_\sigma^{3D}\sigma_z^\infty W^{1-\lambda}}{F_\sigma^{REF}\sigma_z^\infty W^{1-\lambda}}=\frac{\lim_{r\to 0}\left[r^{1-\lambda}\times\sigma_z^{3D,Real}(r)\right]}{\lim_{r\to 0}\left[r^{1-\lambda}\times\sigma_z^{REF,Real}(r)\right]}=\lim_{r\to 0}\left[\frac{r^{1-\lambda}\times\sigma_z^{3D,FEM}(r)}{r^{1-\lambda}\times\sigma_z^{REF,FEM}(r)}\right]=\frac{\sigma_z^{3D,FEM}(0)}{\sigma_z^{REF,FEM}(0)} \quad (1)$$

To discuss the ISSF distribution along the outer interface side, 3D butt joint in figure 1(b) is considered as the new reference solution in the next section under arbitrary material combination. The singularity index along the outer interface side can be obtained from equation (2).

$$\left[\sin^2\left(\frac{\pi}{2}\lambda\right)-\lambda^2\right]^2\beta^2+2\lambda^2\left[\sin^2\left(\frac{\pi}{2}\lambda\right)-\lambda^2\right]\alpha\beta+\lambda^2\left[\lambda^2-1\right]\alpha^2+\frac{\sin^2(\lambda\pi)}{4}=0 \quad (2)$$

Here, α and β are the Dundurs' composite parameters defined by the shearing modulus and Poisson's ratio as

$$\alpha=\frac{G_1(\kappa_2+1)-G_2(\kappa_1+1)}{G_1(\kappa_2+1)+G_2(\kappa_1+1)},\qquad \beta=\frac{G_1(\kappa_2-1)-G_2(\kappa_1-1)}{G_1(\kappa_2+1)+G_2(\kappa_1+1)} \quad (3)$$

where

$$\kappa_j=\begin{cases}\dfrac{3-\nu_j}{1+\nu_j}\ (plane\ stress)\\ 3-4\nu_j\ (plane\ strain)\end{cases}\ (j=1,2). \quad (4)$$

3. The ISSF of 3D reference solution under arbitrary material combinations

By using the similar mesh-independent technique described in section 2, the reference solution in figure 1(a) can be obtained. Figure 3 shows the ISSF ratio of the reference solution obtained from figures 1(b) and 1(b) with fixed displacement in the y-direction (plane strain 2D bonded plate). Here, 2D ISSF K_σ^{2D} was accurately obtained and indicated in [15,16,20]. In figure 3, the maximum ISSF ratio $K_\sigma^{3D}(W/2)/K_\sigma^{2D}(W/2)$ is shown under fixed (α,β) because (α,β) cannot totally control 3D ISSF [19]. For $-0.45\leq\beta\leq 0.45$, the ISSF ratio is in the range $0.537\leq K_\sigma^{3D}(W/2)/K_\sigma^{2D}(W/2)\leq 1.90$.

4. The critical ISSF distribution and critical ISSF of three-dimensional butt joint

The critical ISSF distributions are obtained when the debonding occurs for Suzuki's specimen [13] where the adherents S35C are bonded by adhesive epoxy as shown in figure 1(a). Figure 4 shows the normalized critical ISSF distributions for various h. The critical ISSF distributions are nearly the same for different adhesive thickness h. Figure 5 shows the critical ISSF $K_{\sigma c}^{3D}$ focusing on the middle point y=0 of the outer interface side. The experimental observation shows the debonding starts from the outer interface corner, but the debonding strength can be expressed as a constant critical ISSF $K_{\sigma c}^{3D}(0)$. Figure 5 coincides with the results obtained by using 2D model [20].

IOP Conf. Series: Materials Science and Engineering **369** (2018) 012004 doi:10.1088/1757-899X/369/1/012004

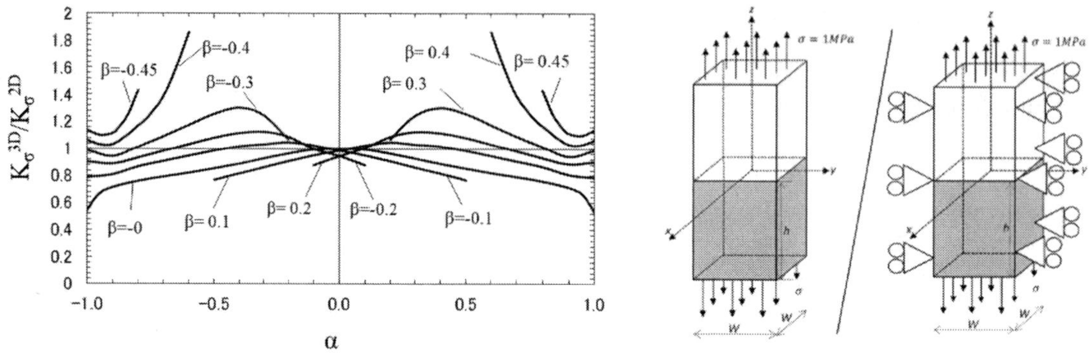

Figure 3. The maximum values of $K_{\sigma}^{3D}(W/2)/K_{\sigma}^{2D}(W/2)$ for various (α, β).

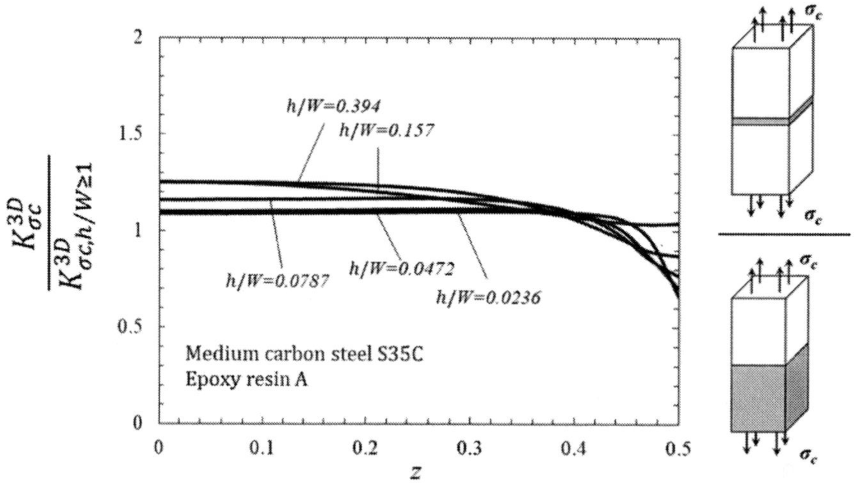

Figure 4. Critical ISSF distribution on the outer edge of the three-dimensional butt joint.

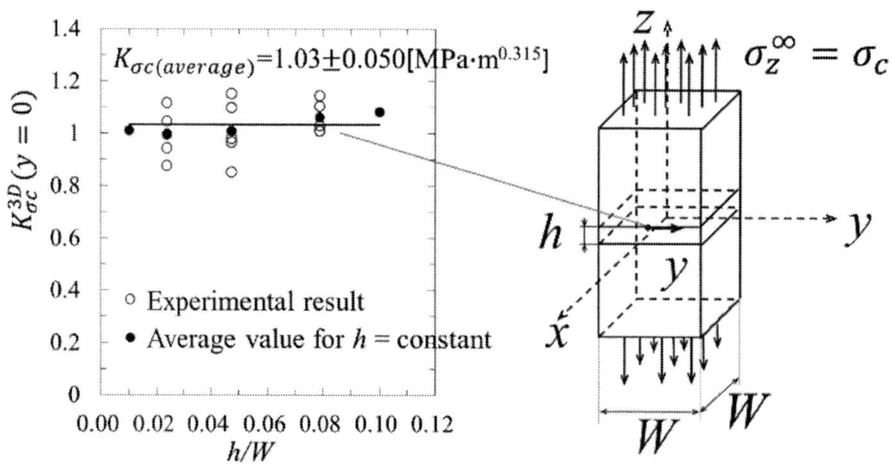

Figure 5. Constant critical ISSF focusing on (y=0).

5. Conclusion

In this study, the three-dimensional butt joint has been analyzed in terms of the intensity of singular stress field. Without considering any defects and residual strain, the elastic and homogenous adhesive is assumed to evaluate the debonding strength. The FEM stress distributions of three-dimensional butt joint are obtained by using different mesh sizes. The results show that FEM stress ratio remains constant near the middle of the outer interface side which is independent of element size. Also, the critical constant ISSF $K_{\sigma c}^{3D} = const$ can be used to express the adhesive strength by focusing on the middle point on the outer interface side. The validity of 2D modelling is confirmed since 3D results coincide with the results obtained by using 2D butt joint model [20].

References

[1] Di Pisa C and Aliabadi M H 2013 An efficient BEM formulation for analysis of bond-line cracks in thin walled aircraft structures *Int. J. Fract.* **179** 129–45

[2] Hasebe N and Kato S 2014 Solution of problem of two dissimilar materials bonded at one interface subjected to temperature *Arch. Appl. Mech.* **84** 913–31

[3] Krishnan A and Xu L R 2013 An experimental study on the crack initiation from notches connected to interfaces of bonded bi-materials Eng *Fract. Mech.* **111** 65–76

[4] Naito K, Onta M and Kogo Y 2012 The effect of adhesive thickness on tensile and shear strength of polyimide adhesive *Int. J. Adhes. Adhes.* **36** 77–85

[5] Hibino Y 1990 Influence of types and surface treatment of dental alloy and film thickness of cements on bond strength of dental luting cement *J. Japanese Soc. Dent. Mater. Dev.* **9** 786–805 (in Japanese)

[6] Asada B, Shinya A and Yokozuka S 1990 Effect of dental adhesive film thickness on bond strength *Adhes. Dent.* **8** 201–26 (in Japanese)

[7] Kagawa F, Yamada J, Suzuki T, Hisamitsu H and Wakumoto S 1991 The relationship between film thickness of resin luting cements and bond strength with the influence of thermal stress *Japan J. Conserv. Dent.* **34** 392–8 (in Japanese)

[8] Ikegami K, Kajiyama M, Kamiko S and Shiratori E 1979 Experimental studies of the strength of an adhesive joint in a state of combined stress *J. Adhes.* **10** 25–38

[9] Yamato T, Shirahama M and Hara S 1957 Effect of adhesive thickness on the strength of chlorinated rubber adhesive *J. Soc. Rubber Sci. Technol.* **30** 842–7 (in Japanese)

[10] Yokoyama T, Nakai K and Ikeda T 2006 Effect of specimen geometry on tensile properties of structural adhesive butt joints In: JSME Annual Meeting pp 861–2 (in Japanese)

[11] Akinmade A O and Nicholson J W 1995 Effect of adhesive layer thickness on the bond strength of a zinc polycarboxylate dental cement *Biomaterials* **16** 149–54

[12] Vallée T, Correia J R and Keller T 2010 Optimum thickness of joints made of GFPR pultruded adherends and polyurethane adhesive *Compos. Struct.* **92** 2102–8

[13] Suzuki Y 1987 Adhesive tensile strengths of scarf and butt joints of steel plates (relation between adhesive layer thicknesses and adhesive strengths of joints) *JSME Int. J.* **30** 1042–51

[14] Akisanya A R and Meng C S 2003 Initiation of fracture at the interface corner of bi-material joints *J. Mech. Phys. Solids.* **51** 27–46

[15] Zhang Y, Noda N A, Takaishi K and Lan X 2010 Effect of adhesive thickness on the intensity of singular stress at the adhesive dissimilar joint *J. Solid Mech. Mater. Eng.* **10** 1467–79

[16] Zhang Y, Noda N A, Wu P Z and Duan M L 2015 A mesh-independent technique to evaluate stress singularities in adhesive joints *Int. J. Adhes. Adhes.* **57** 105–17 (the corrigendum of authorship is published in *Int. J. Adhes. Adhes.* **60**(2015))

[17] Noda N A, Chen X, Sano Y, Wahab M A, Maruyama H and Fujisawa R 2016 Effect of pitch difference between the bolt–nut connections upon the anti-loosening performance and fatigue life *Mater. Des.* **96** 476-89.

[18] Miyazaki T, Noda N A, Ren F, Wang Z, Sano Y and Iida K 2017 Analysis of intensity of

singular stress field for bonded cylinder and bonded pipe in comparison with bonded plate *Int. J. Adhes. Adhes.* **77** 118-37

[19] Noda N A, Suryadi D, Kumasaki S, Sano Y and Takase Y 2015 Failure analysis for coming out of shaft from shrink-fitted ceramic sleeve *Eng. Fail. Anal.* **57** 219-35

[20] Noda N A, Miyazaki T, Li R, Uchikoba T, Sano Y and Takase Y 2015 Debonding strength evaluation in terms of the intensity of singular stress at the interface corner with and without fictitious crack *Int. J. Adhes. Adhes.* **61** 46–64

[21] Shibutani T 2004 Evaluation of crack initiation at interfacial edge on the basis of fracture mechanics concept and application to electronics devices *J. Jpn Inst. Electronics Packag.* **7** 639-44 (in Japanese)

[22] Hattori T, Sakata S, Hatsuda T and Murakami G 1988 A stress singularity parameter approach for evaluating adhesive strength *JSME Int. J. Ser. 1 Solid Mech. Strength Mater.* **31** 718–23

[23] Shiratori M 1994 Problems of joints in packaging of electronic devices *Trans. Japan Soc. Mech. Eng. A.* **60** 1905-12 (in Japanese)

[24] Kitamura T, Shibutan T and Ueno T 2002 Crack initiation at free edge of interface between thin films in advanced LSI *Eng. Frac. Mech.* **69** 1289-99

[25] Shibutani T, Tsuruga T, Yu Q and Shiratori M 2003 Criteria of crack initiation at edge of interface between thin films in opening an sliding modes for an advanced LSI *Trans. Japan Soc. Mech. Eng. A.* **69** 1368-73 (in Japanese)

[26] Takahashi Y, Aihara K, Ashida I, Niguchi K, Yamamoto Y, Arai S, Muto S and Tanaka N 2016 Evaluation of interfacial fracture strength in micro-components with different free-edge shape *Mech. Eng. J.* doi: 10.1299/mej.16-00108

[27] Hirakata H, Takahashi Y, Matsumoto S and Kitamura T 2006 Dominant stress region for crack initiation at interface edge of microdot on a substrate *Eng. Frac. Mech.* **73** 2698-709

[28] Noda N A and Takase Y 2003 Generalized stress intensity factors of V-shaped notch in a round bar under torsion, tension, and bending *Eng. Fract. Mech.* **70** 1447-66

[29] Noda N A, Suryadi D, Kumasaki S, Sano Y and Takase Y 2015 Failure analysis for coming out of shaft from shrink-fitted ceramic sleeve *Eng. Fail. Anal.* **57** 219-35

[30] Noda N A, Uchikoba T, Ueno M, Sano Y, Iida K, Wang Z and Wang G 2015 Convenient debonding strength evaluation for spray coating based on intensity of singular stress *ISIJ International* **55** 2624-30.

[31] Noda N A, Chen X, Sano Y, Wahab M A, Maruyama H and Fujisawa R 2016 Effect of pitch difference between the bolt–nut connections upon the anti-loosening performance and fatigue life *Mater. Des.* **96** 476-89

[32] Noda N A, Shen Y, Takaki R, Akagi D, Ikeda T, Sano Y and Takase Y 2017 Relationship between strain rate concentration factor and stress concentration factor *Theor. Appl. Fract. Mec.* **90** 218–27

[33] Wang Z, Noda N A, Ueno M and Sano Y 2016 Optimum design of ceramic spray coating evaluated in terms of intensity of singular stress field *Steel Research International* **88** doi: 10.1002/srin.201600353.

[34] Miyazaki T, Noda N A, Ren F, Wang Z, Sano Y and Iida K 2017 Analysis of intensity of singular stress field for bonded cylinder and bonded pipe in comparison with bonded plate *Int. J. Adhes. Adhes.* **77** 118-37

[35] Noda N A, Miyazaki T, Li R, Uchikoba T and Sano Y 2015 Dedbonding strength evaluation in terms of the intensity of singular stress at the interface corner with and without fictitious crack *Int. J. Adhes. Adhes.* **61** 46-64

IOP Conf. Series: Materials Science and Engineering **369** (2018) 012005 doi:10.1088/1757-899X/369/1/012005

Rolled plastic film damage simulation considering surface roughness useful for greenhouse roof automatic ventilation

N-A Noda[1,2], H Sakai[1], Y P Huang[1], Y Sano[1] and Y Takase[1]

[1]Department of Mechanical and Control Engineering, Kyushu Institute of Technology Sensui-Cho 1-1 Tobata-Ku, Kitakyushu-Shi, Fukuoka, Japan

E-mail: noda@mech.kyutech.ac.jp

Abstract. Greenhouses are capable of producing a variety of high-value crops year-round. A novel Japanese greenhouse design is gaining popularity because of its automated rolled-up ventilation system that is integrated into the roof. However, due to the frequent movement of the roll-up system, the plastic film deteriorates rapidly and typically lasts for only three or four months. In order to better understand the film deterioration, we studied the mechanics involved at the point of contact between the film and the metal greenhouse frame. We found that film deformation and failure were closely related to stretching and creasing, and these processes were observed at the microscope level. In this paper, the plastic film damage was investigated through the rolling contact analysis focusing on the pipe surface roughness. We found that at the inside film surface the larger strain is caused by the pipe surface roughness. This larger strain generates the wave deformation of the film, crease and line scar whose number increases with increasing the roughness of the arch pipe. The effects of pipe coating and perforated film are discussed to prevent the film damage.

1. Introduction

A novel greenhouse is being used recently since automated rolled-up ventilation system is contained in into the roof to control the temperature. A variety of high-value crops can be produced year-round by providing most suitable temperature. The fully opening automatically ventilated greenhouse can control the inside temperature by automatically rolling up and rolling down the plastic film from the gutter toward the ridge. Until now although partially- opening type green houses are widely used, full-opening type greenhouse are not because the plastic film deteriorates rapidly due to the frequent movement of the roll-up system. It has been requested to investigate the plastic film damage mechanism and to find out how to prevent the damage under rolling contact loading. The several contact stresses, and the rolling contact solutions of elastic film sandwiched between two parallel cylinders were provided by Johnson [1]. Regarding rolling of polypropylene sheet, Nakamura and Takasu investigated the mechanical properties and molecular orientation in the rolled sheets by tensile tests, sonic propagation method, birefringence, and dynamic viscoelastic measurements [2]. Instead of between two parallel pipes, the film deformation was considered between two orthogonal pipes considering rolled-up ventilation [3].

Noda et al summarized the film damage process of automated rolled-up ventilation in the following way [3]. Damage begins on the inside film surface, contacting the greenhouse frame pipes. Compressive rolling stress induces stretching zones with indentations that lead to creases. Eventually, the creases fold over and develop line scars which are perpendicular to the rolling direction. Since the film thickness reduction causes elongation and creasing of the film, these deformations are closely related to film damage [4]. Noda et al also found that temperature severely influences the damage. At 70°C, which are measured pipe surface temperature of greenhouse pipes during the summer, the rolling contact experiment inflicts line scars on the plastic film [4].

However, those previous studies did not consider the pipe surface roughness, which may affect the creation of line scars significantly and therefore the film damage process has not been completely explained yet. In this paper, therefore, different surface roughness will be introduced on the arch pipe and the number of line scar will be studied experimentally for the film sandwiched between the arch pipe and the rolling pipe. The simulation will be also performed by changing the model surface roughness. Moreover, since perforated films are sometimes used to reduce the inside temperature, the effect of holes on the film damage will be also studied. To prevent the film damage contributes the widely use of automated ventilation system.

2. Rolling contact experimental and analytical results considering pipe surface roughness

Figure 1 shows the developed rolling contact machine with which the damage to greenhouse films was investigated [4]. In this apparatus, the pipe that rolls over the film is pulled across and turns under the friction between itself and the film. Each support pipe carries an average load of 15 N. However, the actual load becomes larger when the support pipe deflects. Greenhouse roofs are also subject to another force provided by hard wind and rain and exerted by the holding strap. Therefore, we should assume that loads up to 75 N could be applied to the support pipe.

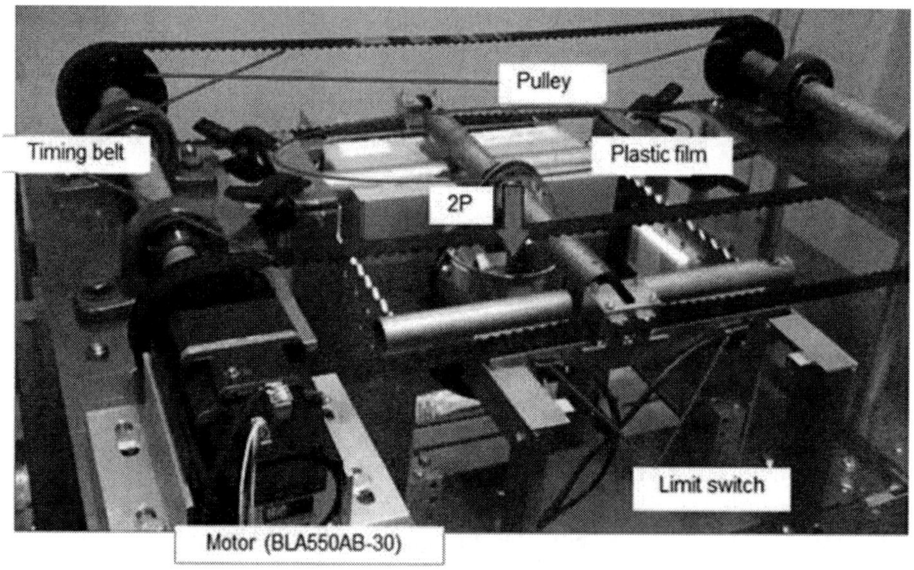

Figure 1. Rolling contact machine to investigate plastic film damage.

Figure 2 shows the number of line scar /cm ($n_{\text{line scar}}$) observed during the rolling contact experiment by changing the support pipe surface roughness R_y. It is seen that with increasing roughness of the support pipe R_y, the number of line scar /cm ($n_{\text{line scar}}$) increases independent of the pipe diameter. The number of line scar /cm ($n_{\text{line scar}}$) can be expressed as the following approximate equation.

$$n_{\text{line scar}}=0.12R_y+3.42 \quad (R^2=0.95). \tag{1}$$

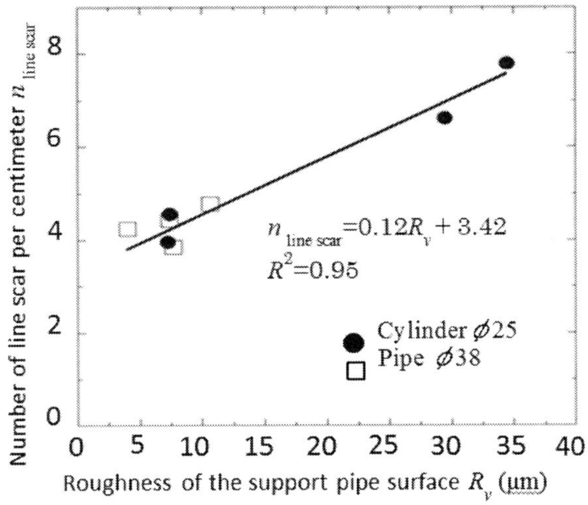

Figure 2. Relationship between the number of line scar /cm and the surface roughness of the support pipe R_y.

The previous FEM study showed that the number of line scar is controlled by the film strain assuming smooth pipe surface [3]. In this study, the effect of surface roughness will be studied by applying the finite element method (FEM). FEM is one of the most used numerical modeling techniques, which can be used for many engineering applications conveniently [3-6]. Figure 3 shows FEM models having an annular protrusion to express the surface roughness. As shown in figures 3(a)-3(d), Modal A has no protrusion to be compared, Model B and Model C have different root radius and Model C and Model D have different heights. The model dimensions are determined by considering real pipe roughness. Figure 4(a) shows the results of Model A. It is seen that the strains are almost the same between the inside and outside surfaces for smooth surfaces. Figure 4(b) shows that even the slight protrusion like Model B causes different strains between the inside and outside surfaces. Figure 4(c) shows that if the root radius becomes a bit smaller like Model C the insider strain becomes larger significantly. Figure 4(d) shows that the higher protrusion height causes larger strain difference between inside and outside surfaces, which may lead the line scar creation at the film inside surface. Figures 4(a)-4(d) shows that the surface roughness affects largely the line scar creation.

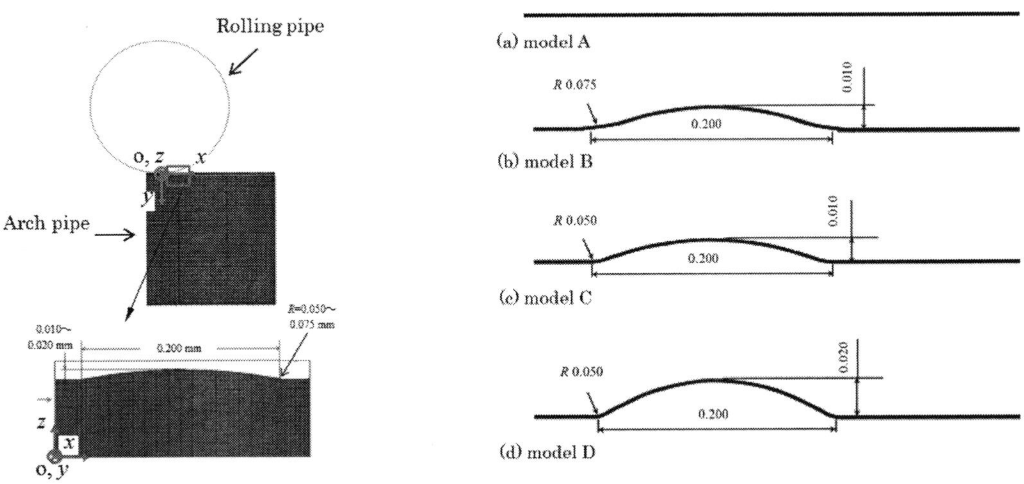

Figure 3. FEM models having an annular protrusion at the arch pipe outer diameter to express the surface roughness.

IOP Conf. Series: Materials Science and Engineering **369** (2018) 012005 doi:10.1088/1757-899X/369/1/012005

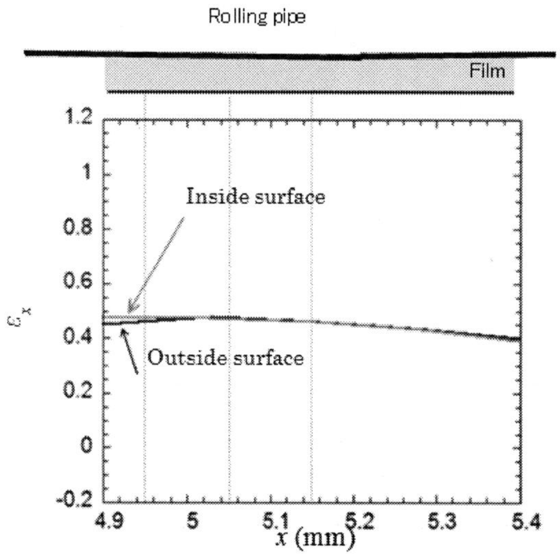

(a) Flat model A

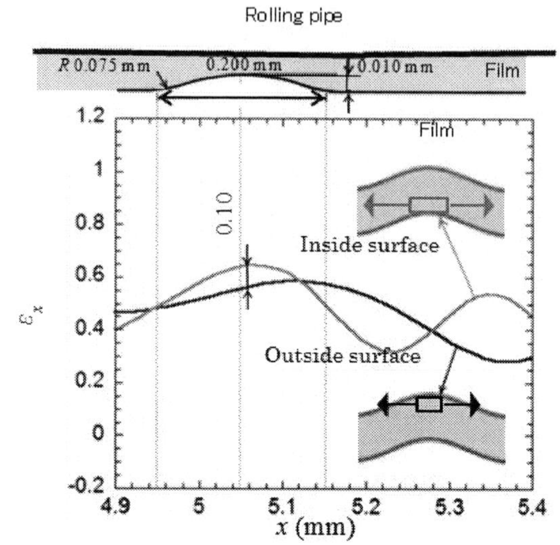

(b) Projection model B

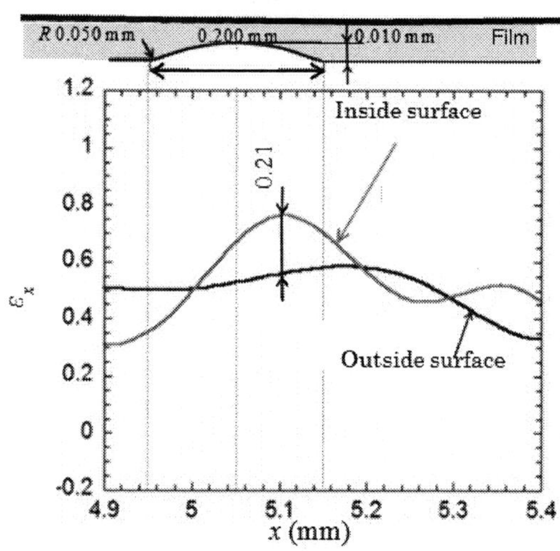

(c) Projection model C

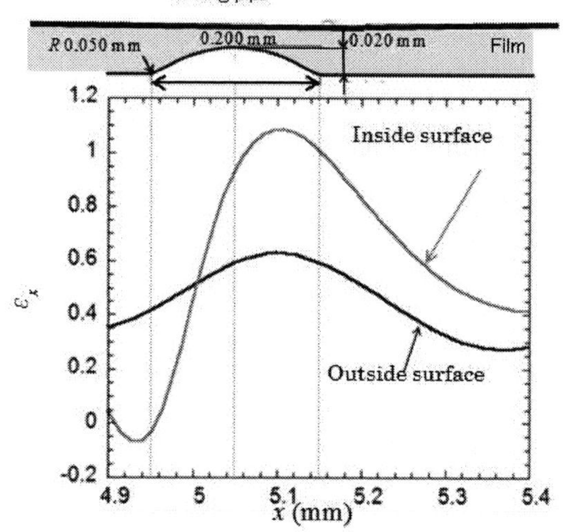

(d) Projection model D

Figure 4. Film elongation along the inside and outside surfaces of the film.

3. Rolling contact analysis and results for coating pipe

As shown in the previous section, the pipe surface roughness leads to the line scar creation. In this section, to prevent the film damage, the pipe coating effect will be discussed. Note that the greenhouse pipe temperature reaches 70°C in summer. The previous study shows that at 70°C the rolling contact stress caused line scars on the plastic film, and therefore, temperature largely affects the damage to plastic films [4]. The pipe coating may be useful for reducing the pipe temperature as well as for reducing the pipe surface roughness. In this study, steel pipes coated with ABS resin whose collars are white, gold, black will be considered with the pipe diameter 38 mm.

Table 1 compares the numbers of line scar and crease for three kinds of coated pipe under the same temperature T=70°C. Compared to no coating, the coated pipes can prevent the line scar and crease creation drastically because of smaller surface roughness. Under the same heating condition, Table 2 compares the numbers of line scar and crease for three kinds of coating pipe. It is seen that when the steel pipe temperature reaches 70°C, the coated surface temperature is higher. Although the coated pipes do not prevent rising temperature, they can decrease the number of line scar under the same heating condition. It may be concluded that the white coating is the most effective to prevent the film damage.

Table 1. Number of line scar/cm and number of crease/cm for three kinds of coating pipe under P=75 N, T=70°C.

Support pipe	Temperature (°C)	Roughness R_y (μm)	Number of creases per centimeter n_{crease}	Number of line scars per centimeter $n_{linescar}$
Steel pipe ϕ38	70	7.38	many	3.85
Coating pipe ϕ38 (white)	70	1.90	0.75	0
Coating pipe ϕ38 (gold)	70	1.08	1.45	0.55
Coating pipe ϕ38 (black)	70	2.30	0.65	0

Table 2. Number of line scar /cm and number of crease/cm for three kinds of coating pipe under the same heating and P=75 N.

Support pipe	Temperature (°C)	Roughness R_y (μm)	Number of creases per centimeter n_{crease}	Number of line scars per centimeter $n_{linescar}$
Steel pipe ϕ38	70	7.38	many	3.85
Coating pipe ϕ38 (white)	70	1.90	0.75	0
Coating pipe ϕ38 (gold)	80	1.08	many	2.43
Coating pipe ϕ38 (black)	90	2.30	many	2.50

Table 3 indicates the stress and deformation of film type A in [4] obtained by FEM when due to the support pipe is changed. Here, steel pipe, resin pipe and steel pipe coated resin pipe are considered. Assume that the coated resin pipe consists of steel and resin layers, each of which has 1 mm thickness. Note that steel Young's modulus is more than 7 times larger than Young's modulus of resin. It is seen that the maximum thickness reduction of the resin pipe is 20% smaller than the one of the steel pipe. The film deformations are almost the same for the steel pipe and the coated pipe within 3%.

Table 3. Stress and deformation of film type A in [4] due to three kinds of support pipes when T=70°C, P=75 N, μ=0.3, and d_1=d_2=38 mm.

	Steel pipe	Resin pipe	Coating pipe
Young's modulu's (MPa)	21000	2900	Outside 2900 Inside 21000
Poisson's ratio	0.3	0.35	Outside 0.35 Inside 0.3
Maximum contact pressure (MPa)	6.81	4.70	7.43

Contact area (mm^2)	10.18	11.56	11.31
Maximum thickness reduction	0.120	0.096	0.117

4. Rolling contact experiment and results for perforated film

Perforated films are often used as a plastic film tunnel which covering whole crops and as a mulch material placed on the soil surface to maintain moisture, reduce weed growth, mitigate soil erosion and improve soil conditions [7,8]. Although perforated films are not often used for traditional greenhouses, they are planed to be used for automated ventilation system to control summer temperatures adequately. However, perforated films look susceptible to the rolling contact damage. In this paper rolling contact machine in figure 1 will be used to investigate the hole effect on the rolling contact damage.

Here 3 mm diameter holes spacing 16.5 mm on the film type D will be considered since this diameter is known to provide good ventilation without losing protection against rain. The previous study [4] showed that film type D is the most desirable film for automated ventilation system because line scars handly appeared at temperatures under 70°C. The temperature was set to T = 70°C, and the same diameter for 38 mm for flame pipe and rolling pipe. The number of cycle N=300. Figure 5 shows two types of rolling paths considered in this study, that is, (a) the path along the hole centres and (b) the path along the hole edges.

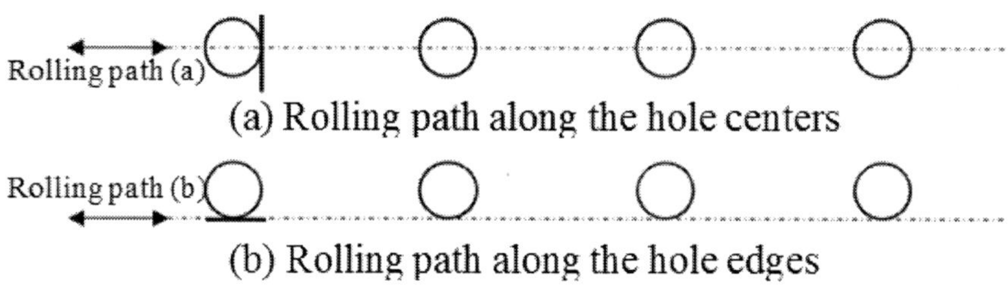

Figure 5. Definition of rolling path (a), (b).

Table 4. Number of line scar/cm of plastic film type D when P=75 N, T=70°C and d_1=d_2=38 mm.

Shape of film	Holes		
Number of line scar	Hole center	Hole edge	Without hole
Number of line scar per centimetre $n_{\text{line scar}}$	0	5	3
	0	0.25	0.15

Table 4 show the number of line scars/cm for film type D. As shown in table 4, no line scar can be seen after the rolling contact of the path (a) in figure 5 along the hole centres. Figures 6(a) and 6(c) show the film after rolling contact when the path is in figure 5(a). It is seen the film damage due to the rolling path (a) in figure 5 is much smaller compared to the film damage without hole shown in figure 7. As shown in figure 6(c), the circular hole is deformed but no line scar can be seen. The line scar can be formed from the wavy creases after they are folded oved by the rolling contact [4]. In perforated films, the creases disappear when the wavy creases are pushed into the holes by rolling contact. This is why no line scar can be seen in figure 6(c). Perforated films may reduce the damage of the film due to the rolling contact.

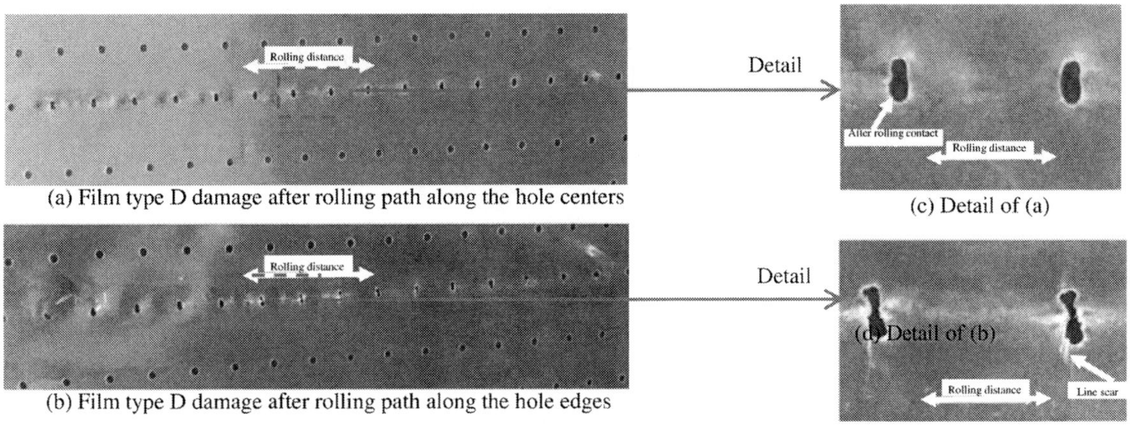

(a) Film type D damage after rolling path along the hole centers

(b) Film type D damage after rolling path along the hole edges

(c) Detail of (a)

(d) Detail of (b)

Figure 6. Perforated films type D in [4].

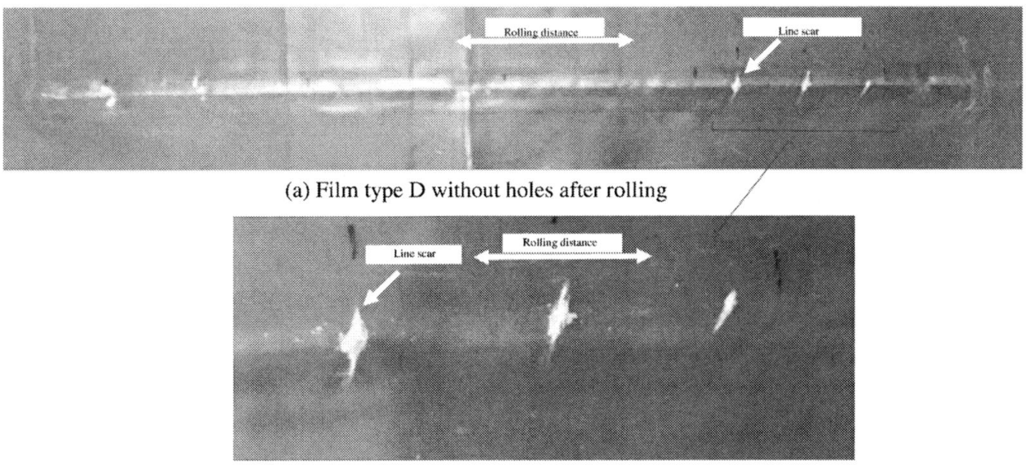

(a) Film type D without holes after rolling

(b) Detail of (a)

Figure 7. Film type D line scars without holes.

Figures 6(b) and 6(d) shows the film damage after rolling contact in figure 5(b) whose path is along the hole edges. The stretched zone can be seen in figure 6(b) similar to figure 7. The line scar also can be seen in figure 6(d). This is because many wavy creases are gathered around the hole edges and they are folded over and lead to the line scar as shown in figure 6(d). Larger hole deformation also can be seen in figure 6(d) compared to the deformation in figure 6(c). Generally speaking, the perforated film damage is smaller than or nearly equal to the damage of the film without hole because the holes reduce the average rigidity of the film and therefore contributes to supress the creation of the wavy creases.

5. Conclusions

A novel greenhouse is being used recently since automated rolled-up ventilation system is contained in the roof to control the temperature. Until now, however, full-opening type greenhouse are not widely used because the plastic film deteriorates in three or four months due to the frequent movement of the roll-up system. To prevent the film damage under rolling contact loading, in this paper, rolling contact simulation and experiment were conducted by focusing on the surface roughness of the flame pipe. Then, the conclusions can be summarized

- FEM rolling contact simulation was performed to clarify the surface roughness on the film damage by assuming a single protrusion on the smooth surface. It was found that larger film strain can be seen at the inner film surface compared to at the outer film surface. With decreasing the root radius of the protrusion and with increasing the height of the protrusion the difference become larger. The larger difference may lead the line scar creation at the film inside surface.

- The experimental study showed that the number of line scar increases with increasing the surface roughness of the pipe. The results can be explained from the FEM simulation considering surface roughness.
- Coated pipe may reduce the film damage. This is because the pipe roughness becomes smaller by the coated layer.
- Perforated film may reduce the damage of the film. This is because the holes reduce the average rigidity of the film and therefore contributes to supress the creation of the wavy creases.

References

[1] Johnson K L 1985 *Contact Mechanics* (Cambridge, UK: Cambridge University Press)
[2] Nakayama K and Takasu H 1993 Effect of Molecular Weight on the Rolling Behaviour of Polypropylene Sheets *J Soc Fiber Sci Tech* **49** 54-9
[3] Wang L, Noda N-A, Luo Q, Nagatomo H, Sano Y and Takase Y 2014 Rolling contact fatigue simulation used for full-open type greenhouse *International Journal of Fracture Fatigue and Wear Proc the 3rd Int Conf Fracture Fatigue Wear* **2** 201-6
[4] Noda N-A, Wang L, Nagatomo H, Sano Y and Takase Y 2016 Contact analysis and simulation of rolled plastic film used for roof ventilation in japanese greenhouses *Trans. of the ASABE* **59** 1421-37
[5] Noda N-A, Chen X, Sano Y, Wahab M A, Maruyama H and Fujisawa R 2016 Effect of pitch difference between the bolt–nut connections upon the anti-loosening performance and fatigue life *Mater Des* **96** 476-89
[6] Noda N-A, Suryadi D, Kumasaki S, Sano Y and Takase Y 2015 Failure analysis for coming out of shaft from shrink-fitted ceramic sleeve *Eng Failure Anal* **57** 219-35
[7] https://www.sswm.info/content/mulching
[8] https://books.google.co.jp/books?id=F1eghGKD6bwC&pg=PA18&dq=perforated+film+agriculture&hl=ja&sa=X&ved=0ahUKEwjvlY3svb7YAhUIlZQKHelPDywQuwUILTAA#v=onepage&q=perforated%20film%20agriculture&f=false

IOP Conf. Series: Materials Science and Engineering **369** (2018) 012006 doi:10.1088/1757-899X/369/1/012006

Synthesis and characterization of a novel polyborosilazane for SiBNC ceramic

C Y Zhang[1], Y Liu[1,2], K Q Han[1], X F Chang[1] and M H Yu[1,2]

[1]State Key Laboratory for Modification of Chemical Fibers and Polymer Materials, Key Laboratory of High-Performance Fiber and Product, Ministry of Education, College of Materials Science and Engineering, Donghua University, Shanghai 201620, P R China

E-mail: yumuhuo@dhu.edu.cn/ liuyong@dhu.edu.cn

Abstract. A novel polyborosilazane (PBSZ) for preparing SiBNC ceramics was successfully synthesized via co-condensation approach using tetrachlorosilan ($SiCl_4$), trichloride (BCl_3) and propylamine ($C_3H_7NH_2$) as starting materials. After pyrolysis of these precursors, amorphous SiBNC ceramics were obtained. The chemical composition, structure and thermal stability of the synthesized PBSZ precursor and SiBNC ceramics were analyzed by using FT-IR, NMR, TGA and XRD methods. The results indicated that the PBSZ contained the major framework of –Si-N-B- and six-membered boron-nitrogen rings. The PBSZ precursor had an approximately ceramic yield of 63 wt% prolyzed at 900°C in nitrogen atmosphere. The SiBNC ceramics shows excellent oxidation resistance and maintained amorphous up to 1600°C.

1. Introduction

Non-oxide amorphous ceramic derived from preceramic polymers have been researched in recent years, such as BN (derived from polycarbosilane), SiC (derived from polyborazine) and SiBNC (derived from polyborosilazane) [1-5]. Among them, the quaternary system SiBNC due to its fine chemical stability, low density, high temperature stability and excellent oxidation resistance, were appropriate for application in high temperature applications and have drawn much attentions [6,7]. However, it is difficult to prepare SiBNC ceramics using traditional powder sintering methods. Up to date, the most feasible approach to prepare SiBNC ceramic (fibers) is the polymer-derived ceramics (PDCs) route [8]. Therefore, to prepare high performance SiBNC ceramics, the key step is to prepare high performance polymer precursor (that is, polyborosilazane, PBSZ) [9-11].

With the aim of improving the performance of PBSZ, many researchers have synthesized many different kinds of PBSZ. For instance, Sneddon *et al* [12] prepared high-yield borazine-substituted hydridopolysilazanes via the reaction of liquid borazine with hydridopolysilazane (HPZ). The boron-modified polymers have higher ceramic yield than the unmodified hydridopolysilazane on account of the structures of borazine-boron in the ceramic. Bernard *et al* [13] prepared $[B(C_2H_4SiCH_3NCH_3)_3]$ through the aminolysis of $[B(C_2H_4SiCH_3Cl_2)_3]$ (tris(dichlorosilylethyl)boranes). Jansen *et al* [14] synthesized precursors by cross-linking of Cl_3Si-CH_2-BCl_2 (trichlorosilyl dichlorobryl methance), $(CH_3)Cl_2Si$-CH_2-BCl_2 (dichlorosilyl dichloroboryl methane) and methylamine. The obtained SiBNC ceramic have high-temperature stability and remarkable high-temperature oxidation resistance in air atmosphere at least up to 1300°C. However, Most synthesis routes are multiple-step, which adding complexity to the process and producing by-product [15,16]. Moreover, both molecular and intermediates are sensitive to moisture, the whole reaction process should keep in sufficiently inert and dry environment and it is difficult to multiple-step processes [17,18].

In this paper, a novel PBSZ was prepared via a copolymerization of BCl_3, $SiCl_4$ and $C_3H_7NH_2$. Moreover, the structure and performance of the obtained PBSZ were studied.

2. Experimental procedures

2.1. Materials

BCl_3 solved in n-hexane were provided by Beijing Multi-Technology Co. n-hexane were provided by Sinopharm Chemical Reagent Co. Ltd. $SiCl_4$ with 99.99% purity were purchased from Bai Lingwei Chemical Co. $C_3H_7NH_2$ with 98.50% purity were obtained from Guoyao Chemical Co. Ltd. Nitrogen with 99.99% purity were used during preparation.

2.2. Polymer precursor synthesis

The glass reaction vessels were dried for 4 h at 100°C. Then vacuumed and filled with high purity nitrogen (N_2). Dried n-hexane (3 L) was cooled to -15 ~ -20°C, and then $C_3H_7NH_2$ (600 ml) dissolved in the pre-dried n-hexane. The mixture of BCl_3 (440 ml) and $SiCl_4$ (50 ml) dissolved in 440 ml n-hexane was gradually added into the mixture solution of $C_3H_7NH_2$ and n-hexane with fully mechanical stirring (the temperature ⩽ -15°C). Then the solution was stirred overnight at room temperature. After the solution standing for an hour, the upper clear liquid was collected and the byproduct was filtered. As a result, about 300 ml oligomers of polyborosilizane could obtained. The oligomers was polymerized at 180-230°C for 50-60 h to gain PBSZ with appropriate viscosity.

2.3. Polymer to ceramic conversion

The obtained PBSZ has lots of active groups, so it was extremely sensitive to air and must be handled under inert atmosphere. The PBSZ was cured from 50 to 350°C in a horizontal Schlenk furnace with a heating rate of 0.25°C/min in flowing nitrogen. Then the PBSZ was pyrolyzed by further heat treated using a heating rate of 0.5°C/min from 350 to 800°C and maintained 2 h at 800°C. Finally, further heated to 1000°C and had a dwelling time for 2 h under nitrogen atmosphere.

2.4. Characterization

Fourier transform infrared (FT-IR) spectra were recorded in the 4000-400 cm^{-1} frequency range on a Nicolet NEXUS-670, using KBr pellets. NMR experiment was performed on a a Bruker Avance 400 spectrometer. ^{1}H and ^{11}B NMR spectra were recorded in $CDCl_3$ with resonance frequencies at 400.13 and 128.37 MHz. Solid-state ^{13}C and ^{29}Si NMR spectra were recorded at 100.61 and 79.49 MHz, respectively, under MAS conditions. Thermo gravimetric analysis (TGA) of the as-obtained samples operating a Netzach 209 F1 Iris instrument was conducted under nitrogen atmosphere. The samples (about 10 mg) were heated from 50°C to 900°C with the heating rate of 10°C/min. Gas Chromatography-Mass Spectrometer (GC-MS) of samples was analyzed by GCMS-QP-2010 instrument. The crystallization of the ceramics was characterized by X-ray diffraction (XRD) analysis (XRD Rigaku, Tokyo, Japan, Cu-K radiation) using graphite crucibles.

3. Results and discussion

3.1. Characterization of PBSZ

Structures of PBSZ were investigated via FT-IR and NMR spectroscopy. As shown in Figure 1, the signal at 3418 cm^{-1} was assigned to N-H bond stretching vibration. The signals at 2959 cm^{-1}, 2931 cm^{-1} and 2872 cm^{-1} were attributed to C-H stretching. The band assigned to B-H bond was located at 1378 cm^{-1} and 1338 cm^{-1}. Assigned to Si-C band in the $Si\text{-}CH_2CH_2CH_3$ was located at 1290 cm^{-1}, 1249 cm^{-1} and 1194 cm^{-1}. A signal peak at 1460 cm^{-1} was attributed to the stretching absorption of N-C and the speaks at 1168 cm^{-1},1072 cm^{-1} and 1023 cm^{-1} were assigned to the stretching vibration band of N-C.

IOP Conf. Series: Materials Science and Engineering **369** (2018) 012006 doi:10.1088/1757-899X/369/1/012006

The characteristic peak of Si-N was observed at 892 cm^{-1} and the peak appeared at 743 cm^{-1} indicated the existence of B-N.

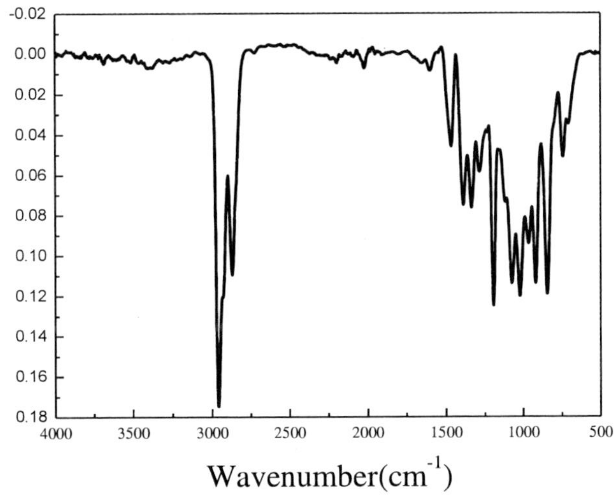

Figure 1. FTIR spectrum of PBSZ co-polycondensed of $SiCl_4/BCl_3/C_3H_7NH_2$.

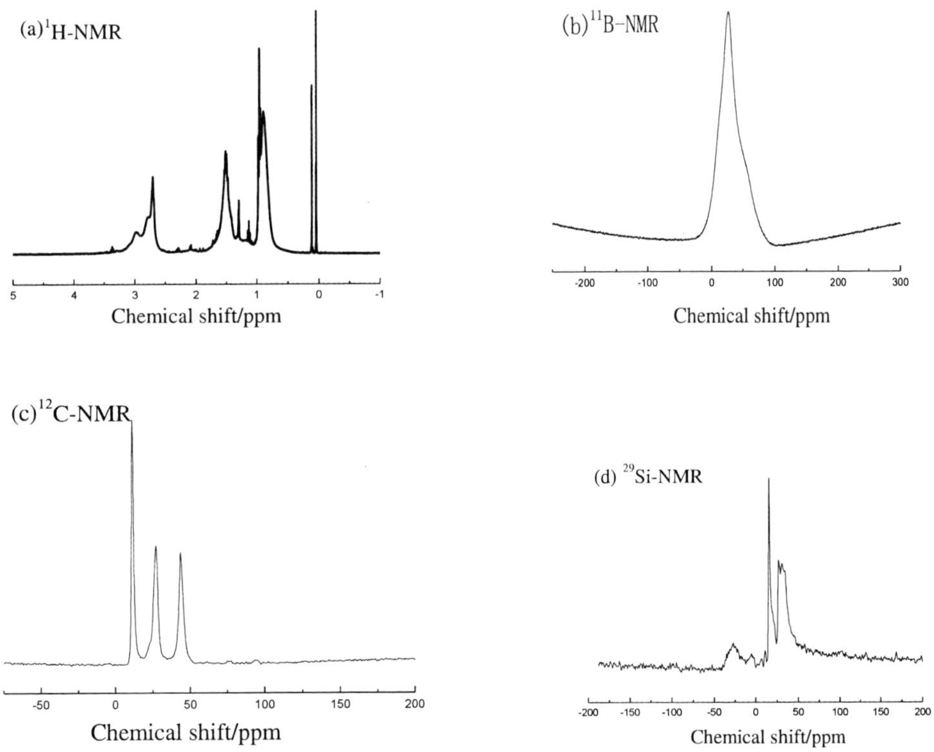

Figure 2. NMR specta of PBSZ copolycondensed of $SiCl_4/BCl_3/C_3H_7NH_2$: (a) ^{1}H-NMR, (b) ^{11}B-NMR, (c) ^{12}C-NMR, (d) ^{29}Si-NMR.

Furthermore, some signals in the NMR spectra were conductive to confirming the structure of the PBSZ. Figure 2 displays the ^{1}H, ^{11}B, ^{29}Si and ^{12}C NMR spectra of the PBSZ. As shown in figure 2(a) of the ^{1}H-NMR spectrum, the signals at 0~0.3 ppm were assigned to Si-$CH_2CH_2CH_3$; the broad peak at 0.5~1.0 ppm was assigned to N-*H*; the multiple peaks at 1.0~1.8 ppm were attributed to B-NCH_2CH_2-Si and the broad peak at 2.5~3.3 ppm was assigned to N-$CH_2CH_2CH_3$. In the ^{11}B-NMR spectrum (figure 2(b)), only a broad peak appeared at 25~27 ppm, which due to BN structural units in six membered borazine rings [19]. In the ^{13}C-MAS-NMR spectrum (figure 2(c)), the signals appeared at 10 ppm, 26 ppm and 44 ppm were attributed to three positional relationships of carbon in N-$CH_2CH_2CH_3$. Also comprise the structure of $SiN(CH_2CH_2CH_3)Si$, $BN(CH_2CH_2CH_3)B$ and $SiN(CH_2CH_2CH_3)B$. The difference of chemical shift of the three kinds of N-$CH_2CH_2CH_3$ was not large is due to the negative induction effect of Si, B to C was delivered by σ bond and the delivery effect of σ band was weak. Therefore, the three peaks in ^{13}C-MAS-NMR spectrum were sharp. In the ^{29}Si-MAS-NMR spectrum (figure 2(d)), the signal at 33 ppm was assigned to Si-N bonds and the signal at 19 ppm was assigned to Si-C bonds.

3.2. Ceramic conversion of PBSZ

The progress of the polymer-to-ceramic conversion of PBSZ was illustrated by TGA from room temperature to 900°C under a flowing nitrogen atmosphere. As shown in Figure 3, the mass loss of PBSZ mostly taken place in three stages. Firstly, from the room temperature to 200°C, the evaporation of low molecular weight molecules leaded to the weight loss of PBSZ. Secondly, the weight loss of PBSZ was attributed to the further cross-linking from 200°C to 500°C, accompany with the release of $C_3H_7NH_2$. At the third stage, B-N and Si-N bonds broken and released $C_3H_7NH_2$ and C_3H_8 from 500°C to 800°C. The final residual mass of PBSZ was about 64.9 wt% at 900°C.

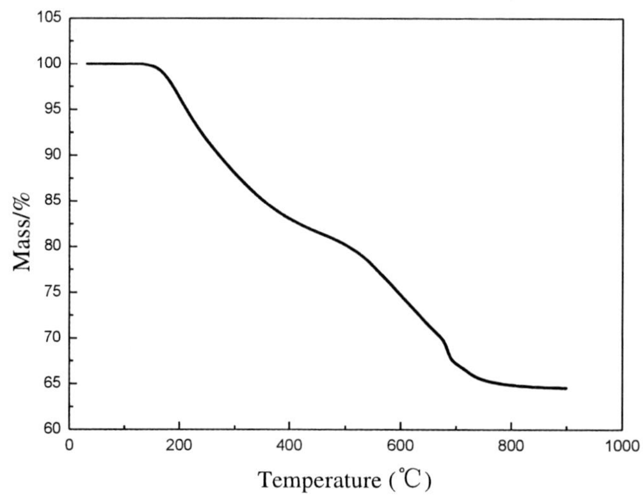

Figure 3. TGA curves of PBSZ copolycondensed of $SiCl_4/BCl_3/C_3H_7NH_2$ (N_2 atmosphere).

To further analysis the pyrolysis process of PBSZ. The released gases at different temperature (200°C, 400°C, 550°C) during the pyrolysis were analyzed by GC-MS. As shown in figure 4, at 200°C, the released gas mainly was the release of small molecules. When the temperature enhanced to 450°C, $C_3H_7NH_2$ and little amount of oligomerss released from the system. The release of $C_3H_7NH_2$ was attributed to the condensation between oligomerss and the release of oligomerss was resulted from the hindrance of cross-linking because of steric-hinerance effect. At 550°C, besides the release of $C_3H_7NH_2$, there was C_3H_8 released from the system. This was due to the broken of B-N and Si-N bonds in nonreduced end of Si-$NHCH_2CH_2CH_3$ and B-$NHCH_2CH_2CH_3$. As we can see, the GC-MS results agreed with the TGA analysis.

Figure 5 shown the SEM microstructure of the SiBNC ceramics pyrolyzed at 800°C. It could be seen that the SiBNC granulation was nearly global and on the surface formed a dense layer.

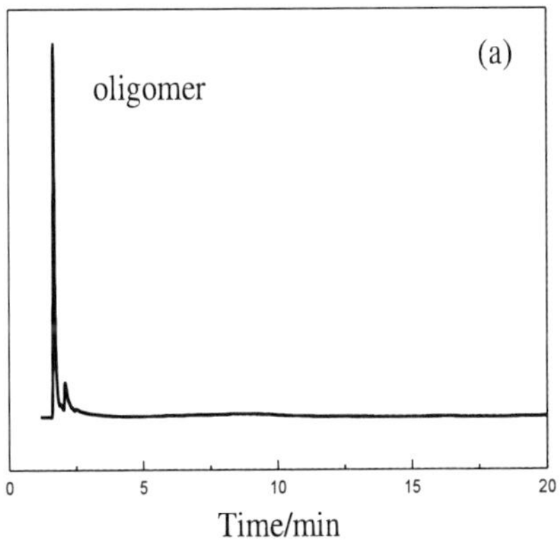

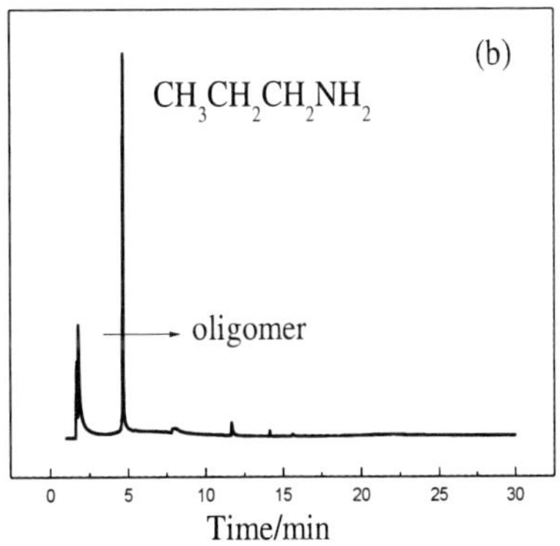

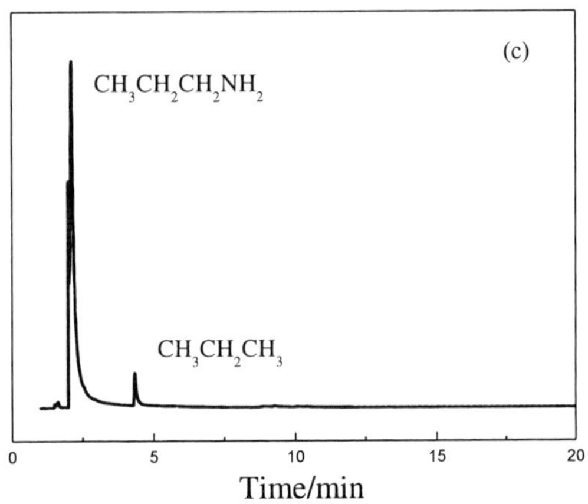

Figure 4. GC-MS spectra of PBSZ pyrolyzed at different time (N_2 atmosphere): (a) 200°C; (b) 450°C; (a) 650°C.

Figure 5. SEM of the SiBNC ceramics pyrolyzed at 800°C.

To investigate crystallization behavior, PBSZ pyrolyzed at different temperatures for 6 h were analyzed by XRD. As shown in figure 6, the pyrolyzed samples from 1000 to 1600°C were all exhibited only an broad peak at 2θ=22.5°, assigned to amorphous silicide, which indicated the high temperature stability and crystallization resistance of SiBNC ceramics.

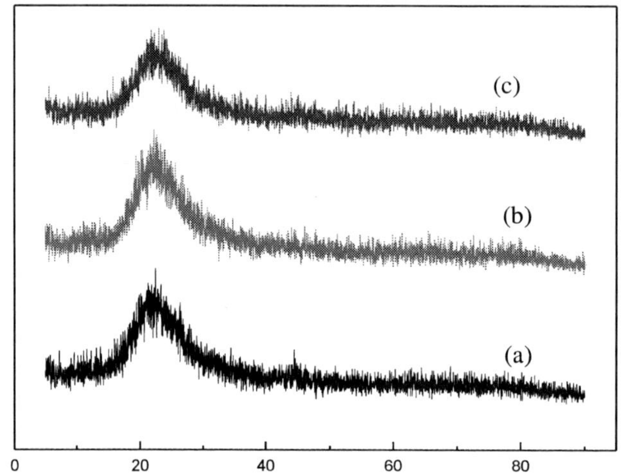

Figure 6. XRD spectra of polymer-derived ceramics pyrolyzed at different temperatures for 6 h: (a) 1000°C; (b) 1400°C; (a) 1600°C.

4. Conclusion

This study demonstrates the possibility of producing high performance SiBNC ceramic derived from a novel polymeric precursor. This polymeric precursor was synthesized in one-step by co-condensation of BCl_3, $SiCl_4$ and $C_3H_7NH_2$. FT-IR and NMR analysis indicated that it contains B-N, Si-N, N-H and N-$CH_2CH_2CH_3$ bonds. The ceramic yield of this polymeric precursor is about 64.9 wt% after annealed at 900°C in nitrogen atmosphere. XRD spectra illustrate that the obtained SiBNC ceramic exhibit high temperature stability and keep amorphous after annealed at 1600°C in nitrogen atmosphere.

Acknowledgments

This work was supported by "the Fundamental Research Funds for the Central Universities"(No. 15D110638), National Science Foundation of Shanghai (No. 15ZR1400400), plan of action for the innovation of Science and Technology in Basic Research of Shanghai (No. 14JC1400100).

References

[1] Lei Y-P and Song Y-C 2012 Boron nitride by pyrolysis of the melt-processable poly [tris (methylamino) borane]: structure, composition and oxidation resistance *Ceramics Int* **38** 271-6
[2] Yajima S, Hayashi J and Omori M 1975 Continuous silicon carbide fiber of high tensile strength *Chem Lett* **4** 931-4
[3] Kusari U, Bao Z, Cai Y, *et al* 2007 Formation of nanostructured, nanocrystalline boron nitride microparticles with diatom-derived 3-D shapes *Chem Commun* **11** 1177-9
[4] Yan M, Song W and Chen Z-H 2011 In situ growth of a carbon interphase between carbon fibres and a polycarbosilane-derived silicon carbide matrix *Carbon* **49** 2869-72
[5] Tang Y, Wang J, Li X, *et al* 2010 Polymer - derived SiBN fiber for high-temperature

structural/functional applications *Chem Eur J* **16** 6458-62
[6] Lee J, Butt D P, Baney R H, *et al* 2010 Synthesis and pyrolysis of novel polysilazane to SiBCN ceramic *J Non-Cryst Solids* **351** 2995-3005
[7] Zhang C, Han K, Liu Y, *et al* 2017 A novel high yield polyborosilazane precursor for SiBNC ceramic fibers *Ceramic Int* **43** 10576-80
[8] Ge Y, Sun S, Wang Q, *et al* 2016 Effect of Fe-contained species on the preparation of α-Si_3N_4 fibers in combustion synthesis *J Am Ceram Soc* **99** 1464-71
[9] Wang Y, Chen L X, Xu T T, *et al* 2017 High char yield novolac modified by Si-B-N-C precursor: Thermal stability and structural evolution *Polym Degrad Stabil* **137** 184-96
[10] Wang S H, Hu X D and Dai Y J 2017 Preparation and electrochemical performance of polymer-derived SiBCN-graphene composite as anode material for lithium ion batteries *Ceramic Int* **43** 1210-6
[11] Zhang Q, Yang Z H, Jia D C, *et al* 2016 Synthesis and structural evolution of dual-boron-source-modified polysilazane derived SiBCN ceramics *New J Chem* **40** 7034-42
[12] Fazen P J, Remsen E E, Beck J S, *et al* 1995 Synthesis, properties, and ceramic conversion reactions of polyborazylene. A high-yield polymeric precursor to boron nitride *Chem Mater* **7** 1942-56
[13] Bernard S, Weinmann M, Cornu D, *et al* 2005 Preparation of high-temperature stable Si B C N fibers from tailored single source polyborosilazanes *J Eur Ceram Soc* **25** 251-6
[14] Jäschke T and Jansen M 2004 Synthesis and characterization of new amorphous Si/B/N/C ceramics with increased carbon content through single-source precursors *C R Chim* **7** 471-82
[15] Yang L J, Zhang Y and Cheng X 2016 Performance and structure of polymer derived SiBCN ceramics *Prog Chem* **28** 308-16
[16] Viard A, Miele P and Bernard S 2016 Polymer-derived ceramics route toward SiCN and SiBCN fibers: from chemistry of polycarbosilazanes to the design and characterization of ceramic fibers *J Ceram Soc Jpn* **124** 967-80
[17] Mou S W, Liu Y, Han K Q, *et al* 2015 Synthesis and characterization of amorphous SiBNC ceramic fibers *Ceramic Int* **41** 11550-4
[18] Al-Ghalith J, Dasmahapatra A, Kroll P, *et al* 2016 Compositional and structural atomisticstudy of amorphous Si-B-N networks of interest for high-performance coatings *J Phys Chem C* **120** 24346-53
[19] Peng Y Q, Han K Q, Zhao X, *et al* 2014 Large-scale preparation of SiBN ceramic fibres from a single source precursor *Ceramic Int* **40** 4797-804

IOP Conf. Series: Materials Science and Engineering **369** (2018) 012007 doi:10.1088/1757-899X/369/1/012007

Rheological characterization of a polyethylene melt, filled with different concentrations of fumed silica at 453 K

D Hadjistamov[1,3] and T Schweizer[2]

[1]DECE GmbH, Helvetierstr. 15, CH-4125 Riehen, Switzerland
[2]ETH Zürich, Institute of Polymers, CH-8093 Zürich, Switzerland

E-mail: dhadjistamov@datacomm.ch

Abstract. Fumed silica is widely used as filler / thickening agent to increase the viscosity or to change the flow behavior type of a suspension. Strong hydrogen bonds between silica-silica and silica-matrix can lead to a suspension with a yield stress. We wanted to know if fumed silica can also form a three-dimensional network in a polyethylene melt at 453 K. The systems with the polyethylene LDPE 1810H (LDPE) filled with different concentration of the fumed silica Cab-o-sil TS720 (silica) were investigated with the MCR300 rheometer from Anton Paar at 453 K. The shear stress τ and the 1st normal stress difference were measured simultaneously in stress growth experiments. The points of the flow curves represent the steady state values from the start-up experiments with a constant shear rate. Just after the steady state values are reached, the stress deformation is stopped, and the stress relaxation begins. The relative residual shear stress is an important value. The absence of residual shear stress means that the structure of the system is destroyed in the stress growth experiment and the system has shear-thinning flow behavior. The systems with high residual shear stress form a strong network that leads to a system with plastic flow behaviour with a yield stress. The shear stress changes insignificantly in the yield stress region with the shear rate. The yield stress can be determined also from the relative residual shear stress curve.

1. Introduction

Fumed silica is used in paints, tooling systems, food, toothpaste etc. [1]. Silica can be added to avoid the sedimentation of the fillers [2]. Systems with fumed silica are studied in polydimethylsiloxane rubber [3,4], in different rubbers [5-7], in the epoxy resin Araldite GY 260 [8], in polydimethylsiloxane [9-18], in polyethylene terephthalate [19]. Recently, the rheological behavior of precipitated silica in polydimethylsiloxane [20] and CPE/NR blends [21] has been investigated. A thermodynamic approach describes the rheological behavior of fumed silica in polystyrene [22,23]. Measurements of the 1st normal stress difference N_1 in colloidal suspensions are scarce, although special effects like negative N_1 have been reported [24]. Cassagnau [25] reports about rheology of silica in organoclay nanocomposites.

No publications exist about the rheological characterization of fumed silica in the LDPE 1810H.

The fumed silica increases the viscosity of a suspension through hydrogen-bonding that builds a three-dimensional network. Depending on the concentration of the fumed silica, the effect can be only an increase in the viscosity or an increase in the viscosity and a change in the flow behavior of the system - to a system with plastic flow behavior [1,8,16]. The main feature of the systems with plastic flow behavior is the existence of a yield stress τ_y or a yield stress region. It can be identified from the flow curve as a region where the shear stress changes insignificantly with increasing shear rates. Many suspensions – e.g. paints, adhesives, pigment dispersions and others show plastic flow behavior.

The main purpose of this paper is to find out whether fumed silica can induce plastic flow behavior in a melt of LDPE at 453 K.

2. Experimental

The LDPE 1810H (BASF AG) is a low-density polyethylene with a weight average molecular weight of 186 kg/mol. The fumed silica Cab-o-sil TS 720 (hydrophobic) has a specific surface area of 100 ±20 m^2/g. The specific gravity is 1.8 g/cm^3.

LDPE granules with different concentrations of silica were mixed on a Brabender kneader in a 30 cm^3 chamber at 30 rpm and 453 K until torque was constant. The resulting mixtures were hot pressed to tablets and investigated with the MCR 300 rheometer from Anton Paar at 453±0.5 K, with a cone-plate measuring system of 4° cone angle.

The points of the flow curves represent the steady state τ_S values from the shear flow start-up experiments with a constant shear rate $\dot{\gamma} = \mathbf{const.}$ (figure 1).

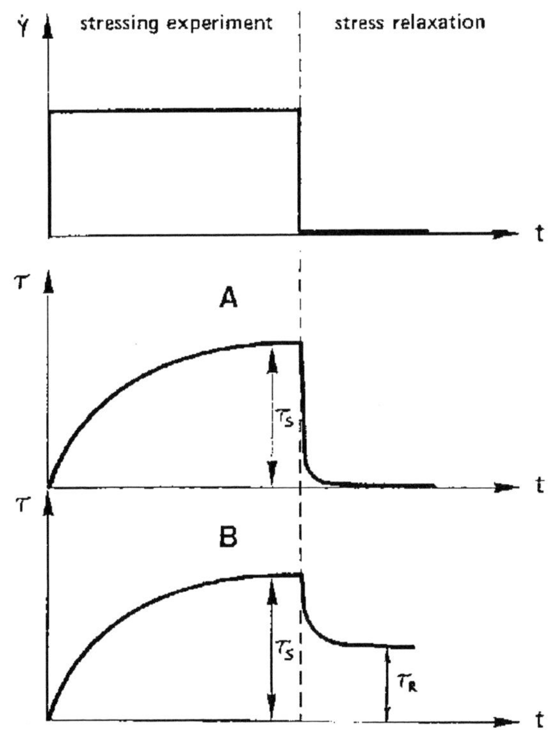

Figure 1. Shear flow start-up experiment – stress relaxation [1].

A shear flow start-up experiment – stress relaxation without residual shear stress $\tau_{\mathbf{R}} = 0$

B shear flow start-up experiment – stress relaxation with residual shear stress $\tau_{\mathbf{R}}$

When steady state flow is reached, the shear deformation is halted, and the stress relaxation is measured. If an existing structure has been destroyed during shear, no residual shear stress $\tau_{\mathbf{R}}$ is observed (figure 1A). If the structure is only partially destroyed after the start-up experiment, a residual shear stress $\tau_{\mathbf{R}}$ is observed (figure 1B). The relative residual shear stress $\tau_{RR} = \frac{\tau_R}{\tau_S}$ is the ratio of residual shear stress to steady state shear stress. It is assumed that the relative residual shear stress represents the part of the structure that has not been destroyed during shear. A system with a better-preserved structure exhibits a larger τ_R and τ_{RR} respectively.

3. Results and discussion

Figure 2 represents the loss modulus curve of LDPE and the flow curve of the system with LDPE, filled with 5wt% silica.

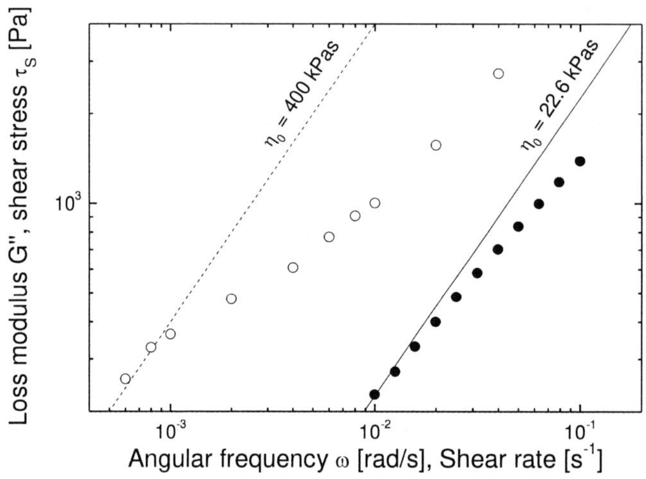

Figure 2. Loss modulus curve of LDPE (•) and flow curve of LDPE, filled with 5wt% fumed silica (o) (MCR 300 rheometer from Anton Paar).

The LDPE melt shows shear-thinning flow behavior with a first Newtonian region up to the onset of shear thinning at 340 Pa. The zero-shear complex viscosity has a value of 22.6 kPas. The addition of 5wt% Cab-o-sil to the LDPE causes only an increase in the shear stress. This melt exhibits shear thinning flow behavior with a zero-shear viscosity of 400 kPas. This value is about 18 times larger than the zero-shear viscosity of unfilled LDPE. The first Newtonian region of the filled melt runs up to the onset of shear thinning of 360 Pa, which is almost equal to the value for the unfilled LDPE. The LDPE, filled with 5wt% fumed silica, as a system with shear-thinning flow behavior, does not indicate any residual shear stress.

The rice of the silica concentration from 5wt% to 10wt% causes at given shear rate a considerable increase in the shear stress (figures 2 and 3), and in the viscosity respectively. The melt with 10wt% silica builds a strong network formation that leads to a change in the flow behavior from shear-thinning (figure 2) to plastic (figure 3).

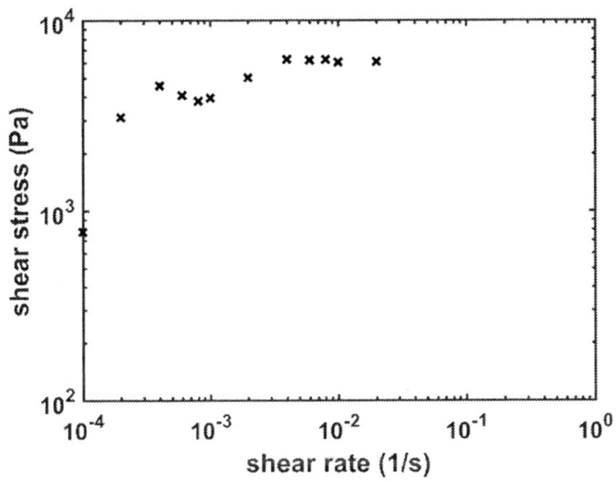

Figure 3. Flow curve of LDPE, filled with 10wt% silica (MCR 300 rheometer, Anton Paar 453 K, CP 50-4).

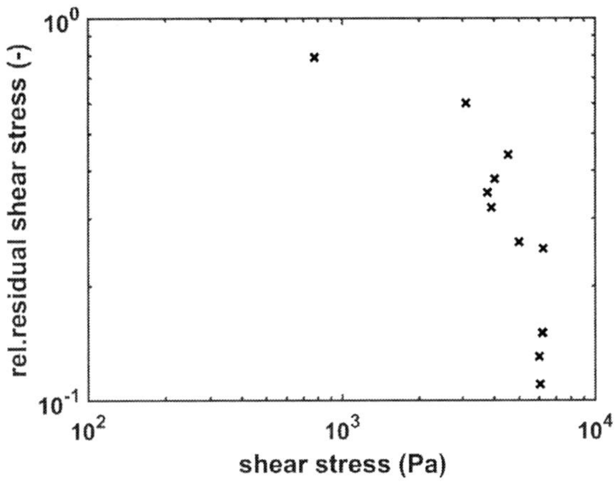

Figure 4. Relative residual shear stress of LDPE with 10wt% silica.

The flow curve of the LDPE melt with 10wt% silica demonstrates two regions where the shear stress changes insignificantly with the shear rate. These sections can be assumed to be yield stress regions at 3.9 kPa and 6.2 kPa. Two yield stress regions have also been experimentally observed for systems with 20wt% and 25wt% Bentone 27 (organoclay) in Araldite GY [1].

The first point of the relative residual shear stress curve has a value of 0.79. About 79% of the three-dimensional structure is still retained or 21% of the structure is destroyed in the shear flow start-up experiment. The values of the relative residual shear stresses (figure 4) decrease in the yield stress regions and consequently a strong distortion of the structure occurs in these regions. Between the first and second yield stress regions there is a transition section (figure 4), where the destruction of the structure and the decrease of the relative residual shear stress slows down with the shear stress.

It is possible to determine the yield stress regions of the filled polyethylene melts also from the relative residual shear stress curve (figure 4). The evaluated yield stress regions are 3.9 kPa and 6 kPa for LDPE, filled with10wt% silica. There is a good agreement between the yield stress values obtained from the flow curve (figure 3) and from the relative residual shear stress curve (figure 4).

The addition of 15wt% silica (figure 5) leads also to a system with plastic flow behavior.

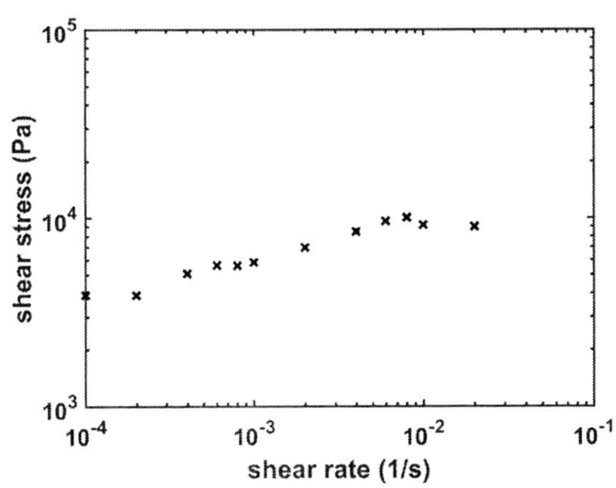

Figure 5. Flow curve of LDPE, filled with 15wt% silica (MCR 300 rheometer, Anton Paar, 453K CP 50-4).

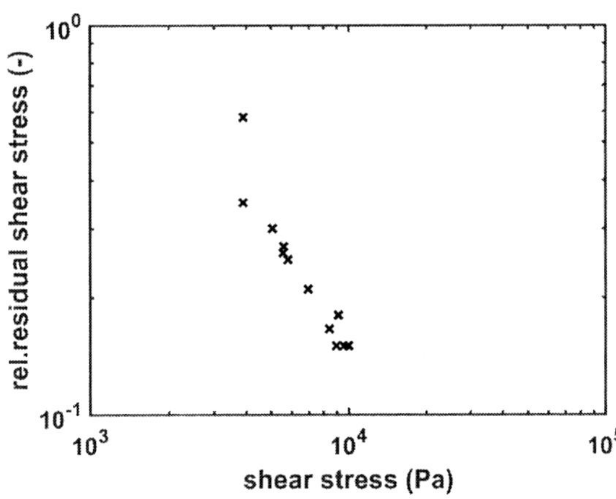

Figure 6. Relative residual shear stress of LDPE with 15wt% silica.

The flow curve of the melt with 15wt% fumed silica (figure 5) shows nearly equal shear stress values compared with the system with 10wt% (figure 3). The higher silica concentration leads to a higher shear stress but does not promote to the expected much bigger strengthening of the structure. The concentration of 10wt% seems to be optimal. The system with 15wt% shows plastic flow behavior with a yield stress section at 3.8 kPa and a second yield stress region at 5.7 kPa (figure 5).

Figure 6 represents the dependence of the relative residual shear stress on the shear stress for the systems with 15wt% silica in LDPE. The first point has a relative residual shear stress of 0.58. About 58% of the three-dimensional structure is still left or 42% of the structure has been destroyed during shear. The relative residual shear stress of the first point has a lower value (figure 6) compared to that of the system with 10wt% silica (figure 4).

Considering the relative residual shear stress curves (figure 6) we can evaluate the yield stress regions of the filled with 15wt% polyethylene melt with 3.9 kPa and 5.6 kPa. There is a good agreement between the yield stress values obtained from the flow curve (figure 5) and from the relative residual stress curve (figure 6).

4. Conclusion

The LDPE shows shear-thinning flow behavior. Silica works at a concentration of 5wt% only as a filler – a large increase of the shear stress is observed (figure 2), but the system retains the shear-thinning flow behavior.

Additions of 10wt% and 15wt% fumed silica form a three-dimensional network which is strong enough to lead to systems with plastic flow behavior with two yield stress regions (figures 3 and 5). The shear stress changes insignificantly with shear rate in these regions, there is a strong decrease in the viscosity respectively. Between the two yield stress regions is the transition section. A rearrangement of the structure takes place in this region. The high values (0.79 and 0.58) of the relative residual shear stress reveal a strong structure in the melt. It is possible to determine the yield stresses from the flow curve and from the relative residual shear stress curve.

The flow behavior of a polyethylene melt with fumed silica depends obviously on the silica concentration. One can now conclude that the fumed silica Cab-o-sil TS720 is able to form a three-dimensional strong structure not only in suspensions with a low viscosity matrix like epoxy resins [8] at room temperature, but also in the polyethylene melt LDPE 1810H at 453 K.

References

[1] Hadjistamov D 2010 *Polym. Eng. Sci.* **50** 811-8
[2] Hadjistamov D 1978 *Colloid Polymer Sci.* **256** 770-83
[3] Southwart D W 1970 *J. Inst. Rubber Ind.* **4** 77-81
[4] Boonstra B *et al* 1975 *Rubber Chem. Technol.* **48** 558-76
[5] Okel T A and Wadell W H 1995 *Rubber Chem. Technol.* **68** 59-76
[6] Shailesh D *et al* 1998 *Rubber World* **218** 1-21
[7] Mongruel A and Cartault M 2006 *J. Rheol.* **50** 115-35
[8] Hadjistamov D 2003 *Appl. Rheol.* **13** 209-11
[9] Cantu T S and Caruthers J M 1982 *J. Appl. Polym. Sci.* **27** 3079-88
[10] Ziegelbaur R S and Caruthers J M 1985 *J. Non-Newtonian Fluid Mech.* **17** 45-68
[11] Kosinski L E and Caruthers J M 1986 *Rheol. Acta* **25** 153-60
[12] DeGroot J V *et al* 1994 *J.Colloid Interface Sci.* **166** 404-13
[13] Piau J-M *et al* 1999 *J.Rheol.* **43** 305-14
[14] Hadjistamov D 1993 *Appl. Rheol.* **3** 113-9
[15] Hadjistamov D 1995 *Appl. Rheol.* **5** 29-33
[16] Hadjistamov D 2002 *Appl. Rheol.* **12** 297-301
[17] Selimovic S *et al* 2007 *J. Rheol.* **51** 325-40
[18] Ma T *et al* 2017 *J. Rheol.* **61** 205-15

[19] Chung S C *et al* 2002 *Macromol. Res.* **10** 221-9
[20] Shim S E and Isayev A I 2004 *Rheol. Acta* **43** 127-36
[21] Phewthongin N *et al* 2006 *J. Appl. Polymer Sci.* **100** 2565-71
[22] Havet G and Isayev A I 2001 *Rheol. Acta* **40** 570-81
[23] Havet G and Isayev A I 2003 *Rheol. Acta* **42** 47-55
[24] Lee M *et al* 2006 *J. Rheol.* **50** 293-311
[25] Cassagnau P H 2008 *Polymer* **49** 2183-96

Modification with polymeric microspheres and its influence on thermal stability, mechanical properties and morphology of polyethylene

A Tor-Świątek

Department of Technology and Polymer Processing, Faculty of Mechanical Engineering, Lublin University of Technology, 36 Nadbystrzycka St., 20-618 Lublin, Poland

E-mail: a.tor@pollub.pl

Abstract. The structure modification of polymers has become essential due to a constant development of their applications, the necessity of changing or enhancing their functional and operational properties as well as higher product requirements. The aim of this study is to identify changes in the thermal, mechanical and structural properties. The article presents research results of low density polyethylene modification with a blowing agent in a form of polymeric microspheres. The temperature of extrudate during extrusion was regards to extrudate properties the morphology and mechanical analysis was carried out.

1. Introduction

The continuous development of the porous materials processing in the scope of the process and construction of machines, tools and devices included in the technological lines determines the enlargement of application areas for these materials. An important reason that causes the use of porous materials is the lowering of the products mass, both for complex or simple systems such as aircrafts, automobiles or packaging [1-3]. In most cases, the costs of manufacturing and their use are reduced. Due to their insulating properties, porous materials are used in construction and refrigeration [4]. They are good soundproofing agents, they can also be used to damp vibrations of machines. A higher interest in two-phase products has led to development of new processing methods such as cellular extrusion and cellular injection molding. A key factor here is the temperature of successive zones in the plasticizing unit. This parameter should be set such that the blowing agent undergoes decomposition in a suitable zone of the machine's plasticizing unit. Another important factor is mixing efficiency which is determined by screw and barrel design [5].

The modification based on the addition of blowing agents involves the introduction of these materials during the processing or directly during their production. These products, decomposed during processing, by forming a mixture of a gaseous substance and a polymer. Porous agents used in the processing of polymeric materials may be in the form of gas, powder, granules, liquids and microspheres [6].

Microcellular blowing agent (porophor) is selected depending on the type of plastic, in the manner ensuring that its decomposition temperature is higher than the plastic melting point, but lower than the plastic extrusion temperature. Cellular plastic in liquid form is not yet a stable system because as a result of interfacial tension at the plastic – gas contact point, and as a result of diffusion, the number of cells in the plastic is reduced, whereas their dimensions enlarge, which is an undesirable effect [7,8]. The obtained microcells continue to enlarge until gas pressure. Favourable structure of a small-cell plastic can be maintained in the product by its immediate cooling and solidification. Microcellular blowing agents used in the extrusion process can be divided into physical and chemical, however this classification is traditional and not exactly accurate [9]. Blowing agents used during microcellular extrusion of plastics may display exothermic or endothermic decomposition characteristics. Porophors used to date usually present exothermic decomposition characteristics. This may be the cause of local overheating and generation of irregular cellular structure of the product. The initiated porophor decomposition process progresses automatically even after energy cut-off. Therefore, products extruded with the use of such agents must be intensely cooled for a long time, in order to prevent strains and keep proper cellular structure [10,11].

As a result of separation of the porous structure, the products obtain new functional and technological properties, such as reducing the weight of the product, saving costs and materials, improvement of damping and insulating properties, reduction in processing shrinkage, increasing stiffness and reduction of flammability and water absorption [12].

Polymer microspheres belong to the new generation of porous agents, containing in their mass liquid hydrocarbon, which boils on the panes of elevated temperature and expands the polymer capsule.

The physical structure of porous products depends on the type of blowing agent used and its degradability and processing conditions and the polymer used [13,14]. To assess the physical structure, the basic geometric features of the pores should be determined, such as, for example, the pore counting ratio, the diameter and the pore distribution. The analysis of the physical structure can be carried out using the computer image analysis method, which allows quantitative and qualitative assessment of the size and distribution of pores in the cross-section of the porous product [15].

2. Experimental

2.1. Materials

LDPE was used, with the trade name Malen E FABS 23D022 (producer Basell Orlen Polyolefins) in the granulated form. The basic properties of the processed LDPE are shown in table 1.

Table 1. Selected properties of LDPE used (producer data).

Property	Value
Density, kg/m^3	921
MFI (190°C, 2,16 kg), g.10min	2
VST(A50, 10N), °C	91
Hardnes Shore (D)	48
Tensile strength, MPa	18
Elongation at break, %	450

As a modificator the polymeric microspheres in the trade name Expancel 951 MB 120 (producer AkzoNobel) was used. Expancel microspheres occur in the form of spherical thermoplastic polymer capsules containing hydrocarbon gas (figure 1). Under the influence of heat, the capsule softens and the expanding hydrocarbon increases its volume, without damaging the very thin shell of the capsule. It is a medium with endothermic nature of decomposition. Spheres can not connect with each other because the capsules retain their barrier properties, which also effectively prevent the release of closed liquid. Expancel 951 MB 120 is a mixture that contains 65% microspheres in an ethylene / vinyl acetate (EVA) copolymer. Properties of microspheres are shown in table 2. The blowing agent in the form of polymeric microspheres was incorporated to LDPE as 0, 0.25, 0.5 and 1% wt.

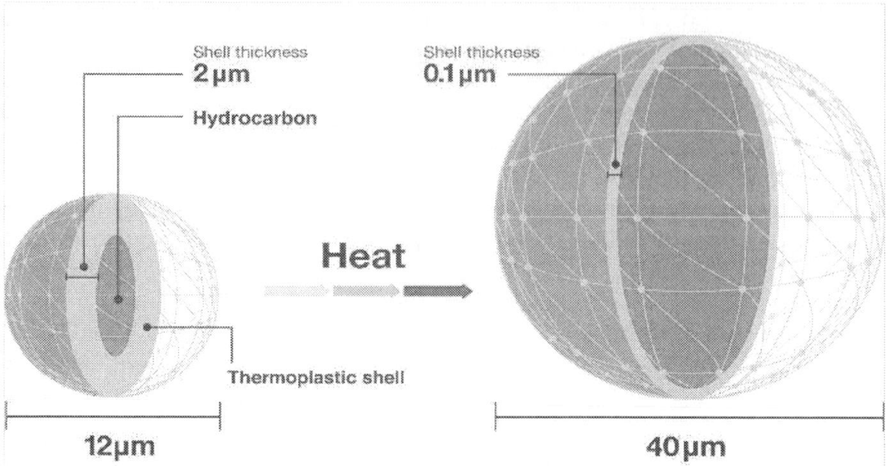

Figure 1. Scheme of the view and essence of action of microspheres (producer data).

Table 2. Selected properties of polymeric microspheres Expancel used (producer data).

Property	Value
Density with T_{max}, kg/m^3	12
Temperature of starting expansion, °C	138-148
Size, µm	28-38
T_{max}, °C	195-210
Processing temperature, °C	140-200

2.2. Processing
To investigate the polymeric microspheres modification effect on the thermomechanical properties and morphology of low-density polyethylene, the specimens were prepared by cellular extrusion process. The process was performed using a profile section extrusion line. This technological line consists of a single-screw extruder type T – 32 – 25 for extruding profiles and pipes. The parameters of the extrusion process was: temperature in plasticizing system zones 170, 180, 190, 200°C, temperature of extruder head 170°C, screw rotational speed 45 rpm, temperature of cooling factor 14°C. As a result, the microcellular extrudate was obtained in the form of a strip (figure 2).

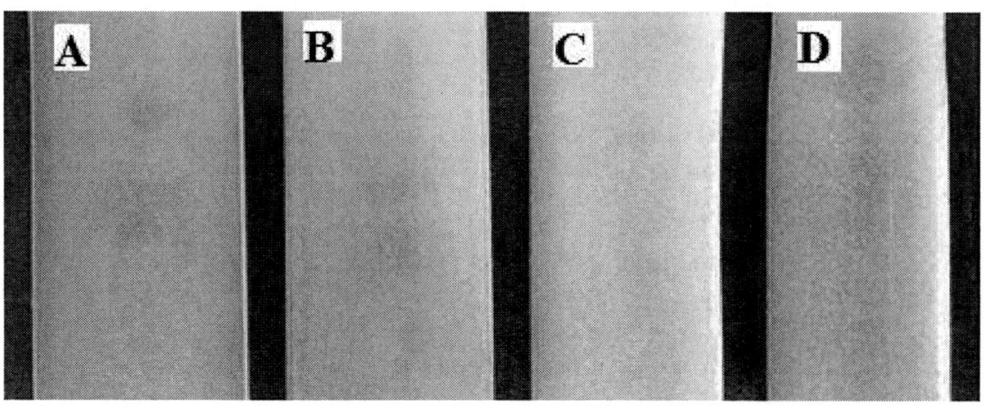

Figure 2. View of extrudate samples: A – 0%, B – 0.25%, C – 0.5%, D – 1%.

The test program developed included measurements of thermal stability of extrudate during extrusion, tensile strength, elongation at break and morphology of extrudates.

The thermal stability was measured with the use of infrared camera V-20, in the section of extrudate including 29 points. Stimulus objects during the ongoing process of microcellular extrusion were nozzle of extruder head and the microporous extrudate leaving the nozzle head. Recorded the infrared radiation emitted by the surface of the examined objects, which after processing has enabled the visualization in the form of thermal images. A detailed analysis of temperature measurement was carried out on the along section of extrudate, which includes 29 points and across section of extrudate, which include 9 measuring points. The change in temperature during the extrusion process of the selected point on the surface of the extrudate at a distance of 40 mm from the nozzle was analyzed.

Measurements of mechanical quantities have been carried out on the Z 010 strength machine at a drawing speed of 50 mm/min. Type 1A specimens were previously excised according to EN ISO 527: 1998.

The morphology examination was performed on a test stand for microscopic examination consisting of an SEM microscope Nova NanoSem 450.

3. Results

The results of thermal stability of the LDPE extrudate during modification with microspheres were shown in figures 3-5.

With the increase of microspheres content, the temperature of the material leaving the nozzle of the extrusion head decreased, while in the tested dosing range it was a decrease of 10.04°C on average. In the tested dosing range, the extrudate temperature dropped uniformly along with the distance from the measuring head. At the same time, in the investigated range, the start of stabilization of the extrudate is visible from point 20 for a sample containing 1% microspheres.

Increasing the dispensing of the microcontroller influenced the cooling of the extrudate. Taking the reference value of the temperature drop in the solid extrudate measuring section (19.30°C), the 0.25% dosing resulted in a relative increase in the extrudate cooling rate by 34,5%, while the microspheres dosage of 0.5% resulted in an increase in the relative rate of extrudate cooling. 56.9%.

The tests of mechanical properties showed a significant effect of the microsphere contents on the tested values, i.e. Young's modulus, tensile strength and elongation at break. The results of individual parameters are shown in figures 6-8.

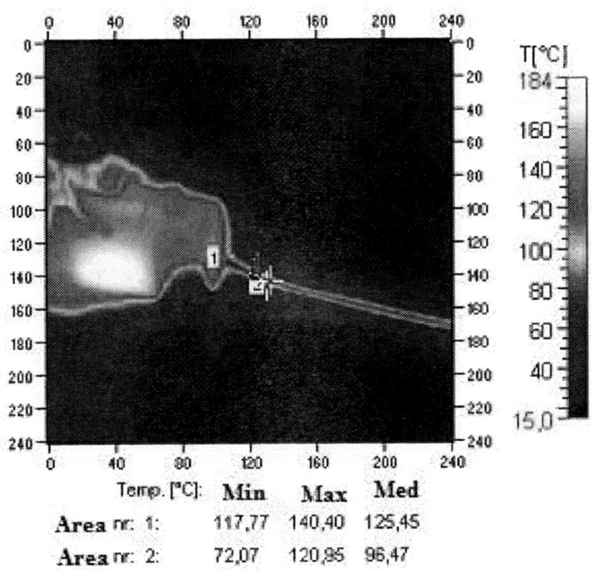

Figure 3. Thermal image with section of extrudate measurements.

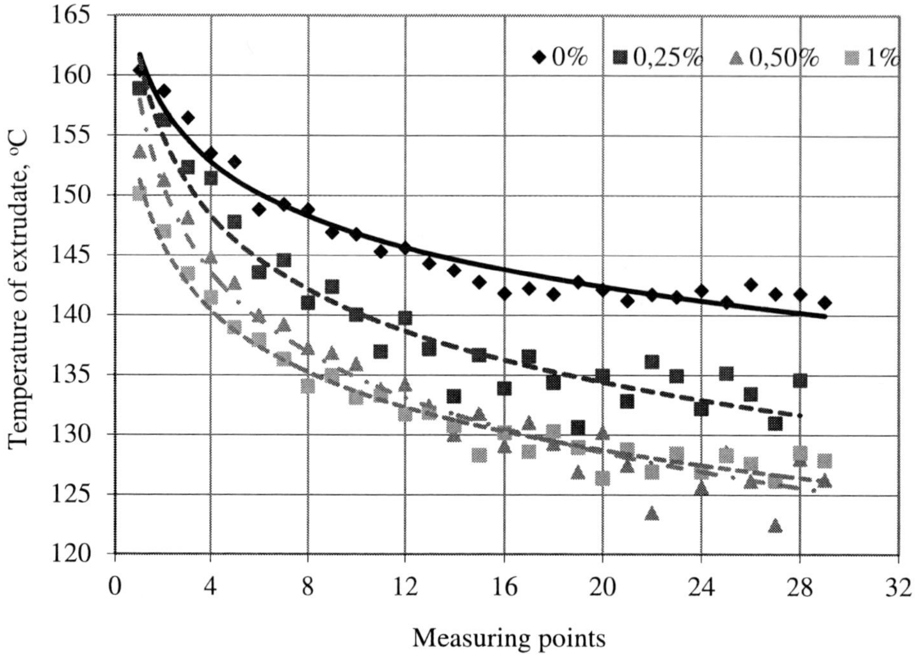

Figure 4. Results of thermal measurements of extrudate temperature, measured along the exrudate section.

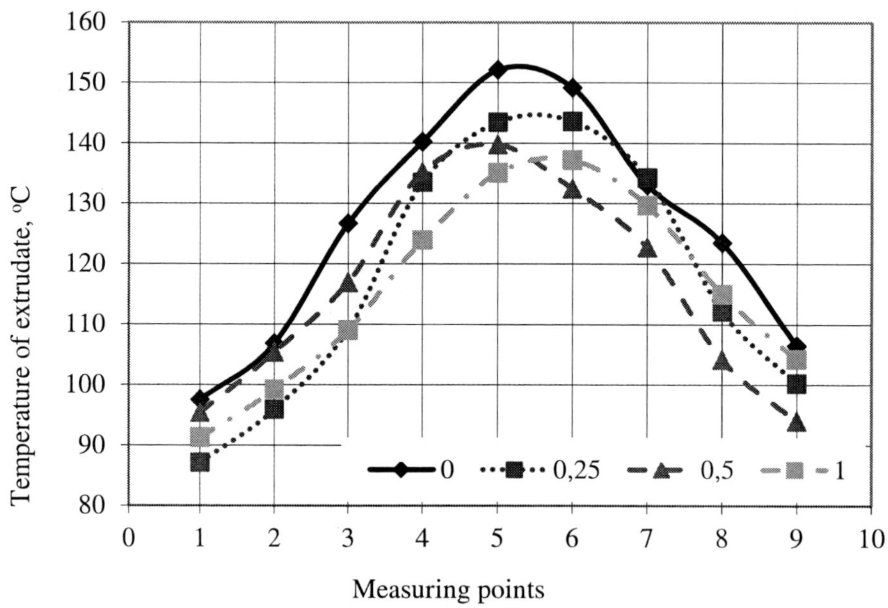

Figure 5. Results of thermal measurements of extrudate temperature, measured across the extrudate section.

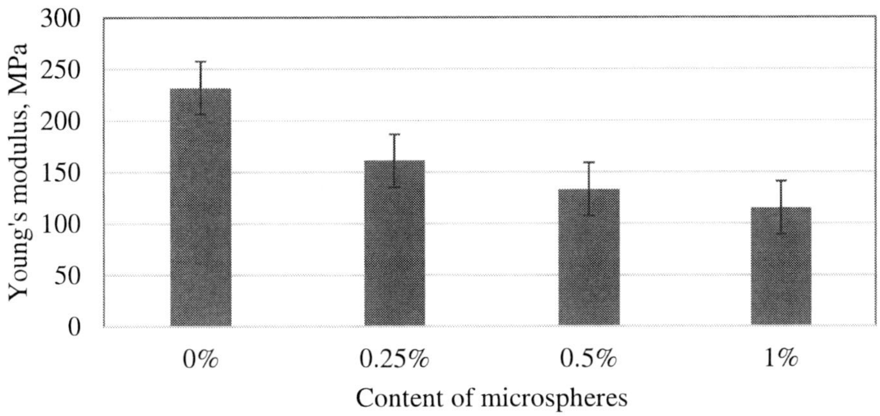

Figure 6. Dependence of Young's modulus on content of microspheres in LDPE extrudate.

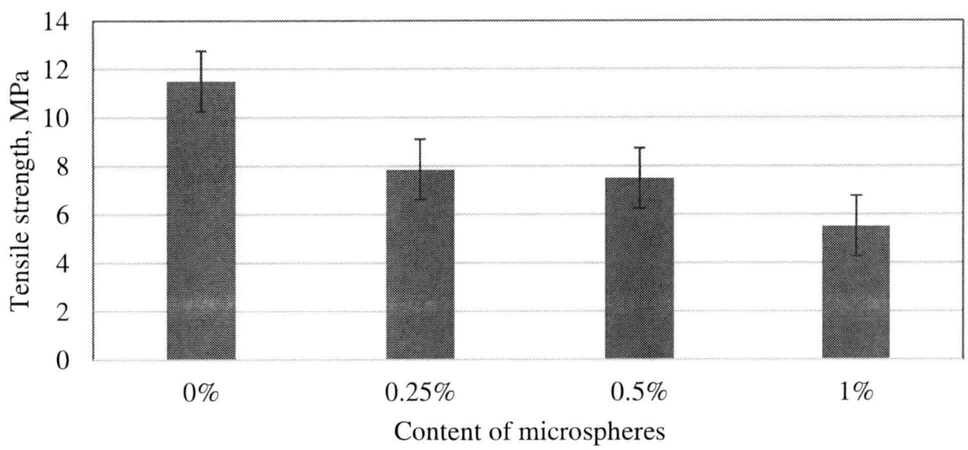

Figure 7. Dependence of tensile strength on content of microspheres in LDPE extrudate.

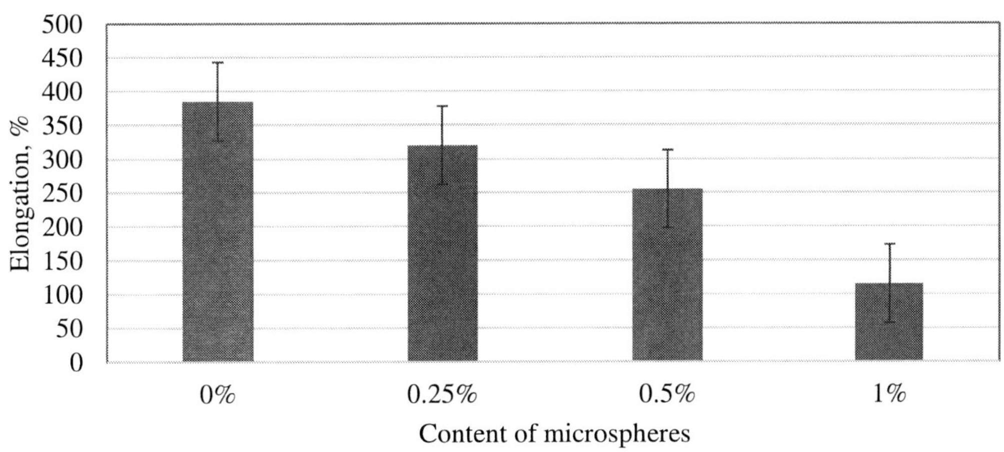

Figure 8. Dependence of elongation on content of microspheres in LDPE extrudate.

Along with increasing the content of microspheres in LDPE, there was a significant decrease in all the mechanical properties tested. The largest decrease in Young's modulus is observed at the content of 0.25% of microspheres and it amounts to 30%, further increase of the content of microspheres by 0.5% and 1% causes further decrease of the modulus by 17% and 13% respectively. In the case of tensile strength also the largest decrease in this property is visible for sample 2 containing 0.25% of microspheres, increasing the content up to 0.5% resulted in a slight decrease in strength by about 4%, subsequent increase in the amount of microspheres in LDPE resulted in further reduction of strength by another 26%. Tested elongation at break shows a similar decrease of 65% of units elongation for samples containing 0.25 and 0.5% microspheres. The highest content of the agent is 1% wt. It caused a further significant decrease of this parameter by another 100% of elongation units.

Morphology studies of LDPE samples modified with microspheres are presented in figure 9. The analysis of the structure of the samples showed that due to the addition of microspheres to the plastic, its structure changes from solid to porous, containing pores of various sizes.

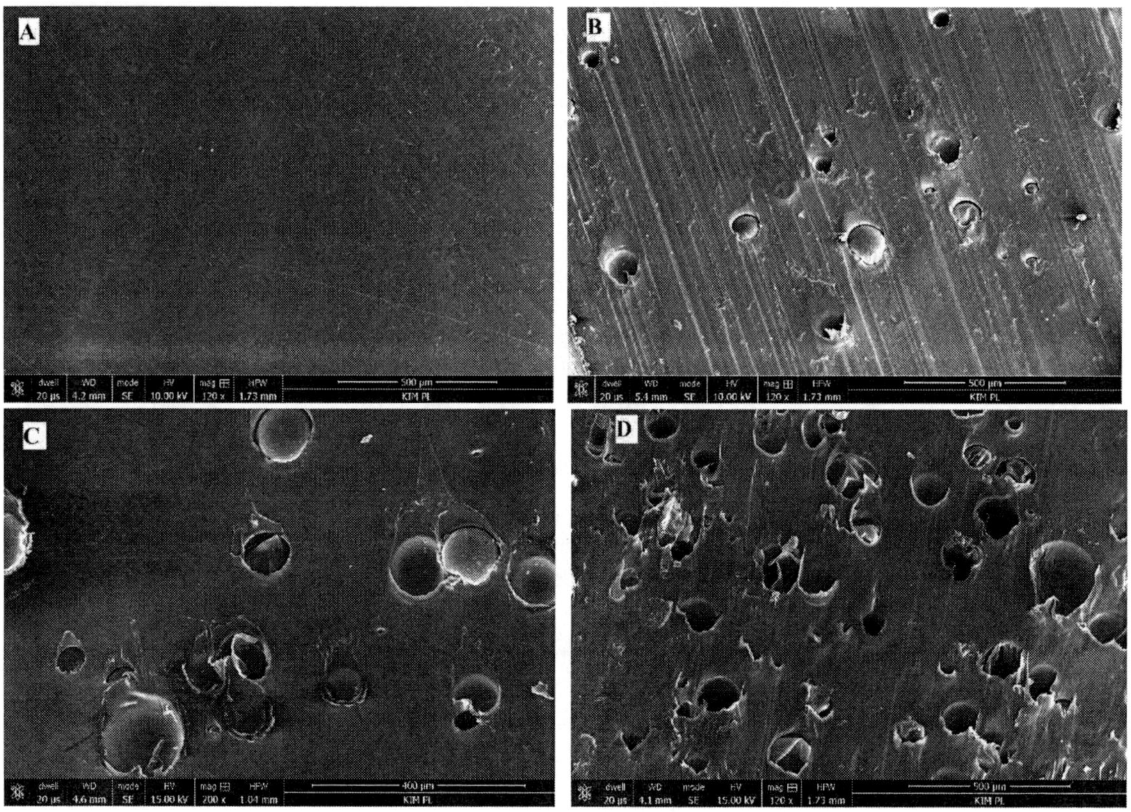

Figure 9. Morphology of LDPE modified with microspheres in amount: A – 0%, B – 0.25%, C – 0.5%, D – 1%.

Along with increasing the content of microspheres in LDPE, the resulting pores are characterized by varying diameter and surface area. The pores B and C visible in figure 9 have a diameter in the range of 80 - 120 μm, and the pore diameter in figure 9D is in the range of 40 - 100 μm. It is worth emphasizing the fact that on the presented structures, there are visible pores, which are empty spaces, as well as whole microspheres with a clear border of the polymer coating, which is a protective layer of the active agent. This confirms the physical expansion of the microsphere under specific conditions without complete decomposition in the polymer matrix.

4. Conclusions

The present study shows that the content of the blowing agent in the form of microspheres significantly affects the thermomechanical properties and the morphology of low density polyethylene. The research found that the inserting of microspheres into LDPE significantly affect the bottom-thermal phenomena in the microcellular extrusion process, especially during the flow in channels and extruder head for cooling the extrudate directly in the extruder head. In the investigated range of microspheres content measure the temperature of extrudate leaving the extruder head attributed to endothermal effect accompanying expansion of the microspheres during the microcellular process. Moreover, we observed that the addition of microspheres significantly affects the mechanical properties of the product, weakening the extrudate by over 60% when dispensing the agent in an amount of 1% wt. in relation to the solid samples. This can greatly limit the applicability of this material. Analysis of the physical structure of the obtained marc used to verify the literature data suggesting complete plasticity of the microsphere in the material [5]. Studies have shown that the structure of the product remains full microspheres with significantly increased sizes, at the same time it should be treated as preliminary studies requiring expansion.

References

[1] Blanco I 2014 *Chinese Journal of Polymer Science* **32** 681-9
[2] Yan F, Yan S and Li Y 2017 *IOP Conference Series: Materials Science and Engineering* **207** 012048
[3] Arumugam V, Mishra R, Militky J, Davies L and Slater S 2018 *Journal of the Textile Institute* **109** 345-57
[4] Duong H M, Liu P and Jewel D 2016 Advanced fabrication and multi-properties of polyimide aerogels and their composites *Polyimides: Synthesis, Applications and Research* (Nova Science Pub Inc) chapter 4 pp 131-62
[5] Garbacz T 2014 *Cellular Polymers* **33** 71
[6] Song K and apfel R E 2004 *Polymer Engineering and Science* **47** 735-42
[7] Tor-Świątek A, Garbacz T and Jachowicz T 2016 *Cellular Polymers* **35** 67-84
[8] Klepka T, Garbacz T and Bociaga E 2014 *Przemysl Chemiczny* **93** 907-10
[9] Błędzki A K and Faruk O 2005 *Journal od Applied Polymer Science* **97** 1090-6
[10] Wong C-M *et al* 2006 *Journal of Cellular Plastics* **42** 153-63
[11] Antunes M *et al* 2009 *Polymer Engineering and Science* **49** 2400-13
[12] Balach J *et al* 2012 *Journal of Microscopy* **246** 274-8
[13] Blanco I, Abate L and Antonelli M L 2011 *Polymer Degradation and Stability* **96** 1947-54
[14] Blanco I, Abate L, Antonelli M L and Bottino F A 2013 *Polymer Degradation and Stability* **98** 2291-6
[15] Tor-Świątek A and Sikora J W 2013 *Przemysl Chemiczny* **92** 538-41

Biopolymer- A beginning towards back to nature

S Gautam[1,2] and A Gautam[1]

[1]Department of Chemical Engineering, Shroff S R Rotary Institute of Chemical Technology, Ankleshwar, Gujarat, 393135 India

E-mail: shinaiitd@gmail.com

Abstract. Biopolymer is regarded as a polymer which can be biodegradable. Polyhydroxyalkanoates (PHAs) is one of the biopolymer which can be recovered from biomass. PHAs are naturally conserved in the cytoplasm of the bacterial cell during the growth. Bacteria/microbes store their energy from carbon sources in the form of hydrocarbons. Intracellular stored compounds are tightly linked with entire cell resulting difficulty of separation. The work aims to extract PHAs from biomass effectively. Chemical and mechanical separation of PHA can be done from biomass. A pretreatment of cells before chemical and mechanical separation is also effective for separation of PHA and has been carried out. Chemical extraction of PHA includes digestion of cell wall in acidic or alkaline medium and releasing PHA in broth, later sedimentation recovers PHA. In recent work different chemical methods were carried out to extract PHA of medium chain length. In one of these, sodium hypochlorite was used to denature the protein and chloroform was used for extraction of purified PHA. A recovery upto 96.6%, PHA by dried weight of cell, was obtained which is quite high comparing to reported literature. Other chemical disruption by sodium chloride, sodium hydroxide and hydrogen peroxide with and without pretreatment have also been carried out.

1. Introduction

Plastics are derived from petrochemical reserves. Common definitions of the term "biopolymer" also include biodegradable plastics from fossil fuels and non-biodegradable plastics from renewable resources.

Biopolymers are made from renewable resources and/or biodegradable waste materials (e.g., waste water, sewage sludge, organic waste) and are fully biodegradable by naturally occurring microorganisms. Typical molar masses of polyhydroxyalkanoate (PHA) vary, depending on the microbial production strain, fermentation conditions, substrate, and isolation method from typically 200 to 2000. Direct extraction of PHA from biomass is highly dependent on polymer structure and chain length of polymer. There have been many pretreatment methods which weakens cell wall of microbe. Later some solvents which oxidize cell wall can be utilize for final PHA extraction.

Biopolymer production is gaining a lot of attention due to biodegradability. The applications of biopolymer are widely in medical sectors. However, the cost of production is limiting its use every common human beings. The cost of production is highly dependent on carbon source which is pure substance for medical applications. Further washing and purification increase cost of production. Another application of biopolymer which is getting focus is, packaging industry [1]. The cost of biodegradable packaging is again not so affordable to medium scale or small scale industries. Therefore, it becomes almost necessary to find cost effective carbon source as well as reduce the use of non-biodegradable polymer. One of such sources can be wastewater from sewage treatment plant, food industry wastewater, process industry wastewater and so forth.

Several pretreatment methods: physical, chemical or biological are applicable to disrupt the microbial cell. Application of heat is one of the pretreatment step, it affects the firmness of cell by denaturation of cell proteins and by destabilization of outer membranes. This can also be done by freeze drying (lyophilization) of biomass [2,3].

A number of methods to release PHA are available in literature, in brief can be divided in solvent extraction, chemical, enzymatic, mechanical operation, supercritical fluids, air classification [4]. Solvent extraction methods are based on that, PHAs are insoluble in water, however, soluble in limited number of solvents. Solvents containing chlorine such a chloroform, 1-2 dichloroethane, methylene chloride and some cyclic carbonates such as propylene and ethylene carbonates are widely used to recover PHAs from biomass. While purifying PHA through solvent extraction carbon chain length may be shorten due to dissolution of hydrocarbon in halogenated solvents. Chemical disruption of cells by sodium hypochlorite is widely used as it has several advantages over others. One of the advantages is the cells need not to be dried and can be used in wet condition. The polymer is not dissolved in hypo solution and remains solid which can be recovered by filtration or centrifugation.

Apart from hypochlorite, disruption of cell can also be done by surfactants. Non-PHA material can be solubilize in surfactant, however, some polymer can also be solubilize in surfactant and results in reduced and short chain length recovery of PHA.

In present work biomass was obtained from common process industry wastewater treatment plant. Biomass was generated in secondary treatment and discarded as waste material. A comprehensive comparison of different PHA recovery methods had been done and several trial and error methods had been approached. The disruption of cells is adapted as per literature. However, substrate and culture used in present work to generate biomass was not found in literature. It is reported here first time.

2. Materials and methods

To extract PHA, biomass was obtained from a local common effluent treatment plant. The wastewater was collected from different process industries like pharmaceuticals, dyes and pigments and specialty chemicals. The wastewater undergoes for treatment by primary, secondary and tertiary methods. In primary treatment physical removal of contamination was done by filtration, coagulation and sedimentation. After primary treatment wastewater was aerated for secondary treatment, where mixed culture consortia were prepared to disintegrate complex hydrocarbons with microbial digestion. After secondary treatment by microbes, sludges were generated which were discarded as solid waste and sent for landfilling (biomass). In the present work this sludge was collected and used to recover PHA.

The biomass received was in the form of slurry. The slurry was washed with distilled water and decanted to recover solids. The process was repeated 4-5 times until supernatant liquid was clear. The sample was measured in volume and the dry weight of that specific volume was carried out in hot air oven at 60°C until biomass is completely dried. To recover PHA, wet biomass was used in all experiments, whilst hypochlorite and chloroform were used to extract PHA. Hypochlorite (30% v/v) with biomass was stirred at 23°C for 2 h. This step weakens walls of cell and release PHA. Cell walls were oxidized and remained in dissolved form in hypochlorite solution then in separating funnel hypochlorite was removed. To increase purity of PHA, chloroform was added in equal volumes to the remaining biomass and stir it for another 1 h. A distinguished layer of PHA was observed in the mixture of biomass, chloroform which was recovered successfully. The recovered PHA was washed with distilled water for 2-3 times to remove sodium hypochlorite or chloroform. Hypochlorite (10% V/V), sodium chloride (99.8%) and chloroform (35%) were procured from Merck, India and were of analytical quality.

3. Results and discussion

Current work demonstrates preliminary findings of PHA recovery from biomass with several methods optimized in literature. The methods applied here were used in literature for different cultures and substrates. As discussed in section 2 substrate used in this work was wastewater from industries. To produce biomass or in other terms to reduce wastewater parameters like total Total Kjeldal Nitrogen, Chemical oxygen demand and total phosphorus, mixed culture consortia was used for secondary treatment.

Results of PHA extracted from biomass with different methods are shown in table 1. In first method sodium hypochlorite was used to disrupt the microbial cell, it dissolves the cell wall and PHA remains undissolved and recovered as a precipitate. The procedure was followed as per Hahn et al., 1994. The recovery of PHA was carried out at dry cell weight basis. The recovery of PHA and purity can be increased by varying sodium hypochlorite solution concentration.

Table 1. Disruption of biomass with different chemical methods and recovery of PHA.

S.N.	Method	Physical appearance	Recovery of PHA based on cell dry weight (%)
1	Disruption by sodium hypochlorite	Clear white	71.8
2	Disruption by hydrogen peroxide treatment	Clear white	49.1
3	Denaturation with Sodium chloride	Dark brown	82.3
4	Denaturation with Sodium hydroxide	Dark brown	96.6
5	Weakening by heating and disruption by Hydrogen peroxide	Clear white	31.3
6	Weakening by heating and disruption by sodium chloride	Dark brown	40.6

High purity P (3HB-*co*-3HV) was recovered by Eutropha R [5]. The procedure includes pretreatment of biomass at 120°C for 60 sec and hydrogen peroxide was added. Recovery of PHA with pretreatment by heat and without heat was observed. Recovery was found to be less with heat treatment. However, purity may have increased.

Similar heat pretreatment was also applied in other experiment and sodium chloride was used for denaturing the cell wall. In these conditions again heat pretreatment depicts reduced recovery of PHA. Sodium hydroxide was used to remove cell walls from biomass and recovery of PHA was more than 96%.

(a) (b)

Figure 1. Images of PHA extracted by two different methods (a) sodium hydroxide, (b) Sodium hypochlorite.

The biomass generated from wastewater treatment was of dark brown color, it was same as of wastewater. Treatment with sodium hypochlorite and hydrogen peroxide oxidized the aromatic compounds which prevails color to the biomass (figure 1). The PHA recovered by this method is white in color. In other cases where oxidizing agent was not used, PHA was brown in color same as the biomass received. The recovery of PHA was calculated by dry weight basis.

The observations and experiments were performed on real effluent treatment plant biomass and represent the first observation in literature for extraction of PHA from biomass of process industries effluent treatment plant. The extraction of PHA from sodium hypochlorite as reported in literature is upto 91%. PHA was produced from glucose and contains a single source of carbon [6]. However, PHA recovered is from heterogeneous source of wastewater from process industries. A common source was the sewage wastewater or any food industry wastewater which contained high cellulose, protein and glucose. The PHA recovery in current work is 71%, the reduced recovery may be the result of reduced sodium hypochlorite concentration. As well as aromatic and cross linked substances present in wastewater which are difficult to disintegrate by mixed culture and may contain less amount of PHA within cell. Hahn *et al* in 1994 utilized single strain culture that is, *Alcaligene eutrophos* which resulted in more purified and good recovery of PHA. PHA recovery can be enhanced with increasing concentration of hypochlorite.

A sewage wastewater treatment plant was used to recover PHA at pilot scale with mixed culture consortia to remove biological nitrogen and reduce chemical oxygen demand. PHA accumulated by microbes was 49% [7]. In other study of glycerol from biodiesel, a side product as a substrate was taken for PHA production. The recovery of PHA was obtained only 24% [8]. Optimization of the process is in process. The heat application as a pretreatment step can reduce the molecular weight of the polymer as well as solubilization of polymer in second step of the chemical treatment. The characterization of biopolymer produced for functional groups and molecular weight is undergoing.

4. Conclusion

It presents a brief explanation of PHA recovered by different methods. The chemical disruption of microbe cells with sodium hypochlorite, sodium hydroxide, sodium chloride and hydrogen peroxide. The application of heat was done to increase the kinetics of the process however resulted in reverse. As per literature available physico-chemical method can be a good combination to extract PHA. It will retain molecular weight of polymer intact as well as it will result in higher yield. In physical separation dissolved air flotation can be a good unit operation where polymer can be separated as hydrophobic particles with bubbles. Ultimate purpose of the work is to optimize the extraction process of PHA from cost effective source so that the generated biopolymer can be used for everyday commodities.

Acknowledgments

Authors would like to thank to Enviro Tech Limited, Ankleshwar Gujarat India for sponsoring the project and permitting to present the work.

References

[1] Blanco I 2014 *Chinese J. Poly. Sci.* **32** 681-9

[2] Kapritchkoff F M, Viotti A P, Alli R C P, Zuccolo M, Pradella J G C, Maiorano A E, Miranda E A and Bonomi A 2006 *J. Biotechnol.* **122** 453-62

[3] Saegusa H, Shiraki M, Kanai C and Saito T 2001 *J. Bacteriol.* **18** 94-100

[4] Mohamed H, Madkour, Daniel H, Mansour A, Shabbaj I and Alexander S 2013 *Biomacromolecules* **14** 2963-72

[5] Liddell J M and Locke T J U S Patent **5** 691 174 1997

[6] Hahn S K, Chang Y K, Kim B S and Chang H N 1994 *Biotech. Bioeng* **44** 256-61

[7] Bengtsson S et al 2017 *New Biotechno.* **35** 42-53

[8] Burniol-Figols A, Varrone C, Egede A D, Balzer Le S, Skiadas I V and Gavala H N 2018 *Water Res.* **128** 255-66

[9] Hu S, Mc Donald A G and Coats E R 2013 *J. Appl. Polym. Online Sci.* **129** 1314-21
[10] Fiorese M L, Freitas F, Pais J, Ramos A M, de Aragao G M F and Reis M A M 2009 *Eng Life Sci.* **9** 454-61

High temperature polymerization monitoring of an epoxy resin using ultrasound

P Maréchal[1,2], N Ghodhbani[1] and H Duflo[1]

[1]Laboratoire Ondes et Milieux Complexes, Normandie Univ, UNIHAVRE, CNRS, LOMC, 76600 Le Havre, France

E-mail: pierre.marechal@univ-lehavre.fr

Abstract. In this study, the real time ultrasonic monitoring is investigated to quantify changes in physical and mechanical properties during the manufacture of composite structures. In this context, an experimental transmission was developed with the aim of characterizing a high temperature polymerization reaction and post-curing properties using an ultrasonic method. First, the monitoring of ultrasonic parameters of a thermosetting resin is carried out in a device reproducing the experimental conditions for manufacturing a composite material with a process known as RTM, that is to say an isothermal polymerization at $T = 160°C$. During this curing, the resin is changing from its initial viscous liquid state to its final viscous solid state. Between those states, a glassy transition stage is observed, during which the physical properties are strongly changing, i.e. an increase of the ultrasonic velocity up to its steady value and a transient increase of the ultrasonic attenuation. Second, the ultrasonic inspection of the thermosetting resin is performed during a heating and cooling process to study the temperature sensitivity after curing. This type of characterization leads to identifying the ultrasonic properties dependence before, during and after the glassy transition temperature T_g. Eventually, this study is composed of two complementary parts: the first is useful for the curing optimization, while the second one is fruitful for the post-processing characterization in a temperature range including the glassy transition temperature T_g.

1. Introduction

The polymerization kinetics is of interest for the thermochemical research community [1]. In the recent decades, the monitoring of viscoelastic parameters of polymers and composite structures from the manufacturing phase to the aging phase has been investigated using various methods: DSC [2], rheology [3], IR [4] or ultrasound [5]. The ultrasound monitoring of resins [6-12] can operate during the fabrication process. In a previous study in reflection mode [13-16], the ultrasonic characteristics were directly related to the reaction kinetics and the reaction conversion. A transmission mode setup would enable to avoid overlapping of echoes during the manufacturing. In the present work, the ultrasonic properties are monitored at high temperature [17-20] in the transmission mode [21-25]. Here, the transmission characterization method is based on the direct transmission of a pulse at the beginning of polymerization (reference state) and enables the monitoring of the ultrasonic properties (instantaneous state) by means of an iterative optimization method [21,22,15]. This paper is divided into two parts: the first describes the theoretical equations and the general approach of the proposed method. The second part presents the experimental setup and associated results with an application of this approach with the high temperature transducer and the RTM6 epoxy resin.

2. Monitoring of the ultrasonic properties

2.1. Low temperature monitoring

As illustrated, Figure 1 shows the different echoes used to the monitoring of the ultrasonic proprieties of an epoxy resin during its polymerization with the transmission method. The structure is composed of two identical aluminum blocks between which the resin liquid is introduced. Multiple echoes can be recorded after various transmission and reflections in the constituting layers. First, the complex spectrum $\underline{S}_1(f)$ corresponds to the direct transmitted echo $s_1(t)$ through the three-layer structure composed of the first aluminum block, the epoxy layer and the second aluminum block (*Alu*/*Epo*/*Alu*). Second, the complex spectrum $\underline{S}_2(f)$ corresponds to the direct transmitted echo $s_2(t)$ through the three-layer structure with a round-trip echo in the epoxy. Third, the complex spectrum $\underline{S}_3(f)$ corresponds to the direct transmitted echo $s_3(t)$ through the structure with a round-trip echo in one of the aluminum blocks. These spectra $\underline{S}_1(f)$, $\underline{S}_2(f)$ and $\underline{S}_3(f)$ are written in the following form:

$$\begin{cases} \underline{S}_1(f) = \underline{S}_0(f)e^{-2j\underline{k}_{Alu}d_{Alu}}e^{-j\underline{k}_{Epo}d_{Epo}}\underline{T}_{Alu/Epo}\underline{T}_{Epo/Alu} \\ \underline{S}_2(f) = \underline{S}_0(f)e^{-2j\underline{k}_{Alu}d_{Alu}}e^{-3j\underline{k}_{Epo}d_{Epo}}\underline{T}_{Alu/Epo}\underline{R}_{Epo/Alu}\underline{R}_{Alu/Epo}\underline{T}_{Epo/Alu} \\ \underline{S}_3(f) = \underline{S}_0(f)e^{-4j\underline{k}_{Alu}d_{Alu}}e^{-j\underline{k}_{Epo}d_{Epo}}\underline{T}_{Alu/Epo}\underline{T}_{Epo/Alu}\underline{R}_{Alu/Epo} \end{cases} \quad (1)$$

where $\underline{S}_0(f)$ is the emission signal; $k_{Alu} = \omega/c_{Alu}(f) + j\alpha_{Alu}(f)$ and $k_{Epo} = \omega/c_{Epo}(f) + j\alpha_{Epo}(f)$ are the longitudinal complex wavenumbers, in aluminum and epoxy resin respectively; d_{Alu} and d_{Epo} are the thicknesses of aluminum and resin layer, respectively; $\underline{T}_{\{Alu/Epo,\ Epo/Alu\}}$ and $\underline{R}_{\{Alu/Epo,\ Epo/Alu\}}$ are the complex transmission and reflection coefficients from aluminum to epoxy (*Alu*/*Epo*) or reversely, from epoxy to aluminum (*Epo*/*Alu*), respectively.

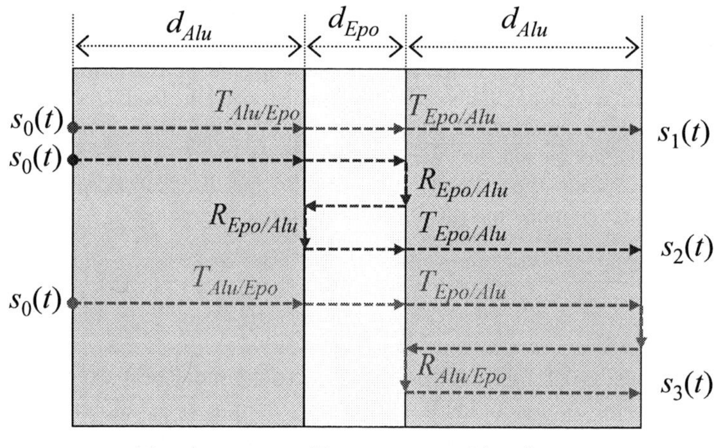

Figure 1. Paths of the transmission signal in three-layer structure composed by a first aluminum block, an epoxy layer and a second aluminum block (Alu/Epo/Alu).

The transfer function $\underline{T}_{21}(f) = \underline{S}_2(f)/\ \underline{S}_1(f)$ contains information on the longitudinal properties of the epoxy resin layer, i.e. the velocity $c_{Epo}(f)$ and attenuation $\alpha_{Epo}(f)$. Consequently, the argument of the transfer function Arg($\underline{T}_{21}(f)$) and its modulus $|\underline{T}_{21}(f)|$ give the following frequency dependencies:

$$\begin{cases} c_{Epo}(f) = \dfrac{-4\pi f d_{Epo}}{Arg(\underline{T}_{21}(f))} \\ \alpha_{Epo}(f) = \dfrac{-1}{2d_{Epo}} \ln\left(\left|\dfrac{\underline{T}_{21}(f)}{\underline{R}^2_{Epo/Alu}}\right|\right) \end{cases} \tag{2}$$

Similarly, the transfer function $\underline{T}_{31}(f) = \underline{S}_3(f)/\ \underline{S}_1(f)$ contains information on the longitudinal ultrasonic properties of the aluminum blocks, i.e. the expressions of the velocity $c_{Alu}(f)$ and attenuation $\alpha_{Alu}(f)$ in the aluminum layer.

The transmission technique consists in calculating the ultrasonic properties of the epoxy relatively to a reference state of the epoxy at the beginning of polymerization. At the reference state, the spectrum $\underline{S}_{1,ref}(f)$, corresponds to the first transmitted echo $s_{1,ref}(t)$:

$$\underline{S}_{1,ref}(f) = \underline{S}_0(f) e^{-2j\underline{k}_{Alu,ref} d_{Alu}} e^{-j\underline{k}_{Epo,ref} d_{Epo}} \underline{T}_{Alu/Epo,ref} \underline{T}_{Epo/Alu,ref} \tag{3}$$

where $\underline{k}_{Alu,ref} = \omega/c_{Alu,ref}(f) + j\alpha_{Alu,ref}(f)$ and $\underline{k}_{Epo,ref} = \omega/c_{Epo,ref}(f) + j\alpha_{Epo,ref}(f)$ correspond to the complex wavenumbers at a reference state, $\underline{T}_{\{Alu/Epo,ref,\ Epo/Alu,ref\}}$ and $\underline{R}_{\{Alu/Epo,ref,\ Epo/Alu,ref\}}$ are the complex transmission and reflection coefficients from aluminum to epoxy (*Alu/Epo*) or reversely, from epoxy to aluminum (*Epo/Alu*) fro the reference state, respectively. The transfer function $\underline{T}_{11}(f) = \underline{S}_1(f)/\underline{S}_{1,ref}(f)$ contains information on instantaneous ultrasonic properties in epoxy. The expression of the velocity $c_{Epo}(f)$ and attenuation $\alpha_{Epo}(f)$ in the epoxy are written as a function of the argument of the transfer function $\mathrm{Arg}(\underline{T}_{11}(f))$ and the modulus $|\underline{T}_{11}(f)|$ in the following form:

$$\begin{cases} c_{Epo}(f) = \dfrac{d_{Epo}}{\dfrac{d_{Epo}}{c_{Epo,ref}(f)} - \dfrac{1}{2\pi f} Arg\left(\dfrac{\underline{T}_{11}(f)}{\underline{T}}\right) + 2d_{Alu}\left(\dfrac{1}{c_{Alu,ref}(f)} - \dfrac{1}{c_{Alu}(f)}\right)} \\ \alpha_{Epo}(f) = \alpha_{Epo,ref}(f) - \dfrac{1}{d_{Epo}} \ln\left(\left|\dfrac{\underline{T}_{11}(f)}{\underline{T}}\right|\right) + 2\dfrac{d_{Alu}}{d_{Epo}}\left(\alpha_{Alu,ref}(f) - \alpha_{Alu}(f)\right) \end{cases} \tag{4}$$

where $\underline{T}$ is a transfer coefficient which is written in the following form:

$$\underline{T} = \frac{\underline{T}_{Alu/Epo} \underline{T}_{Epo/Alu}}{\underline{T}_{Alu/Epo,ref} \underline{T}_{Epo/Alu,ref}} \tag{5}$$

The calculation of the ultrasonic properties during polymerization is summarized by the algorithm procedure in Figure 2. At the beginning of the polymerization reaction, the reference ultrasonic properties in epoxy are calculated from equation (2). These initial properties at the reference state are used to calculate the instantaneous properties in the epoxy resin as a function of time, according to equation (4), including the iterative calculation of transfer coefficient $\underline{T}$ (equation (5)).

2.2. High temperature monitoring

As illustrated by Figure 3, the high temperature transducers with delay lines lead to a five layers structure (*Stl/Alu/Epo/Alu/Stl*), corresponding to the steel delay line of the emission transducer (*Stl*), the aluminum block (*Alu*), the epoxy layer (*Epo*), the aluminum block (*Alu*) and then the delay line of the receiving transducer (*Stl*). The instantaneous signal transmitted through the five layer structure is denoted $s_{1,HT}(t)$ results from the direct transmission transfer function $\underline{T}_d = \underline{T}_{Stl/Alu}.\underline{T}_{Alu/Epo}.\underline{T}_{Epo/Alu}.\underline{T}_{Alu/Stl}$.

The spectra $\underline{S}_{1,HT}(f)$ corresponding to the transmitted echo $s_{1,HT}(t)$ can be written as:

$$\underline{S}_{1,HT}(f) = \underline{S}_0 e^{-2j\underline{k}_{Alu} d_{Alu}} e^{-2j\underline{k}_{Stl} d_{Stl}} e^{-j\underline{k}_{Epo} d_{Epo}} \underline{T}_d \tag{6}$$

IOP Conf. Series: Materials Science and Engineering **369** (2018) 012010 doi:10.1088/1757-899X/369/1/012010

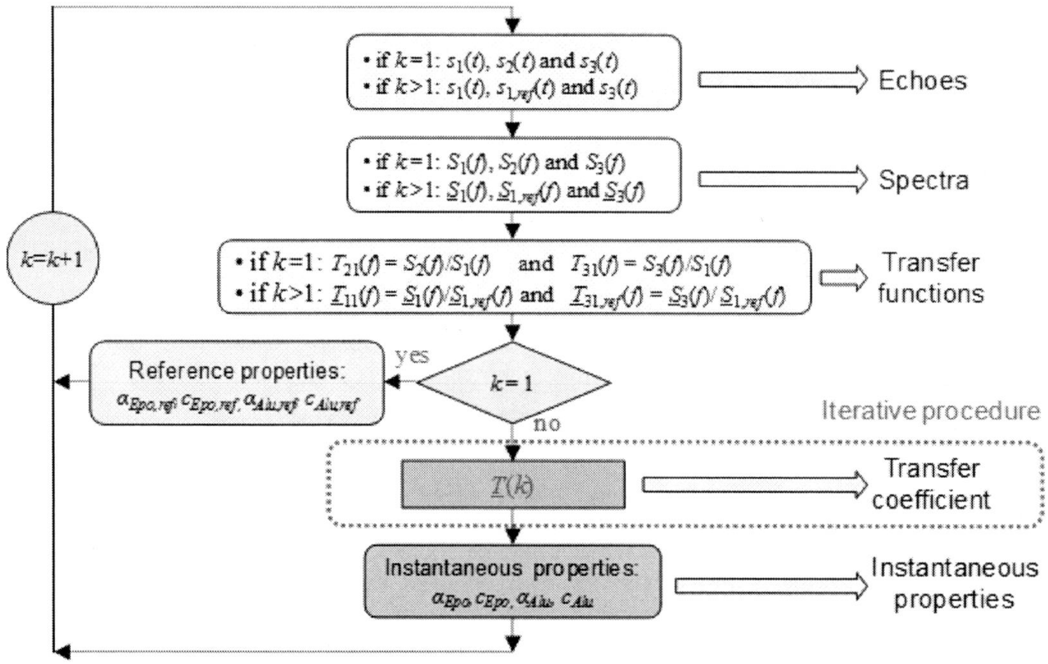

Figure 2. Algorithm for calculating ultrasonic properties.

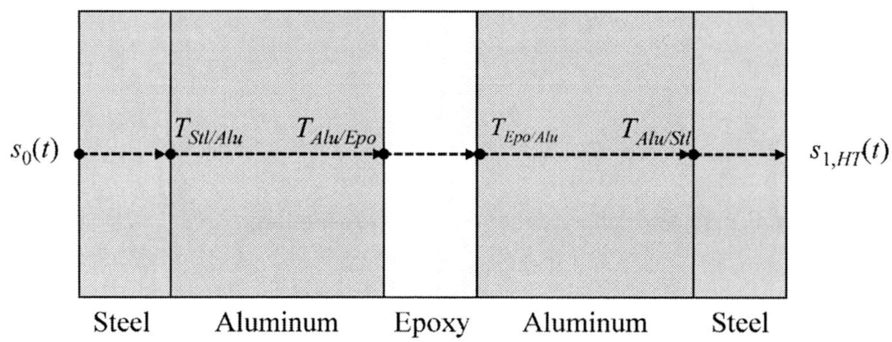

Figure 3. Direct transmission path in a five layer structure ($Stl/Alu/Epo/Alu/Stl$) and associated instantaneous signal $s_{1,HT}(t)$.

Using the procedure described in Figure 2, the ultrasonic velocity c_{Epo} can be written as a function of the time shift between the reference and the instantaneous echo $\Delta t_{11,HT}$, of the time shift in the molds δt_{Alu} and in the delay lines δt_{Stl}. The attenuation α_{Epo} is expressed as a function of the of the ratio between the maximum amplitude of the instantaneous echo $\max(s_{1,HT}(t))$ and that of the reference echo $\max(s_{1,HT,ref}(t))$, and the attenuation variation in the molds $\delta\alpha_{Alu}$ and in the delay lines $\delta\alpha_{Stl}$:

$$\begin{cases} c_{Epo} = \dfrac{d_{Epo}}{\dfrac{d_{Epo}}{c_{Epo,ref}} - \Delta t_{11,HT} + \delta t_{Alu} + \delta t_{Stl}} \\ \alpha_{Epo} = \alpha_{Epo,ref} - \dfrac{1}{d_{Epo}} \ln\left(\dfrac{\max(s_{1,HT}(t))}{\max(s_{1,HT,ref}(t))} \dfrac{1}{\left|\underline{T}_{HT}\right|} \right) + \delta\alpha_{Alu} + \delta\alpha_{Stl} \end{cases} \tag{7}$$

where $\underline{T}_{HT}$ is a transfer coefficient describing the evolution of the direct transmission coefficient $\underline{T}_d$:

$$\underline{T}_{HT} = \frac{\underline{T}_d}{\underline{T}_{d,ref}} = \frac{\underline{T}_{Stl/Alu}\underline{T}_{Alu/Epo}\underline{T}_{Epo/Alu}\underline{T}_{Alu/Stl}}{\underline{T}_{Stl/Alu,ref}\underline{T}_{Alu/Epo,ref}\underline{T}_{Epo/Alu,ref}\underline{T}_{Alu/Stl,ref}} \quad (8)$$

3. Experiment

3.1. Experimental setup

The experimental setup is composed of two contact transducers, one emitter and the other receiver, placed in contact with mold (
(a)). The procedure for preparing the epoxy resin (Resoltech or RTM6) and the instrumentation is detailed in a previous work [13]. The transducer emits longitudinal waves using a pulse generator. The second transducer receives the echoes (
(b)) then acquired using a digital oscilloscope. The temperature is programmed and regulated by a heating oven. A Pt100 temperature sensor is immersed in the epoxy resin in order to record the temperature changes during the polymerization. The acoustic signals and the temperature are automatically recorded and processed with an acquisition period of 10 minutes, using an acquisition program on Matlab.

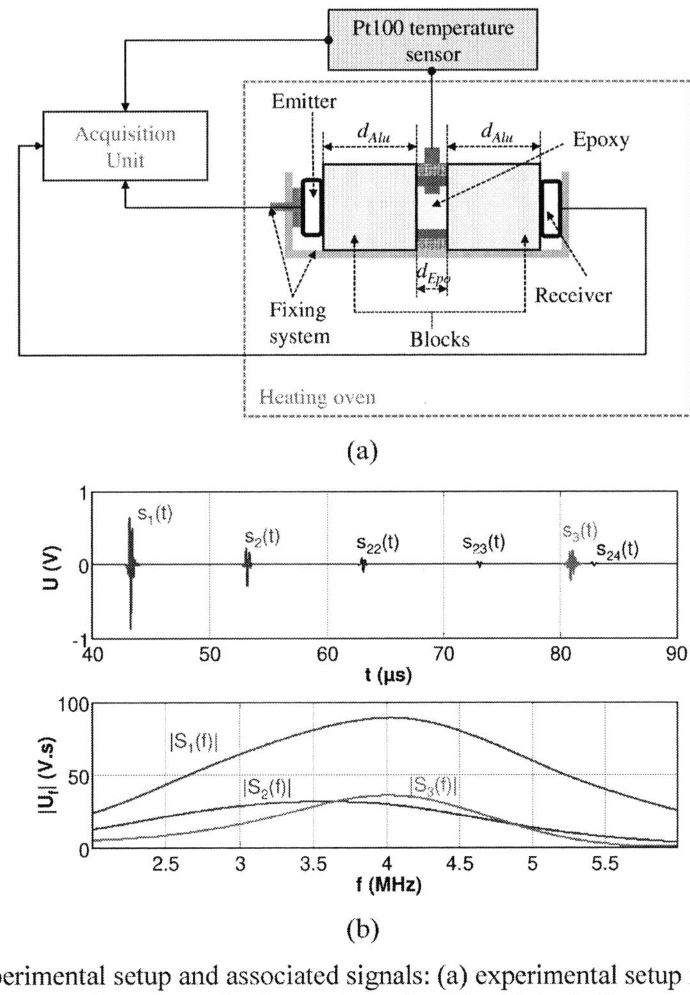

Figure 4. Experimental setup and associated signals: (a) experimental setup for the ultrasonic monitoring; (b) processed transmitted echoes $s_1(t)$, $s_2(t)$ and $s_3(t)$, including round-trip echoes $s_{22}(t)$, $s_{23}(t)$ and $s_{24}(t)$ in the epoxy layer, and associated spectra in the three-layer structure (*Alu*/*Epo*/*Alu*).

3.2. Low temperature processing

As illustrated Figure 5, the ultrasonic properties of the epoxy layer were evaluated while the polymerization was monitored as a function of the acquisition time t_{acq}, at a setpoint temperature $T = 40°C$. These results are illustrated in Figure 5(a) for the velocity $c_{Epo}(t_{acq})$ and in Figure 5(b) for the attenuation $\alpha_{Epo}(t_{acq})$.

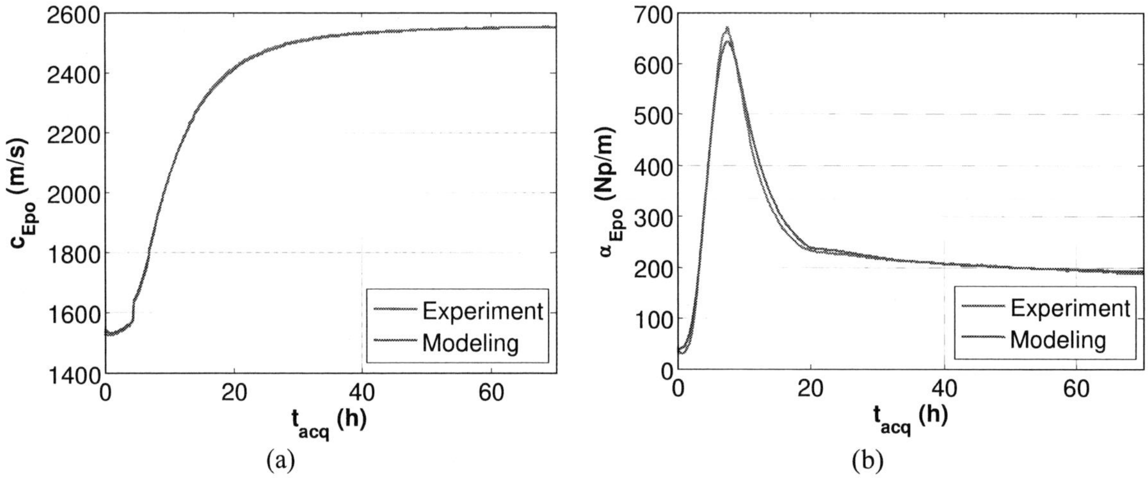

Figure 5. Ultrasonic properties of the epoxy resin: (a) velocity $c_{Epo}(t_{acq})$, and (b) attenuation $\alpha_{Epo}(t_{acq})$, while the polymerization was monitored at a setpoint temperature $T = 40°C$.

3.3. High temperature processing

A high temperature (HT) piezoelectric transducer (GE Measurement and Control) is specifically used for continuous high-temperature transmission measurements (up to 180°C). The center frequency of the transducer is $f_0 = 4$ MHz. This type of transducer has an integrated steel (*Stl*) delay line used for the protection of the piezoelectric material, and as illustrated in Figure 3, it results in multiple round-trip echoes in the delay line. These delay lines make the emission-reflection configuration complicated by the echoes overlap. The chosen method consists in working with the first transmitted echo provided that we know the ultrasonic properties in the aluminum molds and the integrated steel delay lines as well as their evolutions as a function of the temperature. This is what was done with what we called the reference state.

The polymerization monitoring at high temperature of the RTM6 resin was carried out according a precise protocol. The resin was degassed during 30 minutes. Meanwhile, the molds were heated at 80°C, and once the resin was flowed in the mold, the setpoint temperature was set at 160°C. At that time $t_{acq} = 0$, the polymerization monitoring started and the transmitted signals through the five layer structure were recorded periodically. As illustrated by

(a), the liquid viscoelastic phase ranges from $t_{acq} = 0$ and 1.25 h and shows a small variation, mainly due to the temperature increase of the resin. The second phase ranging from $t_{acq} = 1.25$ to 2.25 h is that of the glassy transition, characterized by a strong variation of the echoes amplitude and time-of-flight. The third phase defined by $t_{acq} > 2.25$ h, corresponds to the stabilization of the echoes properties, i.e. solidification of the epoxy. As a result, in

(b), the associated wave velocity c_{Epo} (m/s) and attenuation α_{Epo} (Np/m) in the epoxy layer were extracted using the previously described algorithm. These deduced properties of the epoxy layer clearly show the glassy transition around 1.75 h.

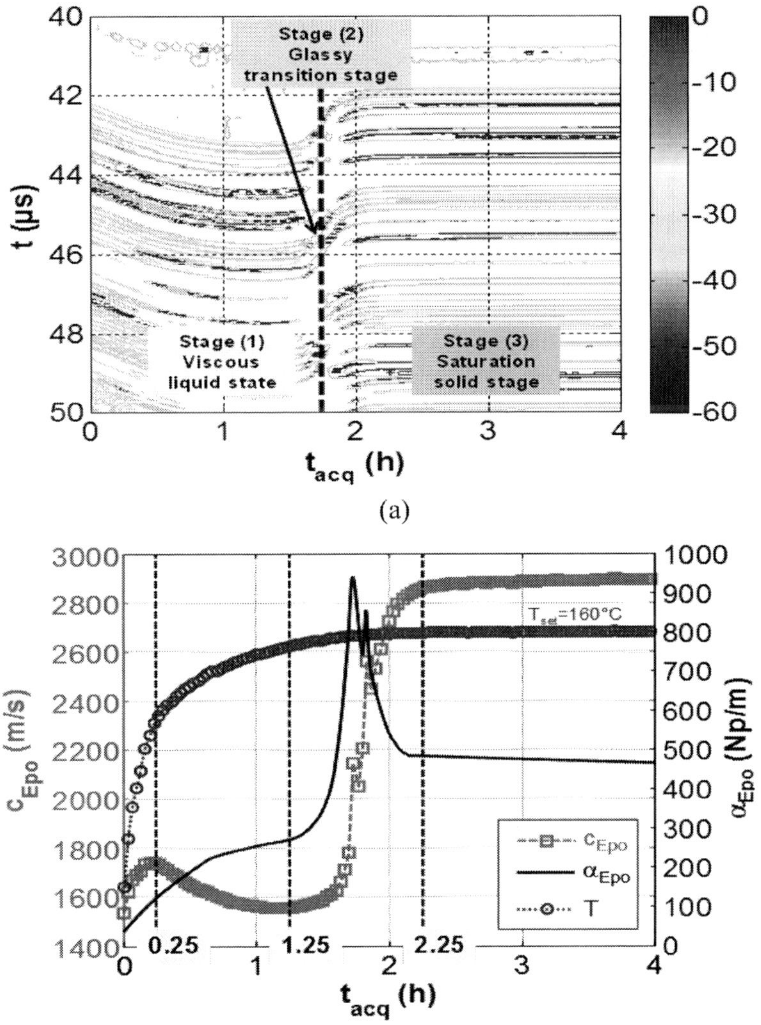

(a)

(b)

Figure 6. (a) Ultrasound monitoring of the transmitted signals through the five layers studied configuration, in the case of a RTM6 epoxy resin with a couple of high temperature transducers and (b) associated velocity c_{Epo} (m/s) and attenuation α_{Epo} (Np/m) in the epoxy layer.

4. Conclusion

The reference properties were estimated using different spectral and temporal methods. The transfer coefficient $\underline{T}$ has been determined in an analytical form and via an iterative optimization based on an inverse approach. Since the processing method has been described and validated, then the ultrasonic properties can be studied at high temperature with transducers integrating delay lines. In this five layer configuration, the ultrasonic properties of the epoxy resin $\{c_{Epo}, \alpha_{Epo}\}$ can be calculated from those evaluated at the reference state $\{c_{Epo,ref}, \alpha_{Epo,ref}\}$, including also those of the aluminum molds $\{c_{Alu}, \alpha_{Alu}\}$ and delay lines $\{c_{Sil}, \alpha_{Sil}\}$ and the appropriate temperature compensations through the transfer coefficient $\underline{T}$. Polymerization monitoring of a RTM6 epoxy resin was carried out at high temperature, leading to the determination of the thermal and phase transformations through ultrasound.

References

[1] Vyazovkin S, Burnham A K, Criadoc J M, Pérez-Maqueda L A, Popescud C and Sbirrazzuoli N 2011 ICTAC Kinetics Committee recommendations for performing kinetic computations on thermal analysis data *Thermochimica Acta* **520** 1-19

[2] Ren R, Xiong X, Ma X, Liu S, Wang J, Chen P and Zeng Y 2016 Isothermal curing kinetics and mechanism of DGEBA epoxy resin withphthalide-containing aromatic diamine *Thermochimica Acta* **623** 15-21

[3] Hu J, Shan J, Zhao J and Tong Z 2016 Isothermal curing kinetics of a flame retardant epoxy resin containing DOPO investigated by DSC and rheology *Thermochimica Acta* **632** 56-63

[4] Duemichen E, Javdanitehran M, Erdmann M, Trappe V, Sturm H, Braun U and Ziegmann G 2015 Analyzing the network formation and curing kinetics of epoxy resinsby in situ near-infrared measurements with variable heating rates *Thermochimica Acta* **616** 49-60

[5] Ndiaye E B, Duflo H, Maréchal P and Pareige P 2017 Thermal aging characterization of composite plates and honeycomb sandwiches by electromechanical measurement *J. Acoust. Soc. Am.* **42** 3691-702

[6] Papadakis E P 1974 Monitoring the moduli of polymers with ultrasound *J. Appl. Phys.* **45** 1218-22

[7] Lindrose A 1978 Ultrasonic wave and moduli changes in a curing epoxy resin *Experimental Mechanics* **18** 227-32

[8] Rokhlin S, Lewis D, Graff K and Laszlo A 1986 Real-time study of frequency dependence of attenuation and velocity of ultrasonic waves during the curing reaction of epoxy resin *J. Acoust. Soc. Am.* **79** 1786-93

[9] Challis R, Alper T, Cocker R, Holmes A and White J 1991 Ultrasonic absorption and velocity dispersion measurements in thin adhesive layers *Ultrasonics* **29** 22-8

[10] Sidebottom D 1993 Ultrasonic measurements of an epoxy resin near its sol-gel transition *Phys. Rev. E* **48** 391-9

[11] Faiz B, Maze G, Decultot D, Moudden A, Aassif E and Ezzaidi M 1999 Ultrasonic characterization of the quality of an epoxy resin polymerization *IEEE Trans. Ultrason. Ferroelect. Freq. Control* **46** 88-196

[12] Lionetto F and Maffezzoli A 2013 Monitoring the cure state of thermosetting resins by ultrasound *Materials* **6** 3783-804

[13] Ghodhbani N, Maréchal P and Duflo H 2015 Curing and post-curing viscoelastic monitoring of an epoxy resin *Physic Procedia* **70** 106-9

[14] Ghodhbani N, Maréchal P and Duflo H 2016 Ultrasound monitoring of cure of an epoxy resin: reaction kinetics, frequency and temperature dependence *Polymer Testing* **56** 156-66

[15] Ghodhbani N 2016 Control and monitoring of viscoelastic properties of a composite during manufacturing (in french) Ph.D. thesis (University of Le Havre)

[16] Ghodhbani N, Maréchal P and Duflo H 2017 Real-time polymerization monitoring of a thermosetting resin around its glassy transition temperature *Proc. Mtgs. Acoust.* **26** 045004

[17] Trelin Y, Vasil'ev I and Roshchupkin V 1961 Measurement of the ultrasound velocity in molten alkali metals *The Soviet Journal of Atomic Energy* **9** 955-7

[18] Yoshimoto N, Shibata H, Yoshizawa M, Suzuki K, Shigematsu K and Kimura S 1996 Sound velocity measurements of molten germanium *Jpn. J. Appl. Phys.* **35** 2754

[19] Greenberg Y, Yahel E, Ganor M, Hevroni R, Korover I, Dariel M P and Makov G 2008 High precision measurements of the temperature dependence of the sound velocity in selected liquid metals *Journal of Non-Crystalline Solids* **354** 4094-100

[20] Kono Y, Park C, Sakamaki T, Kenny-Benson C, Shen G and Wang Y 2012 Simultaneous structure and elastic wave velocity measurement of SiO_2 glass at high pressures and high temperatures in a Paris-Edinburgh cell *Review of Scientific Instruments* **83**

[21] Messineo M G, Frontini G L, Eliçabe G E and Gaete-Garretón L 2013 Equivalent ultrasonic impedance in multilayer media. A parameter estimation problem *Inverse Problems in*

Science and Engineering **21** 1268-87

[22] Messineo M G, Rus G, Eliçabe G E and Frontini G L 2016 Layered material characterization using ultrasonic transmission. An inverse estimation methodology *Ultrasonics* **65** 315-28

[23] Mouhtadi A E 2011 Ultrasonic characterization of viscoelastic homogeneous and composite plates (in french) Ph.D. thesis (University of Le Havre)

[24] Wei S, Z. Yunsheng and Jones M 2014 Using the ultrasonic wave transmission method to study the setting behavior of foamed concrete *Constr. Build. Mater.* **51** 62-74

[25] Trtnik G, Turk G, Kavčič F and Bosiljkov V B 2008 Possibilities of using the ultrasonic wave transmission method to estimate initial setting time of cement paste *Cement and Concrete Research* **38** 1336-42

IOP Conf. Series: Materials Science and Engineering **369** (2018) 012011 doi:10.1088/1757-899X/369/1/012011

Kinetics of curing process in carbon/epoxy nano-composites

J Abenojar[1,2,3], J C del Real[2], Y Ballesteros[2] and M A Martinez[1]

[1]Materials Science and Engineering Department, Álvaro Alonso Barba Institute, Universidad Carlos III de Madrid, Av. Universidad 30, Leganés, 28911
[2]Technological Research Institute/Mechanical Engineering Department, Universidad Pontificia Comillas, C/ Alberto Aguilera 25, Madrid, 28015

E-mail: abenojar@ing.uc3m.es

Abstract. Epoxy resins are routinely used as adhesives, coatings, encapsulates, casting materials, and as matrix in composites. Their high crosslinking supposes too much rigidity, resulting in low toughness. Nano-filled addition could modify toughness, but it could also modify the curing process of the resin. Quality requirements of nano-composite (NC) imply a control of their curing process to ensure the mechanical properties of the material. Therefore, the aim of this work is to study the influence of carbon-based nano-fillers in the kinetics of curing of an epoxy resin. Three different methods are used in this work, Kissinger, Kamal and Model Free (MFK). The methods are complementary and provide information on the curing process of the nano-composites.

1. Introduction

Polymer matrix composite (PMC) are materials which provide a good compromise between properties and weight. These properties depend on the filler used, so additives can be found in several geometries, for example natural fibres [1], carbon fibres [2], glass fibres [3], natural particles [4] or inorganic particles as carbon black [5], silicon carbide [6], boron carbide [7], silica [8], polyhedral oligomeric silsequioxanes (POSS) [9], etc. All these additives can change the mechanical, tribological, thermal, electric or magnetic properties. This design strategy allows obtaining materials with on-demand properties.

Epoxy resins are the most versatile of the structural adhesives being strong but also brittle. They are usually used as polymer matrix to manufacture composites. Epoxy resins are a thermosetting polymer which possesses an oxirane group. This oxirane group is opened and it accelerates curing reaction. Polyaddition reactions with amines or polyamides are well known curing reactions, through which a primary or secondary amines added to the epoxy produce tertiary amine. The presence of particles or nanoparticles can promote or prevent the curing reactions [10,11] since other types of hydroxyl groups with different reactivity can activate the oxirane rings [12]. Amine type affects to curing reactions, so aliphatic amines cure epoxies at room temperature, whereas aromatic amines require higher temperatures.

The main purpose of this work is to study the curing process in an epoxy resin that cures at room temperature (Araldite 2020) modified with nano-carbon particles. For this purpose graphene, graphene oxide, multi-walled carbon nano-tubes and chemically exfoliated graphite nano-platelets were added in wt. 0.5% to the epoxy resin.

Differential scanning calorimetry (DSC) is an efficient method to analyse curing processes of thermoset resins. Kinetic parameters were calculated by non-isothermal scans.

2. Materials and experimental procedure

2.1. Materials

Epoxy resin used to manufacture the nanocomposite was Araldite 2020 by Hunstman (Barcelona, Spain). The rate epoxy: hardener was 100:30 by weight. This resin has a low viscosity and cure at room temperature for 24 hours. The used nano-carbon materials were graphene (G) of 1-2 layers by Avanzare Innovación Tecnológica S.L. (Logroño, Spain), graphene oxide (GO) and graphite nanoplatelets (GKOH) from Nanoinnova Technologies (Madrid, Spain); and multi-walled carbon nanotubes (CNT) by Nanocyl (Sambreville, Belgium).

2.2. Experimental process

A differential scanning calorimetry (DSC) 822e instrument was used and 40 μl aluminium crucibles were filled with approximately 8 mg of composite for each test. Different types of nano-carbon materials were added in 0.5 wt% and mixed mechanically for 10 min to 120 rpm, and then 8 mg approximately of this mixture was brought into pan to carry out the test by DSC. Nitrogen (80 ml/min flow) was used as purge gas. The non-isothermal scans were programmed from 0 to 200°C with a heating rate of 5, 10 and 20°C min^{-1}. For each composition, test was repeated three times.

2.3. Kinetic analysis

To analyse the cure kinetics of specimens, with and without nano-particles, non-isothermal (dynamic) DSC tests were performed. The kinetic analysis is based on the curing rate and it is proportional with the rate of heat flow at any time. So the rate of reaction can be described as equation (1).

$$d\alpha/dt = dH/dt \; x \; 1/\Delta H = k(T)f(\alpha) \quad (1)$$

Being ΔH the total heat of reaction determined from integration of the DSC curve, H the total heat released between start of reaction and time t, α the conversion degree, and $d\alpha/dt$ and dH/dt stand for curing rates and the heat flow. The $k(T)$ is a temperature-dependent parameter and $f(\alpha)$ is a conversion degree function.

Kinetics and calculi of curing activation energy were carried out under three different methods. Empirical methods were performed to calculate the kinetic parameters. At the first method the experimental data were adjusted to a model for heterogeneous reactions, using two kinetic mechanisms: autocatalytic and *n*-order. The autocatalytic mechanism is characterized by a maximum reaction rate at about 30–40% reaction curing progress. For the calculation of the kinetic parameters the Kamal's method was used [13], according to equation (2) where the two mechanisms are included:

$$d\alpha/dt = (k_1 + k_2\alpha^m)(1-\alpha)^n \quad (2)$$

Kamal M R and Sourour [13] explained these parameters. While the chemical kinetic controls the reaction, the rate constants (k_1 and k_2) depend on temperature following an Arrhenius dependence, according to equation (3).

$$K = Ae^{-(E_a/RT)} \quad (3)$$

The pre-exponential constant or frequency factor (A), in the Arrhenius equation, is an empirical relationship between temperature and rate coefficient. It depends on the quantity of molecules or groups in the reaction and their orientation [14].

The model free kinetic (MFK) [15,16] gives reliable predictions about conversions degree of chemical reactions under simulated conditions and allows calculate an E_a which depends on the conversion degree. The E_a was obtained using STARe Software of Mettler Toledo GmbH.

In addition other empirical models (such as Kissinger and Flynn–Wall–Ozawa models) can be used for dynamic kinetic studies. They have already been used to study epoxy based nanocomposites [17]. This model keeps in mind the curing peak temperature and heating rates, it is applying an Arrhenius type equation, according to equation (4), where β is heating rates.

$$\beta/T_p^2 = Ce^{-(E/RT_p)} \tag{4}$$

3. Results and discussion

When non-isothermal scans are integrated at three heating rates, the curing heat (ΔH) and the temperature (T_p) of curing peak are known. These values are shown in table 1. As it is logical whereas rate increases the curve is moved to the right and temperature increases with the heating rate. The addition of nano-particles produces a descent of the curing heat, this effect is more significant when GKOH or GO is added, mainly with GO. In the nano-composite with GO a decrement of the peak temperature is also noted.

Table 1. Influence of heating rate on peak temperature (*Tp*) and total reaction heat (*ΔH*)

β (°C/min)	5	10	20	$\Delta H \pm 1$ (J/g)
	$T_p \pm 1$ (°C)			
Araldite	99	114	130	346
+0.5CNT	99	115	131	338
+0.5G	97	115	131	339
+0.5GKOH	97	113	130	319
+0.5GO	95	111	127	289

According to equation (1) the conversion degree for all heating rates of each nano-composite and clear resin is calculated.

3.1. Kamal kinetic model

From the DSC curves, the conversion degrees versus the temperature or time can be obtained; figure 1 shows conversion degree at three heating rates for nano-composite with 0.5% CNT. Conversion degree plot is the average of three scans and its error is not more than ±1%. Figure 2 shows the curves (figure 1) derived with respect to time and plotted versus the conversion degree. Fitting these curves (figure 2) according to Kamal model equation (2) (see figure 3) then the kinetic parameters are obtained.

In this case reaction constants (k_1 and k_2) are obtained, which provide information about the reaction rates of *n*-order and autocatalytic respectively, together the order reaction (*n* and *m*). The table 2 shows the kinetic parameters for all nano-composites and clear resin.

The reaction constants increase with heating rate, k_2 is greater than k_1, so the autocatalytic reaction is always faster than the *n*-order. Additions of nano-particles decrease the reaction rate thus slowing the curing reactions consequently. Reaction order also decreases, which implies fewer particles present in the medium to produce the curing reaction.

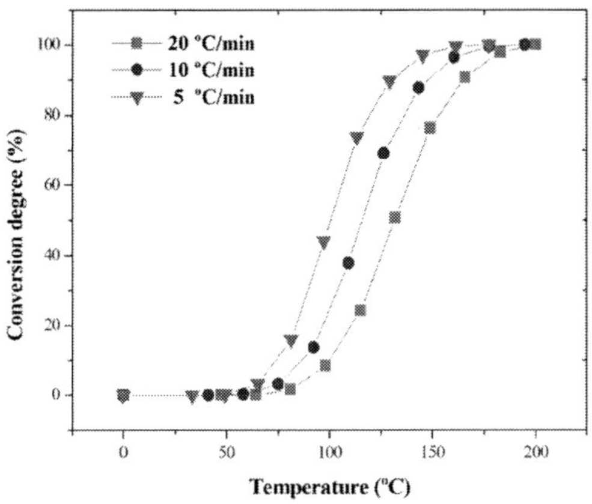

Figure 1. Conversion degree versus temperature of Araldite + 0.5CNT.

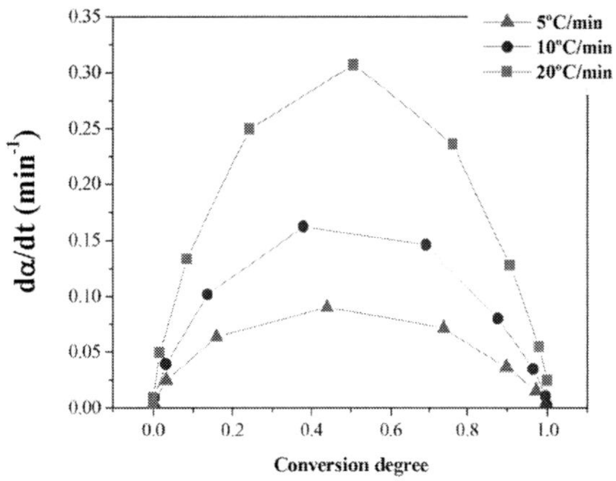

Figure 2. Reacting rates (dα/dt) versus conversion degree Araldite + 0.5CNT.

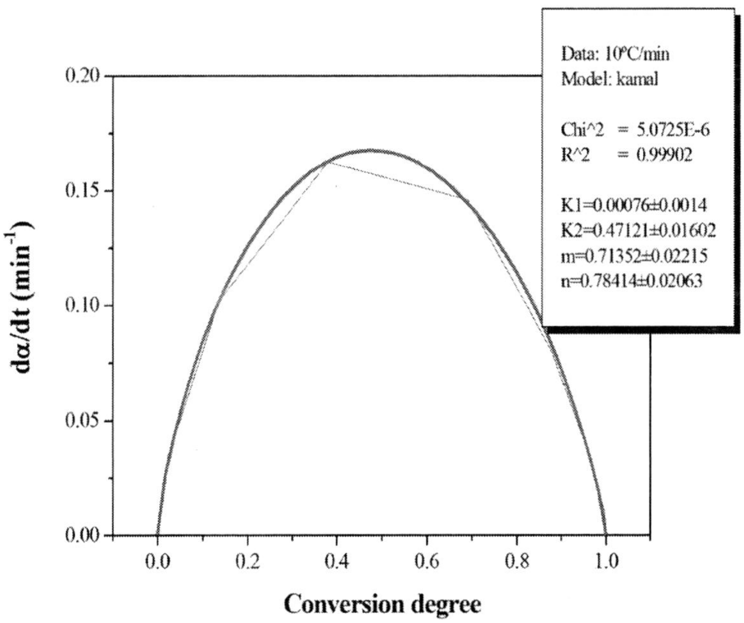

Figure 3. Fit to Kamal model: Reacting rates (dα/dt) versus conversion degree Araldite + 0.5CNT at 10°C/min.

Table 2. Kinetic parameters obtained by kamal model equation (1).

	K_1 (min^{-1})			K_2 (min^{-1})			m+n			R^2		
β (°C/min)	5	10	20	5	10	20	5	10	20	5	10	20
Araldite	0.0102	0.0229	0.0345	0.6010	1.1039	1.6574	2.6	2.6	2.4	0.9919	0.9868	0.9650
+0.5G	0.0004	0.0022	0.0021	0.2662	0.3703	0.8607	1.5	1.3	1.5	0.9993	0.9970	0.9941
+0.5CNT	0.0005	0.0008	0.0034	0.2456	0.4712	0.8270	1.5	1.5	1.4	0.9991	0.9990	0.9937
+0.5GKOH	0.0014	0.0016	0.0035	0.2459	0.4633	0.7712	1.4	1.5	1.4	0.9977	0.9989	0.9923
+0.5GO	0.0008	0.0019	0.0021	0.2292	0.4690	0.6905	1.5	1.5	1.5	0.9987	0.9989	0.9954

3.2. *Activation energy*

The activation energy was calculated from k_1 and k_2 and equation (3), and from heating rates and equation (4). Figure 4 shows how the equations are applied, plotting Ln (k or β/T^2_p) versus $1000/T$ then the data is adjusted to a line whose slope allows to calculate the activation energy and table 3 shows linear regression (R^2) of these adjustments.

In figure 5 the activation energy values are observed, for clear resin, and nano-composites with graphene and nano-tubes. The activation energy from k_1 (constant of n-order reaction) is greater than k_2, being greater for nano-composites. As the *n*-order reaction initially has a maximum rate, the curing reaction of these nano-composites needs more energy at the beginning and in the end than clear resin. However nano-composites with graphene oxide and graphite nano-platelets need less energy than the clear resin. This effect may be due to the presence of OH groups which could catalyse the reaction that favours the beginning of the reaction. The OH groups on the surface of nano-fillers could produce a catalytic effect [18], then the oxirane ring does not need to open, the catalytic reaction would start earlier than the autocatalytic, which has a maximum rate at about 30-40% of the reaction progress. In this case, activation energy from k_2 is similar at all nano-composites.

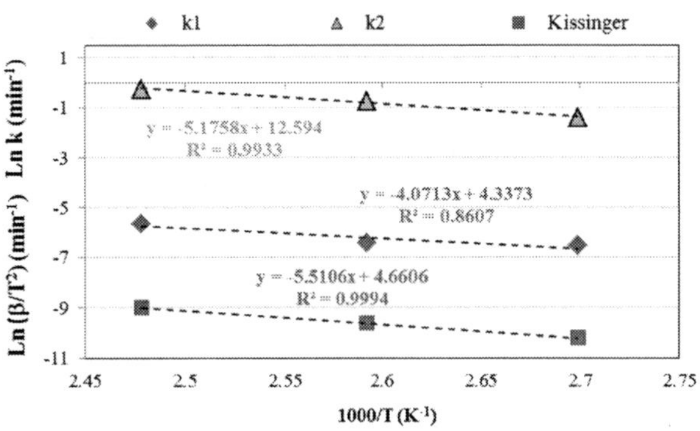

Figure 4. Adjustment for calculation of the activation energy from equations (3) and (4) for nano-composite Araldite + 0.5%GKOH.

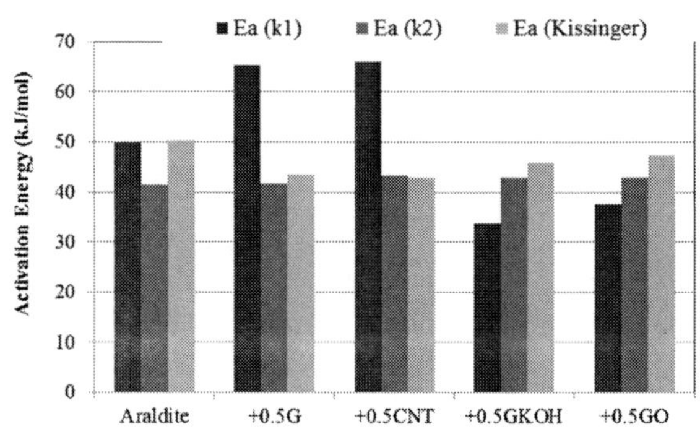

Figure 5. Activation energy obtained from reaction constants of Kamal model (equation (3)) and Kissinger model (equation (4)).

Table 3. R^2 values of the adjustments to a line of Arrhenius equation.

	R^2		
Araldite	0.9574	0.9983	1
+0.5G	0.8787	0.9100	0.959
+0.5CNT	0.8545	0.9994	0.9942
+0.5GKOH	0.8607	0.9933	0.9994
+0.5GO	0.8889	0.9807	0.9987

When Kissinger model is used, activation energy is always higher for the clear resin than the nano-composites, in this case according to equation (4), the heating rates are only taken into account; therefore the reaction mechanisms do not have influence.

When MFK is used, the activation energy depends on conversion degree. Taking into account only the beginning of the reaction, from figure 6 it is clearly appreciable that the clear resin (Araldite) needs more energy than the nano-composites. Activation energy is higher for graphene and nano-tubes than for the other nano-fillers. Therefore at the beginning the nano-fillers favour the curing reaction, specially GO and GKOH. However, in the middle of reaction clear resin and nano-composites have similar activation energy, although nano-composite with graphene needs slightly less energy than the rest. At the end of reaction nano-composite with graphene needs less energy to finish its curing reaction than any other materials; on the other hand the nano-composite that contains graphene oxide needs more energy than every else. This effect could be due to a fast initial reaction in which the graphene oxide could react with the OH groups of oxirane rings. After that, the crosslinking between the chains would be slowed and the last 5% curing takes much more time than in the other nanocomposites.

Others researchers explained these types of effect, thus kinetic models are based on a fast-pre-equilibrium step which happens due to hydrogen bonding between the oxygen of the epoxide ring and OH groups, as proton donor [19-24]. The reaction is also participated by amine-hydroxyl complexes [19]. Nevertheless in amine-hydroxyl complexes the amine is the proton acceptor, thanks to the greater dipolar moment of OH groups. According to Rozenberg [20] when amine loses its nucleophilic character, the oxirane ring reaction does not happen. Beside a compensation effect amine-amine and hydroxyl-hydroxyl reaction can be slighted.

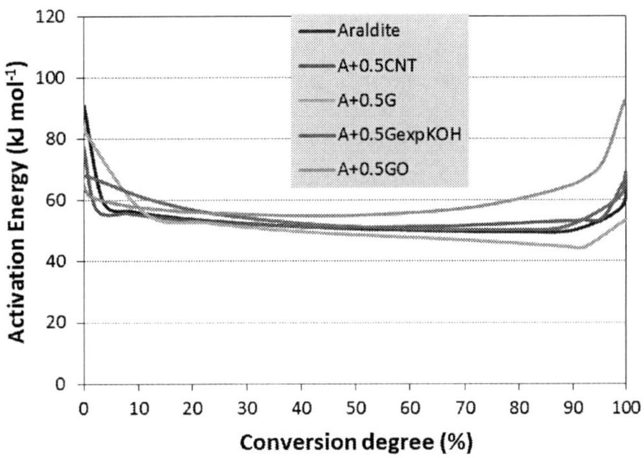

Figure 6. Activation energy obtained by MFK.

4. Conclusions

The empiric models complement each other and give information about the curing mechanisms that are being produced.

The Kamal and MFK models provide similar information, noting that there are two mechanisms: *n*-order at the beginning and end of the reaction and autocatalytic when oxirane group is opened and following while these are present in the medium.

The nano-carbon particles have an influence on the curing of the studied composites, this influence is greater when the nano-particles contain OH groups. These groups catalyse the reaction, but they can also delay its completion.

References

[1] Enciso B, Abenojar J and Martinez M A 2017 Influence of plasma treatment on the adhesion between a polymeric matrix and natural fibres *Cellulose* **24** 1791-1801

[2] Galvez P, Quesada A, Martinez M A, Abenojar J, Boada M J L and Diaz V 2017 Study of the behaviour of adhesive joints of steel with CFRP for its application in bus structures *Compos. Pt. B-Eng.* **129** 41-6

[3] Pihtili H 2009 An experimental investigation of wear of glass fibre–epoxy resin and glass fibre–polyester resin composite materials *Eur. Polym. J.* **45** 149-54

[4] Barbosa A Q, Figueiredo M, da Silva L, Öchsner A and Abenojar J 2017 Toughness of a brittle epoxy resin reinforced with micro cork particles: effect of size, amount and surface treatment

Compos. Pt. B-*Eng.* **114** 299-310

[5] Carbas R J C, da Silva L F M and Andrés L F S 2017 Functionally graded adhesive joints by graded mixing of nanoparticles *Int. J. Adhes. Adhes.* **76** 30-7

[6] Abenojar J, Martinez, M A, Pantoja M, Velasco F and del Real J C 2012 Epoxy composite reinforced with nano and micro SiC particles: curing kinetics *J. Adhes.* **4-6** 418-34

[7] Abenojar J, Martinez M A, Velasco F, Pascual-Sánchez V and Martín-Martínez J M 2009 Effect of boron carbide filler on the curing and mechanical properties of an epoxy resin *J. Adhes.* **85** 216-38

[8] Abenojar J, Tutor J, Ballesteros Y, del Real J C and Martinez M A 2017 Mechanical and thermal properties of polymer matrix nanocomposites reinforced by silica. *Compos. Pt.* B-*Eng.* **120** 42-53

[9] Blanco I and Bottino F A 2016 Kinetics of degradation and thermal behaviour of branched hepta phenyl POSS/PS nanocomposites *Polym. Degrad. Stab.* **129** 374-9

[10] Abenojar J, Encinas N, del Real J C and Martinez M A 2014 Polymerization kinetics of boron carbide/epoxyresin composite *Thermochim. Acta* **575** 144-50

[11] Barbosa A Q, da Silva L F M, Abenojar J, del Real J C, Paiva R M M and Öchsner A 2015 Kinetic analysis and characterization of an epoxy/cork adhesive *Thermochim. Acta* **604** 52-60

[12] Perrin F X, Nguyen T M H and Vernet J L 2007 Chemico-diffusion kinetics and TTT cure diagrams of DGEBA–DGEBF/amine resins cured with phenol catalysts *Eur. Polym. J.* **43** 5107-20

[13] Kamal M R and Sourour S 1973 Kinetics and thermal characterization of thermoset cure *Polym. Eng. Sci.* **13** 59-64

[14] Chang R 2005 *Physical Chemistry for the Biosciences* (Sausalito, CA: University Science)

[15] Vyazovkin S and Wight C A 1998 Isothermal and non-isothermal kinetics of thermally stimulated reactions of solids *Int. Rev. Phys. Chem.* **17** 407-33

[16] Vyazovkin S and Wight C A 1999 Model-free and model-fitting approaches to kinetic analysis of isothermal and non-isothermal data *Thermochim. Acta* **340–341** 53-68

[17] Wang Y T, Wang C S, Yin H Y, Wang L L, Xie H F and Cheng R S 2012 Carboxyl-terminated butadiene-acrylonitrile-toughened epoxy/carboxyl-modified carbon nanotube nanocomposites: thermal and mechanical properties *Express Polym. Lett.* **6** 719-28

[18] Xie H, Liu B, Yuan Z, Shen J and Cheng R 2004 Cure kinetics of carbon nanotube/tetrafunctional epoxy nanocomposites by isothermal differential scanning calorimetry *J. Polym. Sci. B Polym. Phys.* **42** 3701-12

[19] Mijovic J, Fishbain A and Wijaya A 1992 Mechanistic modelling of epoxy-amine kinetics. 2. Comparison of kinetics in thermal and microwave fields *Macromolecules* **25** 986-9

[20] Rozenberg B A 1986 Kinetics, thermodynamics and mechanism of reactions of epoxy oligomers with amines *Adv. Polym. Sci.* **75** 113-65

[21] Smith I T 1961 The mechanism of the crosslinking of epoxide resins by amines *Polym.* **2** 95-108

[22] Xu L, Fu J H and Schlup J R 1994 n Situ near-infrared spectroscopic investigation of epoxy resin−amine cure mechanisms *J. Am. Chem. Soc.* **116** 2821-6

[23] Flammersheim H J 1998 Kinetics and mechanism of the epoxide–amine polyaddition *Thermochim. Acta* **310** 153-9

[24] Riccardi C C, Fraga F, Dupuy J and Williams R J J 2001 Cure kinetics of diglycidylether of bisphenol A– ethylenediamine revisited using a mechanistic model *J. Appl. Polym. Sci.* **82** 2319-25

Infiltration behaviour of liquids over fibres or woven

M A Martinez[1,2], J Abenojar[1], B Enciso[1] and S Lopez de Armentia[1]

[1]Materials Science and Engineering Department. Álvaro Alonso Barba Institute, Universidad Carlos III de Madrid. Av. Universidad 30, Leganés, 28911

E-mail: mamc@ing.uc3m.es

Abstract. The high porosity of fabrics and fibres have hindered the study of the interaction between fluids and those kind of materials. In order to understand penetration mechanisms of polymeric matrices or woven sealing, some properties such as wettability or capillarity must be analysed. The fluid speed through some woven could be compared with metallic meshes in those is easy to determine pores size. In this work it is tried to solve these problems from a theoretical point of view by using hydrostatic laws and capillarity effect.

1. Introduction

Capillarity phenomena together with the permeability to air and steam water are the most important factors in all types of porous materials including fibres, especially regarding natural materials. For example, in the textile industry it is necessary that fabrics allow the evacuation of hot air in order to permit the perspiration of the skin. In this way more comfortable clothes are obtained. This can be achieved by using tighter fibres with a greater number of yarns. Another example of natural material is cork. There are some studies of cork used as a reinforcement of adhesives to improve the impact resistance, the porosity of the cork allows the adhesive to get inside the pores [1], although the material has a low surface energy [2].

In other industries, such as construction, capillarity in concrete is an important drawback because it could produce humidity problems in buildings and, as a consequence, deterioration of the construction [3]. In the automotive industry metallic filters for air and liquids like oil are used. The efficiency on these filters will depend on capillary size and interconnection between pores.

Nowadays, the applications of polymeric composite materials reinforced with fibres like glass, carbon, aramid, etc. are very numerous and continue growing. These materials can be applied in different industries such as aviation, aerospace, sports or automotive, and in conclusion, in all those applications where weight reduction is mandatory without compromising mechanical properties [4,5].

When the fabric spaces are large enough and had the right interconnections to be invaded by fluid molecules, several processes take place. Consequently, according to the imbalance among different atoms can be produced capillarity processes [6-8].

The models by which a liquid is keeping in contact with a porous material independent of its nature has already been studied [9], where some important concepts are defined. For example the ability of some materials to absorption of liquid as hygroscopicity, that is influenced by combination effect of environment conditions. Besides, the nature, physical dimensions and form of the pores affect to vapour amount absorbed.

When a liquid and a solid are in surface contact state the liquid can be absorbed by the solid by means of capillarity. The capacity takes place because of the interactions among the solid, liquid and gas phases. Through the pores, liquid movements are produced by suction phenomena, so it can be considered that the penetration and speed of the process depends on the nature, geometry and roughness of the fibers in this case.

By different equations and laws as Young's equation and Jurin law, the relationships between some measurable concepts are defined: contact angles (θ), the liquid height (h) as a result of pressure variation [10,11]. In accordance with this, the necessary pressure (ΔP), to move the liquid across a capillary of radius r, is obtained by the capillary effect and hydrostatic pressure of the liquid used to measure, according to equation 1 (figure 1). It is very easy to realize that an external pressure (as the one produced in a hot plate press or autoclave) will increase this pressure and therefore the hydrostatic term can be considered negligible.

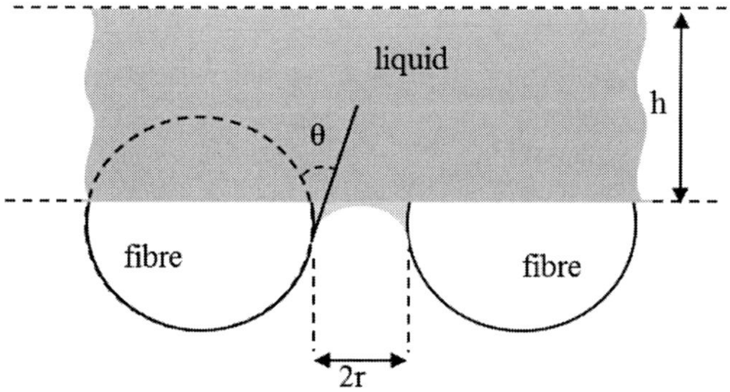

Figure 1. Scheme of liquid infiltration between two fibres (own).

$$\Delta P = \frac{2\gamma_{LG}cos\theta}{r} + h\rho g \quad (1)$$

where: γ_{LG} = liquid surface tension (N/m).

There is an inversely proportion between the contact angle on the fibres or the entry pressure of the liquid with the wettability (Kw), when one increases the other decreases. Besides, the wettability depends on the material type. As well as if Poiseuille's law is considered, speed and pressure are connected by equation 1. In the case of a laminar flow is possible to define the flow rate Q (m^3/s) and fluid speed V (m/s) by equations (2) and (3) where other parameters, as the fluid viscosity μ (Pa.s) and woven thickness L (m), are required.

$$Q = \frac{\pi r^4 \Delta P}{8\mu L} \quad (2)$$

$$V = \frac{r^2 \Delta P}{8\mu L} \quad (3)$$

By Darcy's equation the permeability can be evaluated according to equation (4), that is defined in this case as an ability of the woven fibres because it depends on its interconnected pores.

$$Q = \frac{KA\Delta P}{\mu L} \quad (4)$$

where: A = total area of wet woven

The pore section is not really constant and the length is larger. The uneven shape of the pores produces a length increment and there are also changes in the capillary section. These different sections require higher pressure to make the liquid pass across it, for this reason the measure is more difficult. The measure is easier on metallic meshes with a defined mesh size than woven fibres [12]. Porosity parameters are shown in table 1 and figure 2. The average radius value (r) has been calculated taking into account the mesh dimensions and the capillary diagonal. Thickness (L) was measured with all four layers together. Wet area (A) and pores number that water drop covers (n) were measured during the experiment.

Table 1. Metallic mesh with different pore size [12].

Mesh	r (mm)	L (mm)	A (cm^2)	Pores Number
150	0.181	3.88	15.21	243
110	0.133	2.28	23.80	690
90	0.108	1.60	17.37	920

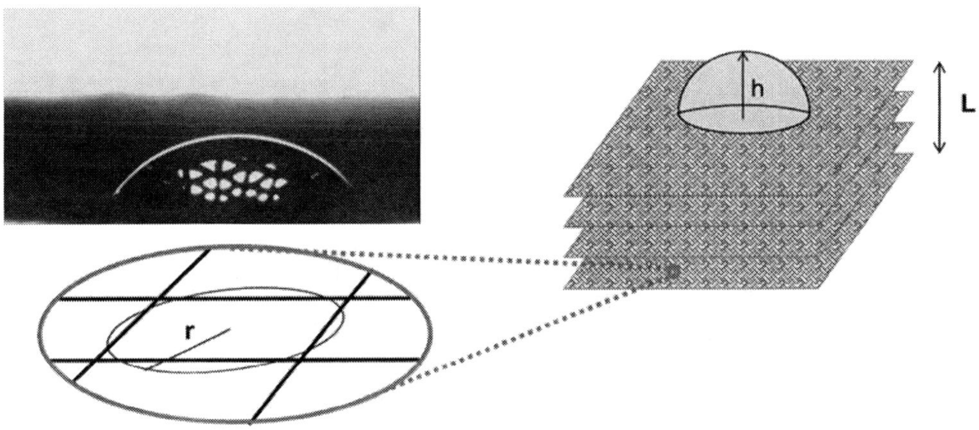

Figure 2. Mesh porosity parameters (own).

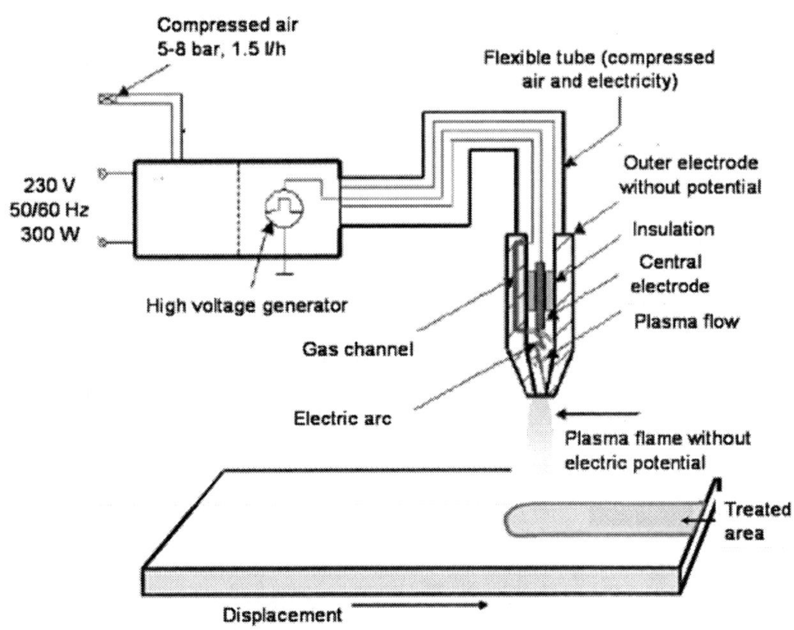

Figure 3. Atmospheric plasma torch treatment device (own).

When a cold plasma treatment is applied, ions and electrons are reorganized and act on the surface materials modifying its chemical and physical nature without affecting volume properties. Atmospheric Pressure Plasma Torch (APPT) mainly effects are cleaning, pollutants decomposition, grafting and polar groups, as -OH, -COOH, -CO and -NH, cause the activation of the surface. Therefore, surface energy increases in a significantly way and it is traduced in polymers adhesion properties improvement [13,14]. As seen in figure 3 the torch is coupled on a table with displacement to control the rate, the torch can also be moved, this way the sample distance is modified. Both parameters need to be optimized to obtain an optimal treatment.

Figure 4. (a) Glass fibre woven picture; (b) Micrograph of untreated Glass fibre; (c) Micrograph of Glass fibre treated with APPT.

2. Experimental study

By measuring contact angles and flow rate calculation, as it was done with metallic meshes, a glass fibre woven with an unknown porosity (Vetrotex-P-204-2400 Tex) provided by Saint-Gobain (Aachen, Germany) was studied (figures 4(a) and 4(b). The process was then carried out on the glass fibre woven which was treated with atmospheric plasma torch (APPT) to raise the wettability (figure 4(c)). The woven was treated with an atmospheric pressure plasma torch (APPT) device with operating conditions 50/60 Hz, 230 V and 16 A (PLASMATREAT GmbH, mod. FG 3001, Steinhagen, Germany). The optimized parameters of the treatment are distance between torch nozzle and woven sample (8 mm) and speed of sample under the plasma flow (0.08 m/s).

Optical goniometer OCA manufactured by Dataphysics of Instruments GmbH (Filderstadt, Germany) was used to calculate the contact angles and drop penetration into the woven fibre. This device has a movie mode which is used to analyse the water drop penetration through several filters. This mode allows determine the liquid flow rate as a function of time. The values obtained for the fibres are compared with those obtained for metallic mesh [12]. So is possible to know the woven fibre capillarity.

3. Results and discussion

The study of the effect of APPT treatment on glass is already carried out in other works [15]. Due to the high intrinsic surface energy of glass the effect of APPT is only cleaning but not activation. In this work the glass fibre woven is covered with a layer of thermoplastic material, as all commercial fabrics have, so plasma treatment acts on the thermoplastic layer instead of directly on glass. For this reason contact angles measured with polar liquids (distilled water and glycerol) decreases and those measured with a dispersive liquid (diiodomethane) increases after treatment, as it is shown in table 2. According to Encinas *et al* [13], APPT treatment produces a surface energy increment and, as a result, a wettability increment.

Table 2. Contact angles (°) on untreated and APPT treated woven glass fibres using different test liquids.

Woven Fibres	Distilled Water (°)	Diiodomethane (°)	Glycerol (°)
GF	112	28	118.0
GF + APPT	42	52	55.3

Figure 5 shows filtration rates of glycerol through untreated and APPT treated woven glass fibres. Results are compared with those obtained with water through metallic meshes. The liquids were changed for their different viscosities, searching in this way the same effect on the woven that on the mesh, and therefore the slopes of the corresponding curves coincide.

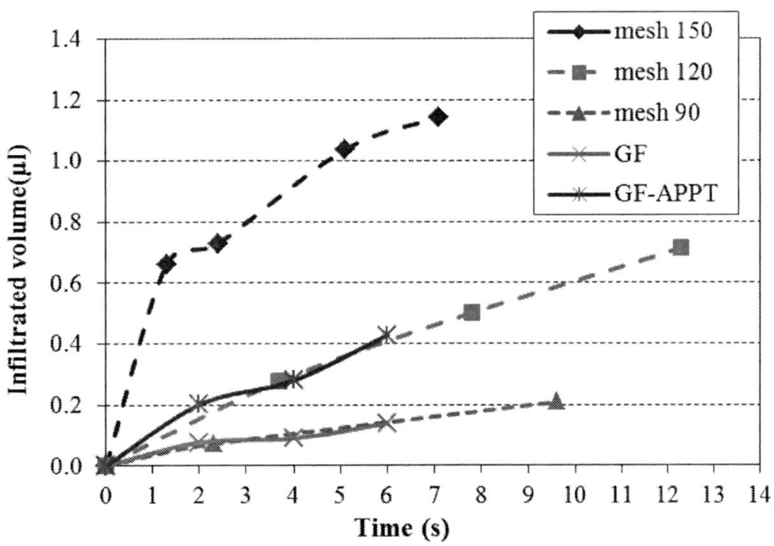

Figure 5. Infiltrated drop volume variation versus time.

Once is verified that the infiltrated drop volume through metallic meshes is similar to the one obtained on glass fibre woven, all infiltration parameters can be calculated with equations from (1) to (4). If surface tension, viscosity and density of distilled water is known (γ_{LG} = 72.75 mJ/m^2, μ= 0.001 Pa.s y ρ = 997.13 kg/m^3 a 25°C) infiltration parameters of the water through metallic meshes can be obtained. Regarding pressure, the hydrostatic component is negligible due to the small size of the drop. Table 3 shows contact angle measurements using water as test liquid on metallic meshes and other infiltration parameters.

Table 3. Infiltration parameters of water obtain on metallic meshes.

Material	Contact Angle (°)	Flow Rate (µl/s)	Pressure (Pa)	Speed (mm/s)	Permeability (m^2 x10^{-3})
Mesh 150 µm	80	0.161	1483	1565	2.77
Mesh 120 µm	120	0.060	1114	1080	0.56
Mesh 90 µm	110	0.022	659	600	0.31

Glycerol infiltration flow on untreated woven glass fibres is the same as the one obtained for 90 µm metallic mesh and glycerol infiltration flow on APPT treated woven glass fibres is very similar to 120 µm metallic mesh. These results could be explained by the effect of the APPT, which increases wettability and therefore flow rate is higher (as in a bigger metallic mesh). Accordingly, the rest of the glycerol infiltration parameters can be obtained if surface tension, viscosity and density are known (γ_{LG} = 63.10 mJ/m^2, µ= 1.5 Pa.s and ρ =1261 kg/m^3) and woven thickness is 0.91 mm. Results are shown in table 4.

Table 4. Infiltration parameters of glycerol on glass fibre woven obtain with and without APPT treatment.

Material	Total Area (cm^2)	Contact angle (°)	Flow rate (µl/s)	Pressure (kPa)	Speed (mm/s)	Permeability (m^2 x10^{-10})
Glass woven	0.75	28.1	0.023	562	0.30	7.45
Glass woven +APPT	4.20	52.2	0.071	667	0.16	3.46

4. Conclusions

The obtained data allow to know the theoretical parameters of the woven fibres.

If drop size and flow rate are known is easy to determine the theoretical parameters that define porous materials, this are porosity, fluid pressure, wettability and permeability values.

In this study has been proved that the more wettability the more flow rate of liquids through glass fibre woven. This wettability increment is due to the effect of APPT treatment which makes that the glass fabric behaviour could be resembled to the one obtained for metallic meshes with bigger pore size.

Acknowledgments

The authors thank MEIC for financing the project TRA2014-56471-C4-2-R.

References

[1] Barbosa A Q, da Silva L F M, Carbas R C, Abenojar J and del Real J C 2012 Influence of the size and amount of cork particles on the toughness of a structural adhesive *J. Adhes.* **88** 452-470

[2] Abenojar J, Barbosa A Q, Ballesteros Y, del Real J C, da Silva L F M, and Martínez M A 2014 Effect of surface treatments on cork: surface energy, adhesion and acoustic insulation *Wood Sci Technol* **48** 207-224

[3] Castro J, Bentz D and Weiss J 2011 Effect of sample conditioning on the water absorption of concrete *Cement Concrete Comp* **33** 805-813

[4] Chawla K K 2012 *Composite Materials. Science and Engineering* (New York: Springer)

[5] Barbero E J 2011 *Introduction to Composite Materials Design* (Boca Raton, Florida: Taylor & Francis Group)

[6] Gregg S J and Sing K S W 1982 *Surface Area and Porosity Adsorption* (London: Academic Press)
[7] Lowell S and Shields J E 1991 *Powder Surface Area and Porosity* (London: Champman & Hall)
[8] Adler P M 1992 *Porous Media: Geometry and Transports* (Oxford: Butterworth-Heinemann)
[9] Liu P S and Chen G F 2014 *Porous Materials Processing and Applications* (New York: Elsevier Inc)
[10] Gilles de Gennes P, Brochard P F and Quéré D 2003 *Capillarity and Wetting Phenomena* (New York: Springer)
[11] Irima D 2008 Capillary force valves *Encyclopedia of Microfluidics and Nanofluidics* (New York: Springer)
[12] Martinez M A, Velasco F and Abenojar J 2014 Behaviour of fluids in porous materials *Mater Sci Forum* **802** 303-8
[13] Encinas N, Díaz-Benito B, Abenojar J and Martínez M A 2010 Extreme durability of wettability changes on polyolefin surfaces by atmospheric pressure plasma torch *Surf. Coat. Technol.* **205** 396-402
[14] Caiazzo F, Canonico P, Nigro R and Tagliaferri V 1996 Electrode discharge for plasma surface treatment of polymeric materials *J. Mater. Process. Technol.* **58** 96-9
[15] Abenojar J, Martínez M A, Velasco F and Encinas N 2013 Modification of glass surfaces adhesion properties by atmospheric pressure plasma torch *Int. J. Adhes. Adhes.* **44** 1-8

IOP Conf. Series: Materials Science and Engineering **369** (2018) 012013 doi:10.1088/1757-899X/369/1/012013

Physical properties and morphology of electrospun composite fiber mats of polyhydroxyalkanoate containing nanoclay and tricalcium phosphate additives

N Tanadchangsaeng[1,2,3] and S Boonyagul[1]

[1]Faculty of Biomedical Engineering, Rangsit University, 52/347 Lak-Hok, Pathumthani 12000 Thailand
[2]Industrial Biomaterials Research Center, Rangsit University, 52/347 Lak-Hok, Pathumthani 12000 Thailand

E-mail: nuttapol.t@rsu.ac.th

Abstract. Recently, nanofiber research has gained substantial attention from scientists. In this study, the main component of the nanofiber sheet is polyhydroxyalkanoate (PHA) polymer, which is strong, ductile, flexible and adhesive to human skin. Two major additives of nanofiber sheet that we applied are nanoclay and tricalcium phosphate. The additives are generally synthetic substances that can be chemically synthesized and compatible with tissues body. Nanoclay has a low density, strong, durable to compressive strength and humidity. While, tricalcium phosphate is a calcium phosphate ceramic that is biocompatible to human tissue. From the reasons above, we proposed to choose both nanoclay and tricalcium phosphate for adding into PHA nanofibers for film formation. Thus, this study aims to investigate the morphological and mechanical properties of the fiber mat by using PHA added with various amount of nanoclay and tricalcium phosphate at 0.1%, 1% and 10% by weight, and fabricate nanofiber samples by electrospinning technique. The tested results of scanning electron microscope (SEM) morphology show that the fibers have a uniformed pattern. The PHA containing nanoclay of all additive contents exhibited micrometer diameter distributions, while PHA loaded with 1% tricalcium phosphate still had the nano-scale diameter range, and might be the optimum additive load for further nanometer medical applications. A tensile test was performed to determine the effect of nanoclay and tricalcium phosphate contents on the mechanical properties of the electrospun PHA films, and reflect the level of modularity. With nanoclay components being integrated into the polymer matrix, subsequent reduction in fiber crystallinity was occurred after addition of nanoclay with an increase of modulus value. The results confirmed that PHA fiber mat containing 1% nanoclay may have a potential for using as a rigid scaffold which bearing force loading in human organ system. Whereas, it can be indicated that PHA fiber mat containing 1% tricalcium phosphate might be employed as a flexible scaffold for biomedical materials application due to a high elongation at break value.

1. Introduction

Electrospinning, as shown in figure 1, is a simple and versatile method for the fabrication of polymer nanofibers. Nanofibers have been investigated as components in wound dressings, protective clothing, filtration, biomedical devices and as reinforcement fibers in composites [1]. Their high surface areas are one factor dictating their potential in tissue engineering applications, especially if the material used is bioactive, biocompatible or nanoporous [2]. Electrospinning can also be used for processing non-woven nanofibers for advanced medical research, such as the fabrication of biocompatible nanofiber scaffolds with three-dimensional extracellular matrices for improving the function, or treatment of, damaged tissues or organs [3,4]. Scaffolds play a crucial role in commanding stem cells differentiation, therefore, electrospun scaffolds stands for a promising way for bone tissue engineering [5].

Until now, a variety of polyesters have been studied in biomedical applications due to their biocompatibility, bioresorbility, biodegradability and high mechanical propertiey [6-8]. Among them, polyhydroxyalkanoate (PHA), which is a family of biopolyesters produced by various microorganisms, can be processed similar to petroleum-based plastics, which their physical properties are of substantial attractiveness to polymer manufacturer [9,10]. The excellent biodegradability and biocompatibility of PHA bioplastics are desired properties for many short-life goods and applications in agriculture and health care [11,12].

Nanoclays have exhibited to improve physical properties of polymer morphology and structures [13,14]. Nanoclays are greatly employed to reinforce thermoplastics for hybrid composites. Incorporation of nanoclay into polymer matrices is still well-known to produce composite scaffolds. Tricalcium phosphate are bioactive and osteoconductive if fabricated with appropriate geometry [15]. The tricalcium phosphate can improve the mechanical properties of polymeric materials, and enhance cell adhesion for inducing the differentiation of cells. Due to a difference of mechanical properties of PHA scaffolds and natural bone, polymer matrix scaffolds made of inorganic materials are too brittle. The addition of these two inorganic additives to polymeric matrix can control the physical properties of the nanocomposite matrix [16,17]. Here, we proposed to choose nanoclay and tricalcium phosphate to add into nanofibers for film formations and assessed the morphological and physical properties of nanoclay/tricalcium phosphate-enriched electrospun PHA scaffolds. This work aims to investigate the strength of the nanofiber sheet after adding nanoclay and tricalcium phosphate by electrospinning nanofiber fabrication.

2. Experiments

2.1. Materials

Polyhydroxyalkanoate (PHA) copolymer was produced from glycerol and levulinic acid by *Ralstonia eutropha* as described in a previous work [18]. Synthetic nanoclay (clay Cloisite 20A) was obtained from Southern Clay Products, Inc. Tricalcium phosphate was purchased from Sigma Aldrich (USA). Dichloromethane (DCM, analytical grade) was also purchased from Sigma Aldrich (USA).

2.2. Polymer solution preparation

For preparation of polymer solution, the polymer mixture of PHA was dissolved in dichloromethane with concentration at 8% (w/v). Then, nanoclay or tricalcium phosphate was blended at 0%, 0.1%, 1.0%, and 10% by weight, respectively, and mixed together with temperature at 60°C for 15 min.

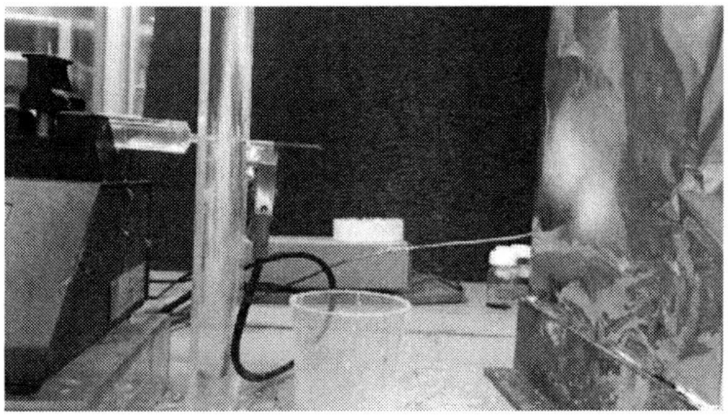

Figure 1. Experimental set up during electrospinning process.

2.3. Fabrication of PHA–composite electrospun scaffolds by electrospinning

A single nozzle electrospinning setup was used to prepare electrospun nanofibers as shown in figure 1. A high voltage power supply (ES60P-10 W, Gamma High Voltage, Florida, USA) was attached to a metal nozzle to create a voltage difference between the metal nozzle and a grounded collector of 15–25 kV. A charged metal nozzle with 0.57 mm outer diameter was placed 15 cm directly above the center of the 20 × 20 cm aluminum flat grounded collector. A mechanical pump (NE-300, New York, USA) was used to deliver the solution from a plastic syringe to the metal nozzle at a rate of 1 ml/hr.

2.4. Morphology characterization

Scanning Electron Microscope (SEM) (JSM-6400, JEOL Ltd., Tokyo, Japan) was used to characterize electrospun fibers characteristic and size. Samples of the electrospun nanofibers were sputter coated with platinum. Images were captured with an accelerating voltage of 5 kV. Image J software (www.imagej.nih.gov) was used for measuring fiber diameter. The average fiber diameter and Standard Deviation (SD) were determined.

2.5. Mechanical property testing

The mechanical properties of electrospun nanofibers were evaluated using a texture analyzer (TA.XT plus, Stable Micro Systems, UK) with a 5 kg load cell equipped with tensile grips. Specimens tested had dimensions 0.5 cm x 3 cm, and were tested at a rate of 5.0 mm/min until fracture occurred.

3. Results and discussion

3.1. Morphological property of electrospun scaffolds of PHA–composite with nanoclay and tricalcium phosphate

Non-woven nanofibers via electrospining were processed by high voltage on the needle and zero voltage on the plate collector (grounded), accumulated electrostatic on the PHA/DCM droplet was pushed by high voltage on the needle to the collector, therefore the droplet became to a solution jet that rapidly elongate the solution into nanosize till the DCM solvent was completely evaporated before the nanosized fiber drop on the collector as shown in figure 1. In this study, we selected 8% PHA solution to be investigated because the 8% PHA nanofiber via electrospinning exhibited a smooth cylindrical, uniformity and diameter is at 0.83 ± 0.14 μm (less than 1 μm) of fibre formation as shown in figures 2A, 3A, and 4.

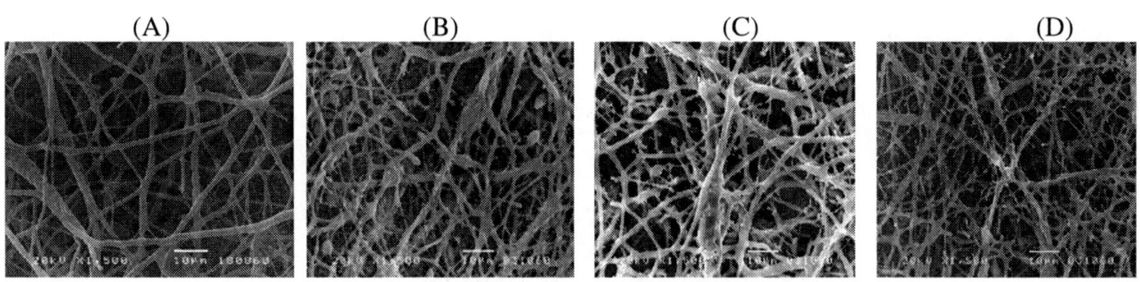

PHA 0.1% 1.0% 10.0%

Figure 2. SEM morphology of PHA fiber mats composited with nanoclay in concentration by polymer of 0% (A), 0.1% (B), 1.0% (C), and 10% (D) by weight, respectively. All the SEM images were captured at 1500x.

Within 1500x microscale, the morphologies of the fibers obtained through electrospinning of PHA/nanoclay solutions (figure 2) are the same (minuscule nanoclay particles attached to fibers) but fiber diameter changes by increasing from nanometer diameter of pure PHA nanofiber to microfiber scale (more than 1 μm). The average diameters of the PHA/nanoclay fibers containing 0.1%, 1.0% and 10% nanoclay were at 2.08 ± 0.14, 2.15 ± 0.91 and 1.91 ± 0.88 μm, respectively as shown in figure 4A. From these results, the PHA containing nanoclay of all additive contents exhibited micrometer diameter distributions.

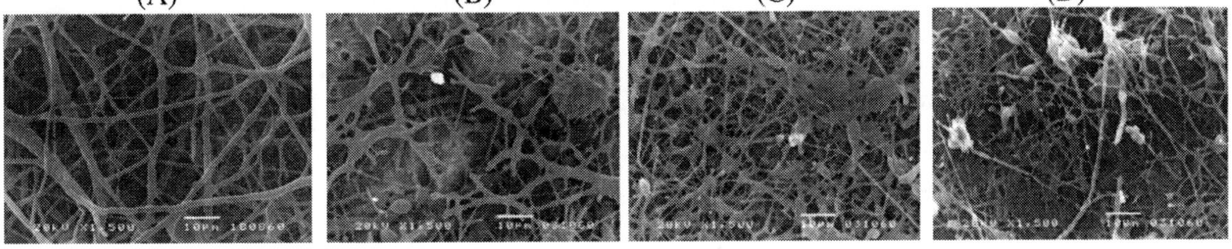

Figure 3. SEM morphology of PHA fiber mats composited with tricalcium phosphate in concentration by polymer of 0% (A), 0.1% (B), 1.0% (C), and 10% (D) by weight, respectively. All the SEM images were captured at 1500x.

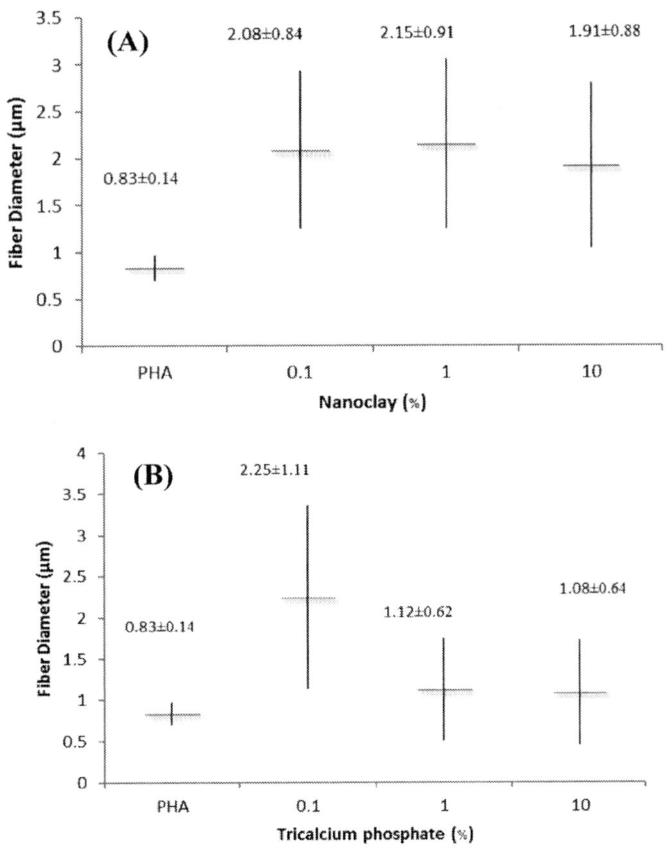

Figure 4. The bar graph between fiber diameter of PHA fiber mats composited with nanoclay (A), and with tricalcium phosphate (B) which are dependent on the amount of each additive.

Whereas, the morphologies of the fibers obtained through electrospinning of PHA/tricalcium phosphate solutions (figure 3) bearing an amount of knots and forming some complex particles attached to the fibers are the same, but fiber diameter still increases from nanometer diameter of pure PHA nanofiber to microfiber scale (more than 1 μm) when loading the additive. The average diameters of the PHA/ tricalcium phosphate fibers containing 0.1%, 1.0% and 10% additive were at 2.25 ± 1.11, 1.12 ± 0.62 and 1.08 ± 0.64 μm, respectively as shown in figure 4B.

For tissue engineering work, the ECM matrix should be in nanoscale scaffold for suitable application. As a result, we confirmed that PHA fibers loaded with 1% tricalcium phosphate still had the nano-scale diameter range, and might be the optimum additive load for further nanometer medical applications.

3.2. Mechanical property of electrospun PHA fibers containing nanoclay and tricalcium phosphate additives

A tensile test was performed to determine the effect of nanoclay and tricalcium phosphate contents on the mechanical properties of the electrospun PHA films. The films was performed at room temperature using a tensile test machine with a strain rate of 5 mm/min. Figure 5 shows typical tensile curve depicting relative between tensile force and elongated distance of PHA fiber mats composited with nanoclay and with tricalcium phosphate, which are dependent on the amount of each additive. Table 1 shows a summary of mechanical data of PHA fibers loaded with each additive.

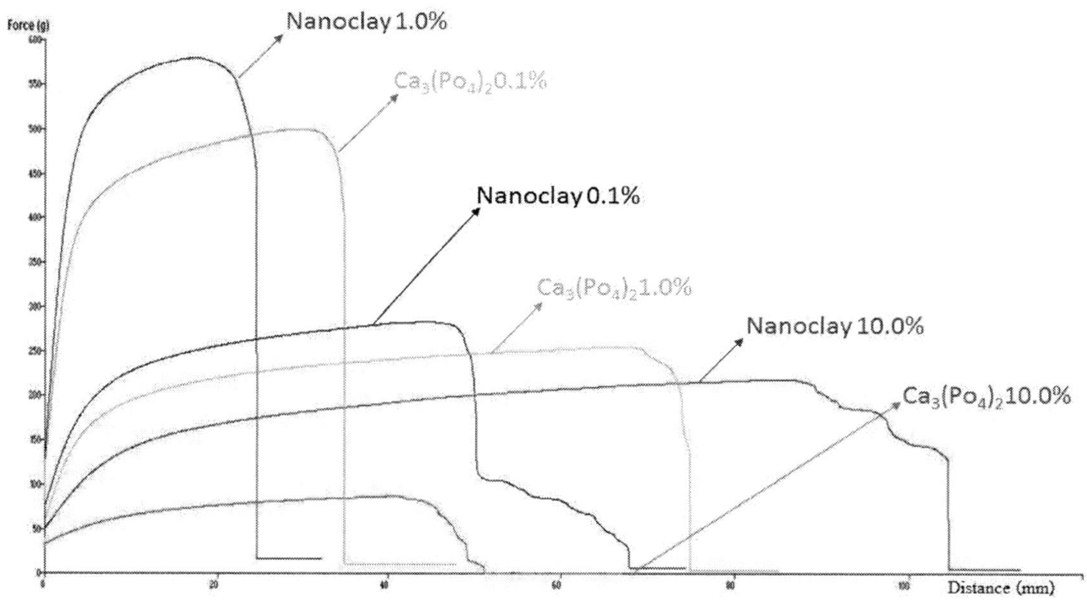

Figure 5. The tensile curve depicting relative between tensile force and elongated distance of PHA nanofibers composited with nanoclay and with tricalcium phosphate, which are dependent on the amount of each additive.

The mechanical property results for PHA fibers containg 1.0% nanoclay indicate a Young's modulus of 50 MPa that is higher than 3 folds compare to 0.1% nanoclay loaded. But the elongation at break value is reduced from 70% to 25% as we decreased the amount of nanoclay. These results are consistent with nanoclay components being integrated into the polymer matrix, and subsequent reduction in fiber crystallinity after addition of nanoclay. As a result, it can be confirmed that PHA fiber mat containing 1% nanoclay may have a potential for using as a rigid scaffold which bearing force loading in human organ system.

Table 1. Mechanical Properties of PHA fiber mats composited with nanoclay and with tricalcium phosphate, which are dependent on the amount of each additive.

Sample	Modulus	% Elongation
Nanoclay 0.1%	15.31 MPa	70%
Nanoclay 1.0%	49.40 MPa	25%
Nanoclay 10.0%	6.55 MPa	105%
Tricalcium phosphate 0.1%	32.5 MPa	36%
Tricalcium phosphate 1.0%	11.19 MPa	76%
Tricalcium phosphate 10.0%	2.74 MPa	53%

For tricalcium phosphate loading in PHA fibers, the mechanical property results for PHA fibers containg 1.0% tricalcium phosphate shows a Young's modulus of 11 MPa, while the elongation at break is at 76%. As a result, it can be indicated that PHA fiber mat containing 1% tricalcium phosphate might be employed as a flexible scaffold for biomedical materials application such as suture and mesh.

4. Conclusion

In this study, we investigated the morphological and mechanical properties of the fiber mat by using polyhydroxyalkanoate (PHA) added with various amount of nanoclay and tricalcium phosphate at 0.1%, 1% and 10% by weight, and fabricate nanofiber samples by electrospinning technique. The tested results of scanning electron microscope (SEM) morphology show that the fibers have a uniformed pattern. The PHA containing nanoclay of all additive contents exhibited micrometer diameter distributions, while PHA loaded with 1% tricalcium phosphate still had the nano-scale diameter range, and might be the optimum additive load for further nanometer medical applications. A tensile test was performed to determine the effect of nanoclay and tricalcium phosphate contents on the mechanical properties of the electrospun PHA films, and reflect the level of modularity. With nanoclay components being integrated into the polymer matrix, subsequent reduction in fiber crystallinity was occurred after addition of nanoclay with an increase of modulus value. The results confirmed that PHA fiber mat containing 1% nanoclay may have a potential for using as a rigid scaffold which bearing force loading in human organ system. Whereas, it can be indicated that PHA fiber mat containing 1% tricalcium phosphate might be employed as a flexible scaffold for biomedical materials application due to a high elongation at break value.

Acknowledgment

The authors appreciate financial support from the Research Institute of Rangsit University (contract number 58/2557). We thank National Nanotechnology Center (NANOTEC) for use of the tensile testing equipment and Thailand Institute of Scientific and Technological Research for SEM experiment. We wish to thank Thanawat Saengsunanon, Phimphichaya Sanmanot, Acharawan Panlomso and Stayu Kitmongkolpaisarn from Rangsit University for laboratory assistance in this study.

References

[1] Hu X, Liu S, Zhou G, Huang Y, Xie Z, *et al* 2014 *J Control Release* **185** 12-21
[2] Ko F K and Wan Y 2014 *Introduction to Nanofiber Materials* (Cambridge University Press)
[3] Sun B, Long Y, Zhang H, Li M, Duvail J, *et al* 2014 *Prog Polym Sci* **39** 862-90
[4] Gaharwar A K, Mukundan S, Karaca E, Dolatshahi-Pirouz A, Patel A, *et al* 2014 *Tissue Eng Part A* **20** 2088-101
[5] Jang J H, Castano O and Kim H W 2009 *Adv Drug Deliv Rev* **61** 1065-83
[6] Agrawal C M and Ray R B 2001 *J Biomed Mater Res* **55** 141-50
[7] Patel A, Gaharwar A K, Iviglia G, Zhang H B, Mukundan S, *et al* 2013 *Biomaterials* **34** 3970-

83
[8] Zhang H B, Patel A, Gaharwar A K, Mihaila S M, Iviglia G, *et al* 2013 *Biomacromolecules* **14** 1299-310
[9] Lenz R W and Marchessault R H 2004 *Biomacromolecules* **6** 1-8
[10] Luzier W D 1992 *Proc Natl Acad Sci* **89** 839-42
[11] Sudesh K, Abe H and Doi Y 2000 *Prog Polym Sci* **25** 1503-55
[12] Chen G-Q 2009 *Chem Soc Rev* **38** 2434-46
[13] Bordes P, Pollet E and Averous L 2009 *Prog Polym Sci* **34** 125-55
[14] Wu C J, Gaharwar A K, Schexnailder P J and Schmidt G 2010 *Materials* **3** 2986-3005
[15] Kang Z, Zhang X, Chen Y, Akram M Y, Nie J, *et al* 2017 *Mater Sci Eng: C* **70** 1125-31
[16] Gaharwar A K, Kishore V, Rivera C, Bullock W, Wu C J, *et al* 2012 *Macromol Biosci* **12** 779-93
[17] Schexnailder P J, Gaharwar A K, Bartlett R L, Seal B L and Schmidt G 2010 *Macromol Biosc* **10** 1416-23
[18] Tanadchangsaeng N and Yu J 2013 *J Appl Polym Sci* **129** 2004-16

Molecular dynamics study on glycolic acid in the physiological salt solution

S Matsunaga

National Institute of Technology, Nagaoka College, 940-8532 Nagaoka, Japan

E-mail: matsu@nagaoka-ct.ac.jp

Abstract. Molecular dynamics (MD) study on glycolic acid in the physiological salt solution has been performed, which is a model of a biofuel cell. The structure and charge distribution of glycolic acid in aqueous solution used in MD is beforehand optimized by Gaussian09 utilizing the density functional theory. MD is performed in the NTV constant condition, *i.e.* the number of particles, temperature, and volume of MD cell are definite. The structure difference of the glycolic acid and oxalic acid is detected by the water distribution around the molecules using the pair distribution functions, $g_{ij}(r)$, and the frequency dependent diffusion coefficients, $D_i(\nu)$. The anomalous dielectric constant of the solution, i.e. about 12 times larger than that of water, has been obtained, which may be attributed to the ion pair formation in the solution.

1. Introduction

Recently much attention has been riveted on the biodegradable polymers. They are used in the broad area, especially in the medical treatment, e.g. surgical suture, cartilage support, and capsules carrying drugs to the affected part. Poly glycolic acid (PGL) is one of the promising candidates which can be produced from biological origin resources. The degradation process of PGL by hydrolysis has been examined by experiment and molecular dynamics (MD). However, the physical properties of resolved PGL, or glycolic acid in the physiological salt solution is not sufficiently investigated [1].

Besides, according to the recent experimental reports of the CREST (Core Research for Evolutionary Science and Technology) project, a sort of rechargeable biofuel cell has been constructed using glycolic acid and oxalic acid aqueous solution without CO_2 emission [2]. Although the extensive experimental and ab initio studies have been reported regarding this system, the detailed structure and transport coefficients related to electric properties have not been examined yet by classical MD on the stand point of multi-component organic molecular solutions, to the best of our knowledge. Out of these motives, we perform the MD study on the glycolic acid in NaCl aqueous solution to simulate a physiological salt solution in a human body.

2. Simulation procedure and results

As mentioned in the previous section, the electric power circulation method based on the glycolic acid ($HOOC\text{-}CH_2OH$) and oxalic acid ($(COOH)_2$) redox couple has been proposed. The electric power charge has been achieved using TiO_2 in the potential region of -0.5 to -0.7 V vs. the reversible hydrogen electrode (RHE). The reaction in the cell is summarized as,

$$\text{Cathode: } HOOC\text{-}CH_2OH + H_2O \rightarrow (COOH)_2 + 4H^+ + 4e^- \quad (1)$$

$$\text{Anode: } O_2 + 4H^+ + 4e^- \rightarrow 2H_2O \quad (2)$$

$$\text{Total:} \quad HOOC\text{-}CH_2OH + O_2 \rightarrow (COOH)_2 + H_2O \tag{3}$$

Although, these reactions are formed in alkaline solutions, we perform MD in a physiological salt solution to examine the hydration of glycolic acid and oxalic acid in detail. The optimized structure and charges of used molecules have been prepared using Gaussian09 by the density functional theory (DFT) in advance. The calculations are carried out at the B3LYP/6-311++G(d,p) level of the theory in a cavity in the IEF-PCM solvent field. The obtained molecules and ions are shown in figures 1(a)-1(d).

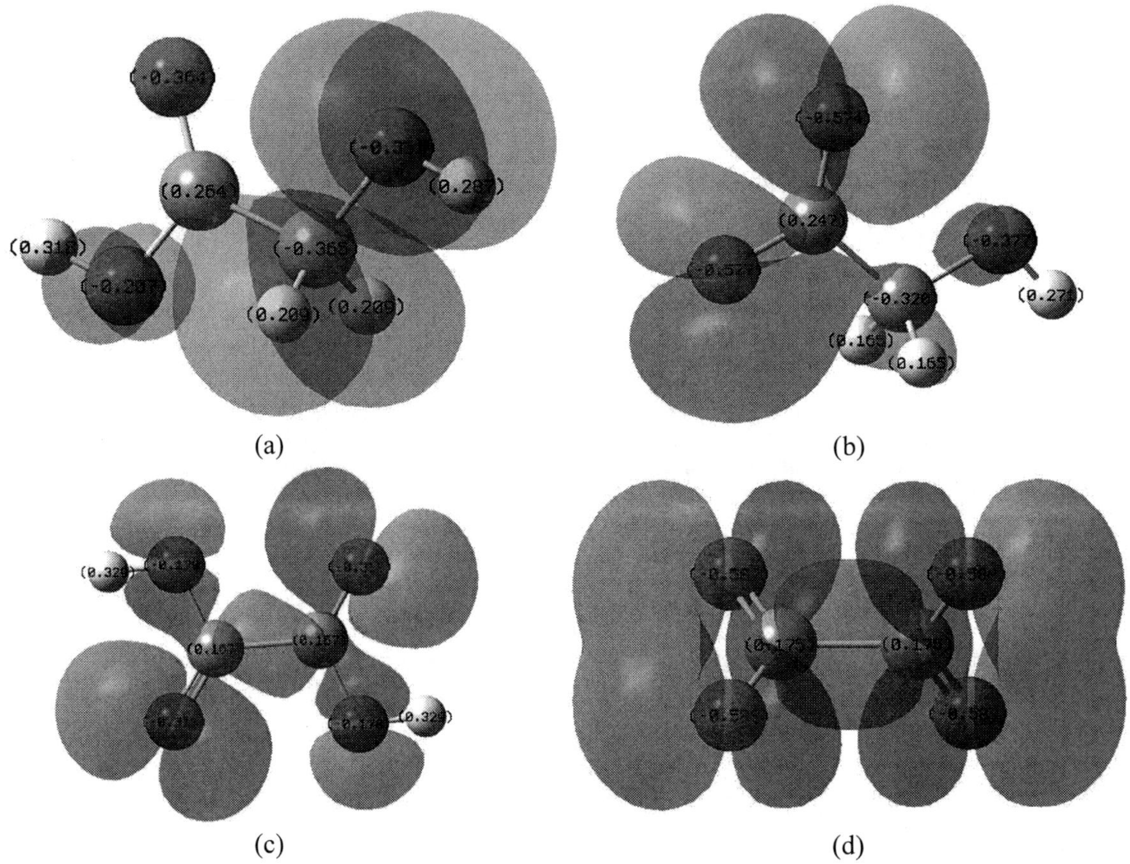

Figure 1. (a) Glycolic acid, (b) Glycolic acid anion, (c) Oxalic acid and (d) Oxalic acid anion.

In MD calculation, glycolic acid and oxalic acid are assumed to be ionalized in the solution, i.e. they are in the form of anion. Therefore the constituents of MD are glycolic acid anion, oxalic acid anion, Na^+, Cl^-, and water molecule.

The procedure of MD simulation is same as our previous works [3-5]. We briefly give outline below. The rigid body water model, TIP4P, is used. The charged Lennard-Jones (L-J) type potentials are used for the potential function for water molecules as [6,7],

$$\phi_{ij}(r) = \frac{z_i z_j e^2}{r} + 4\varepsilon\left\{\left(\frac{\sigma}{r}\right)^{12} - \left(\frac{\sigma}{r}\right)^{6}\right\} \tag{4}$$

From now on, the symbols i and j in the equations stand for the constituent ions; z_i is the charge for the constituent ions i; and e is the elementary charge. The parameters of equation (4) are taken from the literature. The interactions between Na^+-Na^+, Na^+-Cl^-, Cl^- - Cl^-, TIP4P - Na^+, and TIP4P - Cl^- are described as [8],

$$\phi_{ij}(r) = \frac{z_i z_j e^2}{r} + \frac{C}{r^9} - \frac{D}{r^6} \tag{5}$$

The softness of the interactions is represented by the power 9 of the second repulsive term. MD is performed in the NPT constant condition, using about 2100 molecules, i.e. 1.5 wt% glycolic acid anion and 1.8 wt% oxalic acid anion in a physiological salt solution, at 1atm, 310 K. The one timestep is 0.2 fs. Ewald method is used for charge interaction. SEIGRESS package (Fujitsu) is used for calculation [9].

The partial pair distribution functions, $g_{ij}(r)$, are obtainable from the simulation data, which represents configurations of ions in the solution, defined as [10],

$$n_{ij}(r) = \sum_{n<r/\Delta r} (N_j/V) g_{ij}(n \cdot \Delta r). \tag{6}$$

where $n_{ij}(r)$ stands the integrated coordination number. The obtained $g_{ij}(r)$'s are shown in figures 2(a)-2(b). The disparities of $g_{ij}(r)$'s reflect the structural difference between glycolic acid anion and oxalic acid anion, and the water distribution around them. In figure 2(a), $g_{ij}(r)$ for C(glycolic acid anon) - O(water) and C(glycolic acid anon) - H(water) have plateaus around 3.9 Å and 4.5 Å, respectively.

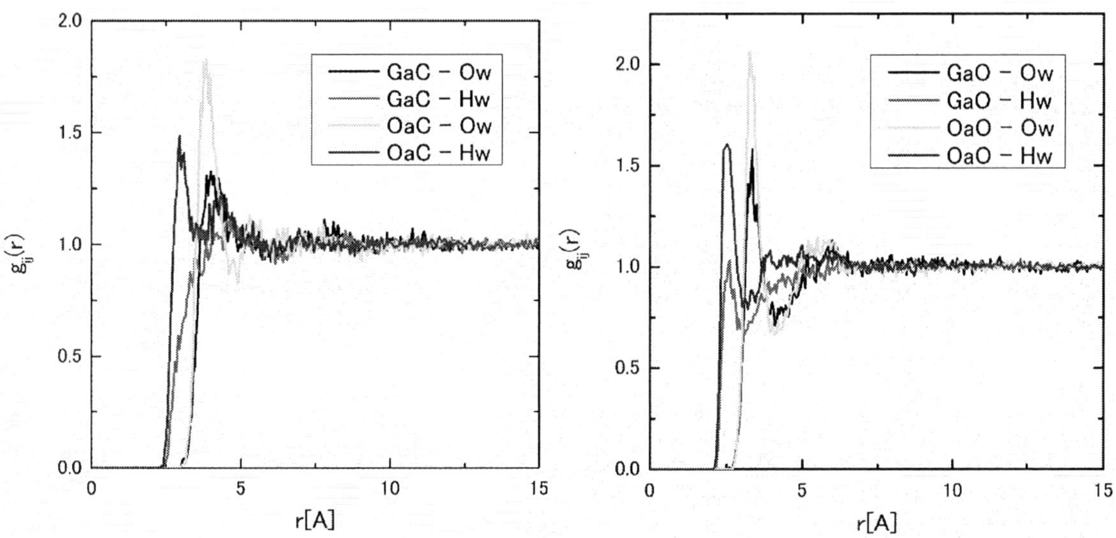

(a) (b)

Figure 2. (a) $g_{ij}(r)$ for C(glycolic acid anon) - O(water), C(glycolic acid anon) - H(water), C(oxalic acid anon) - O(water), and C(oxalic acid anon) - H(water), and (b) $g_{ij}(r)$ for O(glycolic acid anon) - O(water), O(glycolic acid anon) - H(water), O(oxalic acid anon) - O(water), and O(oxalic acid anon) - H(water).

On the other hand, $g_{ij}(r)$ for C(oxalic acid anon) - O(water) and C(oxalic acid anon) - H(water) have pronounced peaks at 3.7 Å and 3.0 Å, respectively. The difference of the peak heights of $g_{ij}(r)$'s around a glycolic acid anion and an oxalic acid anon might be attributed to the difference of charges of anions, i.e. a glycolic acid anon -1, an oxalic acid anon -2. The four C in an oxalic acid anon attract H of water most, which causes the peak of $g_{ij}(r)$ for C(oxalic acid anon) - H(water) at 3.0 Å. In figure 2(b), the difference of the peak heights of $g_{ij}(r)$'s around a glycolic acid anion and an oxalic acid anon are also attributed to the difference of charges of anions. Regarding the peak positions, $g_{ij}(r)$ for O(glycolic acid anon) - O(water) and O(glycolic acid anon) - H(water) have peaks at 3.4 Å and 2.6 Å, respectively. On the other hand, $g_{ij}(r)$ for O(oxalic acid anon) - O(water) and O(oxalic acid anon) -

H(water) have high peaks at 3.2Å and 2.5Å, respectively. Therefore, the structural difference between glycolic acid anion and oxalic acid anion are detected most by the distribution of H in water around them.

Next we examine the transport properties, i.e. the shear viscosity and the diffusion coefficients. The major advantages of the biofuel cell are its safety and small size. To design a small sized cell, the transport properties are particularly important, because the small sized fluid is treated in the cell. Therefore, on the stand point of physical properties, the similar transport features as water are desirable.

The shear viscosity is calculated using the Green-Kubo's formula as [11],

$$\eta = \frac{V}{k_B T} \int_0^{\infty} \langle \sigma^{xz}(t)\, \sigma^{xz}(0) \rangle dt. \tag{7}$$

where $\sigma^{xz}(t)$ and $\sigma^{xz}(0)$ are the off diagonal elements of the stress tensor; k_B is the Boltzmann's constant: V is the volume of the system. The estimated shear viscosity is 1.72 cP at 1atm, 310 K, which is about 2 times larger than that of pure water 0.89 cP at 1atm 298 K.

The diffusion coefficient for i-th constituent can be estimated using the mean square displacement (MSD) as,

$$D_i = \lim_{t\to\infty} \frac{1}{6t} \langle |\boldsymbol{r}_i(t) - \boldsymbol{r}_i(0)|^2 \rangle. \tag{8}$$

The calculated diffusion coefficients for constituents are listed in table 1. The obtained D_i's values for glycolic acid and water are comparable to those in literatures [12,13]. D_i for oxalic acid anion is much smaller than that of glycolic acid and water.

Table 1. Diffusion coefficients estimated from MSD.

constituent	D_i ($\times10^{-5}$ cm^2/s)
glycolic acid anion	3.04
oxalic acid anion	1.87
Na^+	2.63
Cl^-	2.34
water	4.68

The dynamical properties of glycolic acid anion and oxalic acid anion are estimated by the frequency dependent diffusion coefficient $D_i(\nu)$, which is expressed as [10],

$$D_i(\nu) = \frac{1}{3} \int_0^{\infty} \langle \mathbf{v}_i(t) \cdot \mathbf{v}_i(0) \rangle \cos(2\pi\nu t)\, dt, \tag{9}$$

where $\langle \mathbf{v}_i(t) \cdot \mathbf{v}_i(0) \rangle$ stands for the velocity auto-correlation function (VAF) for the ith atom species.

As seen in figure 3, the lower frequency distribution in the infrared region is attributed to the cage effect by the surrounding water molecules. The structural difference between glycolic acid anion and oxalic acid anion is observed most in middle and higher frequencies. Although the bond length is fixed and only bending is allowed in these molecules in MD, the high frequency distributions around 900 [1/cm] are obtained. These results may be yielded by the interaction between atoms in the different molecules, i.e. the inter molecular interaction. In figure 3, the frequency distribution of water and glycolic acid are mainly in lower region, and the higher frequencies are seen most in oxalic acid anion, which may prevent the diffusion of oxalic acid anion. These facts suggest that the cell should be operated in glycolic acid rich region around 3 wt%.

Finally, we would like to refer to the static dielectric constant of the solution estimated by MD, ε, which is defined as [11],

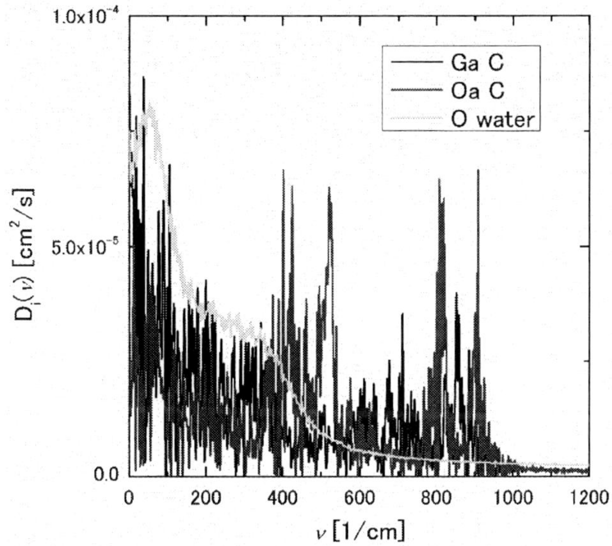

Figure 3. $D_i(\nu)$ for C(glycolic acid anion), C(oxalic acid anion) and O(water).

$$\frac{1}{\varepsilon - 1} = \frac{3k_B\varepsilon_0 TV}{M^2} - \frac{1}{2\varepsilon_s + 1} \tag{10}$$

where k_B is the Boltzmann's constant, ε_0 is the static dielectric constant in a vacuum, T is the temperature, V is the volume of the MD cell. The term M^2 is expressed as,

$$M^2 = \langle \left| \sum_{i=1}^{N} \vec{\mu_i} \right|^2 \rangle. \tag{11}$$

where $\vec{\mu_i}$ stands for the dipole moment of the i-th molecule. The static dielectric constant, ε_s, on the outside of the cell is assumed to be infinity under the periodic boundary condition. Then, ε is expressed as,

$$\varepsilon = \frac{M^2}{3k_B\varepsilon_0 TV} + 1 \tag{12}$$

The estimated static dielectric constant of the solution is 9.6×10^2, which is about 12 times larger than that of pure water. According to the experiment, for example, alkali halides water solution yields decrease in ε by the formation of hydration shell around ions, which prevents creating the effective dipole moments in the solution. On the other hand, in the case of onium salts (e.g. Bu_4NBr etc.) water solution, ε increases as the concentration of solute increases, because the ions may associate to form ion pairs [14]. The present results of increasing ε also might be explained by the large dipole moment formation by ion pairs in the solution. If the relation of ε and the concentration of solute, c, follows the Onsager's equation, dε/dc is anticipated to be proportional to M^2 [15]. However, the detailed analysis is left for further study.

3. Conclusion

MD simulations have been performed on the glycolic acid and oxalic acid in NaCl aqueous solution as a model of human body. The obtained $g_{ij}(r)$'s suggest that oxalic acid anions attract water molecule around them to form a hydration shell. The transport properties for glycolic acid anion and oxalic acid anion are quite different, which may be attributed to the structural difference. The anomalous dielectric constant has been obtained. This fact may be explained by the large dipole moment formation in the solution, however the further analysis are left for the future study.

Acknowledgment
I would like to express my thanks to Professor S. Tamaki for his helpful comments and encouragement on this study. This study was supported by JSPS KAKENHI Grant Number 15K05136. Part of the results in this study was obtained using the supercomputing facilities at Research Institute for Information Technology, Kyushu University.

References

[1] Ratner B D 2004 *Biomaterials Science: An Introduction to Materials in Medicine* 2nd ed. (Amsterdam, The Netherland: Elsevier Academic Press)
[2] Watanabe R, Yamauchi M, Sadakiyo M, Abe R and Takeguchi T 2015 *Energy Environ. Sci.* **8** 1456-62
[3] Matsunaga S and Tamaki S 2014 *J. Sol. Chem.* **43** 1771-90
[4] Matsunaga S 2016 *Int. J. Mol. Sci.* **17** 45-1-18
[5] Matsunaga S 2017 *J Mol. Liq.* **226** 90-5
[6] Jorgensen W L, Chandrasekhar J, Madura J D, Impey R W and Klein M L 1983 *J.Chem. Phys.* **79** 926-35
[7] Mahoney M W and Jorgensen W L 2000 *J. Chem. Phys.* **112** 8910-22
[8] Peng Z, Ewig C S, Hwang M-J, Waldman M and Hagler A T 1997 *J. Phys. Chem.* A **101** 7243-52
[9] FUJITSU Technical Computing Solution SCIGRESS http://www.fujitsu.com/global/solutions/business-technology/tc/sol/scigress/index.html (Accessed 5 March 2018)
[10] Hansen J-P and McDonald I R 1986 *Theory of Simple Liquids* 2nd ed. (London: Academic Press)
[11] Allen M P and Tildesley D J 1987 *Computer Simulation of Liquids* (Oxford: Clarendon Press)
[12] Casalini T, Rossi F, Santoro M and Perale G 2011 Int. J. Mol. Sci. **12** 3857-8370
[13] Idrissi A, Sokoli F and Perera A 2000 *J. Chem. Phys.* **112** 9479-88
[14] Wang P a nd Anderko A 2001 *Fluid Phase Equilibria* **186** 103-22
[15] Onsagaer L 1936 *J. Am. Chem. Soc.* **58** 1486-93

IOP Conf. Series: Materials Science and Engineering **369** (2018) 012015 doi:10.1088/1757-899X/369/1/012015

Electrochemical studies of the composite polymer-containing coating on the 1579 aluminium alloy with welded joint

A S Gnedenkov[1,3], S L Sinebryukhov[1], D V Mashtalyar[1,2], V S Egorkin[1,2], I E Vyaliy[1] and S V Gnedenkov[1,2]

[1]Institute of Chemistry of FEB RAS, 159 Pr. 100-letiya Vladivostoka, Vladivostok, 690022, Russia
[2]Far Eastern Federal University, 8 Sukhanova St., Vladivostok, 690950, Russia

E-mail: asgnedenkov@mail.ru

Abstract. The features of corrosion process on the surface of the 1579 Al alloy with welded joint area were investigated using SIET (scanning ion-selective electrode technique) and SVET (scanning vibrating electrode technique) in 3% NaCl. The development of corrosion was realized along the weld interface. Methods of composite polymer-containing layer generation based on plasma electrolytic oxidation (PEO) sufficiently improve the anticorrosion behavior of the Al alloy. Corrosion of Al alloy with protective layers are described in details. Composite coatings have the best anticorrosion properties and increase the stability of material in the corrosive medium.

1. Introduction

Al alloys are attractive material to be used in such fields as automobile, aerospace and marine technology. At the same time, the intermetallic particles presence in the aluminium alloys composition, makes these materials highly liable to localized corrosion, especially in marine environment [1–3].

For the industrial implementation of aluminium alloys, a reliable anticorrosion protection of these materials is required. Plasma electrolytic oxidation (PEO) method is very promising way to generate a protective and anticorrosion layer on the Al alloy surface [4–13].

The method of composite polymer-containing layer formation via superdispersed polytetrafluoroethylene (SPTFE) treatment of base PEO-coating was proposed for Mg alloys [14,15]. SPTFE treatment can be used to increase the anticorrosion protection of Al alloys. The use of SPTFE allows us to seal the PEO pores with the polymer and to make a reliable barrier that will sufficiently reduce ions penetration to the surface of the substrate, and upgrade the sample behavior in the corrosive environment [14,15].

In this work, the way of anticorrosion layer generation on the 1579 Al alloy surface is proposed. The polymer treated PEO-layer can be characterized by increased anticorrosion properties. The protective and electrochemical behavior of the formed coatings were studied.

In accordance with our works [4,16–19], the reliable methods to study the mechanism and kinetics of the corrosion include localized electrochemical techniques [11,20–26], such as Scanning Ion-Selective Electrode Technique (SIET) and Scanning Vibrating Electrode Technique (SVET).

The study of the corrosion on the welded joint area surface of the 1579 aluminium alloy by means of localized techniques is presented. The formation and study of composite coating on the sample surface, which increases protective properties of the working material, have been examined as well.

2. Experimental

2.1. Samples

In this work, the 1579 Al alloy with welded joint (6.701 wt. % Mg; 0.639 wt. % Zn; 0.138 wt. % Cu; 0.548 wt. % Mn; 0.150 wt. % Fe; 0.126 wt. % Zr; 0.107 wt. % Sc; 0.116 wt. % Cr; 0.095 wt. % Ni; 0.021 wt. % Ti; balance – Al) was a specimen for study.

To obtain the element distribution on the aluminium alloy surface and welded joint area the EDX-800HS (Shimadzu, Japan) was used.

The PEO-layer was obtained on the Al alloy surface at 0.5 A/cm^2. The contents of the electrolyte were: NaF, 0.3 g/l, $C_4H_4O_6K_2 \cdot 0.5H_2O$, 15 g/l. The time of PEO was about 160 s. The PEO-layer thickness was about 12 μm.

For polymer-containing layer formation, the subsequent treatment of the surface of PEO-coating using Forum® SPTFE was proposed to upgrade its anticorrosion behavior. The way of polymer-containing layer generation, suggested in this experiment, was presented in our previous works [14,15]. 10 wt. % SPTFE suspension in the ethanol was used to obtain the polymer layer. The immersion period of the sample in the suspension was 30 s. Then the material was dried at 25 °C for 0.5 hours. The time and temperature of treatment of the sample with polymer-containing layer after immersion were 20 minutes and 350 °C, correspondingly.

2.2. Electrochemical studies

To investigate the development of the corrosion process on the welded joint surface of the Al alloy and to study the anticorrosion properties increase after generation of the composite coating, the SVET – SIET equipment (Applicable Electronics, USA) was used.

The SVET probe was a Pt–Ir insulated wire. The tip of the probe was covered with a Pt black. This probe was located at 100 μm above the studied surface. The amplitude of probe vibration was equal to 15 μm. In this work, the SIET electrode was sensitive to the H^+ ions. The composition of the SIET microelectrode membrane was described in [22]. 0.01 M KH_2PO_4 in 0.1 M KCl was the inner reference solution. The SIET microelectrode was located at 40 μm above the studied surface.

Specimens were investigated in 3% NaCl solution under open circuit potential conditions.

3. Results and discussion

To obtain the element distribution on the surface of the studied material the EDX analysis was used. The EDX data show the homogeneous element distribution in the surface and bulk layers of the Al alloy and welded joint zone of the specimen (figure 1). This result shows the probability of a homogeneous occurrence of the electrochemical processes on the sample surface, which will be investigated by localized techniques.

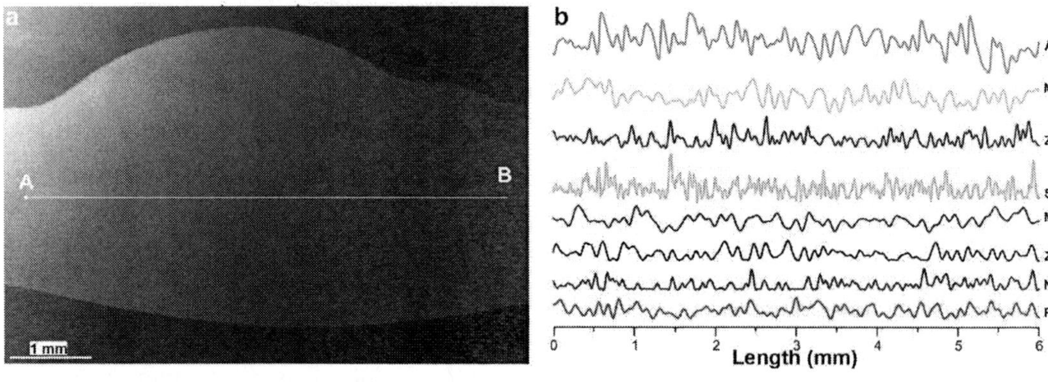

Figure 1. SEM image of the Al alloy specimen with welded joint zone (a) and element distribution along the line AB, according to the EDX analysis (b).

IOP Conf. Series: Materials Science and Engineering **369** (2018) 012015 doi:10.1088/1757-899X/369/1/012015

Figure 2 depicts the optical image of the studied welded joint zone before SVET/SIET measurements. This investigated zone is outlined by frame (figure 2). After 0.5 hours of the specimen exposure, the starting of the corrosion process happened. There is a generation of the anodic area on the welded joint zone.

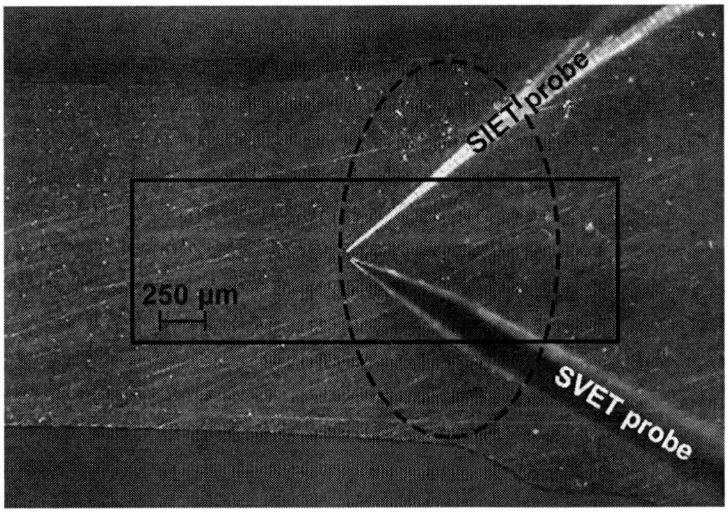

Figure 2. The optical image of the welded joint zone of the Al alloy specimen before SVET/SIET measurements. The weld interface is marked by dashed line.

Lower pH in the anodic area for the Al alloy can be described by the following reaction, which makes the pit medium more acid, and, therefore, the corrosion is realized:

$$Al + nH_2O \rightarrow Al(OH)_n^{3-n} + nH^+ + 3e^-, n = 1 - 3. \quad (1)$$

Analysis of the SVET and SIET data indicated the development of corrosion process from the edge of the investigated zone to the bulk layers of the weld interface during the specimen immersion from 1.5 to 9 hours (figure 3(a)). The weld interface underwent the intensive corrosion destruction in accordance with the SVET/SIET results. Microdefects in the morphological structure are the reason of high electrochemical activity of the welded joint. In the welded joint zone, there are areas with different potential, which result in the occurrence of the corrosion process.

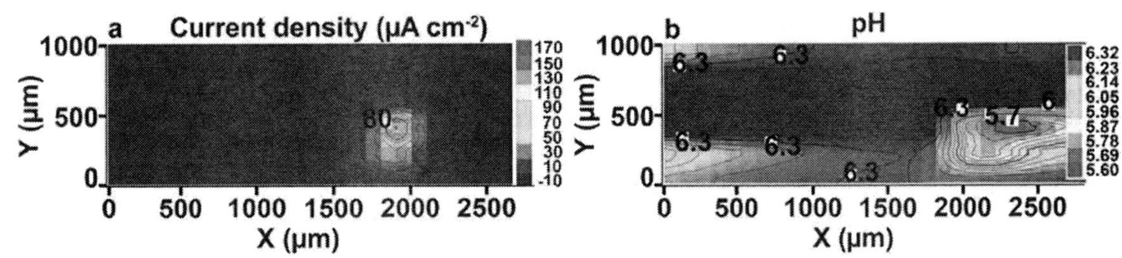

Figure 3. Maps of the current density (a) and pH distribution (b) on the surface of the Al alloy specimen with welded joint zone after 9 hours of the immersion in 3% NaCl.

It was ascertained that weld interface under the influence of chloride environment, accelerated the corrosion degradation of the investigated sample. Thereby, the anticorrosion protection of Al alloy is necessary.

SVET and SIET detected the high corrosion in the bulk layers of the welded joint area after 9 hours of the sample immersion in chloride-containing medium (figures 3(a) and 3(b)). In the anodic zone, the current density reached 170 μA/cm^2. The pH in the anodic zone reduced down to 5.6 (figure 3(b)), in accordance with the reaction (1). Therefore, it was established that welded joint zone of 1579 Al alloy was the area of the corrosion process passing, according to SVET/SIET maps (9 hours).

Figure 4 shows the optical image of the studied surface after PEO and SPTFE treatment. The studied area is limited by a frame (figure 4). Since the electrochemical activity of the polymer treated specimen practically did not change during all experimental time SVET and SIET maps presented after 500 hours of specimen exposure to 3% NaCl. The SVET/SIET methods registered a small corrosion activity changes on the surface of the studied specimen (figure 5). After polymer-containing layer generation on the sample surface, the welded joint area became a cathodic one (figure 5).

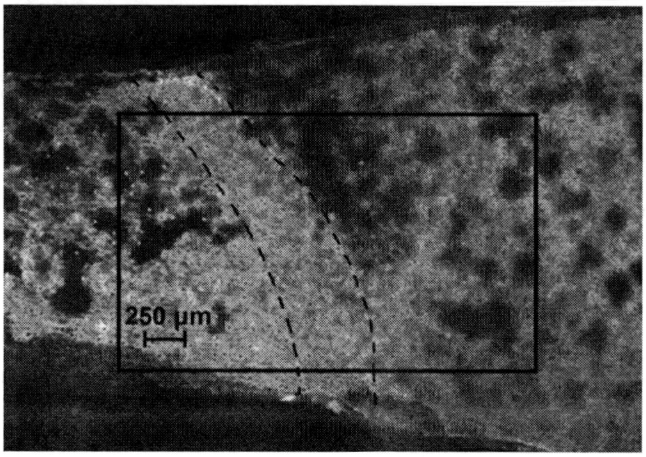

Figure 4. The optical image of the studied welded joint area of the Al alloy specimen with composite coating after 500 hours of specimen exposure. The weld interface of sample is marked by dashed line.

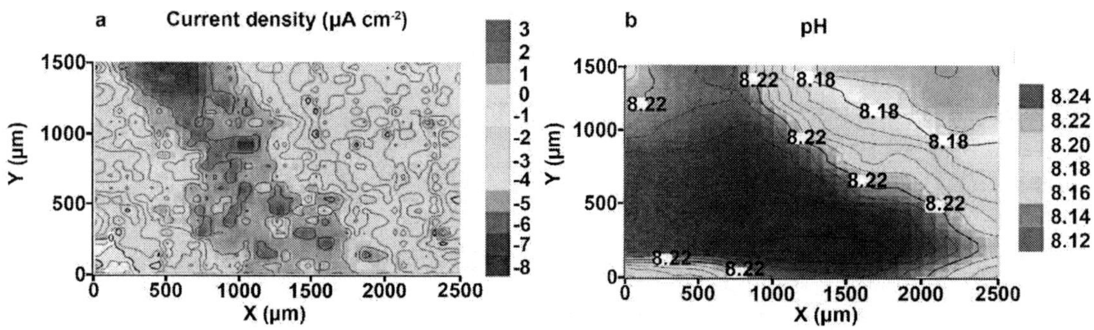

Figure 5. Maps of the current density (a) and pH distribution (b) on the weld interface of the Al alloy specimen with composite layer after 500 hours of the immersion.

In contrast to the homogeneous distribution of elements in the material investigated by the EDX analysis, the SVET/SIET data show the passing of different electrochemical processes on the sample surface. Corrosion activity of the welded joint area is connected with the microdefects in this zone, which were registered by SVET and SIET methods. After generation of the composite coating, these defects were covered, and the weld interface became a cathodic area (figure 5). The other part of the specimen is the anodic area with low corrosion activity; anodic current density is 3 μA/cm^2 (figure 5(a)).

Unopened pores of PEO-layers, which is one of the features of such type of coatings [14–16,27–30], is the reason of low anodic current density on the surface of the studied material. The current density gradient values for the bare Al specimen after 9 hours of the sample immersion in 3% NaCl solution (figure 3(a)) was about 180 μA/cm^2. This parameter for the specimen with composite layer was 11 μA/cm^2 (figure 5(a)). Cathodic zone at the welded joint area was also detected by SIET (figure 5(b)). Higher pH in the cathodic zone is connected with the oxygen reduction reaction [26]. The pH in cathodic and anodic areas for the specimen with the composite layer is 8.2 and 8.1, correspondingly (figure 5(b)) and 6.3 and 5.6 for the bare specimen (figure 3(b)). This is the result of corrosion process inhibition by means of polymer treatment of the specimen, thus indicating the high anticorrosion properties of the sample with composite layer.

Figure 6 shows the kinetic development of the Al alloy corrosion for the bare sample and for the sample with composite coating. The results confirm the anticorrosion behavior increase due to the coating generation. The total current, calculated from the SVET data, significantly reduced for the polymer treated sample.

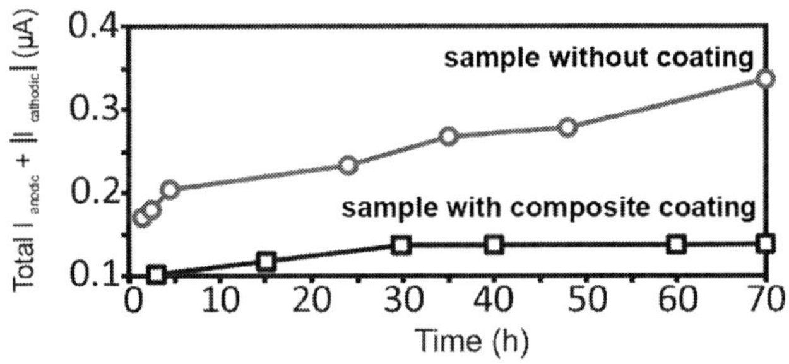

Figure 6. The total current changing with time for 1579 aluminium alloy samples with welded joint.

This result revealed a sufficient protection of the Al alloy specimen against the corrosive environment via the PEO process and polymer treatment. Such layers reduce the corrosion activity on the investigated sample surface. The weld interface became a cathodic area, and the corrosion stability of the studied sample was increased.

4. Conclusion

Features of the corrosion of 1579 Al alloy with welded joint zone have been studied. Corrosion process passing along the weld interface was investigated by SVET/SIET methods. The high electrochemical activity of the welded joint area is connected with microdefects in its morphological structure.

The composite coating reduces the corrosion intensity of the investigated specimen. The current density gradient values between anodic and cathodic area on the specimen surface decreased from 180 μA/cm^2 down to 11 μA/cm^2 as a result of PEO and SPTFE treatment. These methods enable us to sufficiently improve anticorrosion properties of the 1579 Al alloy with welded joint area in accordance with the SVET/SIET data.

Acknowledgments

This work was supported by the Grant of Russian Science Foundation (project no. 17-73-10132).

References

[1] Moreto J A, Marino C E B, Bose Filho W W, Rocha L A and Fernandes J C S 2014 *Corros. Sci.* **84** 30–41

[2] Grilli R, Baker M A, Castle J E, Dunn B and Watts J F 2010 *Corros. Sci.* **52** 2855–66

[3] Jinlong L, Hongyun L and tongxiang L 2016 *Mater. Res. Bull.* **83** 148–54

[4] Gnedenkov A S, Sinebryukhov S L, Mashtalyar D V and Gnedenkov S V 2014 *Solid State Phenom.* **213** 143–8
[5] Němcová A, Skeldon P, Thompson G E, Morse S, Čížek J and Pacal B 2014 *Corros. Sci.* **82** 58–66
[6] Gnedenkov S V, Khrisanfova O A, Sinebryukhov S L, Puz A V and Gnedenkov A S 2008 *Mater. Manuf. Process.* **23** 879–83
[7] Sinebryukhov S L, Gnedenkov A S, Khrisanfova O A and Gnedenkov S V. 2009 *Surf. Eng.* **25** 565–9
[8] Gnedenkov S V, Sinebryukhov S L, Puz A V, Gnedenkov A S, Vyaliy I E, Mashtalyar D V and Egorkin V S 2014 *Solid State Phenom.* **213** 149–53
[9] Sun M, Yerokhin A, Bychkova M Y, Shtansky D V, Levashov E A and Matthews A 2016 *Corros. Sci.* **111** 753–69
[10] Ivanou D K, Yasakau K A, Kallip S, Lisenkov A D, Starykevich M, Lamaka S V, Ferreira M G S, Zheludkevich M L and Aparicio M 2016 *RSC Adv.* **6** 12553–60
[11] Mohedano M, Serdechnova M, Starykevich M, Karpushenkov S, Bouali A C, Ferreira M G S and Zheludkevich M L 2017 *Mater. Des.* **120** 36–46
[12] Gnedenkov S V, Khrisanfova O A, Zavidnaya A G, Sinebryukhov S L, Egorkin V S, Nistratova M V, Yerokhin A and Matthews A 2010 *Surf. Coatings Technol.* **204** 2316–22
[13] Egorkin V S, Gnedenkov S V, Sinebryukhov S L, Vyaliy I E, Gnedenkov A S and Chizhikov R G 2018 *Surf. Coatings Technol.* **334** 29–42
[14] Gnedenkov S V, Sinebryukhov S L, Mashtalyar D V, Egorkin V S, Sidorova M V and Gnedenkov A S 2014 *Corros. Sci.* **85** 52–9
[15] Gnedenkov S V, Sinebryukhov S L, Mashtalyar D V, Nadaraia K V, Gnedenkov A S and Bouznik V M 2016 *Corros. Sci.* **111** 175–85
[16] Gnedenkov A S, Sinebryukhov S L, Mashtalyar D V and Gnedenkov S V 2016 *Corros. Sci.* **102** 269–78
[17] Gnedenkov A S, Sinebryukhov S L, Mashtalyar D V and Gnedenkov S V 2016 *Solid State Phenom.* **245** 89–96
[18] Gnedenkov A S, Sinebryukhov S L, Mashtalyar D V and Gnedenkov S V 2013 *Surf. Coatings Technol.* **225** 112–8
[19] Gnedenkov A S, Sinebryukhov S L, Mashtalyar D V and Gnedenkov S V 2012 *Phys. Procedia* **23** 98–101
[20] Williams G, McMurray H N and Grace R 2010 *Electrochim. Acta* **55** 7824–33
[21] Kirkland N T, Williams G and Birbilis N 2012 *Corros. Sci.* **65** 5–9
[22] Alvarez-Pampliega A, Lamaka S V, Taryba M G, Madani M, De Strycker J, Tourwé E, Ferreira M G S and Terryn H 2012 *Electrochim. Acta* **61** 107–17
[23] Yan M, Gelling V J, Hinderliter B R, Battocchi D, Tallman D E and Bierwagen G P 2010 *Corros. Sci.* **52** 2636–42
[24] Falcón J M, Otubo L M and Aoki I V 2016 *Surf. Coatings Technol.* **303** 319–29
[25] Nazarov V A, Taryba M G, Zdrachek E A, Andronchyk K A, Egorov V V and Lamaka S V 2013 *J. Electroanal. Chem.* **706** 13–24
[26] Nardeli J V, Snihirova D V, Fugivara C S, Montemor M F, Pinto E R P, Messaddecq Y and Benedetti A V 2016 *Corros. Sci.* **111** 422–35
[27] Gnedenkov A S, Sinebryukhov S L, Mashtalyar D V and Gnedenkov S V 2016 *Corros. Sci.* **102** 348–54
[28] Sinebryukhov S L, Gnedenkov A S, Mashtalyar D V and Gnedenkov S V 2010 *Surf. Coatings Technol.* **205** 1697–701
[29] Gnedenkov S V, Sinebryukhov S L, Mashtalyar D V, Imshinetskiy I M, Gnedenkov A S, Samokhin A V and Tsvetkov Y V 2015 *Vacuum* **120** 107–14
[30] Gnedenkov S V, Khrisanfova O A, Zavidnaya A G, Sinebrukhov S L, Kovryanov A N, Scorobogatova T M and Gordienko P S 2000 *Surf. Coatings Technol.* **123** 24–8

Dielectric properties of metallic alloy FeCoZr-dielectric ceramic PZT nanostructures prepared by ion sputtering in vacuum conditions

O Boiko

Electrical Engineering and Computer Science Faculty, Lublin University of Technology, 38A Nadbystrzycka Street, 20-618 Lublin, Poland

E-mail: oleksandr.boiko@pollub.edu.pl

Abstract. The main objective of the research was investigation of dielectric properties of $(FeCoZr)_x(PZT)_{(100-x)}$ granular nanocomposites and determination the influence of isochronous annealing in temperatures of 398 K-573 K on them. The impedance spectroscopy methodology was used. The measurements of electrical parameters, such as: phase shift angle φ, dielectric loss factor tgδ, capacity C and conductivity σ of $(FeCoZr)_x(PZT)_{(100-x)}$ nanocomposites have been performed. Frequency dependencies of these parameters were obtained for the ambient temperature range 98 K-373 K for the frequencies ranging from 50 Hz to 10^5 Hz. It was established, that the conductivity σ of the tested materials before the percolation threshold demonstrates non-linear dependence on frequency. Furthermore, it increases when the ambient temperature is increasing, which indicates a dielectric type of the material. The two types of electrical conduction: capacitive (phase shift angle φ takes negative values) and inductive (φ takes positive values) have been observed. It was concluded that the hopping conductivity dominated in the nanocomposites. Voltage and current resonances phenomena are observed in the materials. The isochronous annealing intensifies the dielectric properties of $(FeCoZr)_x(PZT)_{(100-x)}$ nanocomposites.

1. Introduction

Ultralight soft structures with high temperature capability are a very important class of materials with very high implementation potential [1,2]. High values of Curie temperatures for iron (1043 K) and cobalt (1388 K) formed the basis for the development and broadening of the research spectrum of magneto-soft materials containing these elements or their alloys [3,4]. However, thin films FeCo demonstrate high values of coercivity H_c (100-200 Oe) and magneto-crystalline anisotropy, which makes it difficult to find some potential applications. One of the solutions is adding (alloying) some additional metalloid component ("semi-metal") to the structure of FeCo alloy. The examples of such materials are: CoFeZr [5], CoFeB [6], CoFeNi [7], CoFeAl [8] etc.

Studies of electro-magnetic properties carried out over the last twenty years have shown that ferromagnetic materials in combination with piezo or ferroelectric components form structures, properties of which are changing under the influence of both electric and magnetic fields [9,10]. A frequently used ferroelectric for this purpose is Lead Zirconate Titanate $PbZrTiO_3$, which is a typical material with piezoelectric properties [11,12]. It belongs to the group of ceramic due to the structure and manufacturing methods.

The materials been tested in the present paper are nanostructures contains FeCoZr ferromagnetic alloy nanoparticles randomly arranged in Lead Zirconate Titanate ferroelectric matrix $PbZrTiO_3$ (abbreviated as PZT). The dielectric matrix of PZT was chosen for samples production due to its specific features: it's less corrosion-resistant matrix than alumina (according to previous investigations of $(FeCoZr)_x(Al_2O_3)_{(100-x)}$ nanocomposites) [13], and having crystalline structure it demonstrate ferroelectric or piezoelectric properties [14], which broaden the spectrum of its investigations. Furthermore, $(FeCoZr)_x(PZT)_{(100-x)}$ nanofilms are interesting structures to study percolation and resonant phenomena [15], hopping conductivity and relaxation mechanisms [16] in bulk materials.

The paper contains new experience in describing of dielectric properties of novel materials with use of dielectric spectroscopy method. Furthermore, the article presents a big number of nanocomposites potential applications due to demonstrated properties and original information about modelling and experimental results.

2. Experimental

The main objective of the research – granular nanomaterials of $(FeCoZr)_x(PZT)_{(100-x)}$, where x – metallic concentration (at.%), were produced using ion sputtering technique in a vacuum chamber. The sputtered beam during the process of nanocomposites synthesis included mixture of argon Ar and oxygen O_2 ions with its partial pressures of $P_{Ar} = 7.4\times10^{-2}$ Pa and $P_{O2} = 3.0\times10^{-3}$ Pa, respectively. Sputtering target consisted of the metallic alloy FeCoZr plate and ferroelectric ceramic PZT ribbons (with different distances between them) attached to it, which allowed to obtain series from 11 nanocomposite samples with various ratios of metallic and dielectric phases x ranging from 29.0 at.% to 90.0 at.% during one production cycle. Glass-ceramic substrates were used for nanolayers deposition at temperature of 373 K. All of the details related to films production process are described in [15]. The chemical compositions of the $(FeCoZr)_x(PZT)_{(100-x)}$ nanolayers were investigated by using of energy-dispersive X-ray spectroscopy and Rutherford Backscattering technique. Samples thicknesses were verified by SEM with accuracy of approx. 3 % and got values from the range of 0.4 µm-2 µm. To examine the structure and phase state of the nanocomposites Mössbauer spectroscopy was used. All of the details related to structure and chemical composition investigations of tested nanomaterials are described in [17,18].

Table 1. Samples specifications and parameters.

x (at.%)	***h*** (µm)	***a*** (cm)	***b*** (cm)	***R*** (MΩ)	**σ**(kΩ·cm)$^{-1}$	***C*** (F)	**tgδ** (a.u.)
29.0	0.4016	0.46	0.20	168.00	0.137	2.76E-12	6.87
49.8	1.3067	0.20	0.20	71.54	0.321	6.18E-12	7.20
64.4	1.0899	0.46	0.21	1.79	127.362	7.68E-09	2.32
80.0	1.5692	0.46	0.20	0.15	977.304	1.1E-07	1.94
90.0	0.8400	0.47	0.20	2.97	49.321	1.09E-08	0.99

Some specifications related to samples dimensions and electrical parameters at room temperature (f = 50 Hz) are presented in table 1. The table includes: a – length, b – width, h – thickness, R – resistance, σ – conductivity, C – capacity and tgδ – loss factor. In order to investigate the dielectric properties of tested materials the impedance spectroscopy was used. Measurements of electrical parameters (phase shift angle φ, conductivity σ, capacity C and loss coefficient tgδ) have been conducted using experimental station described in [19]. Frequency dependencies of $\varphi(f)$, $\sigma(f)$, $C(f)$ and $tg\delta(f)$ have been obtained for ambient temperatures range of $98\ K \leq T_p \leq 373\ K$ for frequencies of 50 Hz to 10^5 Hz. To determine the impact of thermal treatment on dielectric properties $(FeCoZr)_x(PZT)_{(100-x)}$ films were subjected to 15-min. isochronous annealing in tubular heater (air atmosphere) in temperatures of $398\ K \leq T_a \leq 573\ K$ with a step of 25 K. The publication presents the results of AC dielectric parameters measurements obtained for nanocomposites with x_1 = 49.8 at.% and x_2 = 64.4 at.% ratios.

3. Results and discussion

Figure 1 presents selected conductivity (case 1a) and phase shift angle (case 1b) versus frequency characteristics for a nanocomposite sample of $x_1 = 49.8$ at.%. The measurements have been performed directly after samples production. From the figure 1(a) can be observed that conductivity σ increases (two orders of magnitude) when measurement temperature increase. Such behaviour is typical for dielectric materials [20]. In this case the influence of temperature in conductivity in suitable temperature ranges can be described by the relationship:

$$\sigma = \sigma_0 \exp\left(-\frac{E_g}{2kT}\right) \quad (1)$$

where σ_0 – fixed value for each dielectric material, E_g – band gap energy, k – Boltzmann constant.

On the other hand, σ demonstrates a weak dependence on frequency. Figure 1(b) illustrates that phase angle φ takes negative values from approx. 0° to approx. -90° as the frequency *f* increases. This means, that the nanocomposite indicates capacitive type of current conduction. Capacitive properties of $(FeCoZr)_x(PZT)_{(100-x)}$ nanocomposites are described in [21]. The materials exhibit such kind of properties can find potential application in the industry as low-dimensional capacitors or nanosensors.

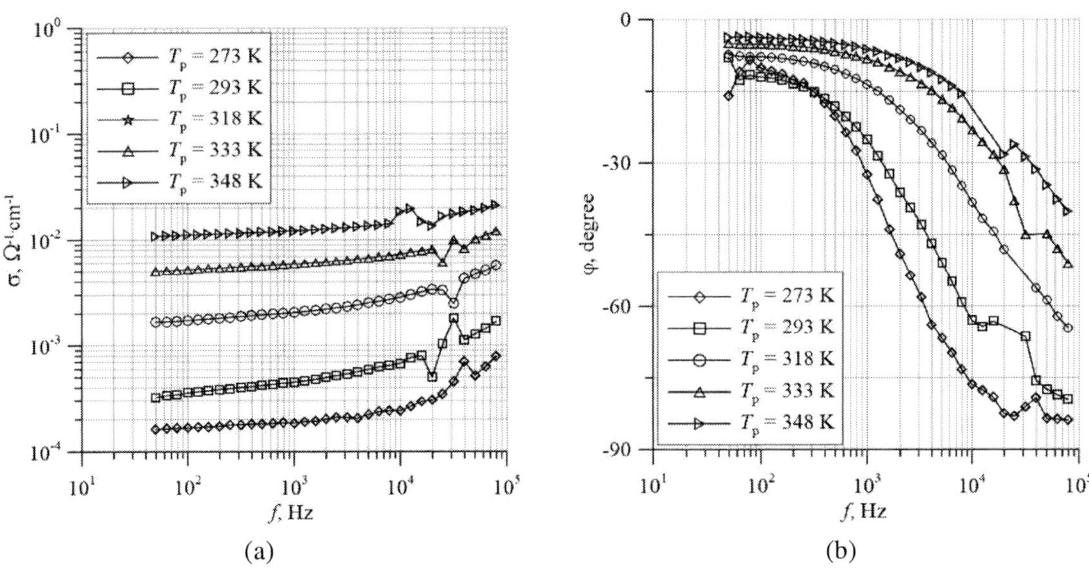

Figure 1. Frequency dependencies of: (a) conductivity σ and (b) phase shift angle φ of $(FeCoZr)_{49.8}(PZT)_{50.2}$ nanocomposite measured directly after preparing for selected ambient temperatures T_p.

To investigate the influence of isochronous annealing on dielectric properties $(FeCoZr)_{49.8}(PZT)_{50.2}$ sample was heated at temperature of $T_a = 523$ K. This time we can observe completely different situation, presented in figure 2. The materials conductivity σ at temperatures of $T_p < 318$ K exhibits nonlinear behaviour (figure 2(a)). Furthermore, we can notice the occurrence of its local minima for frequencies of f_{RP}. From figure 2(d) we can also observe minimal values of tgδ(*f*) dependencies at the same frequencies. It can be explained in the next way: at this situation f_{RP} is a resonant frequency and we can ascertain that the current resonant phenomenon appears in the material. The resonant frequency can be found using formula:

$$f_{RP} = \frac{1}{2\pi (L_P C_P)^{1/2}} \quad (2)$$

where L_P and C_P – inductance and capacitance in parallel circuit, (as shown in figure 3(a)). This electrical phenomenon are widely observed in electronics and can be presented using equivalent circuit diagram of parallel connection of *RLC* elements, as shown in figure 3(a). At this case, $\underline{U}$ – voltage across the open terminals, current through the capacitor C_p equals the current through the inductor $L_p \rightarrow \underline{I}_C = \underline{I}_L$, but they are in opposite phases which means that the resulting current equals to the current through reactance $R_p \rightarrow \underline{I} = \underline{I}_R$. It should be admitted, that it is one of the rear cases when current resonant is observed with nonzero value of phase angle φ (in our case φ = +90°, see figure 2(b)).

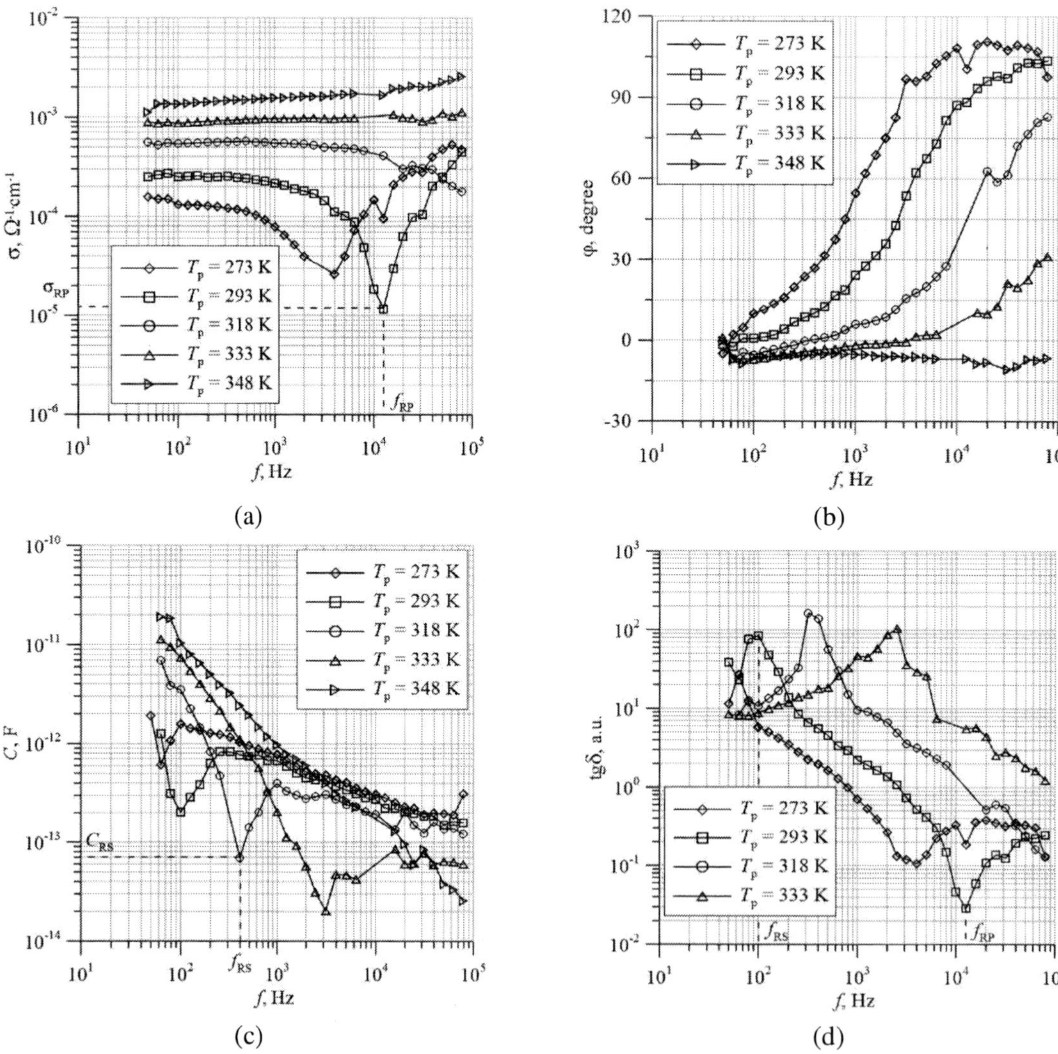

Figure 2. Frequency dependencies of: (a) conductivity σ, (b) phase shift angle φ, (c) capacity *C* and (d) loss coefficient tgδ of $(FeCoZr)_{49.8}(PZT)_{50.2}$ nanocomposite measured after 15-min. thermal treatment in temperature of T_a = 523 K.

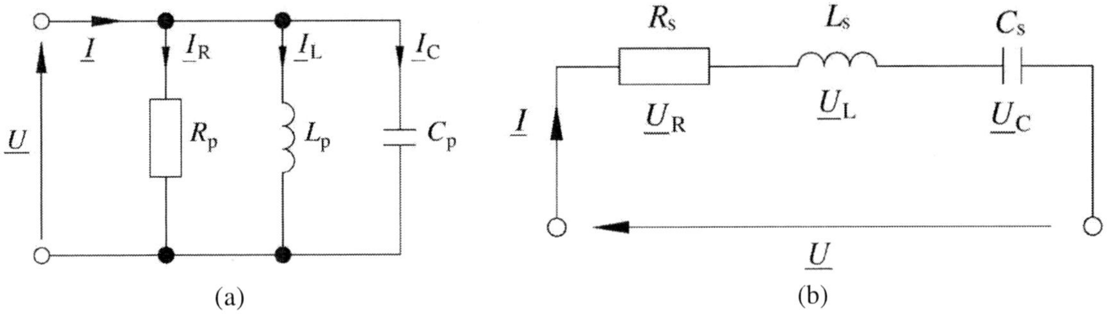

Figure 3. Equivalent circuit diagram of for both resonant cases: (a) current resonance: and (b) voltage resonance phenomena.

Figure 2(b) shows us positive values of phase shift angle φ for the temperatures of $T_p \leq 318$ K. The isochronous annealing in 523 K caused changes in the type of electrical conduction from capacitive to inductive. At this case, we can observe interesting electrical property of the nanocomposite, so-called non-coil inductance. Such sudden change in the conduction type can be related to the additional oxidation of metallic nanoparticles within ferroelectric volume which intensifies dielectric properties. Oxide shell around nanoparticle forms additional potential barrier which the charges should overcome to take part in electrical conduction [18]. Non-coil inductance means that nanocomposite sample, compared to conventional inductor, doesn't have a coil and a core from the material with magnetic properties. In case of $(FeCoZr)_x(PZT)_{(100-x)}$ nanolayers last can be replaced by FeCoZr nanoparticles because of it ferromagnetic nature.

Frequency dependence of capacity C is presented in figure 2(c). We can notice that, $C(f)$ for the temperatures $T_p \leq 318$ K demonstrates some minimal values at the frequencies f_{RS}. At the same frequencies tgδ characteristics take maximum values and phase shift angle achieves zero value. Such situation clearly indicates voltage resonance phenomenon in conventional *RLC* circuits, so-called series resonance. At this case current $\underline{I}$ is the same on each elements, the voltages $\underline{U} = \underline{U}_R$ and inductive reactance X_L and capacitive reactance X_C of a circuit are equal:

$$X_L = X_C \tag{3}$$

than

$$\omega_{RS} L_S = (\omega_{RS} C_S)^{-1} \tag{4}$$

Due to the fact that angular frequency $\omega_{RS} = 2\pi f_{RS}$, resonant frequency f_{RS} can be found as follows:

$$f_{RS} = \frac{1}{2\pi (L_S C_S)^{1/2}} \tag{5}$$

where L_S and C_S – inductance and capacitance in series circuit, (as shown in figure 3(b)).

Observation these phenomena indicates possible applications of $(FeCoZr)_x(PZT)_{(100-x)}$ nanofilms in electronics in radio transmitters, tuned circuits, filters, oscillators and other electrical devices working in wide frequency spectrum.

4. Hopping charge exchange in nanocomposites

Metal-dielectric nanocomposites with granular structure in a number of cases exhibit electrical conduction of hopping type [15]. It is related to the non-linear exponential dependence of conductivity on frequency. Same situation occurs in the case of FeCoZr-PZT nanocomposites below the percolation threshold as was established in [13] and is shown on figure 4(a). Figure 4(a) demonstrates conductivity versus frequency characteristics for nanocomposite sample with $x_2 = 64.4$ at.% subjected to thermal treatment in tubular heater at temperature of $T_a = 573$ K. To examine the mechanism of electrical conduction of $(FeCoZr)_x(PZT)_{(100-x)}$ nanolayers the model of hopping conductance was used. The detailed description, main assumptions and its verification can be found in [13]. Moreover, some simulations of electrical parameters of tested materials were conducted in [16].

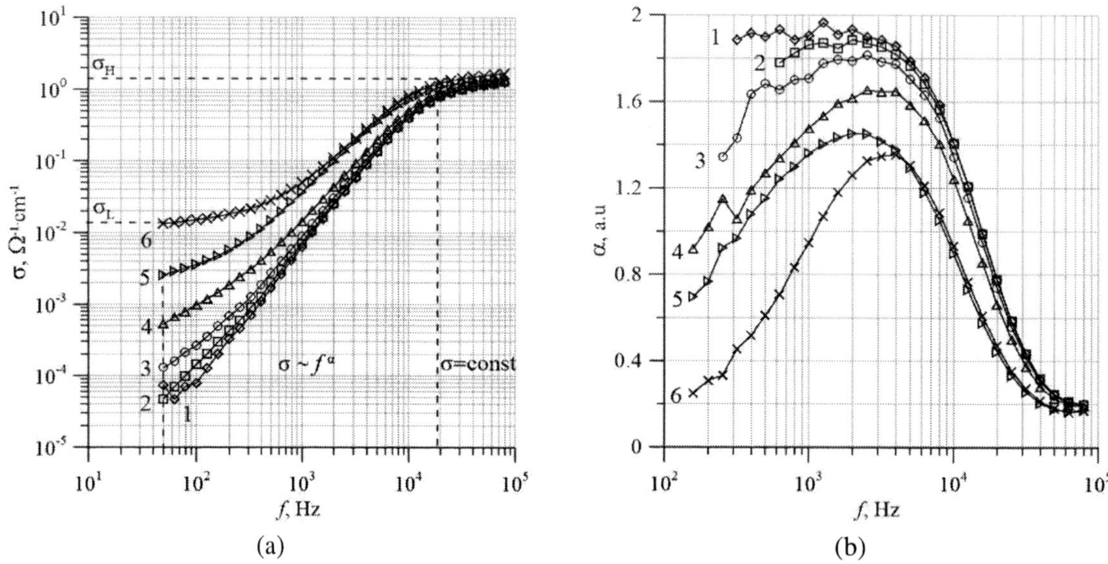

Figure 4. Frequency dependencies of: (a) conductivity σ and (b) frequency factor α of nanocomposites sample x = 64.4 at.% measured after 15-min. thermal treatment in temperature of $T_a = 573$ K for temperatures: *1* – 98 K, *2* – 153 K, *3* – 218 K, *4* – 273 K, *5* – 313 K, *6* – 373 K.

In the case of sample x_2, the $\sigma(f, T_p)$ dependencies for nanocomposites in dielectric regime are characterized by a sigmoid-like shapes, displaying three frequency areas (curve 6 on figure 4(a)): low-frequency area where $\sigma_L \approx$ const, medium frequencies with non-linear behaviour, high-frequency area with $\sigma_H \approx$ const. From figures 2(a) (temperature range $T_p \leq 293$ K) and 4(a) (temperatures of 98 K-373 K) we can observe strong conductivity versus frequency dependence:

$$\sigma(f) \sim \sigma_0 f^{\alpha(p)} \tag{6}$$

where $\alpha(p)$ is the frequency coefficient dependent on probability of electrons hops p. Electrical conduction in the material is carrying out in a way of hopping charge exchange (on different distances) between the potential wells (FeCoZr nanoparticles). The probability of such hops p and the return hops (1-p) can be described using following relation:

$$p = \frac{\sigma_L}{2\sigma_H} \tag{7}$$

In the case of tested samples parameter p takes values from 2×10^{-5} ($T_p = 98$ K) to 5×10^{-3} ($T_p = 373$ K). Frequency dependencies of exponent α are shown in figure 4(b). According to the model, α should take values from the range of $0 < \alpha < 2$ what we can notice in figure 4(b). The complex analyses of the results of electrical measurements contained in the section can be concluded that $(FeCoZr)_x(PZT)_{(100-x)}$ nanocomposites demonstrates hopping conductivity in alternating current at frequencies ranging from 50 Hz to 10^5 Hz.

5. Conclusions

AC electrical parameters measurements of $(FeCoZr)_x(PZT)_{(100-x)}$ nanolayers with $x_1 = 49.8$ at.% and $x_2 = 64.4$ at.% have been performed. The frequency and temperature dependencies of capacity C, conductivity σ, phase angle φ, loss coefficient tgδ and exponent α have been obtained. Dielectric properties of the materials measured directly after producing and subdued to thermal treatment at temperatures of 523 K and 573 K have been investigated. It was concluded that:

- Non-annealed nanolayer $(FeCoZr)_{49.8}(PZT)_{50.2}$ demonstrates dielectric type of conduction. Conductivity σ of the material is almost independent on frequency but shows temperature dependence, and phase shift angle φ takes negative values in whole frequency range;
- After the annealing of the sample in temperature of 523 K the materials character changes from capacitive to inductive due to the positive values of φ. Current and voltage resonances at resonant frequencies f_{RP}/f_{RS} and non-coil inductance phenomena in whole frequency range can be observed. Isochronous annealing intensifies the dielectric properties of the layer;
- Nanofilm $(FeCoZr)_{64.4}(PZT)_{35.6}$ annealed in temperature of 573 K demonstrates nonlinear exponential dependence of conductivity versus frequency. In both cases (figures 2(a) and 4(a)) the mechanism of charge carriers of the materials are defined as hopping conductivity;
- Potential applications of the tested materials in micro/nanoelectronics and electrical engineering have been considered.

References

[1] Li X, Dong Y, Liu M, Chang C and Wang X M 2017 *J. Alloy. Compd.* **696** 1323-8
[2] Yang B, Li X, Guo R and Yu R 2017 *Mater. Design* **121** 272-9
[3] Hoque S M, Makineni S K, Pal A, Ayyub P and Chattopadhyay K 2015 *J. Alloy. Compd.* **620** 442-50
[4] Le Graet C, Spenato D, Beaulieu N, Dekadjevi D T, Jay J-Ph, Pogossian S P, Warot-Fonrose B and Ben Youssef J B 2016 *EPL* **115** 17002
[5] Zheng F, Ma Z, Gao H, Pan F C, Li S T, Cao J W, Bai J M and Wei F L 2017 *J. Electron. Mater.* **28** 17448-52
[6] Wisniowski P, Dabek M, Wrona J, Cardoso S and Freitas P P 2017 *J. Appl. Phys.* **122** 213906
[7] El Kammouni R, Kurlyandskaya G V, Vazguez M and Volchkov S O 2016 *Sensor. Actuat. A-Phys.* **248** 155-61
[8] Wu D, Zhang Z, Li L, Zhang Z, Zhao H B, Wang J, Ma B and Jin Q Y 2015 *Sci. Rep.-UK* **5** 12352
[9] Nayek C, Murugavel P, Kumar S D and Subramanian V 2015 *Appl. Phys. A-Mater.* **120** 615-22
[10] Shimomura N, Sawada K, Nozaki T, Doi M and Sahashi M 2012 *Appl. Phys. Lett.* **101** 012403
[11] Zhu D P, Jiang Q H and Li Y W 2017 *Mater. Res. Express* **4** 125705
[12] Lu G P, Dong H T, Chen J G and Cheng J G 2017 *J. Sol-Gel Sci. Tech.* **82** 530-5
[13] Koltunowicz T N, Zukowski P, Boiko O, Saad A, Fedotova J A, Fedotov A K, Larkin A V and Kasiuk J 2015 *J. Electron. Mater.* **44** 2260-8
[14] Xiao M, Zhang Z B, Zhang W K and Zhang P 2017 *Appl. Phys. A-Mater.* **124** 8
[15] Boiko O, Koltunowicz T N, Zukowski P, Fedotov A K and Larkin A V 2017 *Ceram. Int.* **43** 2511-6
[16] Zukowski P, Koltunowicz T N, Boiko O, Bondariev V, Czarnacka K, Fedotova J A, Fedotov A K and Svito I A 2015 *Vacuum* **120** 37-43
[17] Kasiuk J V, Fedotova J A, Marszalek M, Karczmarska A, Mitura-Nowak M, Kalinin Y E and Sitnikov A V 2012 *Solid State Phys.* **54** 178-84
[18] Koltunowicz T N, Fedotova J A, Zhukowski P, Saad A, Fedotov A, Kasiuk J V and Larkin A V 2013 *J. Phys. D: Appl. Phys.* **46** 125304
[19] Koltunowicz T N 2015 *J. Appl. Spectrosc.* **82** 653-8
[20] Koltunowicz T N, Zukowski P, Czarnacka K, Boiko O, Fedotova J A and Kasiuk J V 2015 *J. Alloy. Compd.* **650** 262-7

[21] Koltunowicz T N, Zukowski P, Boiko O, Czarnacka K, Bondariev V, Saad A, Larkin A V and Fedotov A K 2016 *J. Mater. Sci.-Mater. El.* **27** 1171-6

IOP Conf. Series: Materials Science and Engineering **369** (2018) 012017 doi:10.1088/1757-899X/369/1/012017

Coarse grain simulations of linear polymer melts for chain lengths near the reptation transition

L L Fan[1] **and L Liu**[1,2]
[1]School of Information Science and Engineering, Dalian Polytechnic University, Dalian, China

E-mail: liu.li@dlpu.edu.cn

Abstract. Melts of linear polymers differing in the lengths of the chains have been studied by coarse graining simulations. Each polymer is represented by a branched string of soft blobs, and uncrossability constraints are added to prevent chain crossings. Their dynamical and rhelogical properties are discussed by comparisons to assess the impact of the chain length on the bulk properties without using extrapolation techniques. A satisfactory quantitative agreement with experiments is observed for the diffusion coefficients and zero-shear viscosities of the linear polymers. The power-law relation between viscosity and molecular mass is indicative of the transition regime between Rouse and reptation dynamics.

1. Introduction

As the third branch of scientific study, computer simulations have become more and more prominent with advancements in modern computing technologies. They offer increased control over system parameters in comparison to experiments, access to measures sometimes inaccessible in experiments and a platform to test theoretical predictions[1, 2, 3]. At first sight, it is tempting to perform a polymer simulation where all chemical structures are included in an effort to capture as accurately as possible the interaction details[4, 5]. However, the fact is that polymers exhibit clear time and length scaling for dynamical behaviors. A direct prediction of the large scale dynamics and rheology from such a detailed method would require a huge computational effort that would be costly in terms of computation time. To reduce the computer consumption for the simulation of polymeric systems, coarse-graining approaches may be applied for the study of rheological responses in such systems [6, 7, 8].

The basic concept in coarse graining is to represent a group of atoms or monomers as a larger group unit that neglects the atomistic details within the unit. By this means, the replaced realistic parts will interact with each other as new repeated coarse grained particles. The coarse grain level is depending on the concern of the properties of interest. The less details are needed, the more monomer units could be assembled into a particle. According to this, particle-based simulations can be performed on different scales. In coarse-grained molecular simulations, one particle represents the center of mass of a few atoms. Accordingly a polymer chain will be substituted by string of hard spheres that with sufficient repulsive potential at short distance to prevent chain crossings [9, 10, 11, 12]. In order to reach even larger time and length scales, one coarse-grained particle may lump together many more atoms,such as every 20 monomeric units can be replaced with a mesoscopic entity, which is called 'blob'. This leads to the soft interactions of the blobs that they are able to pass through each other easily[13, 14, 15, 16]. To prevent the unrealistic crossing, the additional restrictions have to be implemented in the entangled polymer system. Due to the generic nature of polymer, most of their chemical differences are not distinguished in large length scales. This naturally inspires us to develop a super coarse graining model that only keep track of the dynamics of the center of mass of the entire chain. Consequently, each polymer chain is sketched into a single particle that interacts with other coarse grained particles under the action of soft conservative energy and dominate entanglement effects[17, 18, 19]. Of course the structure of particle based program does not depend on the level of coarse graining, it in general follows the same loop: in the box with periodic boundary conditions, forces on the particles are calculated at a certain time, positions and velocities are updated to the next time step, and with the new velocities and positions, new forces are evaluated, etcetera.

Computer simulations of polymer dynamics at the level of atoms or chain segments are being used by a number of groups to probe the above concepts. A complication inherent to these studies is the extremely slow dynamics of highly entangled polymers; this makes the simulations very computational intensive, even when using coarse-grained models. In this work, we present a mesoscopic coarse graining model for studying the entangled polymer system. The entanglement effect is accomplished by introducing the uncrossability constraint. The dynamical and rheological properties of linear chains are studied through the typical blob mean square displacements and stress autocorrelation function. A global overview of the TWENTANGLEMENT algorithm[14, 15], the simulation set-up and the applied analysis methods are provided in Section 2. The simulation results are presented and discussed in Section 3, where comparisons are implemented between linear polymers. Our main conclusions are summarized in Section 4.

2. Model and methods

We will apply the coarse-grained TWENTANGLEMENT model, to simulate linear polyethylene melts. Briefly, each coarse-grained particle or 'blob' represents a polymer fragment of 20 backbone carbon atoms plus attached hydrogen atoms.

2.1. Coarse-grain interactions

Because of the coarse-graining, the equation of motion of the position of the blobs can be described by the simple first order Langevin equation, which involves the conservative forces $\mathbf{F}^C$ derived from the potential of mean force Φ,

$$\mathbf{F}_i^C = -\boldsymbol{\nabla}_i \Phi \tag{1}$$

as well as the blob friction ξ and random forces $\mathbf{F}^R$ with the usual statistical properties,

$$d\mathbf{R}_i = -\frac{1}{\xi}\mathbf{F}_i^C dt + \mathbf{F}_i^R,, \tag{2}$$

the fluctuation-dissipation theorem demands the relation between ξ and $\mathbf{F}^R$ given in Eq 3 as a condition for thermal equilibrium at the temperature of T =450 K.

$$\langle \mathbf{F}_i^R(t) \rangle = 0, \langle \mathbf{F}_i^R(t)\mathbf{F}_j^R(0) \rangle = 6k_B T \xi \delta_{ij} \delta(t). \tag{3}$$

where k_B is the Boltzmann constant.

The potential of mean force can be approximately presented as a sum of three independent parts, ϕ^b, ϕ^θ and ϕ^{nb}, with the usual meaning of bonded potential, angular potential and non-bonded or van der Waals potential separately. Using microscopic simulations[7, 14] the distribution function P between bonded and non-bonded blobs can be determined. Hence, the potential of mean force can be obtained by taking minus $k_B T$ times the logarithm of measured distributions of blobs, i.e.$\Phi = -k_B T \ln P(\mathbf{R}^n)$.

IOP Conf. Series: Materials Science and Engineering **369** (2018) 012017 doi:10.1088/1757-899X/369/1/012017

2.2. Uncrossability

Because each blob represents a flexible collection of bonded atoms, the non-bonded repulsive interactions between blobs are much softer than those between atoms. The repulsions are, in fact, too weak to inhibit crossings of the coarse grained polymers and thus a key property determining the dynamics of polymer melts is lost upon coarse-graining. To prevent these unrealistic crossing events, Padding and Briels developed the TWENTANGLEMENT algorithm that uses elastic forces to contain and reverse bond crossings[14, 15]. In short: an 'entanglement point' $\mathbf{X}$ is introduced at the point where the bond between blobs i and $i+1$ first crosses another bond. The binding energies ϕ^{bond} of these bonds are henceforth calculated using the contour length $L_{i,i+1}$ of the path from blob to blob via the entanglement point. The path between two consecutive blobs can collect multiple entanglement points by 'crossing' other paths, resulting in twisting paths of contour length

$$L_{i,i+1} = |\mathbf{R}_i - \mathbf{X}_1^{i,i+1}| + |\mathbf{X}_1^{i,i+1} - \mathbf{X}_2^{i,i+1}| + \ldots + |\mathbf{X}_p^{i,i+1} - \mathbf{R}_{i+1}|, \tag{4}$$

running from one bead to the next via a sequence of p entanglement points. Beads may slide along an entanglement point, thereby causing the entanglement point to slip from one bond to the next. An entanglement point survives until the two involved bond paths disentangle, at which occasion the entanglement point is annihilated. The positions of the entanglement points are recalculated every time step by minimizing the elastic energy stored in the bonds, subject to fixed blob positions. The forces exerted by the entanglements on the blobs, by means of the bond energy $\phi^{\text{bond}}(L_{i,i+1})$, are included in the equation of motion of the blobs.

2.3. Analysis methods

The dynamics of polymers in the melt will be quantified by the diffusion coefficient, as extracted from the mean square displacement

$$D(t) = \frac{1}{6t}\left\langle |\mathbf{R}_i^{cm}(t) - \mathbf{R}_i^{cm}(0)|^2 \right\rangle, \tag{5}$$

with $\mathbf{R}_i^{cm}(t)$ the center of mass position of the ith polymer at time t, and the pointy brackets denoting an average over all reference times and molecules.

Stress relaxation moduli are calculated from equilibrium simulations by means of the auto correlation

$$G(t) = \frac{V}{6k_BT}\sum_{\alpha\beta}\langle\sigma_{\alpha\beta}(t)\sigma_{\alpha\beta}(0)\rangle, \tag{6}$$

where the sum runs over the six off-diagonal elements and the stress tensor is given by

$$\sigma_{\alpha\beta} = -\frac{1}{V}\sum_{\langle i,j\rangle}(Y_{i,\alpha} - Y_{j,\alpha})\, f_{ij,\beta}. \tag{7}$$

Here $Y_{i,\alpha}$ denotes the α-component of the position $\mathbf{Y}_i$ of the ith entity – either a blob or an entanglement point – in the box of volume V, $f_{ij,\beta}$ is the β-component of the force exerted on this entity by entity j, and the sum runs over all entity pairs[15].

3. Results and discussions

In this section the dynamical and rheological properties of melts of polymers will be studied through comparisons. We note that advances in hardware and software have made it possible to simulate the linear polymers over considerably longer time scales than in previous TWENTANGLEMENT studies[15]. This has now made possible a direct evaluation of the dynamical properties of the longer chains, without resorting to the procedure of estimation.

IOP Conf. Series: Materials Science and Engineering **369** (2018) 012017 doi:10.1088/1757-899X/369/1/012017

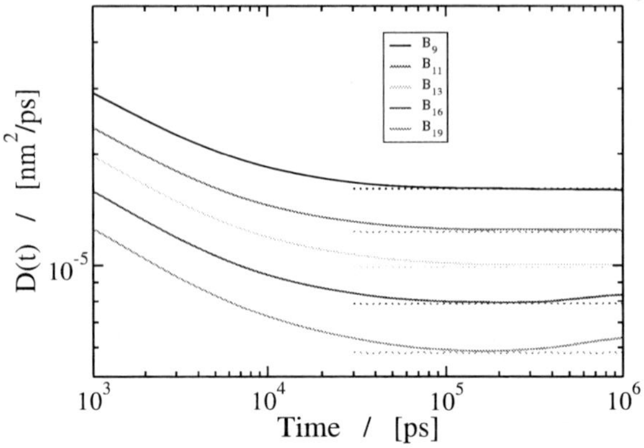

Figure 1. Time-dependent diffusion coefficients for linear polymers of various lengths, as extracted from the mean squared displacement of the polymer center of mass.

3.1. Diffusion

From theory and experimental observations it is known that the diffusion coefficient in a melt of linear polymers decays with increasing molecular mass M, with $D \propto M^{-1}$ for short polymers in the Rouse regime and $D \propto M^{-2}$ for long polymers in the reptation regime[20, 21, 22]. Figure 1 shows the time-dependent center-of-mass diffusion coefficients as extracted from the simulations by Equation 5. After a transient decay on the nanosecond time scale, the functions are seen to reach a plateau value that defines the long time diffusion coefficient D. For the longest chain, which does not reach a clear plateau, D is approximated by the minimum value of the $D(t)$ curve. The terminal diffusion coefficients are plotted in Figure 2 against the molecular weight, showing a satisfactory quantitative agreement with the experimental data by Pearson *et al.*[23]. Fitting the data with a power law yields an exponent of -1.27, indicating that the chains

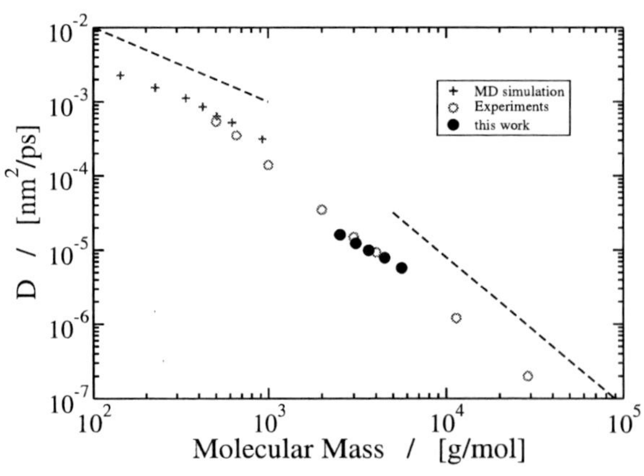

Figure 2. Long time diffusion coefficients of linear polymers as a function of molecular mass. The black dots are the results of this study, the red circles represent experimental data from Pearson *et al.*[23], and the blue pluses show molecular dynamics simulation results obtained by Mondello *et al.*[24, 25]. The dashed lines illustrate the slopes expected for the Rouse regime (left, $D \propto M^{-1}$) and the reptation regime (right, $D \propto M^{-2}$).

IOP Conf. Series: Materials Science and Engineering **369** (2018) 012017 doi:10.1088/1757-899X/369/1/012017

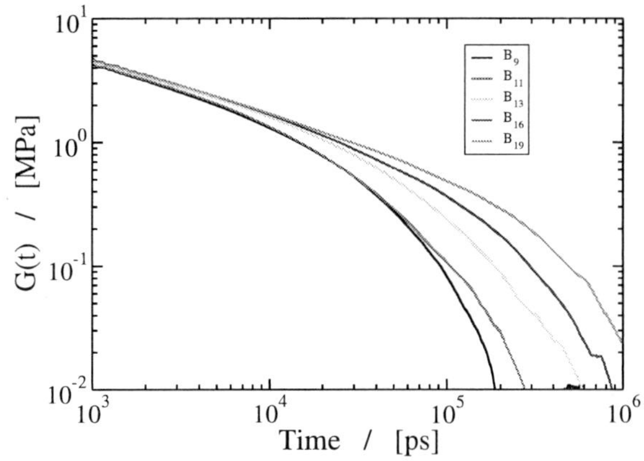

Figure 3. Block-averaged stress relaxation moduli for five melts of linear polymers of various chain lengths.

under investigation are in the transition region between the Rouse and reptation regimes. In comparison with the experimental data, the simulated transition occurs at a higher molecular mass and over a wider mass range; we suspect that this is a consequence of the softness of the coarse-grained system, in which the beads are less strongly confined by their neighbours and the uncrossable bonds than are the atoms in an experimental melt.

Both Figure 1 and Figure 2 suggest that the simulations with the largest linears have not yet reached the terminal diffusion coefficients. Besides this, the diffusion coefficient may have an additional regime between Rouse and reptation as suggested by Harmandaris *et al.*[20].

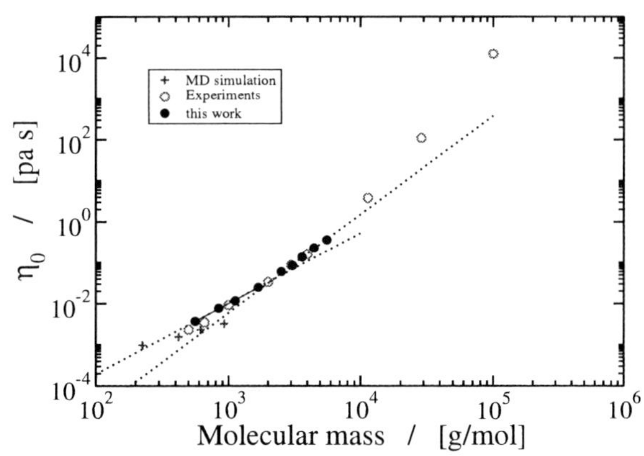

Figure 4. Zero shear viscosities of melts of linear polymers as a function of the molecular mass. The black dots are the results of this study, the red circles represent experimental data from Pearson *et al.*[23], and the blue plusses show simulation results obtained by Mondello *et al.*[24] using molecular dynamics. The dashed lines highlight the experimental scaling laws at low mass ($\eta \propto M^{1.8}$) and high mass ($\eta \propto M^{3.5}$).

3.2. Stress relaxation moduli
Stress relaxation moduli for five linear polymers are collected in Figure 3. The curves almost coalesce for short time and fan out for large time, with the longer polymers relaxing more slowly than the shorter ones. The resulting viscosities at vanishing shear rate are plotted in Figure 4, supplemented with the zero-shear viscosities of three shorter chains, downto B_2. The results indicate that the simulations are in the transition regime between Rouse and reptation dynamics. The power of 1.71 ± 0.05 extracted from the first four points is appreciably lower than the power of 2.4 ± 0.2 extracted from the next four points, confirming that the simulated melts are in the transition regime. Comparison with the experimental viscosities of Pearson *et al.*[23] reveals a good quantitative agreement. The location and width of the transition regime of the simulated melts appear to agree with those of the experimental melts.

4. Conclusions
The diffusive and rheological properties of linear polyethylene polymers of various lengths have been studied by computer simulations using the highly coarse-grained TWENTANGLEMENT technique. Comparing the simulated linear polymers, highlights the impact of the chain length on the dynamical properties of the melt. By calculating the autocorrelation function of the chains, it is found that the longer arm has a slower relaxation decay. Comparison with the experimental viscosities gives good quantitative agreement and a regime for chains lies between the Rouse and reptation is investigated and confirmed from the scaling exponents. In future work, longer chains and new topologies of polymer will be investigated by this model.

Acknowledgments
Authors acknowledge the financial support of National Natural Science Foundation of China(Grant:21704010) and Natural Science Foundation of Liaoning Province(Grant:2015010217-301).

References
[1] Maurel G, Schnell B, Goujon F, Couty M and Malfreyt P 2012 *J. Chem. Theory Comput.* **8** 4570–9
[2] Ramirez-Hernandez A, Müller M and de Pablo J J 2013 *Soft Matter* **9** 2030–6
[3] Snijkers F, Pasquino R, Olmsted P and Vlassopoulos D 2015 *J. Phys.: Condens. Matter* **27** 473002–28
[4] Anogiannakis S D, Tzoumanekas C and Theodorou D N 2012 *Macomolecules* **45** 9475–92
[5] Pérez-Aparicio R, Colmenero J, Alvarez F, Padding J T and Briels W J 2010 *J. Chem. Phys.* **132** 024904:1–11
[6] Peters L B, Salerno M K, Agrawal A, Perahia D and GrestS G S 2017 *J. Chem. Theory Comput.* **13** 2890–6
[7] Reith D, Pütz M and Müller-Plathe F 2003 *J. Comput. Chem.* **24** 1624–36
[8] Padding J T and Briels W J 2011 *J. Phys. Condens. Matter* **23** 233101
[9] Harmandaris V A, Mavrantzas V G and Theodorou D N 1998 *Macromolecules* **31** 7934–43
[10] Tjornhammar R and Edholm O 2014 *J. Chem. Theory Comput.* **10** 5706–15
[11] Jabbarzadeh A, Atkinson J D and Tanner R I 2003 *Macromolecules* **36** 5020–31
[12] Kremer K and Grest G S 1990 *J. Chem. Phys* **92** 5057–86
[13] Goga N, Rzepiela A J, de Vries A H, Marrink S J and Berendsen H J C 2012 *J. Chem. Theory Comput.* **8** 3637–49
[14] Padding J T and Briels W J 2001 *J. Chem. Phys.* **115** 2846–59
[15] Padding J T and Briels W J 2002 *J. Chem. Phys.* **117** 925–43
[16] Steenbakkers R J and Schieber J D 2015 *J. Chem. Phys.* **143** 024905
[17] Sprakel J, Padding J T and Briels W J 2011 *Europhys. Letters* **93** 58003:1–6
[18] Padding J T, Mohite L V, Auhl D, Briels W J and Bailly C 2011 *Soft Matter* **7** 5036–46
[19] de Oliveria I S S, den Otter W K and Briels W J 2012 *J. Chem. Phys.* **137** 204908
[20] Harmandaris V A, Mavrantzas V G, Theodorou D N, Kröger M, Ramírez J, Öttinger H C and Vlassopoulos D 2003 *Macromolecules* **36** 1376–87
[21] Pearson D S, Fetters L J and Graessley W W 1994 *Macromolecules* **27** 711–9
[22] Rouse P E 1953 *J. Chem. Phys.* **21** 1272–80
[23] Pearson D S, Strate G V, Meerwall E V and Schilling F C 1987 *Macromolecules* **20** 1133–41
[24] Mondello M and Grest G S 1997 *J. Chem. Phys.* **106** 9327–36

[25] Mondello M, Grest G S, Webb III E B and Peczak P 1998 *J. Chem. Phys.* **109** 798–805

Formation of hybrid nanocomposites polymethylolacrylamide/ silver

L G Kolzunova[1,3], E V Shchitovskaya[1,2,3] and I G Rodzik[1]

[1]Russian Academy of Science, Far Eastern-Branch, Institute of Chemistry, Vladivostok, Russia
[2]Far Eastern Federal University, Vladivostok, Russia

E-mail: kolzunova@ich.dvo.ru/ shchitovskaya@gmail.com

Abstract. In this study, polymethylolacrylamide/silver composites have been formed by incorporating silver nanoparticles into the pre-electrosynthesized polymer film. The composites were formed in a two-step process involving the sorption of silver nitrate by a polymer matrix followed by chemical reduction of Ag-ions. The presence of crystalline silver phase in the polymer was confirmed by X-ray phase analysis (XRD), plasmon resonance and scanning electron microscopy (SEM). The small-angle X-ray scattering (SAXS) method has obtained the distribution functions of silver particles over radii. It is established that the content of silver in composites without chitosan is 10-15 times higher than with its additive. The dependences of cyclic voltammetry in pure phosphate buffer (pH 6.86) and in the presence of hydrogen peroxide were obtained. It has been shown that polymer/silver composites exhibit selectivity to hydrogen peroxide.

1. Introduction

Hybrid materials containing inclusions of metal nanoparticles in the polymer matrix are of particular interest among the polymer composites formed by the method of electropolymerization. Nanoparticles (NPs) are promising for use in catalysis, biology, computer technology, solar batteries, in medicine and optoelectronic devices [1-4]. For example, silver NPs (Ag) are widely used to increase the sensitivity of a surface in certain spectroscopic measurements, such as fluorescence, Raman scattering and second harmonic generation, various optical materials, configurable plasmon structures, biocidal coatings, and also in photochemistry, electrochemistry, etc. [5-8]. Recently, polymers that react to various changes in the environment (pH, temperature, ionic strength, the presence of certain substances, the electric field, etc.) have been developed. That so called "smart" polymers have found wide application in biotechnology and medicine as chemical and biological sensors [9,10].

At present the method for the electrochemical formation of polymers based on acrylamide and its derivatives has been thoroughly investigated [11]. Studies are under way to create hybrid polymer/metal materials on their basis. In this vein, it is of interest to investigate the possibilities of forming polymethylolacrylamide/silver composites for use as biosensors or electrocatalysts [10,12-14].

The aim of this investigation was to study the method of incorporating silver particles into the polymer matrix formed by the electro(co)polymerization of acrylamide, formaldehyde, N,N'-methylene-bis-acrylamide and chitosan.

2. Experimental

Formation of polymer films was carried out from an aqueous solution of acrylamide (AA), formaldehyde, N, N'-methylenebisacrylamide (MBA), chitosan (Chit) and zinc chloride. The substances were taken at the required concentrations in potentiostatic electrolysis at E = -1.16 V with respect to saturated silver chloride electrode [15].

The electropolymerization time was 3 minutes. Synthesis of polymers and electrochemical studies of composites were carried out using potentiostat/galvanostat IPC-Pro 8.0, "Cronas" (Russia). The working electrode (cathode) was a rod of stainless steel AISI 304, platinum plate S = 7.5 cm^2 was an anode. Stainless steel electrodes were previously degreased with a fine magnesium oxide paste, rinsed thoroughly with deionized water. Electrolysis was carried out in glass cells with a volume of 10 cm^3 without separating of the anode and cathode spaces and without preliminary solution degassing. The zinc metal sublayer was removed in 0.5 M sodium sulfate solution for 48 hours and washed with deionized water.

The introduction of silver ions into the polymer matrix was carried out by immersing the electrode with a polymethylolacrylamide film for 96 hours in a solution of $AgNO_3$ of various concentrations. Silver chemical reduction was carried out in a freshly prepared 0.01M solution of sodium tetrahydroborate ($NaBH_4$), by placing composite 25-30 seconds and then thoroughly washed with deionized water [5].

X-ray phase analysis (XRD) was performed on an Advance D-8 diffractometer (Bruker-AXS, Germany), PDF-2 data base was used.

Scanning field emission electron microscope S5500 (Hitachi, Japan) was used to investigate the structure, morphology, and element analysis of polymer films.

The distribution of silver particles was carried out by small-angle X-ray scattering (SAXS) method using a HECUS S-3 diffractometer (Micro-Pix, Austria).

The absorption spectra of the composites were taken in the far ultraviolet and visible regions in the range 300-900 nm, at room temperature on a Shimadzu UV-2550 (Japan) spectrophotometer with a film holder.

3. Results and discussion

One way to modify polymer films is to immobilize various particles in them to create composite materials with different properties. In this paper we have applied the chemical reduction method for silver particles implantation into the polymer matrix.

The composites were labeled according to the concentration of $AgNO_3$ ($mol \cdot L^{-1}$) solution in which immobilization of silver ions was carried out: 10-3–(I-1); 5·10-3–(I-2); 10-2–(I-3); 5·10-2–(I-4)–for composites without Chit and 10-3–(II-1); 5·10-3–(II-2); 10-2–(II-3); 5·10-2 (II-4)–for composites with Chit.

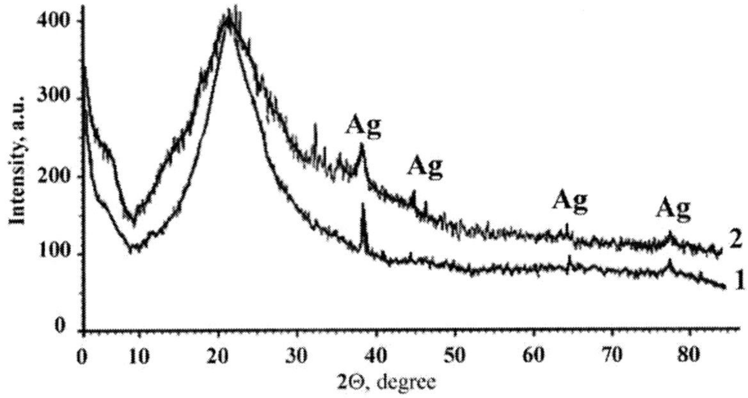

Figure 1. X-ray diffraction patterns of polymers with silver particles I-1 – (1) and II-3 – (2).

To confirm the inclusion of silver particles in the composite, the structure of the film surface was studied and X-ray phase analysis (XRD) was performed. The study of thin films with silver was carried out with grazing incidence X-ray diffraction, when the signal is formed mainly in the film material, thereby significantly increasing the intensity of the diffraction lines. X-ray diffractograms (figure 1) of nanocomposites differentiate the amorphous halo of the polymer matrix and the intense Bragg reflections in the relevant fields, which corresponds to the presence of zero-valence metallic silver (face-centered cubic crystals) [16]. Silver is fixed in all composites.

Scanning electron microscopy (SEM) was used to investigate the morphology of the surface of composites and to confirm the shape, size, and distribution of silver particles both on the external surface of the polymer and in the near-electrode layer. Figure 2 shows SEM images (with different magnifications) of the outer surface of polymer/silver composites without chitosan (I-2)–(a, c) and with chitosan (II-2)–(b, d).

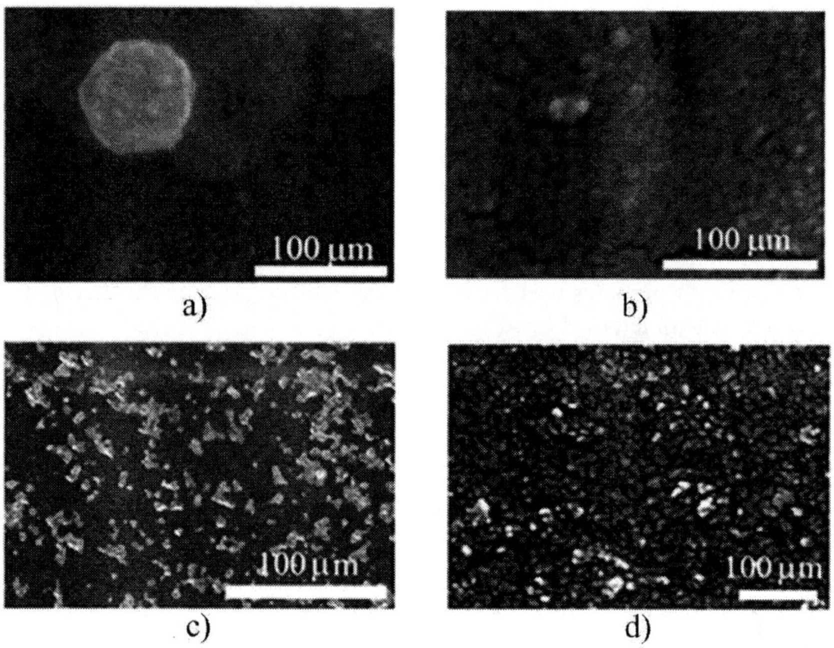

Figure 2. SEM images of the outer surface of composites without chitosan I-2 (a, b) and with chitosan II-2 (c, d).

SEM images (figure 2) shows the silver particles are found in spherical forms on the surface of the polymethylolacrylamide/Ag composite which does not contain Chit (a, b). That aggregates are 80-90 μm in size, and consisting of smaller particles with a diameter less than 50 nm. The surface of the composite polymethylolacrylamide/Ag with the addition of Chit is denser, silver is detected as individual fine particles (c, d).

An elemental analysis can be carried out simultaneously with scanning electron microscopy, as addition to the surface morphology study [16].

On the outside of the composite without Chit, the amount of silver is 15 times larger (Ag = 51.42 ± 1.42 at.%) than on the surface of the chitosan containing composite (Ag = 3.36 ± 0.57 at.%), which may be due to the formation of a complex of silver with Chit.

According to the SEM data, no silver was found on the near-electrode side of the polymer, both in the composite without chitosan and with the addition of Chit. Perhaps this is due to the fact that the solution of silver nitrate, in which the samples were kept, did not penetrate to the surface of stainless steel due to the increased density of the electrode near the electrode layer, in agreement with literature

[17]. As a result, after the reduction of silver ions by NaBH4, the silver particles were localized in the volume and near the outer surface of the polymer.

As revealed by the SEM, the aggregation of particles in the polymer/silver composite is observed, leading to a change in their size and shape, and therefore we obtained the distribution functions of the silver particles along the radii for all polymer films by the SAXS (figure 3). It is established that the silver particles have a radius from 120 Å to 145 Å. The polymer substrate does not significant affect the SAXS results since the magnitude intensity is 3 orders lower and the average polymer particles radius is less than 10 Å.

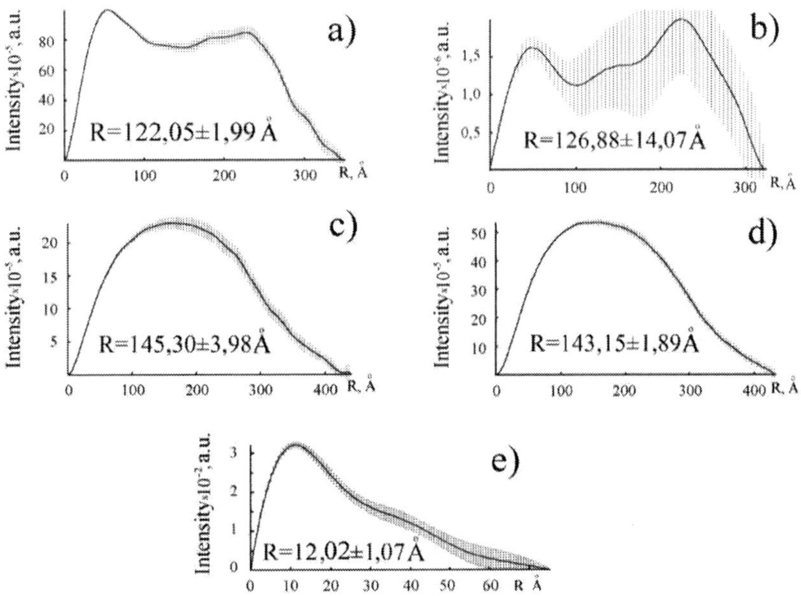

Figure 3. The distribution functions of Ag particles along radii for composites (I) without Chit – (a, c) and (II) with the addition of Chit – (b, d) obtained in Ag (NO_3 ($mol \cdot L^{-1}$): 10^{-3} – (a, b), 5 10^{-2} – (c, d), the blank polymer substrate (e).

For composites formed from a low-concentration $AgNO_3$ solution, two peaks are observed, which corresponds to the shape of two spherical particles connected to each other. As the $AgNO_3$ concentration increases, the amount of silver particles in the composite increases and the shape of the particles changes markedly. At the maximum concentration of $AgNO_3$, the particle distribution function corresponds to the shape of one large spherical particle consisting of smaller particles.

Figure 4 shows the absorption spectra of wet films in the far ultraviolet and visible regions, taken for all samples. It can be seen from the figure that most of the spectra have a pronounced absorption band in the range of 380-420 nm, in which plasmon resonance of silver nanoparticles is detected.

The shape of the bands indicates a wide range of nanoparticle size distribution. According to the literature [18-20], the position of the maximum of the plasmon resonance band at 410 nm and higher in the long-wavelength region of the spectrum indicates that the diameter of the obtained nanoparticles is more than 20 nm. Indirectly, the brown color of the films indicates the sufficiently large size of the obtained particles. Analysis of the intensity of radiation absorption in the spectra shows that very few nanoparticles are formed at low concentrations of $AgNO_3$ (samples 1, 2), while an increase in the $AgNO_3$ concentration by a factor of 10 in composites 3 and 4 significantly increases the amount of silver nanoparticles. This is evidenced by the high intensity of the absorption bands in the spectra. A similar conclusion about the effect of the experimental conditions can be made in composites with chitosan (II) in which, in comparison with composites without it (I), a more complete reduction of silver is observed together with the formation of a larger number of nanoparticles in the bulk of the polymer.

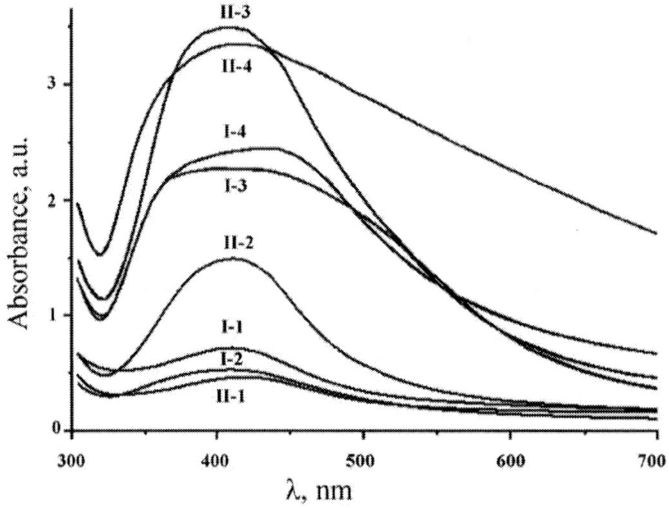

Figure 4. Plasmon resonance absorption spectra in the far ultraviolet and visible regions for composites without Chit – (I), with Chit – (II), obtained in a solution with different content of Ag(NO)$_3$, mol/l: 10^{-3}–(1), 5 10^{-3}–(2); 10^{-2}–(3); 5 10^{-2}–(4).

The electrochemical properties of composites were studied by using cyclic voltammetry (CVs) of the hydrogen peroxide reduction [21]. A comparison of the reduction rate of H_2O_2 on polymer/silver composites without Chit (I-2) and with Chit (II-2) in 0.1 M phosphate buffer saline (PBS) at pH 6.86 are shown in figures 5 and 6.

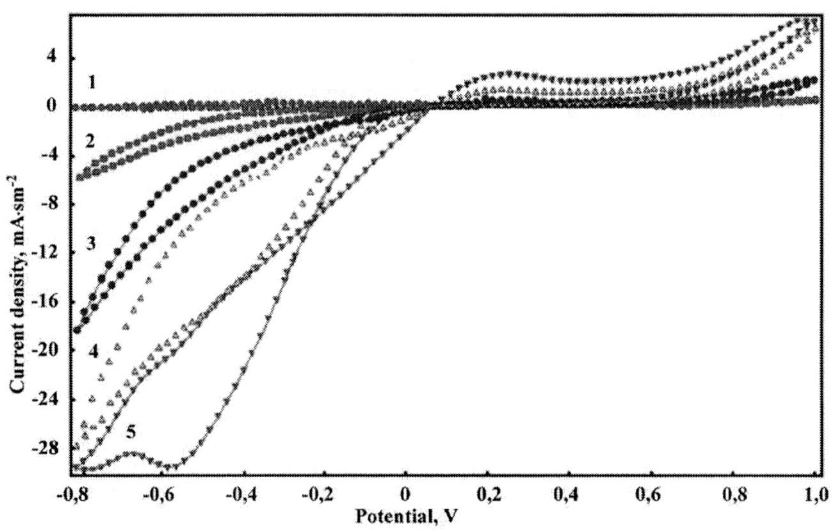

Figure 5. CVs for the composite I-2 in 0.1 M PBS-(1) and in PBS with addition of H_2O_2 (mol·L^{-1}): 0.12–(2), 0.24–(3), 0.36–(4), 0.48–(5), (scan rate: 100 mV/s).

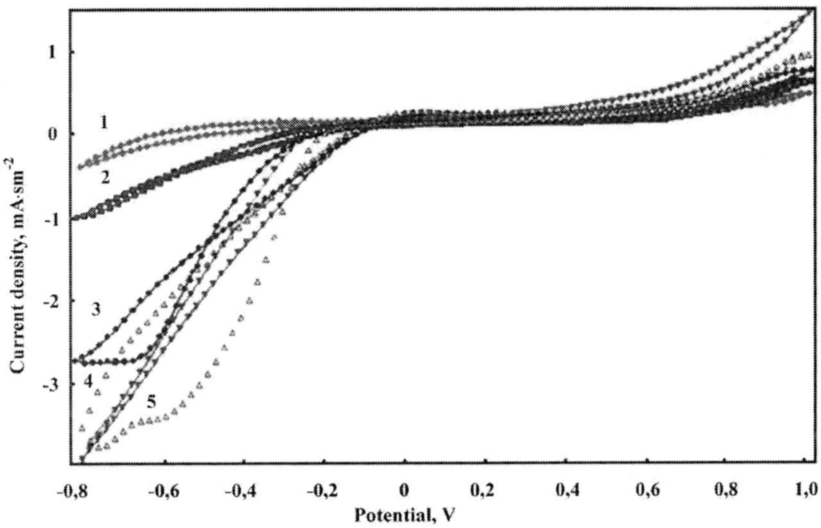

Figure 6. CVs for composite II-2 in 0.1 M PBS-(1) and in PBS with addition of H_2O_2 (mol·L^{-1}): 0.12–(2), 0.24–(3), 0.36–(4), 0.48–(5), (scan rate: 100 mV/s).

The addition of hydrogen peroxide leads to an increase in the values of the anodic and cathodic peaks of the current density for all the studied composite electrodes. An increase of the hydrogen peroxide concentration leads to an increase of the hydrogen release rate for polymer/silver I-2 composites (without Chit), which have the hydrogen peroxide current density at -0.6 V almost 5 times higher than polymer/silver II-2 composites (with Chit) (figure 7).

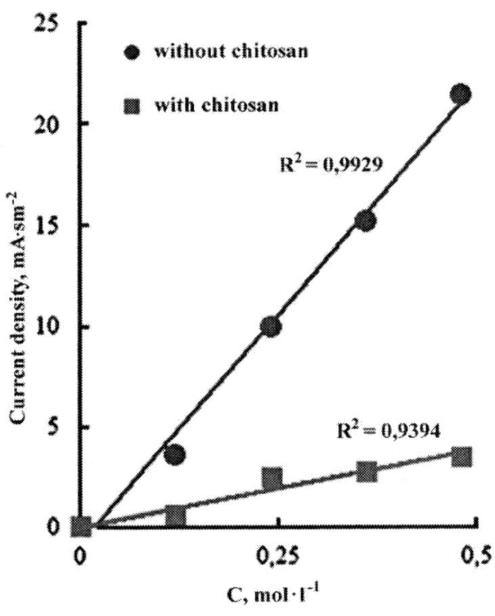

Figure 7. Cathodic current density dependence (at E = -0.6 V) on hydrogen peroxide concentration in 0.1 M PBS for polymer/silver composites.

Most likely, such a result is associated with the morphology of the surface of composites with chitosan, coarsening of the particles, which are included in the volume of the polymer.

4. Conclusions

Thus, it has been shown that the introduction of silver particles into a matrix performed by the electro(co)polymerization method is possible in a two-step process involving the sorption of silver nitrate by a polymer matrix followed by its chemical reduction. The presence of a crystalline silver phase in the composite was confirmed by X-ray diffraction analysis. By the methods of SAXS and SEM it was established that the content of silver particles with average sizes from 10 to 20 nm in composites without chitosan is 10-15 times higher than with chitosan. Using cyclic voltammetry, it has been found that polymer/silver composites exhibit selectivity to the presence of hydrogen peroxide in solutions.

Acknowledgments

The research was partial supported by the Russian Government State (0265-2014-0001) and by grant (№ 18-3-031) of Program "The Far East".

References

[1] Gao Y, Munroe N, Kong X and Jones K 2009 *J. Power Sources* **189** 935-42
[2] Schrand A M, Braydich-Stolle L K, Schlager J J, Liming D and Hussain S M 2008 *Nanotechnology* **19** 235104
[3] Xu R, Wang D, Zhang J and Li Y 2006 *Chemistry - An Asian J* **1** 888-93
[4] Blanco I 2018 *J Nanomed* **1** 1002-4
[5] Krutyakov Y A, Kudrinsky A A, Olenin AYu and Lisichkin G V 2008 *Uspekhi Khimii* **3** 242-69
[6] Bogle K A, Dhole S D and Bhoraskar V N 2006 *Nanotechnology* **17** 3204-8
[7] Guzman M, Dille J and Godet S 2012 *Nanomed. Nanotech. Biol. Med.* **8** 37-45
[8] Kim J S, Kuk E, Yu K N, *et al* 2007 *Nanomed. Nanotech. Biol. Med.* **3** 95-101
[9] Mahouche-Cherguia S, Guerrouachea M, Carbonniera B and Chehimib M M 2013 *Colloids and Surfaces A: Physicochem. and Engineering Aspects* **439** 43-68
[10] Perelshtein I, Applerot G, Perkas N, Guibert G, Mikhailov S and Gedanken A 2008 *Nanotechnology* **19** 245705
[11] Kolzunova L G, Kalugina I Y and Kovarsky N Y 1996 *Zh. Prikl. Chemistry* **69** 135-41
[12] Ni K, Chen L and Lu G 2008 *Electrochem. Commun.* **10** 1027-30
[13] Ruparelia J P, Chatterjee A K, Duttagupta S P and Mukherji S 2008 *Acta Biomater.* **4** 707-16
[14] Sakardina E A, Kravchenko T A, Zolotukhina E V and Vorotyntsev M A 2015 *Electrochim. Acta* **179** 364-71
[15] Karpenko M A and Kolzunova L G 2015 *Vestnik FEB RAS.* **4** 79-83
[16] Balasubramanian C, Godbole V P, Rohatgi V K, Das A K and Bhoraskar S V 2004 *Nanotechnology* **15** 370-3
[17] Karpenko M A, Kolzunova L G and Karpenko A A 2006 *Russian Journal of Electrochemistry* **42** 89-97
[18] Brukhanov V V, Tikhomirova N S, Gorlov R V and Slezhkin V A 2013 *Izv. KSTU* **6** 115-22
[19] Slezhkin V A and Gorlov R V 2011 *Izv. KSTU.* **1** 115-22
[20] Wang Y, Guo X, Pan R, Han D, Chen T, Geng Z, Xiong Y and Chen Y 2015 *Mater. Science and Engineering* **53** 222-8
[21] Lu W, Luo Y, Chang G and Sun X 2011 *Biosensors and Bioelectronics* **26** 4791-7

IOP Conf. Series: Materials Science and Engineering **369** (2018) 012019 doi:10.1088/1757-899X/369/1/012019

Research on the improvement of mixed titania and Co(Mn) oxide nano-composite coatings

G Yar-Mukhamedova[1,3], M Ved[2], A Karakurkchi[2], N Sakhnenko[2] and R Atchibayev[1]

[1]Physics Department, Al-Farabi Kazakh National University, IETP, 71 Al-Farabi Avenue, Kazakhstan
[2]Physical Chemistry Department, National Technical University "Kharkiv Polytechnic Institute", 2 Kyrpychova Street, 61002 Kharkiv, Ukraine

E-mail: gulmira-alma-ata@mail.ru

Abstract. The structure and the properties of the oxide films formed on titanium in the diphosphate based electrolytes by means of plasma electrolytic oxidizing at direct current density of 2–2.5 A·dm^{-2} have been studied. Oxide layers of different composition and content of alloying elements were obtained by modification of electrolytes and variation in current density. The interelectrode voltage during PEO, chemical and phase composition, topography and microstructure of the formed layers depend on the electrolyte composition and applied current density. The spark-discharge regime was shown to be reached at inter-electrode voltage 100 to 130 V depending on the composition of electrolyte. The effect of chemical composition and surface morphology formed mixed oxide films on the corrosion resistance and catalytic activity has been discussed.

1. Introduction

Titanium is the most common oxide in heterogeneous catalysis and photo-catalysis for the purification gas and liquid media from toxicant [1,2]. This fact is attributed to the fairly high chemical stability under different operating conditions, no toxicity, and relatively low cost of this material. At the same time it was found that, in most cases, catalyst materials based on mixed two or three component oxide systems exhibit high activity and selectivity not only in heterogeneous red-ox reactions [3–5] but also in photo-catalytic ones [6–8]. Synthesis of nano-composite catalytic disperse systems based on titania is carried out by various methods, including sol-gel processes [6,9,10].

The formation different kind of oxide layers has been achieved by the anodic oxidation and plasma surface treatment in the sparking regime (plasma electrolytic oxidizing PEO) seems to be the very promising since it does not need the sophisticated facilities, allows to form various types of titania and to incorporate different species into the layer by modification of electrochemical parameters and of electrolyte chemistry [11–14].

Titania layers containing manganese oxides obtained by PEO in acetate-borate electrolyte was noted to be catalytic active in CO conversion to CO_2 [15]. Cobalt-containing oxide coatings on titanium are obtained from a silicate electrolyte with cobalt acetate addition [16]. However, the increase in the catalytic activity of above materials in CO oxidation reaction was achieved by additional impregnation followed by annealing. In [17] catalytic materials on titanium and aluminum doped with transition metal oxides (Mn, Fe, Co, Ni) were obtained in one stage by the PEO method, and also in two stages combining the PEO treatment, followed by impregnation in solutions. The need for additional technological operations was due to the low content of dopants and the uneven distribution of catalytic components in the surface layers.

Our positive experience [18,19] related to the development of electrolytes and single step PEO regimes for nano-composite manganese-containing oxide coatings forming at valve metals from diphosphate bath. This approach was extended to the synthesis of cobalt-containing layers [20]. Nevertheless the study of cobalt and manganese effect on morphology, and, consequently, the catalytic activity and corrosion resistance of mixed oxide systems is topical and new. In this work some results concerning the chemical composition, topography, corrosion resistance and catalytic properties of the oxide layers formed on Ti by plasma electrolytic treatment and alloying by cobalt, manganese and phosphorus has been presented.

2. Materials and experimental procedure

Titanium alloy sheets of composition, wt.%: Ti – 99.2–99.7, and impurities – 0.3–0.8; were subjected to the plasma electrolytic oxidation. The samples' pre-treatment procedure included mechanical cleaning with sandpaper from contaminants, degreasing in a 0.2–0.3 M NaOH, and etching in a mixture of a 0.1–0.3 M HF and 0.3–0.9 M HNO_3, and thorough rinsing with running water. The PEO treatment of the samples was carried out in the diphosphate solution [19,20], with or without the addition of Co and Mn sulfates (table 1). Anodic polarization of samples was carried out from a stabilized current source by a direct current of different densities (table 1) until the sparking regime was achieved. The treatment lasted 30 minutes after attaining the sparking mode to obtain a uniform oxide layer with dopants. The Ti alloy anode was placed into the electrochemical cell (volume 200 ml) the form of the graphite cathode provided the formation of the uniform modified layer on the both sides of the sheet. The temperature of oxidation (293–300 K) was stabilized by water shield. The thickness of the coatings was at least 15–20 μm.

Table 1. The treatment parameters and the code of the studied samples.

Sample code	Electrolyte composition, $mol \cdot dm^{-3}$	Current density i, $A \cdot dm^{-2}$	Addition of alloying elements
1	$K_4P_2O_7$ – 0.5		–
2	$K_4P_2O_7$ – 0.5; $CoSO_4$ – 0.1	2.0	Co
3	$K_4P_2O_7$ – 0.5; $MnSO_4$ – 0.1		Mn
4		2.5	

The surface of the anodized specimens was observed on scanning electron microscope Hitachi S 4200 at magnification up to 10,000 with attached ESEM-50 Philips EDS analyzer. The backscattering electron image and the energy dispersion analysis were done. The X-ray diffraction was done by a Brucker AXS D8 diffractometer, using Cu Kα radiation and Ni filter. The obtained spectra were analyzed using the CPDF PDF – 2/2001 database.

The susceptibility to corrosion was checked in the free aerated 2 M NaOH and Ringer (composition, $mol \cdot dm^{-3}$: NaCl – 0.15, KCl – 0.004, $CaCl_2$ – 0.004) solutions by comparing the values of open circuit potentials (E_{oc}). The potentials presented in the paper are given relative to the standard hydrogen electrode (SHE). The catalytic properties of the oxide systems were studied in the model reaction oxidation of carbon (II) oxide to CO_2 as in [18,19].

3. Results and discussion

The stages of the titanium alloy oxidizing at direct current of 2 $A \cdot dm^{-2}$ in different electrolytes one can see at chronograms of inter-electrode voltage in figure 1. Section I corresponds to a linear change in the inter-electrode voltage U up to 60 V at first 2 min of oxidizing. This section known as pre-spark stage is almost the same for chronograms in all electrolytes and is attributed with formation of phase titania.

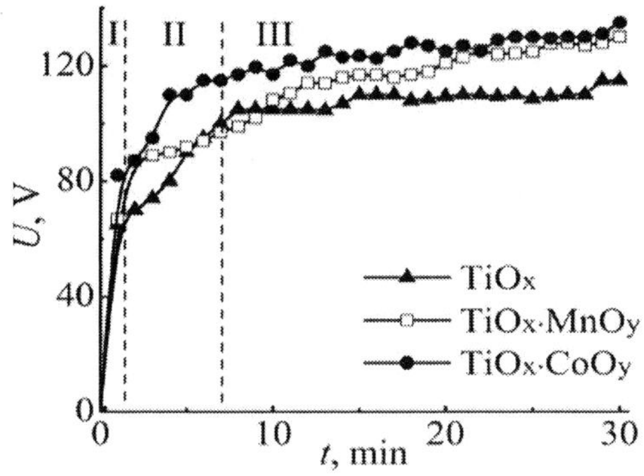

Figure 1. Chronograms of inter-electrode voltage at applied anodic i=2 A·dm^{-2} polarization of Ti alloy in diphosphate electrolyte with or without addition of cobalt or manganese sulfates.

Voltage oscillations are observed in section II of the chronograms, when local sparking begins at the anode. This is due to competitive processes of titania formation / dissolution, and thermo-chemical parallel reactions with electrolyte components. At this stage, the dopants (Co, Mn, P, K) are incorporated into the matrix of titania, forming oxides that heal defects. The dependence of the spark voltage U_s on the nature of the dopants is explained by differences in the specific electrical resistivity of oxides (table 2). It is quite natural that the spark voltage U_S increases in the series TiO_2 < $TiO_2 \cdot MnO_x$ < $TiO_2 \cdot CoO_x$.

Table 2. Effect of the current density on the PEO voltage and composition of oxide coatings on Titanium.

Electrolyte	Current density i, A·dm^{-2}	Voltage, V		Specific electrical resistance of oxide, ρ (Ω·cm)	Content of dopants in surface layer, at%				P to Ti ratio
		U_S	U_O		Co	Mn	P	K	
$K_4P_2O_7$	2.0	67–73	105	TiO_2 – 10^{13}	–	–	22±1	13.2±0.5	0.53
	2.5		110		–	–	27±1	12.0±0.5	
$K_4P_2O_7$, $CoSO_4$	2.0	82–85	120	CoO – 10^6–10^{10}	5.7±0.2	–	35±1	10.2±0.5	0.76
	2.5		130	Co_3O_4–10^3–10^5	6.3±0.2	–	36±1	7.5±0.5	0.72
$K_4P_2O_7$, $MnSO_4$	2.0	80–85	125	MnO – 10^9–10^{12}	–	4.6±0.2	37±1	9.1±0.5	0.63
	2.5		130	Mn_2O_3–10^5, Mn_3O_4 – 10^4–10^5 MnO_2 –10^{-1}–10^{-3}	–	5.0±0.2	38±1	7.6±0.5	0.79

Section III at chronograms corresponds to the micro-arc mode of PEO, and the voltage stabilization is due to the increase in the thickness of the mixed oxide coatings. The operating voltage U_O also depends on the composition of the electrolyte and, consequently, the formed oxides, especially their thermal resistance. The operating voltage in cobalt and manganese containing solutions after 25 min PEO is practically the same, which is due to identical specific electrical resistances MnO and CoO that are the products of intermediate oxides thermal decomposition.

Surface layers morphology depends on the applied current density and composition of oxide coatings. The size of the rounded (doughnut like) grains observed for titanium in diphosphate electrolyte as shown in figure 2(a) increases for Co or Mn containing mixed coatings as shown in figures 2(b) and 2(c) obtained at the same current density i 2 A·dm^{-2}, but it decreases at increasing i up to 2.5 A·dm^{-2} as shown in figure 2(d). The micro cracks of the doughnut like and of the acicular grains, have been seen as shown in figure 2(a). The presence of Co ions considerably modified the appearance of the oxide layer as shown in figure 2(b) due to the healing of cracks and defects when dopant is incorporated into the surface layers. On the other hand, the effect of the current density can be seen in the case of layer formation in the Mn containing solution (cf. as shown in figures 2(c) and 2(d)). The structure consisting of the doughnut like grains filled with the small grains has been also observed as shown in figure 2(b). The size of the small grains slightly decreases with the increased *i* and thus higher Mn content (sample 4).

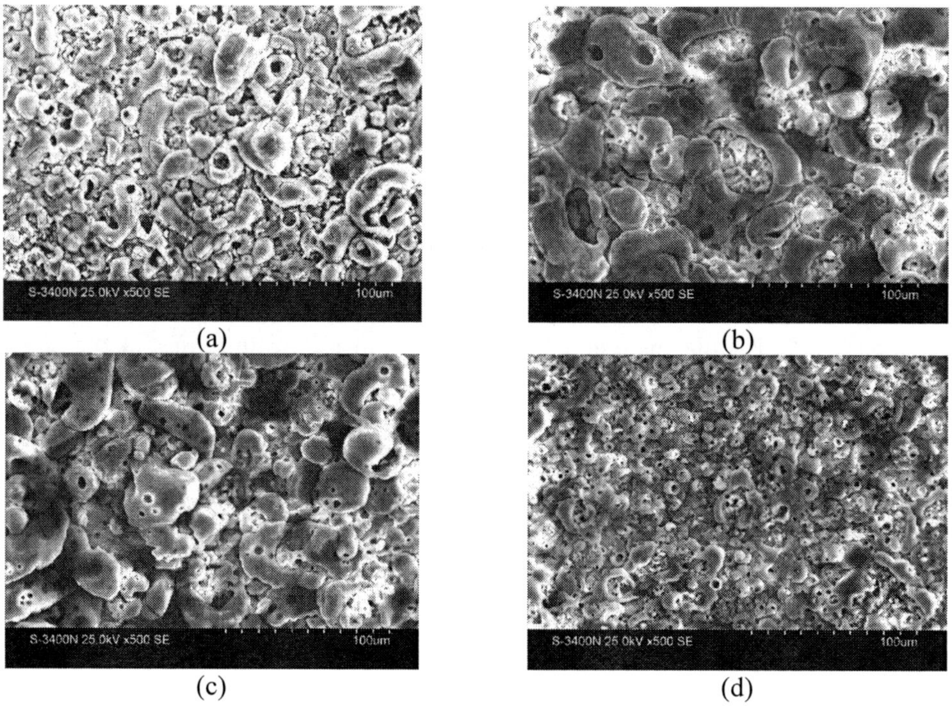

Figure 2. Appearance of the surface of Ti after the PEO treatment at different conditions. Samples: 1 (a), 2 (b), 3 (c), 4 (d).

As follows from the EDS results as shown in figure 3 the X-ray characteristic spectrum taken from the surface reveals the presence of other elements beside Ti and O in the layer. The atomic % of P, K, Co and Mn evaluated from the obtained spectra not taking into account the presence of O, H and B and the ratio of P/Ti are shown in table 2. It is seen that with increasing current and consequent operating voltage the content of Mn, Co and the amount of P and the P/Ti ration increase, whereas the amount of K decreases. The concentration of titanium in the surface valleys of specimen 1 (spectrum 3) is naturally higher than on the peaks (spectrum 2) as shown in figure 3(a). For the other oxide systems this difference is not so noticeable, but at the same time, the content of dopant metals on the peaks (spectrum 2) of the relief is higher than in valleys (spectrum 3) as it follows from figures 3(b) and 3(c).

A series of diffraction lines for α-Ti and TiO_2 on X-ray diffraction patterns for specimens 1, 2, 4 was obtained as shown in figure 4 which is in accordance with [21]. It should be noted that the analysis of the X-ray spectra obtained by the routine procedure shows the presence of the α-P_3Ti_5 in the 1 and 4 specimens. Furthermore, one can find small peaks at angles $2\theta \sim 37°$ and change in intensity double peaks at $2\theta \sim 76°$ on X-ray diffraction patterns of systems $TiO_x \cdot CoO_y$ and $TiO_x \cdot MnO_y$. Taking into account the data from the EDS, it may be concluded that this difference may be attributed with Co or Mn incorporation in titanium oxide matrix. At the same time the low concentration of these elements does not allow to determine the phase composition with high accuracy.

IOP Conf. Series: Materials Science and Engineering **369** (2018) 012019 doi:10.1088/1757-899X/369/1/012019

The results of testing the catalytic activity of the mixed oxide coatings with show that the ignition temperature T_i corresponding to the top efficient operation of a catalyst is in the range from 510 K to 520 K for $TiO_x{\cdot}MnO_y$ systems, obtained by PEO. Parameter T_i is slightly higher than for platinum-based catalysts. The complete conversion of CO at $TiO_x{\cdot}MnO_y$ surface is achieved at T_c=670 K (table 3). As can be seen from the data in table 3 mixed oxides $TiO_x{\cdot}CoO_y$ inferior in catalytic activity the $TiO_x{\cdot}MnO_y$ system as it was observed in [18,19]. This is due to variability of manganese oxidation numbers and more developed and a rough surface of mixed manganese containing oxide. Obtained results are in accordance with the data [15–17] and advantage of this work is that observed in this paper coatings do not need additional impregnation. Thus above data confirm the efficiency of materials as catalysts for the CO conversion and gaseous wastes neutralization.

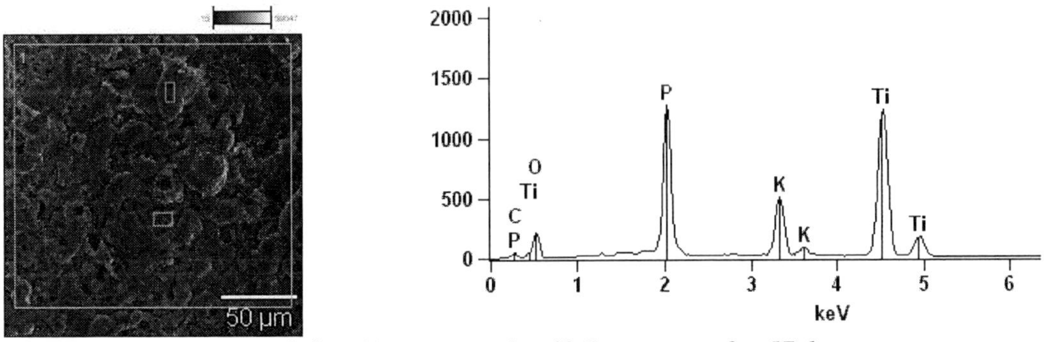

ω(Ti), at%: spectrum 2 – 49.5; spectrum 3 – 57.6

(a)

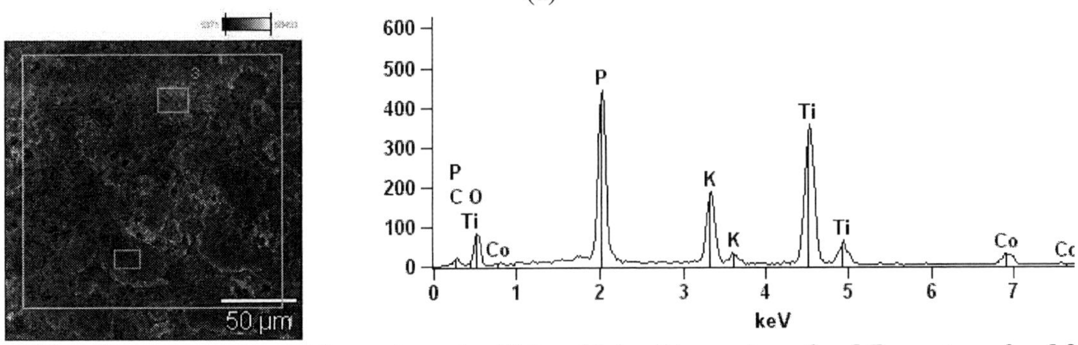

ω(Ti), at%: spectrum 2 – 40.9; spectrum 3 – 51.2; ω(Co), at%: spectrum 2 – 5.7; spectrum 3 – 5.2

(b)

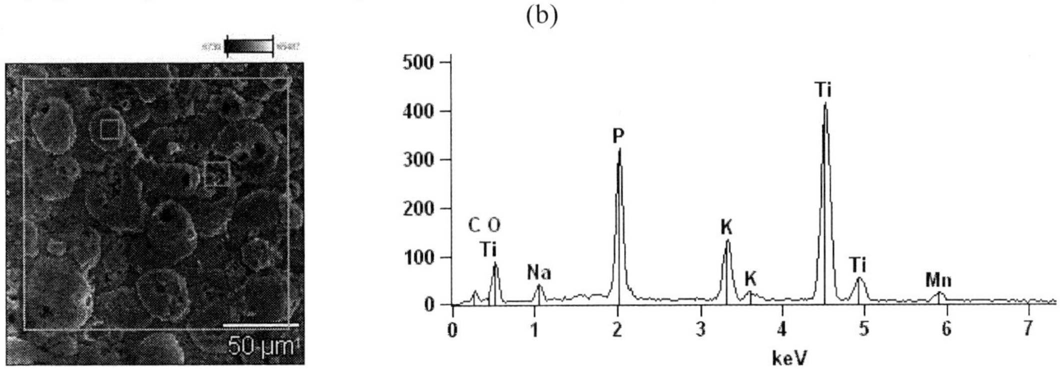

ω(Ti), at%: spectrum 2 – 46.6; spectrum 3 – 46.4; ω(Mn), at%: spectrum 2 – 5.0; spectrum 3 – 4.6

(c)

Figure 3. Surface morphology, X-ray spectra taken from the surface and composition of oxide coatings at hills (spectrum 2) ad valleys (spectrum 3) for samples: 1 (a), 2 (b), 4 (c).

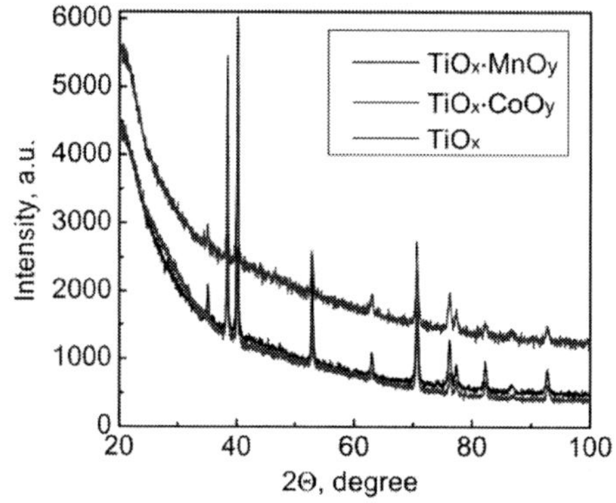

Figure 4. Diffraction patterns of PEO coatings.

Table 3. The catalytic activity of the materials in the CO oxidation reaction.

Catalyst material	The content of the active ingredient ω, at.%	Ignition temperature T_i, K	Temperature of complete conversion T_c, K
Pt [22]	100	490	635
Pt_{exp}	100	495	650
$TiO_x{\cdot}CoO_y$	Co – 6.3	565–570	700–710
$TiO_x{\cdot}MnO_y$	Mn – 5.0	510–520	670–675

Table 4 shows the results of the open circuit potential E_{oc} measurements of samples immersed in 2M NaOH and Ringer solutions. It is seen that PEO substantially increases the corrosion resistance, as follows from the shift of the E_{oc} into the positive direction. The lowest corrosion resistance in both solutions reveals $Ti{\cdot}TiO_x$ system (sample 1) although it is significantly higher than unoxidized titanium (sample 0).

Table 4. Open circuit potential of oxide systems in different media.

Sample		E_{oc}, mV in solutions	
Code	Composition	2M NaOH	Ringer solutions
0	Ti	−600	−480
1	$Ti{\cdot}TiO_x$	370	100
2	$TiO_x{\cdot}CoO$	580	900
4	$TiO_x{\cdot}MnO_y$	450	580

Passivity of doped oxide systems in an alkaline medium is explained by the basic nature of cobalt and manganese oxides. At the same time, it should be noted that manganese oxides are less resistant in chloride-containing media compared to cobalt oxides. Some conclusions may be drawn from the comparison of corrosion behavior data with the surface layers topography. The highest positive open circuit potential of sample 2 may be associated with the closely packed doughnut-like grains filled with small grains, cf. as shown in figure 3(b). The lowest corrosion resistance of sample 1 may be associated with the cracking of the formed layer, which promotes the penetration of electrolyte to the substrate. Specimen 4 has the loosest structure, which may provide lower corrosion resistance as compared with sample 2.

4. Conclusions

- The oxide layers containing the alloying elements might be formed on Ti by means of the anodic polarization in the spark-discharge regime (at current densities 2–2.5 $A \cdot dm^{-2}$).
- The chemical and phase composition as well as the topography, the microstructure of the formed layers can be varied by modification of the electrolytes and by the altering the applied current density.
- The high corrosion resistance of mixed oxide systems in alkaline and in Ringer solutions has been estimated.
- Mixed oxide coatings are characterized by the developed surface and high catalytic activity in the carbon (II) oxide conversion reaction.

References

[1] Bagheri S, Muhd J N and Bee A H S 2014 *Sci. World J.* **1** 1-21
[2] Anpo M and Kamat P 2010 *Environmentally Benign Photocatalysts: Applications of Titanium Oxide-based Materials* (Springer Science) pp 234-97
[3] Rudnev V, Morozova V, Kaidalova T and Nedozorov P 2007 *Russ J Inorg Chem* **52** 1350-4
[4] Lukiyanchuk I, Rudnev V and Tyrina L 2016 *Surf Coat Technol* **307**(C) 1183-93
[5] Lukiyanchuk I, Chernykh I, Rudnev V, Ustinov A, Tyrina L, Nedozorov P and Dmitrieva E 2014 *Prot Met Phys Chem Surf* **50** 209-17
[6] Zaleska A 2008 *Recent Patents on Engineering* **2** 157-64
[7] Khairy M and Zakaria W 2014 *Egyptian Journal of Petroleum* **23** 419-26
[8] Sakhnenko N, Ved' M and Bykanova V 2014 *Func mater* **21** 492-7
[9] Ma X, Zhou W and Chen Y 2017 *Materials* **10** 631
[10] Yar-Mukhamedova G, Belyaev D, Mussabek G and Sagyndykov A 2017 *SGEM 2017* **17** 107-14
[11] Gupta P, Tenhundfeld G, Daigle E and Ryabkov D 2007 *Surf Coat Tech* **201** 8746-60
[12] Wu H, Lu X and Long B 2005 *Mater Lett* **59** 370-5
[13] Yerokhin A, Nie X and Leyland A 1999 *Surf Coat Technol* **122** 73-93
[14] Parfenov E, Yerokhin A, Nevyantseva R, Gorbatkov M, Liang C and Matthews A 2015 *Surf Coat Tech* **269** 2-22
[15] Rudnev V, Vasilyeva M, Kondrikov N and Tyrina L 2005 *Appl Surf Sci* **252** 1211-20
[16] Vasilyeva M, Rudnev V, Ustinov A, Korotenko I, Modin E and Voitenko O 2010 *Appl Surf Sci* **257** 1239-46
[17] Lukiyanchuk I, Rudnev V, Tyrina L and Chernykh I 2014 *Appl Surf Sci* **315** 481-9
[18] Sakhnenko N, Ved M, Karakurkchi A and Galak A 2016 *Eastern-European Journal of Enterprise Technologies* **3** 37-43
[19] Sakhnenko N, Ved' M, Androshchuk D and Korniy S 2016 *Surf Eng Appl Electrochem* **52** 145-51
[20] Yar-Mukhamedova G, Ved' M, Karakurkchi A and Sakhnenko N 2017 *IOP Conference Series: Mater Sci and Eng* **213**
[21] Rokosz K, Hryniewicz T, Raaen S, Chapon P and Dudek L 2017 *Surface and Interface Analysis* **49** 303-15
[22] Snytnikov P, Belyaev V and Sobyanin V 2007 *Kinet Catal+* **48** 93-102

Degradable fiber and its application in horizontal well acid-fracturing

H W Zhang[1,3], X M Yan[1], M Y Cui[1], C Liang[1], J Yan[1], Y P Sun[1], H B Li[1] and S Z Zhang[2]

[1]Research Institute of Petroleum Exploration & Development, Beijing 100083, China
[2]PetroChina Tarim Oilfield Company, Xinjiang Korla, 841000, China

E-mail: etczhanghw@163.com

Abstract. In order to acidize whole horizontal well interval uniformly, temporary plugging technology was chosen as one of the most effective methods. Laboratory tests were also carried out to evaluate their diverting properties. The results show that clean self-diverting acid could increase its viscosity during reaction with carbonate, and which made it divert to other layers. And for fiber, it has good temporary plugging and degradable performance, which can meet the demand of uniform acidification and lower damage to oil reservoir. According to the analysis of fiber characteristic, some using principles were also given. This technology was applied on ten wells, and temporary plugging and diverting display could be clearly observed by observing pumping pressure change. Furthermore, final production rate after stimulated was increased obviously. All these results showed that this technology could get a better application effect in horizontal well acidification.

1. Introduction

When performing acidification on the carbonate reservoir, the corrosion rate of acid liquid is high on the core no matter the concentrate of the used hydrochloric acid (15% or 20%), for the high content of calcite. Under the same pressure system, the acidification usually effects along high-permeability zone and large pore path, and can hardly works on reservoirs with low permeability and heavy pollution [1]. Especially for horizontal wells of the long well section, the acid rock reaction is fast, the horizontal section is long, anisotropy exists and a lot of wells are completed with open wells and screen tubes, and general acidification fails to evenly form the whole horizontal section [2]. From 1930s, the research and application of diverting acidification and fracture technology has been started domestically, including mechanic diverting, foam diverting, particle temporary plugging and high polymer gel diverting. In most technologies, the special property of diverting agent is used under specified conditions for temporary plugging of high-permeability zone, so the working liquid is forced to divert into the low-permeability zone to increase the reservoir production ratio and well production rate [3-5].

Commonly used methods onsite are liquid chemical diverting and solid particle temporary plugging diverting. The chemical diverting works by increasing viscosity of acid liquid for temporary plugging [6,7]. Besides, adding solid particle into the pumped liquid is a commonly used method, too. Currently, fiber agent is commonly used for the following advantages: good compatibility with formation fluid and rock, no secondary pollution on stratum, good chemical stability in pumping process, good compatibility with the treating fluid, degradable in stratum, no harm to the stratum [8-11].

In this paper, lab experiments on the performance of temporary plugging diverting, degradation and reservoir protection of these two kinds of materials are shown and construction parameters are optimized by combining the experiment results. The onsite results prove the obvious effects of temporary plugging and production increase if achieved.

2. Clean self-diverting acid system

The viscosity of common diverting liquid mainly comes from the gel of polymer, the interaction between polymer molecules of HMW (amount to millions). This acid system usually brings harm to the reservoir and unsatisfying acidification effect. Along with the development of chemical synthesis and application of surfactant, diverting acid technology of viscoelasticity surfactant started to appear.

The viscosity of surfactant comes from the micellar structure generated by gathering of surfactant molecules. After entering the stratum, the acid liquid reacts with the rock and generates $CaCl_2$ brine. The surfactants interact with the brine and turn into gel aggregate from individual ions, while long spiral micellars are generated. When the solution property changes suddenly, the number of micelle increases obviously and huge layered micelle, gel, is formed. The gel can increase the flow resistance of the acid in the pore path and plug its flowing and force to acid liquid turn to other pore paths with less flow resistance. After acidification, the produced oil gas or mutual solvent contact with long micellars and turn them into spherical micellars, which apparently decreases the viscosity of the residue acid and facilitate the rapid flowback of residue acid and effective clean of the acid-ething paths.

As shown in the process of viscosity increase and gel breaking, the fresh has low viscosity less than 3 mPa.s. However, with the proceeding of the acid-rock reaction, the pH value increases and the viscosity of the residue acid is high and can even reach 1400 mPa.s when pH=2.5, where the pH value keeps increasing but the viscosity remains changes. See figure 1 for the test results.

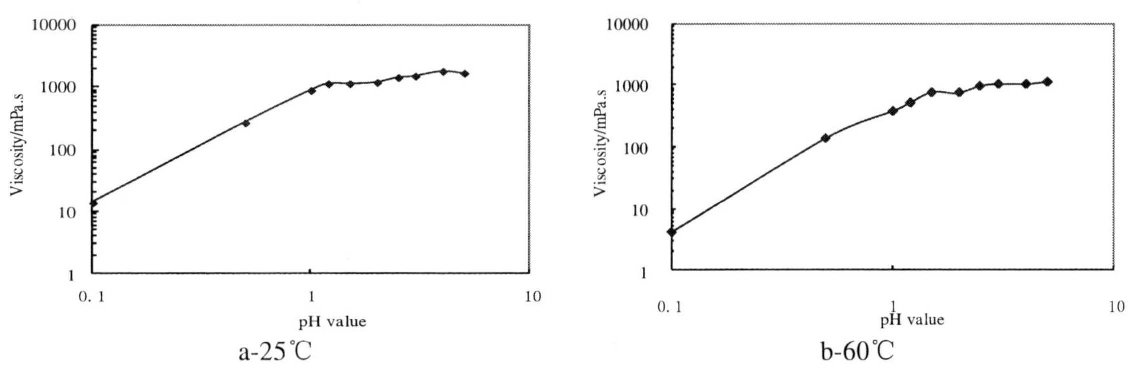

Figure 1. Relationship between viscosity and residual aicd concentration under 25℃ and 60℃.

To evaluate the diverting performing of the acid liquid, the three-core parallel connection equipment in figure 2 is used as shown in figure 2. The used rock core has a diameter of 2.5 cm and length of 6cm with permeability ratio of adjacent rock cores of 0.5. The test liquid is injected under uniform pressure. Steps for test are: saturating the rock core by brine, measuring the permeability of the rock core (the actually measured permeability are 0.5 md, 1.0 md and 2.0 md), injecting the acid liquid (conventional acid and diverting acid) until the rock core breaks, saturating the rock core by brine again, measuring the permeability of the rock core after acid ething, perform nuclear magnetism scanning on rock core section and observe the wormhole pattern the diverting effect.

As the viscosity of the diversity acid increases after entering the high-permeability rock core, the injecting pressure of acid liquid raises apparently (figure 3), indicating that the diverting acid works well on increasing the flow resistance and diverting the acid liquid.

IOP Conf. Series: Materials Science and Engineering **369** (2018) 012020 doi:10.1088/1757-899X/369/1/012020

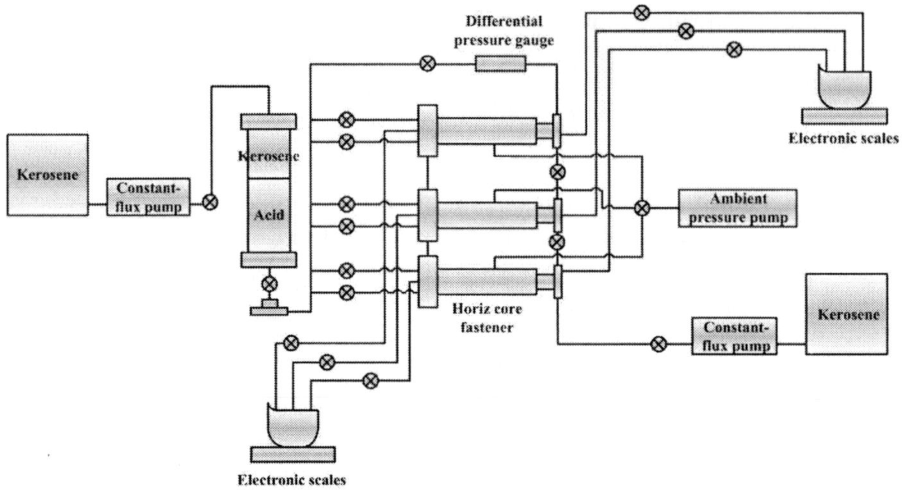

Figure 2. Experimental equipment of self-diverting acid flowing in core.

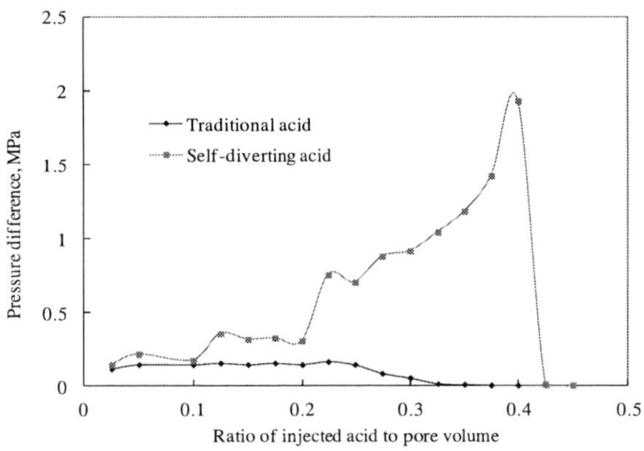

Figure 3. Experimental results of injection pressure with different acid.

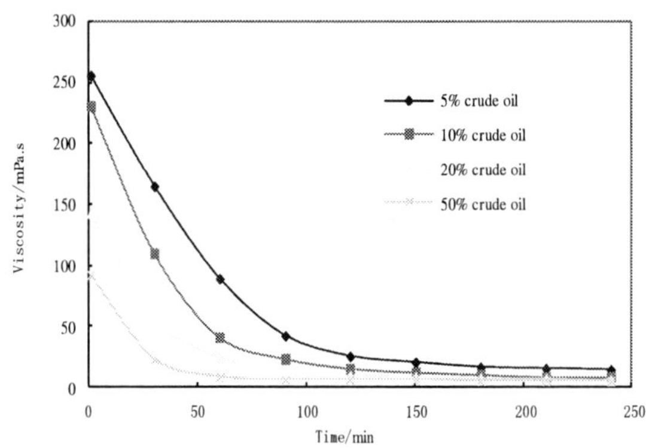

Figure 4. Variation of residual acid viscosity.

By using the cone-plate test system, the relationship between viscosity and time for residue acid after gel broken is measured under the shearing rate of 170 s^{-1}. After acidification, the residue acid gel breaks once it contacts with the oil gas. Add different quantity of light crude into the gel of the diverting residue acid and mix them together (figure 4). As for the gel generated by the cleaning diverting residue acid shall break its structure completely after meeting with the crude but keep its structure when meeting with water except that a large amount of water is used for attenuation and damage its structure and discharge its viscosity, the acid system has a combined function of water controlling.

3. Biodegradable fiber

The leak-off control fiber (LCF) is a kind of artificial fiber, which is made of organic polymer by hot melt, drawing, surface coating, chopping and other processes. It is cylindrical with an average length of 4-8 mm and a diameter of 10-20 μm. Table 1 showed its basic properties. This thin material is flexible and can form compact three-dimensional network structure after entering fractures and pores. The action principle is: the fiber itself has the function of bridging into a network and flexible transformation; therefore as the temporary plugging fiber enters an acid-ething fracture or wormhole, a compact "filter screen" takes shape. As a result, the flow resistance of acid liquid is increased, and the purposes of temporary plugging and increasing the diverting efficiency of acid liquid are completed.

Table 1. Leak-off control fiber properties.

Property classification	Test results	
	70℃	90℃
Strength (cN/tex)	4～4.4	6.3～8.8
Extension (%)	12～36	10～28
Elastic Modulus (cN/tex)	22～62	55～114
Flexible degree of recovery (%)	70～85	72~85
Collusion strength (cN/tex)	2.6～4.6	4.4~5.1
Knot strength (cN/tex)	2.1～3.5	4.0~4.6

Disperse the fibers into the water base liquid evenly in onsite construction, and a strong three-dimensional network structure is formed in the dropout zone to control the loss so that the fractures and high-porosity stratum with large dropout can be effectively plugged in the acid fracturing and acidification. Based on different types of fracture width and pores, the temporary plugging system is used for the plugging experiment. See figure 5 for the experiment equipment.

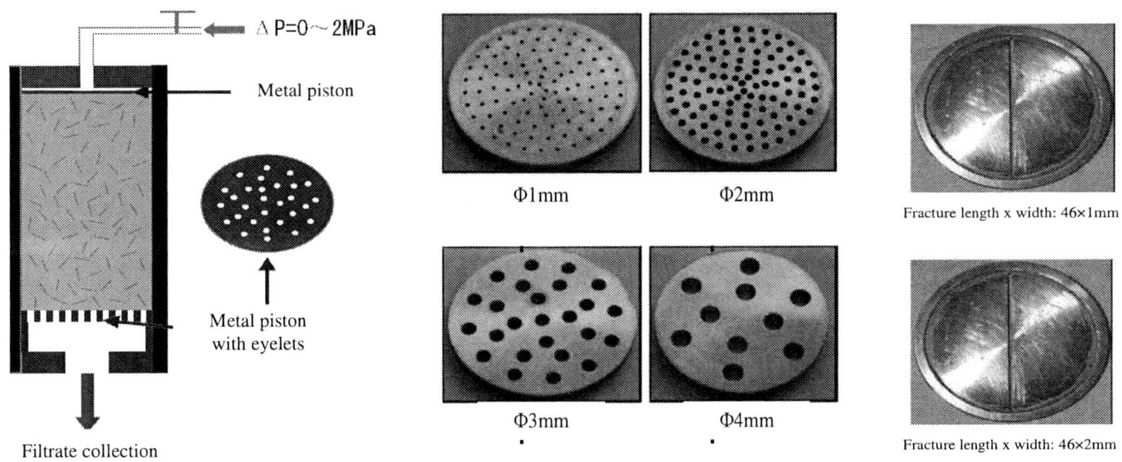

Figure 5. API leakage apparatus and different types of metal pistons.

The leakoff volumes in 30 minutes are measured under normal temperature and differential pressures of 0.5 MPa, 1.0 MPa and 2.0 MPa (table 2).

Table 2. Leak off of base fluid and fluid with fiber.

Fracture length × width (mm×mm)	Pressure difference/MPa			
	0.5	0.5	1	2
	Leakoff volume of base fluid (ml)	Leakoff volume of fiber temporary plugging liquid (ml)		
46×0.5	25 800	26.5	11.2	4.5
46×1	42 500	46.5	18.8	8.5
46×2	86 500	56.7	24.5	11.4
Φ1mm pore plate	39 800	32.9	15.6	7.3

As shown in the result, fiber has obvious plugging effect and the increase of differential pressure can also increase the plugging performance of fiber. In analysis, the increase of differential pressure accelerates the formation of temporary plugging layer and reduces further leakoff of liquid. To verify this conclusion, experiments on temporary plugging layer formed under differential pressures of 0.5 MPa, 1.0 MPa and 2.0 MPa were carried out. The experiment result shows that high differential pressure can accelerate the leakoff of temporary plugging liquid and forms compact plugging layers at the opens of fractures and pores rapidly. Therefore, in actual acidification, after injecting of fiber temporary plugging liquid, the construction displacement shall be increased rapidly, so as to make the base liquid leak rapidly and form a compact plugging layer so that the acid liquid is diverted to other stratum.

a-before degradation b- after degradation

Figure 6. Fiber dispersion in different kinds of fluid and experimental photos after degradation.

To assess the compatibility and degradation of fiber in different kinds of liquid, five types of liquid are prepared: clean water, temporary plugging liquid, guar gum, clean self-diverting acid and gelled acid with same amounts of fiber added (figure 6). The test result shows, the liquid after degradation are all basically clean, indicating that fiber has good biodegradability. However, the compatibility of fiber is best in clean self-diverting acid and temporary plugging liquid and is second in clean water followed by that in guar gum and gelled acid. As for the guar gum and gelled acid has polymer, there is residue separated out after degradation.

By using residue liquid degraded by the above five kinds of liquid, rock core experiment method is used to assess the permeability damage on rock core by residue liquid degraded by fiber in different liquid mediums so as to verify its performance on stratum protection. The result shows that the recovery rate of permeability of rock core after damaged by residue liquid is 78.32%-95.71%, where the value for clean self-diverting acid is 95.71% as the biggest, indicating that this fiber shows good protective performance after degraded in clean self-diverting acid.

Based on the results, following principles shall be observed in field application: ① add fiber by agitator tanks of the fracturing blender truck; ② mix the fiber with a dosage of 1% to 1.5%; ③ determine the adding speed based on the usage dosage and displacement of fiber-carried fluid; ④ the pumped displacement of subsequent acid liquid shall be larger than twice that of fiber and the fiber-carried liquid; ⑤ multistage injection of fiber-carried liquid and acid liquid can be realized based on the section number of stratum and the degree of heterogeneity, so as to achieve temporary plugging diverting.

4. Field application

Oil field K in Kazakhstan is buried with an average depth of 4335 m and a maximum depth of 6000 m. This oil field has high limestone content and an average calcite content of 75% and bad stratum physical property. As its average porosity is 7.2% and the average permeability is 0.098×10^{-3} μm^2, the stratum is the type of low porosity and extra-low permeability. With strong heterogeneity, the stratum is divided into high-productive zone and low-productive zone. The micro pore structure contains mainly throat path and natural fractures. The original stratum temperature is 91℃, which belongs to mesotherm stratum. The formation pressure coefficient of the original stratum is as high as 1.8 to 1.9, which however decreases rapidly, now being lower than 1.0. Horizontal wells and screen tubes are mainly used for horizontal wells of this oil field. The length of horizontal wells is 106 m to 711m and the average horizontal section of 33 horizontal wells is 413 m.

For the rapid reaction between hydrochloric acid and carbonatite and the acid liquid tends to flow in the direction with the least resistance, so general acidification can only changes local spaces close to the horizontal section bottom. For any well with a horizontal section not more than 100 meters, uniform acidification and pollution remove of the whole well section can not be achieved. The test result shows that the diverting acid can works efficiently for rock core with a permeability ratio more than 2. Therefore, for Oil Field K with an average horizontal section length of 413 m and average maximum single-well permeability ration over 100, the temporary plugging diverting technology is an effective solution for uniform transformation.

In 2015, clean self-diverting acid and temporary plugging fiber were used in the guiding experiment on Well A, the first horizontal well. By using screen pipe completion, the average section has a length of 669 m. All together, there were 34 effective sections with and effective thickness of 120 m and a maximum permeability ratio of 160 in the well logging interpretation. Two temporary plugging were proposed to be performed for the horizontal section based on the physical property of stratum. Then the displacement is immediately increased after pumping temporary plugging liquid with fiber is completed based on the lab experiment result, so as to achieve fast leakoff of the base liquid and formation of fiber plugging layer.

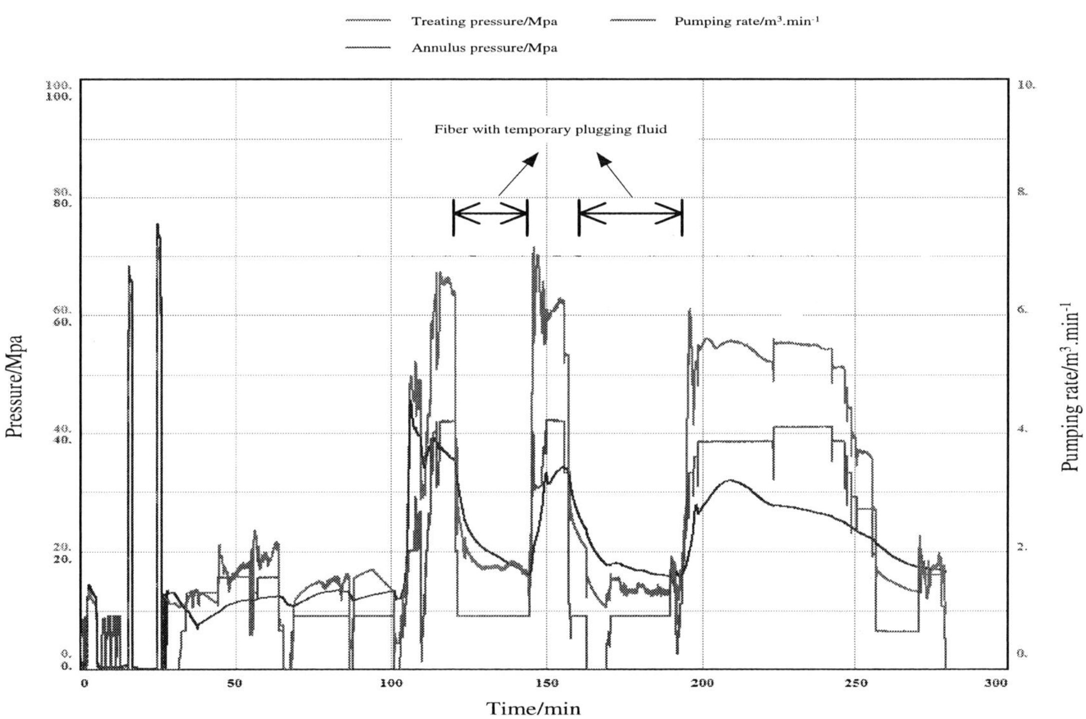

Figure 7. Acid-fracturing curve of Well A temporary plugged by clean self-diverting acid and fiber.

On Oct. 27, 2015, the construction was completed; temporary plugging fiber of 0.68 t was pumped in by a fracturing blender truck twice with a pumped displacement of 0.9 m^3/min. After pumping, the displacement was increased to 4 m^3/min. See figure 7 for construction curve of temporary plugging acid fracturing of Well A. For the applying of oil casing for connecting and closing the casing valve and monitoring the pressure change, the well bottom pressure that the casing presented was more accurate. As shown in the construction curve, after the pumped fiber temporary plugging liquid entered into the stratum, the casing pressure kept increasing and dropped after a certain point, indicating that the plugging diverting of liquid and connecting to the new stratum is successful.

After acid fracturing, the well blowing daily output is 183 t/d and cumulative increase of oil is 45 974 t in 300 days with average increasing crude of 153.2 t/d, indicating that good experiment effect is obtained. After the success of horizontal well A, this process is promoted for vertical and horizontal wells of open-hole completion and screen completion. In 2015, the transformation was applied to 6 wells in all with cumulative increasing oil of 23 292 t, average daily increasing oil of single well of 51.64 t/d; in 2016, the transformation was applied to 4 wells in all with cumulative increasing oil of 15 486t, average daily increasing oil of single well of 42.31 t/d, indicating that significant economic benefits were obtained.

Meanwhile, promotional experiments on wells with great physical property difference and uneven production profile and acid-fracturing transformation has become one of the major transformation processes of Oil Field K and related oil fields in Kazakhstan.

5. Conclusions

- By experiment, the clean self-diverting acid is featured in favorable diverting, retardance and low damage, which fit the carbonate reservoir acidification transformation of Oil Field K in Kazakhstan with strong anisotropism and long horizontal well section.
- Degradable fiber can form compact barrier bed, which is an effective method for realizing the intrastratal and interlayer diverting and even transformation. As shown in the experiment result, fibers can reduce the leakoff and form the barrier bed; the greater the displacement is, the greater the pressure difference is, the faster the leakoff of base liquid for temporary plugging fibers, the easier the plugging works and the acid liquid is diverted. Therefore, in actual acid pressure construction, the working displacement shall be increased rapidly once the injection of temporary plugging liquid is injected.

References

[1] Zhang X H, Wang J Q and Qi F L 2008 Study and application of temporary plugging acidizing in heterogeneous reservoir Drilling & Production Technology **30** 58-60, 64

[2] Yuan B, Su Y L, Feng Z T, *et al* 2013 Productivity distribution and flow characteristics of volume-fractured horizontal wells *Journal of Shenzhen University of Science and Engineering* **30** 545-50

[3] Wu Y, Chen F and Cheng N 2008 Enhancing re-fracturing effect with artificial temporary plugging technology *Drilling & Production Technology* **31** 59-61

[4] Gerard G, Valerie Y, Robert R, *et al* 2010 Fluid-diversion monitoring: the key to treatment optimization *SPE European Formation Damage Conference* (Scheveningen)

[5] Kalfavan L J and Martin A N 2009 The art and practice of acid placement and diversion: history, present state and future *SPE Annual Technical Conference and Exhibition* (New Orleans)

[6] Leal J J A, Malik A R, Nunez G W, *et al* 2010 Field trials of a novel fiber-laden self-diverting acid system for carbonates in Saudi Arabia *SPE Deep Gas Conference and Exhibition* (Manama)

[7] Ai K, Li Q D and Yuan Z P 2008 Application of acid fracturing technology with visco-elastic surfactant acid in Tahe oilfield *Oil Drilling & Production Technology* **30** 71-4

[8] Cohen C E, Tardy P M J, Tim Lesko, *et al* 2010 Understanding diversion with a novel fiber-laden acid system for matrix acidizing of carbonate formations *SPE Annual Technical Conference and Exhibition* (Florence)

[9] Martin F, Quevedo M, Tellez F, *et al* 2010 Fiber-assisted self-diverting acid brings a new perspective to hot, deep carbonate reservoir stimulation in Mexico *SPE Latin American & Caribbean Petroleum Engineering Conference* (Lima)

[10] Jairo L J, Ataur R M, Solares J R, *et al* 2011 Successful application of novel fiber laden self-diverting acid system during fracturing operations of naturally fractured carbonates in Saudi Arabia *SPE Middle East Oil and Gas Show and Conference* (Manama)

[11] Tomislav B, Nihat M G, Jairo L J, *et al* 2012 Stimulation strategies to guard against uncertainties of carbonate reservoir *SPE Saudi Arabia Section Technical Symposium and Exhibition* (Al-Khobar)

IOP Conf. Series: Materials Science and Engineering **369** (2018) 012021 doi:10.1088/1757-899X/369/1/012021

Aging behaviour and properties evaluation of high-density polyethylene (HDPE) in heating-oxygen environment

B Zhao[1], S Zhang[1], C Sun[1], J Guo[1], Y X Yu[1] and T Xu[1]

[1]China Special Equipment Inspection and Research Institute, Hepingjie, Xiyuan, Chaoyang District, Beingjing, 100029, P R China

E-mail: zhaobo19840626@163.com

Abstract. The high-density polyethylene (HDPE) samples aged in heating-oxygen environment were prepared. The mechanical properties were determined by the tensile test and small punch test (SPT), and the chemical properties were determined by Fourier transform infrared spectroscopy (FT-IR) and gel permeation chromatography (GPC). Meanwhile, microscopically morphology of tensile fracture was observed successively. The result showed that tensile strength increasing continuously with the aging time increasing; meanwhile elongation and maximum load in SPT reached the maximum value at the aging time of 28 days. The carbonyl index on the sample surface increased significantly with the aging time increasing. At the same time, the molecular weight increased until the aging time is 28 days, and then changed slightly. This means that the aging behaviours based on cross-linking increased firstly, and then degradation played a more important role in materials properties change. The microscopically morphology of tensile test samples could prove this point with the amount of crazes decrease, and the brittleness of aged samples increase.

1. Introduction

Over the past years, polyethylene (PE) has been widely used for pipe materials of drinking water and gas, and it has been reached 50% in the construction of medium and low pressure pipe network [1-3]. The high-density polyethylene (HDPE) has become the widely used materials in gas or water deliver pipe since 1950s [4-5]. HDPE pipe has many excellent properties, such as corrosion resistance in soil, economic performance andlow price. But the other fact is that aging effect in multiple environments was one of the most important limitations to it. Nowadays there were many studies on the aging process in ultraviolet (UV) radiation, but fewer studies in heating-oxygen environment [6,7]. Most scholars think that the process of polymer had a low reaction rate and unsustainable change tendency of early stage in heating-oxygen environment [8-12]. However, the mechanism of the variation was still not clear. Meanwhile, most scholars used the full size samples to determine the polymer aging degree. Some articles noted that the aged layer maybe less than 100 μm, and it is very thin comparing with the original substrate , would result for full size samples not satisfy the evaluation condition of the aged polymer properties [13-15]. Therefore, we want to seek a method replaced full size sample, and it could show more information of the properties on aged layer, that is more suitable for the aged polymer evaluation.

2. Experimental preparation

2.1. Materials preparation
The HDPE raw material is M7600. It is a type of PE 80 and produced by SINOPEC Yanshan Company. Bar-extruded method for all samples. The heating-oxygen environment provided by a heating cabinet. The aging condition making reference to others' research was kept at constant temperature of 90±1℃ with air circulating, and 7 days was one cycle for calculating the aging duration. The aging durations are 0 to 8 cycles, and the aging time of 0, 7, 14, 21, 28, 35, 42, 49 and 56 days [16].

2.2. Mechanical properties test

2.2.1. Tensile test. The tensile samples were conducted on dumbbell shape, referring to ASTM D638 (type I), and 5 parallel samples were prepared in each aged case. This experiment was tested in a the universal mechanical tester GWT4503, provide by SANS , and with the extensometer DBX-800, also provided by SANS. Loading speed in this test was 2mm/min, and the room temperature was 25±3℃ in test.

2.2.2. Small punch test （SPT）. SPT was referred to ASTM F2183. The samples were conducted on disk samples, and 4 parallel samples were prepared in each aged case at least. The SPT sample was 10mm in diameters, and 1.0 mm in thickness. The loading system was similar as the tensile test, and loading speed was 2 mm/min. The structure diagram of SPT test device is shown in figure 1. For this device, it was 2.5 mm in punch diameter, and 4mm in diameter of hole in the base with 0.2 mm chamfer radius.

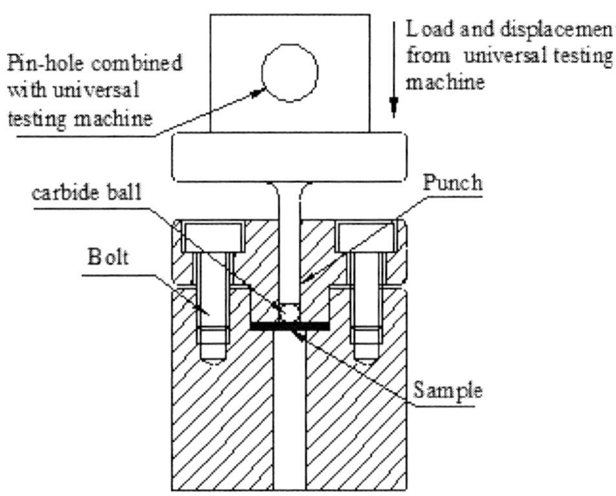

Figure 1. Structure diagram of SPT test device.

2.3. Chemical properties

2.3.1. FT-IR measurement. ICarbonyl index of aged HDPE in 28 d and 56 d were detected to evaluate the aging degree of HDPE We use the Bruker LUMOS IR-Microscope with Attenuated Total Reflection (ATR) mode to test it. For surface evaluation, the detection points were random choosed, and for the depth evaluation, the measurement of depth were point-by-point scanning from heated surface to interior on sample fracture surfaces after sanding. In order to calculate the carbonyl index accurately, the carbonyl consumption is derived from 1730 cm^{-1} to 1765 cm^{-1}..

2.3.2. GPC measurement. The molar mass distribution was determined by the high temperature GPC of Tosoh HLC-8321. The temperature of the injection block and the column compartment making reference to others' research was set to 145℃ [17]. The gel chromatographic columns with dimensions of 7.8 mmI.D.×30 cm*2 were used, and the flow rate of the mobile phase was 1 ml/min. The polymers were dissolved for 4 h in 1,2,4-trichlorobenzene containing butylated hydroxytoluene as antioxidant at a concentration of 0.5 mg/ml, and a temperature of 145℃, 200 ml of the polymer solution for each injection. The sample was then cut from SPT samples with full thickness.

2.4. Microstructure observation

A scanning electron microscope (SEM) of HITACHI SU 3400 was used to observe microscopically morphology with tensile samples. The observation area was the necking region on the tensile samples' surface, with gold sputtering first.

3. Result and discussion

3.1. Mechanical properties

Figures 2 and 3 show the results of tensile tests of HDPE at different aging time, and figure 2 is a stress-strain curves.

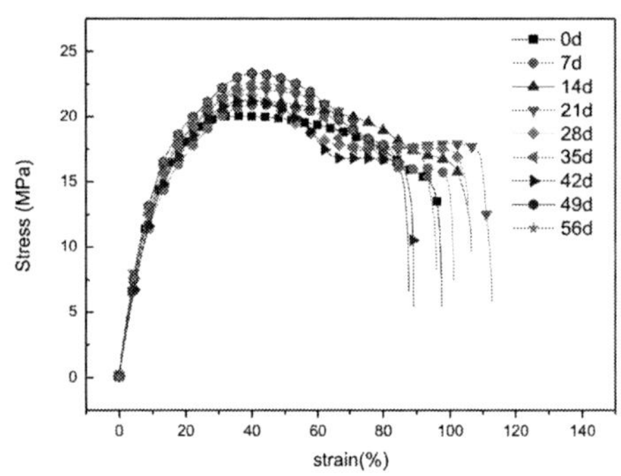

Figure 2. Stress-strain curves at different aging time.

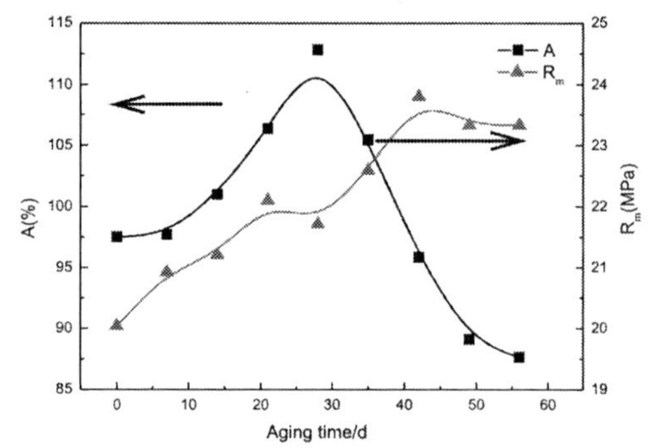

Figure 3. The change of tensile strength and elongation at different aging time.

Figure 2 shows the tensile curves performed very similar at different aging time. There are no obvious yield platform in this curves, and tensile strength (R_m) almost equals to yield strength. The range of plastic deformation is very wide in all of the tensile curves, which show a viscoelastic fracture characteristic.

Figure 3 shows the variation tendency of R_m and elongation at break (A). According to the results, tensile strength (Rm) increased continuously with the increasing of aging time, and it reached 23.33 MPa at the aging time of 49 days and 56 days, which increased about 16% than the original value of 20.04 MPa at 0 day. The variation tendency of elongation was not consistent with the increasing of aging time. At first it was increased, and then decreased slightly with the increasing of aging time at 28 days. The maximum value of the elongation was 112.82% at the aging time of 28 days, which increased about 15% than the 0 day. The minimum value was 87.66% at the aging time of 56 days, which was about 10% decreased. The results show clearly that heating-oxygen environment could cause HDPE aging, with loss of ductility to a certain extent.

Figure 4 shows the load-displacement curves in SPT of aged HDPE samples, and figure 5 shows the maximum load of HDPE with increasing aging time.

It can be seen that the load-displacement curves had similar characteristics for all samples, and the variation tendency of maximum load was regular with increasing aging time.

In this case, the maximum load in SPT increased rapidly firstly, and then declined with the increasing of aging time. The maximum value of maximum load occurred at 28 days, which was 217.5 N, increased about 21% than the 0 day. It must be noted that, the variation tendency of maximum load in SPT is very similar to the elongation in tensile test, and the SPT changed more significantly than tensile tests.It indicated that the SPT might be a more susceptible method to evaluate the HDPE aging degree than the traditional tensile tests in heating-oxygen environment.

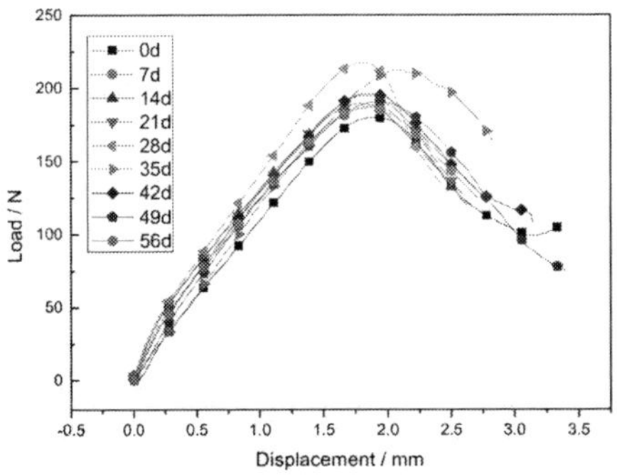

Figure 4. Load-displacement curves in SPT at different aging time.

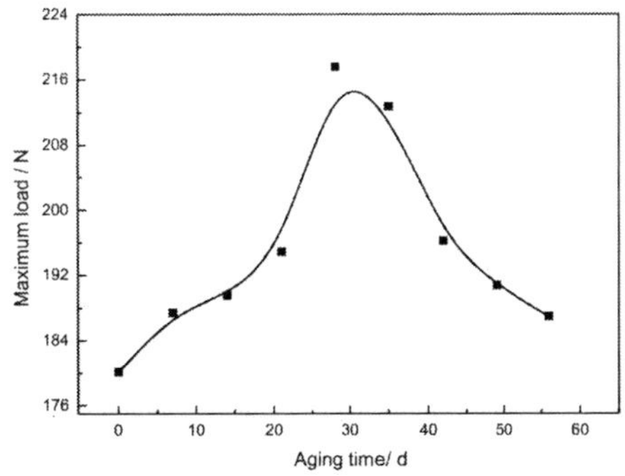

Figure 5. The change of maximum load in SPT at different aging time.

3.2. Chemical properties

Figure 6 shows the number-average molecular weight (Mn) and weight-average molecular weight (M_w) of HDPE at different aging time.

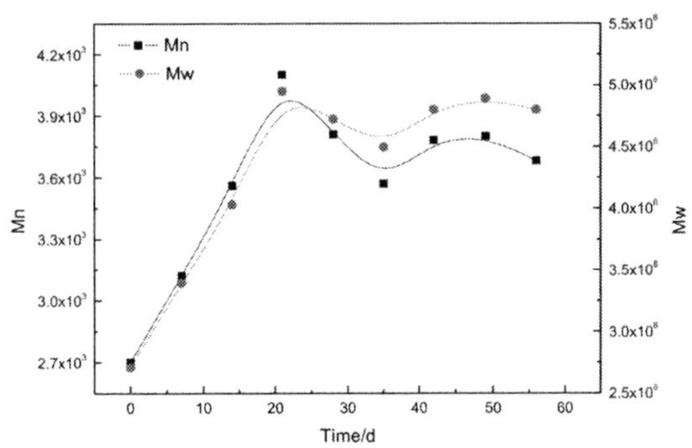

Figure 6. The number-average molecular weight (M_n) and weight-average molecular weight (M_w) at different aging time.

According to the results, HDPE had a high macromolecular chain structure. M_n and Mw performed similar variation tendency with increasing of aging time, and the variation value almost change in the same level. Firstly, the M_n and Mw increased rapidly until the aging time of 28 days, and then basically kept stable. In the stable-stage, M_n kept the value within a range of $3.6\text{-}3.9\times10^3$, and Mw kept in $4.5\text{-}5.0\times10^6$. The results indicated that molecular chains of HDPE may have been broken in this heating-oxygen environment, but not significantly. For the samples made by raw materials without additives, the raise of M_n and M_w may be attributed to the slight cross-linking reaction [18].

Figures 7 and 8 show carbonyl index of HDPE surface at 28 days and 56 days, respectively.

The results showed that carbonyl occurred both on the two HDPE samples' surface with different aging time, and the carbonyl index with 56 days aging (1.53-1.70) was much higher than the 28 days (0.61-1.03). It is well known that carbonyl index can be used as a quota to judge the state of polymer degradation, and is associated with chain scission. So the results indicated that the degree of oxidation and chain scission on HDPE surface would be more and more with the increasing of aging time. As the result shown, the oxygen degradation only occurred on the sample surface, and it could increase the susceptibility of slow-crack-grow, so that elongation was reduced. But for the tensile strength, the force appeared on the cross section, and the sample is hardening for cross-linking all the time. So it increased.

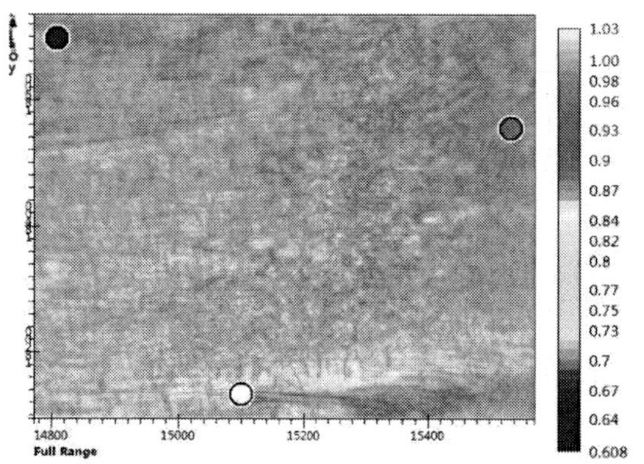

Figure 7. Carbonyl index of HDPE surface at the aging time of 28 days.

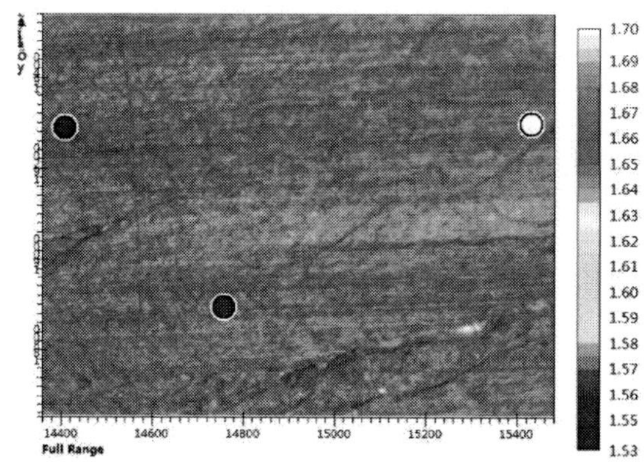

Figure 8. Carbonyl index of HDPE surface at the aging time of 56 days.

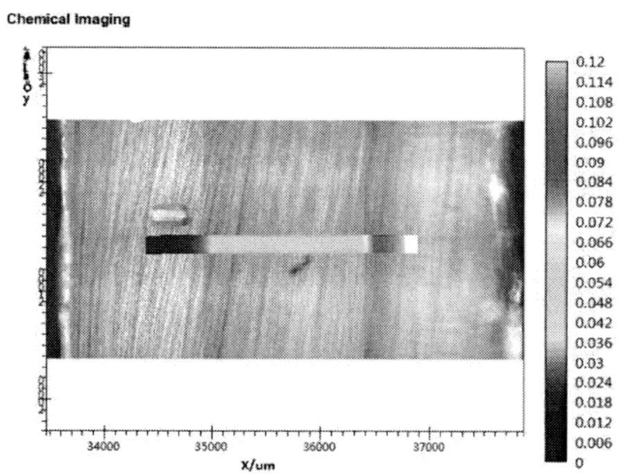

Figure 9. Carbonyl index from surface to interior at the aging time of 56 days.

Figure 9 shows carbonyl index from surface to interior at the aging time of 56 days at HDPE surface.

It can be learned that the maximum carbonyl content location occurred on the sample surface after aging experiment. The carbonyl index declined from the surface to the interior linearly, and nearly disappeared at the center of sample interior. It indicated that the chain scission reaction almost occurred only on a very thin surface layer.

Figure 10 shows the microscopically morphology in necking region of HDPE sample surface in tensile test.

It can be seen that HDPE absorbed energy by an elastic shrinkage perpendicular to the direction of stress at original situation [19,20]. At the same time, it can be seen that a lot of crazes formed in microfibers and cavities, which is perpendicular to the tensile direction, and arranged closely. The density of crazes decreased with the increasing of aging time, and disappeared completely at the aging time of 56 days. At this time the tensile stress causes a brittle fracture on HDPE surface.

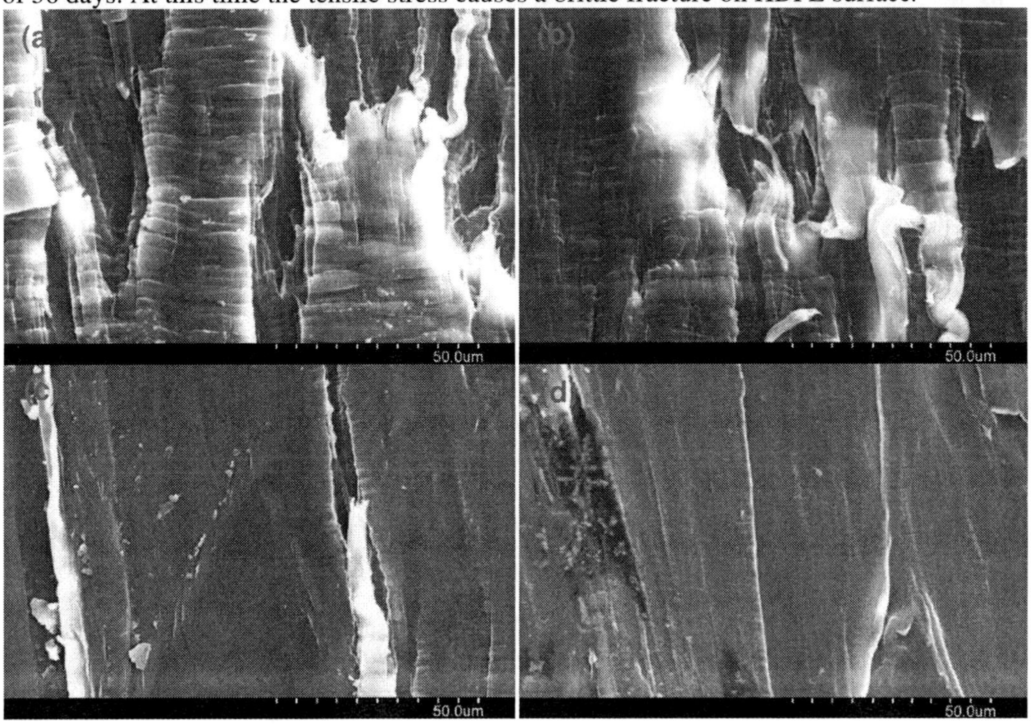

Figure 10. The microscopically morphology in necking region of HDPE surface in tensile test, (a) 0 day aging, (b) 7 days aging, (c) 28 days aging, and (d) 56 days aging.

3.3. Discussion

There are three stages of HDPE aging in the heating-oxygen environment, and the properties changes due to the competition of strengthening and degradation in this environment, as shown in figure 11.

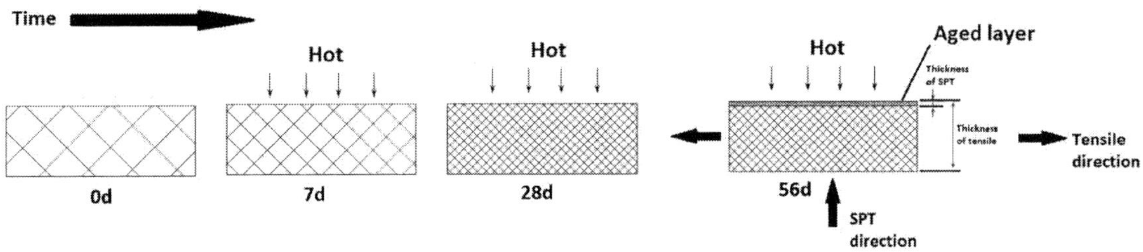

Figure 11. The three stages of HDPE aging in the heating-oxygen environment with the increasing of aging time.

At the aging time of 0 day, HDPE possessed original properties, with low strength and more craze initiation characteristic. After that time, the HDPE aging reaction began in the hot-oxygen environment. For the low thermal activation energy of environment, reactions dominated by radical combination, and cross-linking reaction began in this stage [21]. For the same temperature in the aging process, the reaction had essentially the same strong from sample surface to interior, because it is only relative to the reaction temperature of environment. At the aging time of 28 days, the cross-linking reaction reached the limit in the heating-oxygen environment. Mechanical properties such as elongation at break and maximum load in SPT reached the maximum at this stage, and HDPE possessed properties of high strength and ductility. At the aging time of 56 days, the degradation reactions became explicit on the results. The process of HDPE degradation in heating-oxygen environment can be divide into three stages relating to the reactions of free radical production, bond broken, and recombination., and they were nanmed the radical initiation stage, propagation stage, and termination stage. The chemical reaction show as equations(1)-(6) [22]. The aged layer just situated at the surface, and covered on the original HDPE sample substrate. It could prevent the oxygen degradation spreading into the interior, so that the aged layer was very thin, about only micro-scale. Meanwhile, the aging degree was gradually weakened from the surface to interior of the sample. This phenomenon has been proving by the previous experiments in this article. The chain scission reacted on the thin aged layer, and GPC results came from full thickness samples, so that there was no significant decrease in molecular weight results.

Initiation stage:

$$RH \rightarrow R\cdot \tag{1}$$

Propagation stage:

$$R\cdot + O_2 \rightarrow RO_2\cdot \tag{2}$$

$$RO_2\cdot + RH \rightarrow ROOH + R\cdot \tag{3}$$

Termination stage:

$$R\cdot + R\cdot \rightarrow R—R \tag{4}$$

$$RO_2\cdot + R\cdot \rightarrow ROOH \tag{5}$$

$$RO_2\cdot + RO_2\cdot \rightarrow ROOH + O_2 \tag{6}$$

The result also shows that SPT changed more significantly than tensile tests. The reason of it can be illustrated in figure 11, which shows the effective regions of SPT samples and tensile samples in the test . The aged layer was very thin, which is about micro-scale. As for the SPT test, the thickness of samples is only about 1.0 mm, much thinner than the tensile test samples. The maximum deformation occurred on the age layer in SPT test, and increased the susceptibility of the material crack growth, and its' value decreased rapidly than the traditional tensile test.

4. Conclusion

Mechanical properties of HDPE changed regularly in heating-oxygen environment with increasing of aging time. Tensile strength increased continuously with the increasing of aging time, meanwhile elongation and maximum load in SPT reached the maximum value at the aging time of 28 days for cross-linking reaction. Then degradation of aged layer on the sample surface played a more important role in materials properties, causing it to decrease.

The mechanical properties of aged HDPE varied more significantly in SPT than the traditional tensile test, and it might be a more susceptible method to evaluate the HDPE aging degree than the traditional tensile tests in heating-oxygen environment.

Acknowledgments

The authors thank the China Special Equipment Inspection and Research Institute (CSEI, Government-affiliated institutions of China) and General Administration of Quality Supervision, Inspection and Quarantine (AQSIQ, Government-affiliated institutions of China) for the financial support through projects T2015-CSEI-0003.

References

[1] Whelton A J and Dietrich A M 2009 Critical considerations for the accelerated ageing of high-density polyethylene potable water materials *Polymer Degradation & Stability* **94** 1163-75

[2] Whelton A J, *et al* 2011 Impact of chlorinated water exposure on contaminant transport and surface and bulk properties of high-density polyethylene and cross-linked polyethylene potable water pipes *Journal of Environmental Engineering* **137** 559-68

[3] Dietrich A M, *et al* 2010 Polar and non-polar contaminant transport in new and chlorine aged polyethylene pipes Project 4138 *Water Research Foundation* (Denver) p 131

[4] Ammala A, *et al* 2011 An overview of degradable and biodegradable polyolefins *Progress in Polymer Science* **36** 1015-49

[5] Liu G, *et al* 2012 Innovative photocatalytic degradation of polyethylene film with boron-doped cryptomelane under uv and visible light irradiation *Chemical Engineering Journ*al **213** 286-94

[6] Gijsman P, *et al* 1999 Comparison of the uv-degradation chemistry of polypropylene, polyethylene, polyamide 6 and polybutylene terephthalate *Polymer Degradation & Stability* **65** 433-41

[7] Liauw C M, *et al* 2004 Effect of hindered piperidine light stabilizer molecular structure and uv absorber addition on the oxidation of hdpe. part 2: mechanistic aspects—molecular modeling and electron spin resonance spectroscopy study *Journal of Vinyl & Additive Technology* **10** 159-67

[8] Langlois, *et al* 1995 Thermo-oxidative aging of crosslinked linear polyethylene: stabilizer consumption and lifetime prediction *Polymer Degradation & Stability* **40** 399-409

[9] Tsotsis T K and Lee S M 1998 Long-term thermo-oxidative aging in composite materials: failure mechanisms *Composites Science & Technology* **58** 355

[10] Boukezzi L, *et al* 2008 Observations on structural changes under thermal ageing of cross-linked polyethylene used as power cables insulation *Iranian Polymer Journal* **17** 611-24

[11] Weon J I 2010 Effects of thermal ageing on mechanical and thermal behaviors of linear low density polyethylene pipe *Polymer Degradation & Stability* **95** 14-20

[12] Zhao B, *et al* 2017 A study on small punch test to characterize mechanical property of aged high-density polyethylene (HDPE) Key Engineering Materials **734** 104-15

[13] Jaekel D J, *et al* 2011 Characterization of peek biomaterials using the small punch test *Journal of the Mechanical Behavior of Biomedical Materials* **4** 1275-82

[14] Rodríguez, *et al* 2014 Small punch test on the analysis of fracture behaviour of pla-nanocomposite films *Polymer Testing* **33** 21-9

[15] Lan H Q, *et al* 2016 Thermal oxidative aging laws of PE gas pressure pipes *Natural Gas Industry* **36** 78-83

[16] Kriston I, *et al* 2008 Dominating reactions in the degradation of hdpe during long term ageing in water *Polymer Degradation & Stability* **93** 1715-22

[17] Bi D Z, *et al* 2004 Novel phenomenon of HDPE during oven thermal aging *China Synthetic Rubber Industry* **1** 47

[18] Choi B H, *et al* 2009 Experimental and theoretical investigation of stress corrosion crack (SCC) growth of polyethylene pipes *Polymer Degradation & Stability* **94** 859-67

[19] Ivankovic A, *et al* 2004 Crack growth predictions in polyethylene using measured traction-separation curves *Engineering Fracture Mechanics* **71** 657-68

[20] Han S O, *et al* 2000 Thermal stability of thick tubular HDPE pipe with temperatures and heat transfer fluids *American Chemical Society, Polymer Preprints, Division of Polymer Chemistry* **41** 1106-7

[21] Abdelaal F B and Rowe R K 2014 Effect of high temperatures on antioxidant depletion from different HDPE geomembranes *Geotextiles & Geomembranes* **42** 284-301

Microstructure and critical strain of dynamic recrystallization of 6082 aluminum alloy in thermal deformation

W W Ren[1,2], C G Xu[1], X L Chen[1] and S X Qin[1]

[1]Beijing Research Institute of Mechanical and Electrical Technology, Beijing 100083, China

E-mail: renweiwei2009@126.com

Abstract. Using high temperature compression experiments, true stress true strain curve of 6082 aluminium alloy were obtained at the temperature 460℃-560℃ and the strain rate 0.01 s^{-1}-10 s^{-1}. The effects of deformation temperature and strain rate on the microstructure are investigated; $(-\partial \ln\theta / \partial\varepsilon) - \varepsilon$ curves are plotted based on σ-ε curve. Critical strains of dynamic recrystallization of 6082 aluminium alloy model were obtained. The results showed lower strain rates were beneficial to increase the volume fraction of recrystallization, the average recrystallized grain size was coarse; High strain rates are beneficial to refine average grain size, the volume fraction of dynamic recrystallized grain is less than that by using low strain rates. High temperature reduced the dislocation density and provided less driving force for recrystallization so that coarse grains remained. Dynamic recrystallization critical strain model and thermal experiment results can effectively predict recrystallization critical point of 6082 aluminium alloy during thermal deformation.

1. Introduction

Dynamic recrystallization (DRX) plays an important role in controlling grain size during metal thermal forming. It not only can refine grains, but also control mechanical properties and electrical properties [1,2]. It is now recognized as a powerful tool to control the structure and properties of metals under industrial processing operations [3]. True stress and true strain can be obtained by Gleeble thermal simulation experiment under constant temperature and strain rates. Microstructure and grain size after dynamic recrystallization can be investigated by EBSD. Study of the relationship between the deformation parameters and the microstructure evolution is of great significance for controlling the properties of the material [4].

Deformation temperature and strain rate are two important deformation parameters of materials. The relationship between flow stress and strain is established through $\sigma - \varepsilon$ curve [5]. The occurrence of DRX is indicated by the appearance of a peak in the stress –strain curve. However, DRX is actually initiated before the strain corresponding to the peak stress. This threshold strain is known as the critical strain [6,7]. The difficulty is how to measure the critical strain, where the material begins to recrystallize dynamically. Jonas pointed out that the hardening rate(θ)-stress(σ) curve can be obtained by processing the $\sigma - \varepsilon$ curve. The inflection point of $\theta - \sigma$ curve is the critical stress, the corresponding strain of which is critical strain ε_c. Poliak E I *et al* [8] study shows that the relationship between the hardening rate and the stress is equivalent to the relationship between the logarithm of the hardening rate and the strain. Guo H L *et al* [9] obtained the critical strain model of the dynamic recrystallization of 7075 aluminum alloy through the compression test and simulated the dynamic recrystallization process by Finite Element Method (FEM), the result show the model was verified; Chen X H *et al* [10] studied the effect of temperature and strain rates on microstructure of 7085 aluminum alloy in high temperature and investigated the critical strain model; Shi L *et al* [11] researched on critical strain model and results indicate model is in good agreement with the experiment. The dynamic recrystallization behavior of various aluminum alloy materials has been studied widely. At present there are few studies on the dynamic recrystallization of 6082 aluminum alloy.

The stress-strain curves were obtained alloy based on thermal compression experiment of the 6082 aluminum alloy, $(-\partial\ln\theta/\partial\varepsilon)-\varepsilon$ curve are 2 order polynomial fitted at different temperatures and strain rates, the extremum point of the curve is critical strain of the dynamic recrystallization. According to the experimental data of EBSD grain, recrystallized grain evolution and dynamic recrystallization fraction distribution is researched. The process of dynamic recrystallization can be simulated by FEM base on the experimental data which will be helpful to better control the internal microstructure and mechanical properties of the 6082 aluminum alloy.

2. Experimental procedure

The chemical composition(wt.%) of 6082 aluminum alloy is 0.93 Si, 0.165 Fe, 0.056 Cu, 0.454 Mn,0.65 Mg, 0.117 Cr, 0.002 Zn, 0.03 Ti, The initial average grain size is 60 μm. The specification of raw materials is Φ30 mm×1000 mm, compressed specimens are machined to dimension Φ10 mm×15 mm, both the surface are perpendicular to the axis, the axis is parallel to the compression direction. The forming temperature is 460℃, 500℃, 540℃, 560℃ and the strain rate is 0.01 s^{-1}, 0.1 s^{-1}, 1 s^{-1}, 10 s^{-1}. The operation model of the thermal simulator is Gleeble-3500,heating speed of 1℃. s^{-1}, the specimens are heat preserved for three minutes when up to the set temperature. After the compression is finished, specimens are immediately quenched in water to retain the deformed microstructure. The compressed specimens are sectioned along the axial direction, microstructure is characterized by electro-polishing of the specimens for 10 seconds with voltage 20 V and current 1 A in a solution consisting of 10% perchloric acid and 90% ethanol. By EBSD observation of the internal microstructure, it is grain disorientation greater than or equal to 10° that distinguish the grain boundaries. On account of detection limitation in microstructure [12], the grain misorientation less than 3 degrees is not considered.

3. Results and discussions

3.1. Stress-strain curves

The true stress-true strain curves of 6082 aluminum alloy deformed at different temperatures and strain rates are shown in figure 1. It is seen that the flow stress is increased with increasing strain rate and decreasing temperature. When the strain rates are 0.01 s^{-1} and 0.1 s^{-1}, all the flow curves exhibit a peak and then follow by gradual fall to a steady stress state which is indicative of the occurrence of DRX [13]. In the steady state, the hardening and the softening due to DRX are in a state of balance, flow stress almost remain the same with the increase of deformation. When the strain rates are 1 s^{-1} and 10 s^{-1}, stress-strain curves are in fluctuation. Stress peaks and troughs appears alternately, which shows that the working hardening and the dynamic recrystallization played the dominant role alternately.

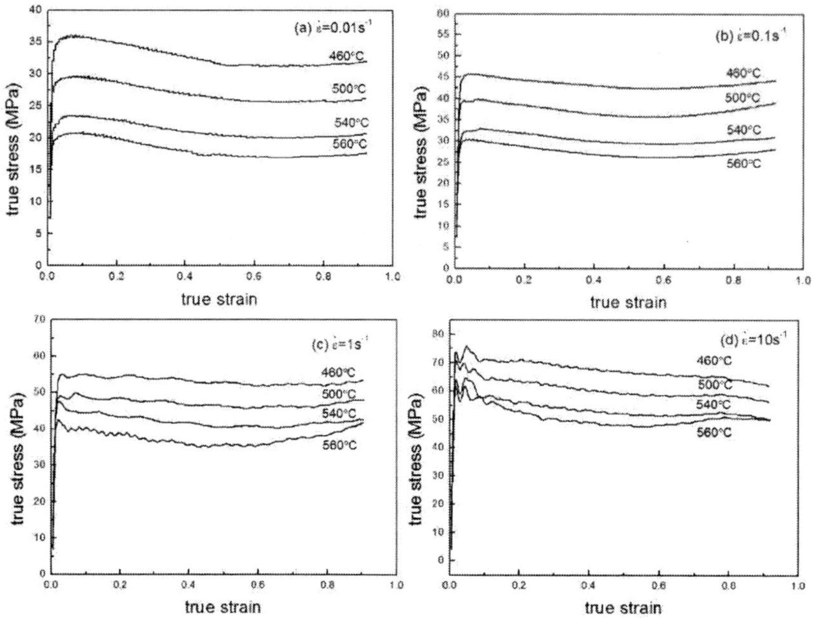

Figure 1. True stress-strain curves of 6082 Al-Si-Mg aluminum alloy at different strain rates o(a)$\dot{\varepsilon} = 0.01$ s^{-1} (b) $\dot{\varepsilon} = 0.1$ s^{-1} (c) $\dot{\varepsilon} = 1$ s^{-1} (d) $\dot{\varepsilon} = 10$ s^{-1}.

3.2. Stress-strain constitutive equations

Arrhenius equation describes the hot deformation behavior of mental [14], and gives equation between Zener–Holloman parameter and flow stress σ. This theory can be expressed as in equations (1)-(3):

$$Z = \dot{\varepsilon}e^{Q/(RT)} \tag{1}$$

$$\dot{\varepsilon} = AF(\sigma)e^{Q/(RT)} \tag{2}$$

$$F(\sigma) = \begin{cases} \sigma^{n_1}, \alpha\sigma < 0.8 \\ e^{\beta\sigma}, \alpha\sigma > 1.2 \\ [\sinh(\alpha\sigma)]^n, \text{for all } \sigma \end{cases} \tag{3}$$

σ is flow stress (MPa); Q is activation energy (kJ mol^{-1}); $\dot{\varepsilon}$ is strain rate (s^{-1}) ;T is temperature (K); R is constant (8.314 J (mol·K)$^{-1}$); α, β and n_1 are material constants , $\alpha=\beta/n_1$.

Based on the figures 2(a) and 2(b), the values β and n_1 are evaluated, β and n_1 are the average value of the slopes of the $\ln\dot{\varepsilon} - \sigma$ curve and $\ln\dot{\varepsilon} - \ln\sigma$ curve.

Activation Q can be calculated as follows:

$$Q = R\left(\frac{\partial\ln\dot{\varepsilon}}{\partial\ln[\sin(\alpha\sigma)]}\right)_T \left[\frac{\partial\ln[\sinh(\alpha\sigma)]}{\partial(1/T)}\right]_{\dot{\varepsilon}} \tag{4}$$

The figures 2(c) and 2(d) illustrates the relationships among $\ln\dot{\varepsilon}$, $\ln[\sin(\alpha\sigma)]$ and T^{-1}. According to the above equation and figure 2, the material constants are listed in table 1.

Table 1. Material constants of Al-Si-Mg alloy.

α(MPa^{-1})	n	A(s^{-1})	Q(kJ · mol^{-1})
0.0269	5.89	4.36×10^{10}	174.9

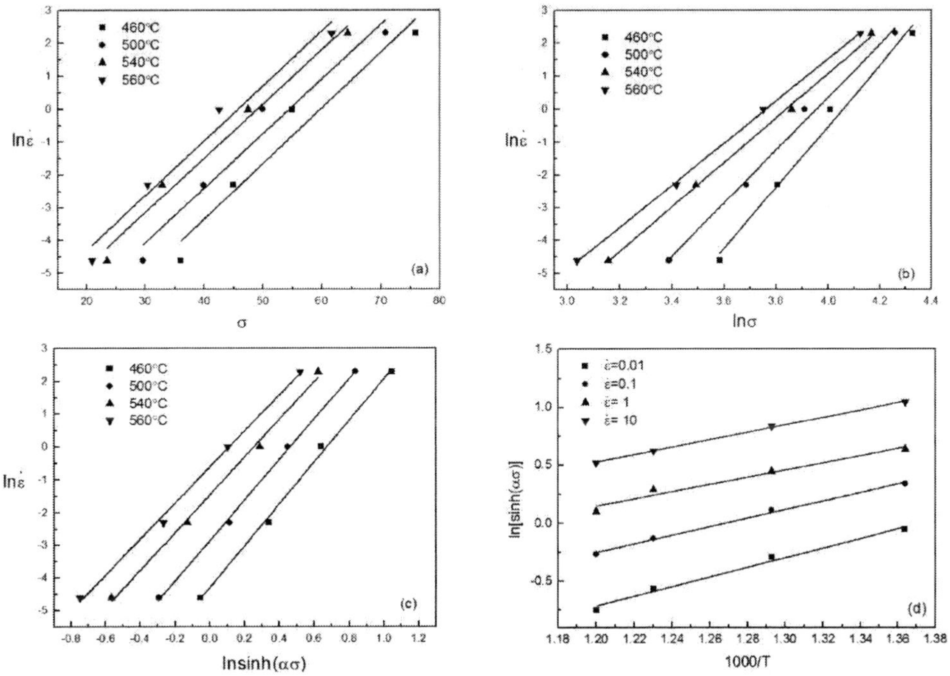

Figure 2. Linear relationships of various parameters of 6082 aluminum alloy: (a) $\ln\dot{\varepsilon} - \sigma$; (b) $\ln\dot{\varepsilon} - \ln\sigma$; (c) $\ln\dot{\varepsilon} - \ln\sinh(\alpha\sigma)$; (d)$\ln\sinh(\alpha\sigma) - T^{-1}$.

So the mathematic model for stress of 6082 Al-Si-Mg alloy is summarized as follows:

$$\sigma = 37.2\ln\left\{\left(\frac{Z}{4.36\times10^{10}}\right)^{\frac{1}{5.89}} + \left[\left(\frac{Z}{4.36\times10^{10}}\right)^{\frac{2}{5.89}} + 1\right]^{\frac{1}{2}}\right\} \tag{5}$$

$$Z = \dot{\varepsilon}e^{\frac{174.9}{8.31T}} \tag{6}$$

3.3. Effect of deformation parameters on microstructure

The figure 3 is EBSD image of grain structure when the temperature 540℃, true strain 0.6 and the strain rate $0.01s^{-1}$, $0.1s^{-1}$, $1s^{-1}$, $10s^{-1}$, Corresponding parameters ln*Z* is 21.3, 23.6, 25.9 and 28.2.

Red line is grain boundary. Dynamic recrystallization occurs at temperature 540℃ and different strain rates from photographs, average grain size of recrystallized grains decreases with the increase of strain rate. It is also conducted that the gains are relatively small under the conditions of high *Z*.

The figure 4 is EBSD image of recrystallization volume fraction when the temperature 540℃, true strain 0.6 and the strain rate 0.01 s^{-1}, 0.1 s^{-1}, 1 s^{-1}, 10 s^{-1}, blue represents dynamic recrystallization region, yellow is the not recrystallization region. Based on the data from EBSD post-processing software, the recrystallization volume fraction in (a) (b) (c) (d) are 78%, 60.3%, 67.8%, 69%. At lower strain rate, the dynamic recrystallization process is adequate, and the recrystallized grains have enough time to grow up (figures 4(a) and 4(b)). The dynamic recrystallization of materials is closely related to the deformation time, and the deformation time at high strain rate is short, and the dynamic recrystallization will be limited. The recrystallization volume fraction is decreased with the increase of strain rates or lnZ.

The results from grain evaluation and recrystallization volume fraction evaluation reveals that high strain rates will refine grains and low strain rates is beneficial to improve the fraction of dynamic recrystallization when temperature keep the same.

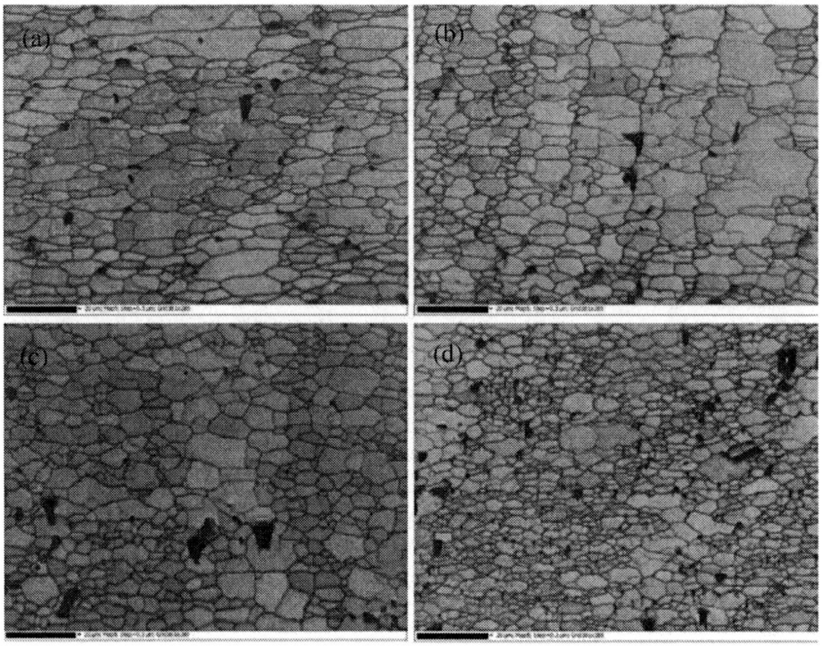

Figure 3. EBSD grain structure of 6082 aluminum alloy at different strain rates: (a) $\dot{\varepsilon}$ = 0.01 s^{-1}(lnZ=21.3); (b) $\dot{\varepsilon}$ = 0.1 s^{-1}(lnZ=23.6); (c) $\dot{\varepsilon}$ = 1 s^{-1}(lnZ=25.9); (d) $\dot{\varepsilon}$ = 10s^{-1}(lnZ=28.2).

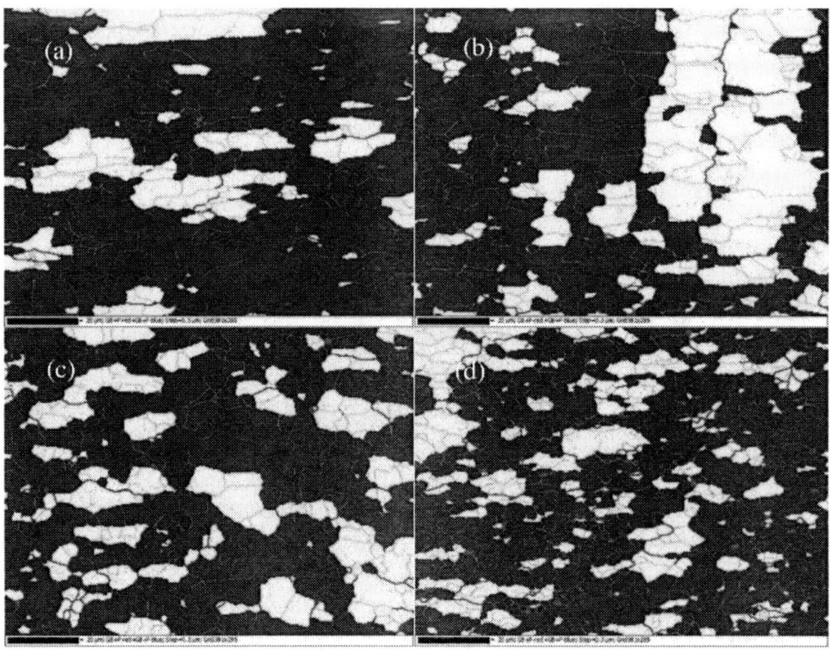

Figure 4. Recrystallization volume fraction of 6082 aluminum alloy at different strain rates: (a) $\dot{\varepsilon}$ = 0.01 s^{-1} (lnZ=21.3); (b) $\dot{\varepsilon}$ = 0.1s^{-1} (lnZ=23.6); (c) $\dot{\varepsilon}$ = 1 s^{-1} (lnZ=25.9); (d) $\dot{\varepsilon}$ = 10 s^{-1} (lnZ=28.2).

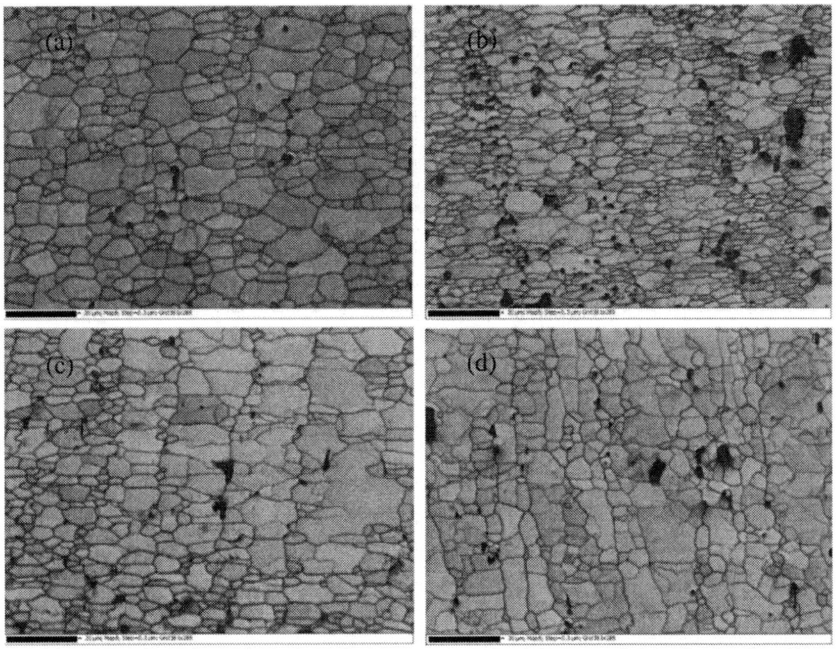

Figure 5. EBSD grain structure of 6082 aluminum alloy at different temperatures: (a) 460℃ (lnZ=26.4); (b) 500℃ (lnZ=25); (c) 540℃ (lnZ=23.6); (d)560℃ (lnZ=23).

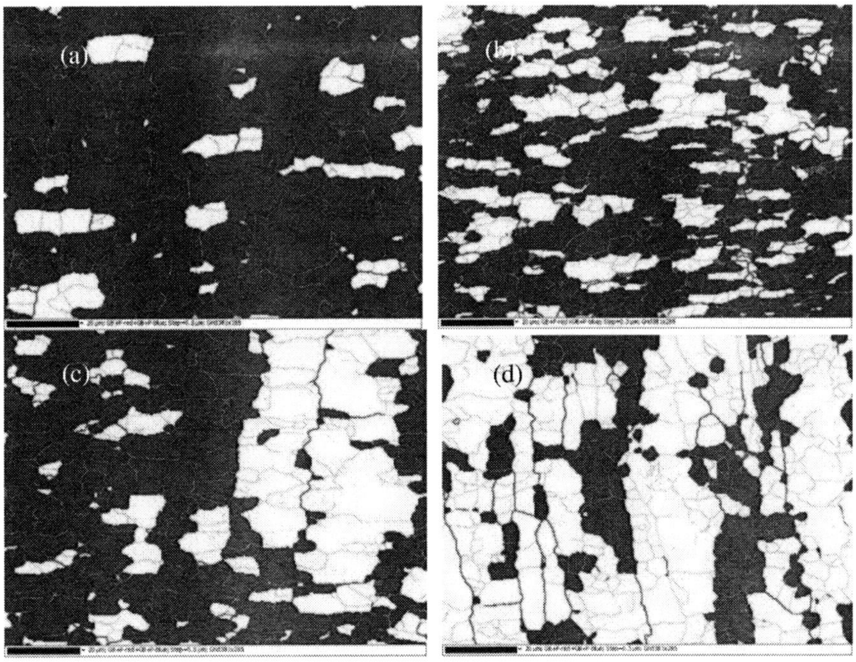

Figure 6. Recrystallization volume fraction of 6082 aluminum alloy at different strain rates: (a) 460℃ (lnZ=26.4); (b) 500℃ (lnZ=25); (c) 540℃ (lnZ=23.6); (d) 560℃ (lnZ=23).

The figure 5 is EBSD image of grain structure when the strain rate is 0.1 s^{-1}, true strain 0.6 and the temperature is 460℃, 500℃, 540℃, 560℃, Corresponding parameters ln$Z$ is 26.4, 25, 23.6 and 23. Red line is grain boundary. Dynamic recrystallization occurs at strain rate 0.1 s^{-1} and different temperatures from photographs. At lower temperatures recrystallization grains are equiaxed grains (figures 5(a) and 5(b)), which gradually transform to columnar grains as the equiaxed grains grow up to annex the small grains (figures 5(c) and 5(d)). Average grain size of recrystallized grains increases with the increase of deforming temperature.

With the increase of deforming temperature, the growth of recrystallized grain boundary enhanced. High temperature reduces dislocation density and provides a less driving force for dynamic recrystallization nucleation. As is shown in figure 6(d), blue represents dynamic recrystallization region.

From the figure 6 above, the deformation keeps similar, amount of recrystallization volume fraction increase as deformation temperature decrease at a given strain rate. The higher the deformation temperature, the smaller recrystallization volume fraction of the dynamic recrystallization. It is also shown that the dynamic recrystallization fraction decrease with the decrease of ln$Z$.

3.4. Critical strain of dynamic recrystallization

By experiment, the true stress-true strain curve is established, but it is difficult to reflect which strain the dynamic recrystallization is initiated from the curve [15]. Based on the Jonas critical strain theory, the critical strain ε_c is acquired as the follow equation [8]:

$$\frac{\partial}{\partial\sigma}\left(\frac{\partial\theta}{\partial\sigma}\right)_{\dot{\varepsilon}} = \frac{\partial}{\partial\varepsilon}\left(\frac{\partial\ln\theta}{\partial\varepsilon}\right)_{\dot{\varepsilon}} = 0 \tag{7}$$

$$\theta = \frac{\partial\sigma}{\partial\varepsilon} \tag{8}$$

Figure 7 shows the relationship between $(-\partial\ln\theta/\partial\varepsilon) - \varepsilon$ curves when the strain rate is 0.01 s^{-1} and true strain is 0.6 at different temperatures. By two order Polynomial fitting, the relationships between $(-\partial\ln\theta/\partial\varepsilon)$ and ε are listed at different temperatures and strain rates in table 2.

Table 2. Relationships between $-\partial\ln\theta/\partial\varepsilon$ and ε at different strain rates and temperatures.

$\dot{\varepsilon}$ (s^{-1})	T(℃)	$-\partial\ln\theta/\partial\varepsilon$-$\varepsilon$ relation
0.01	460	$-\partial\ln\theta/\partial\varepsilon = 251.29 - 9311.04\varepsilon + 129492.7\varepsilon^2$
	500	$-\partial\ln\theta/\partial\varepsilon = 290.24 - 9332.84\varepsilon + 104531.8\varepsilon^2$
	540	$-\partial\ln\theta/\partial\varepsilon = 206.30 - 4201.6\varepsilon + 47829.8\varepsilon^2$
	560	$-\partial\ln\theta/\partial\varepsilon = 286.35 - 6754.54\varepsilon + 95838.4\varepsilon^2$
0.1	460	$-\partial\ln\theta/\partial\varepsilon = 481.5 - 17012.1\varepsilon + 237154.1\varepsilon^2$
	500	$-\partial\ln\theta/\partial\varepsilon = 368.2 - 30703.8\varepsilon + 289440.2\varepsilon^2$
	540	$-\partial\ln\theta/\partial\varepsilon = 463.8 - 13541.2\varepsilon + 131183.9\varepsilon^2$
	560	$-\partial\ln\theta/\partial\varepsilon = 547.6 - 9668.3\varepsilon + 130935.5\varepsilon^2$
1	460	$-\partial\ln\theta/\partial\varepsilon = 349.2 - 105492.6\varepsilon + 713130.2\varepsilon^2$
	500	$-\partial\ln\theta/\partial\varepsilon = 286.4 - 36536\varepsilon + 451408.2\varepsilon^2$
	540	$-\partial\ln\theta/\partial\varepsilon = 527 - 26554.9\varepsilon + 232095\varepsilon^2$
	560	$-\partial\ln\theta/\partial\varepsilon = 208.9 - 9311.04\varepsilon + 129492.7\varepsilon^2$
10	460	$-\partial\ln\theta/\partial\varepsilon = 19320 - 611406\varepsilon + 3417790\varepsilon^2$
	500	$-\partial\ln\theta/\partial\varepsilon = 6043 - 235446\varepsilon + 1622740\varepsilon^2$
	540	$-\partial\ln\theta/\partial\varepsilon = 7982 - 185460.5\varepsilon + 1609900\varepsilon^2$
	560	$-\partial\ln\theta/\partial\varepsilon = 10615 - 209476.1\varepsilon + 2597670\varepsilon^2$

According to the two order fitting polynomial, at the point where $-\partial\ln\theta/\partial\varepsilon$ is the minimum value, the corresponding strain is critical ε_c of dynamic recrystallization.

Peak strain ε_p is obtained from the true stress-strain curves. Critical strain ε_c and peak strain ε_p can be the function of Zenner-Hollomon parameter Z as shown in figure 8.

$$\varepsilon_p = 1.2 \times 10^{-2} Z^{0.069} \quad (9)$$

$$\varepsilon_c = 7.5 \times 10^{-3} Z^{0.073} \quad (10)$$

$$\varepsilon_c/\varepsilon_p \approx (0.678 \sim 0.707) \quad (11)$$

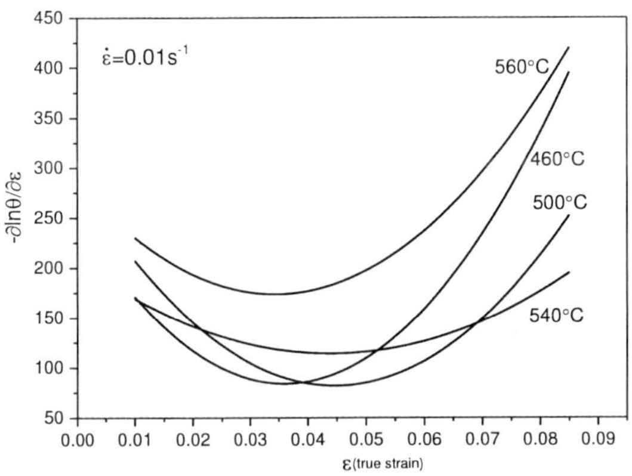

Figure 7. The $(-\partial \ln\theta/\partial\varepsilon) - \varepsilon$ curves of 6082 aluminum alloy at strain rate 0.01 s-1.

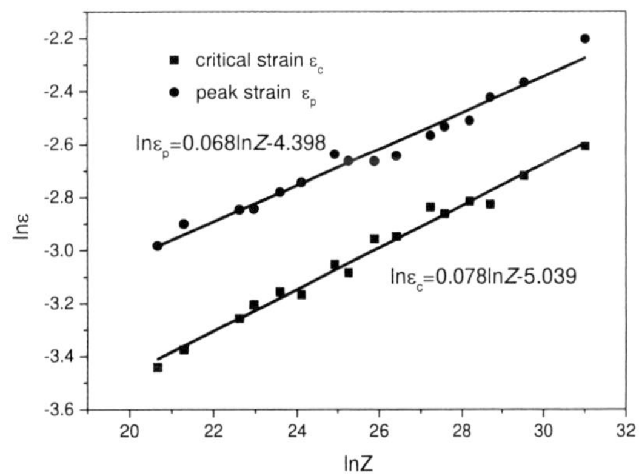

Figure 8. $\ln\varepsilon_p$ versus $\ln Z$ and $\ln\varepsilon_c$ versus $\ln Z$ plot.

4. Conclusions

The dynamic recrystallization behavior of 6082 Al-Si-Mg aluminum alloy is investigated by compression experiment in high temperature. The main results can be summarized as follows:

- At lower strain rates, hardening and softening due to dynamic recrystallization are in balance; at high strain rates, hardening and softening due to dynamic recrystallization alternately dominated.
- The deformation activation energy of 6082 Al-Si-Mg is about 174.9 kJ mol^{-1}.
- Lower strain rates are beneficial to increase the volume fraction of DRX, the average grain size is coarse; High strain rates are beneficial to refine average grain size, the volume fraction of DRX is less than that in low strain rates. High temperature cause coarse grains.
- The relationship between initial critical strain ε_c of dynamic recrystallization and Zenner-Hollomon parameter Z: $\varepsilon_c = 7.5 \times 10^{-3} Z^{0.073}$.
- $\varepsilon_c/\varepsilon_p \approx (0.678 \sim 0.707)$.

References

[1] Ebrahimi G R, Keshmiri, Maldar A R and Momeni A 2012 *Mater. Sci. Technol* **28** 467-73
[2] Liang T, Anderson I, Riedemann T and Russell A 2014 *Acta Materialia* **77** 151-61
[3] Poliak E I and Jonas J J 1996 *Acta Mater* **44** 127-36
[4] Liang T, Anderson I, Riedemann T and Russell A 2017 *Mater. Sci. Eng. A* **690** 348-54
[5] Liang T, Anderson I and Russell A 2017 *Materials Science* **49** 2787-94
[6] Luton M J and Sellars C M 1969 *Acta metal* **17** 1033
[7] Sellars C M 1981 *Metals Forum* **4** 75
[8] Poliak E I and Jonas J J 2003 *ISIJ International* **43** 684-91
[9] Guo H L, Sun Z C and Yang H 2013 *The Chinese Journal of Nonferrous Metals* **6** 1507-14
[10] Chen X H, Chen K H and Yang P X 2013 *The Chinese Journal of Nonferrous Metals* **1** 44-50

[11] Shi L, Yang H and Guo G 2014 *Journal of Plasticity Engineering* **1** 65-70
[12] Feng D, Zhang X M and Liu S D and Deng Y L 2014 *Mater. Sci. Eng. A* **608** 63-72
[13] Mirzadeh H and Najafizadeh A 2010 *Materials and Design* **31** 1174-9
[14] Zenner C and Hollmon J H 1944 *Journal of Applied Physics* **22** 22-32
[15] Peng J, Wang J Y and Zhong L P 2016 *Trans. Nonferrous Met. Soc. China* **26** 945-55

IOP Conf. Series: Materials Science and Engineering **369** (2018) 012023 doi:10.1088/1757-899X/369/1/012023

Effects of the bonded component heights on the stress intensity factors of the edge interface cracks

X Lan[1], Q L Tang[1], X R Li[1] and N A Noda[2]

[1]School of Energy and Power Engineering, Shandong University, Jingshi Road 17923#, Jinan, Shandong province, China
[2]Department of Mechanical Engineering, Kyushu Institute of Technology,1-1 Sensui-cho, Tobata-ku, Kitakyushu-shi, Fukuoka, 804-8550, Japan

E-mail: lanxin@sdu.edu.cn

Abstract. The height of the bonded component of a bonded strip changes with the scope of requirements in practical applications. Therefore, one must understand the effect of the influence of the height of the bonded component. In this research, the stress intensity factors of an edge-cracked asymmetrically bonded strip are systematically computed using the proportional crack tip opening displacement method based on FE analysis. Then, the stress intensity factors are compared for arbitrary sets of material combinations in the whole range of Dundurs' material space with varying the height of the bonded component, and the combined effects of the relative height of the bonded component and material combinations on the stress intensity factors are discussed for the typical engineering materials.

1. Introduction

Due to advances in adhesive materials with improved mechanical properties, bonded systems have found applications in a wide spectrum of industries. Interfacial cracks are usually observed within the vicinity of the bonding region since there exists a high singular stress field. The stress intensity factors (SIFs) are usually employed in predicting the stress state as well as the crack initiation and propagation. The stress intensity factors (SIFs) are usually employed in predicting the crack initiation and propagation for the interface cracks. Many researches were found to determine the stress state of an edge-cracked bonded strip accurately till recently. Lan *et al* [1] investigated the SIFs of a bi-material bonded strip for different material combinations, and the variation of SIFs with varying crack size were then discussed. Noda and Lan [2] computed the SIFs of an edge interface crack in the bonded semi-infinite planes for different material pairs in the $\alpha - \beta$ space. Lan *et al* [3] investigated the energy release rate (ERR) and mode-mixity of an edge-cracked longitudinal symmetry and asymmetry bonded components, and the effect of geometrical configurations were discussed together with the fracture toughness. As known, one may predict the onset of an interfacial crack using the SIF based fracture criterion. Therefore, in this research, the SIFs are computed for various height of the bonded component using the numerical procedure based on the proportional crack-tip opening displacement with FE analysis[4]. And the effects of the bonded component height on SIFs are examined and discussed accordingly.

IOP Conf. Series: Materials Science and Engineering **369** (2018) 012023 doi:10.1088/1757-899X/369/1/012023

2. Analysis method

The coordinate origin is set at the right crack tip and the interface crack lies along the negative x-axis. Two elastic materials are ideally bonded along the x-axis, material 1 is located above the interface and material 2 is located below. The stress singularity field in the plane stress and plane strain conditions can be expressed using equation (1) [5].

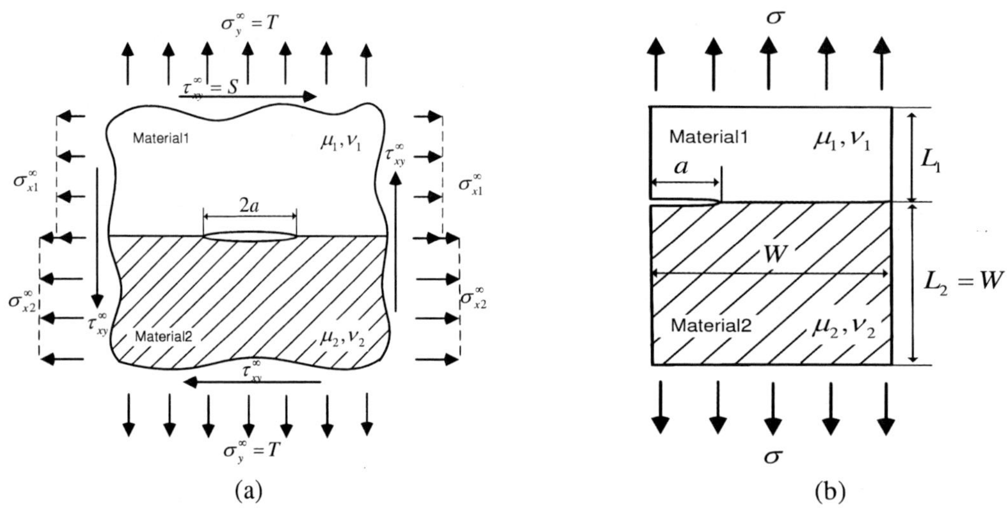

Figure 1. (a) Central interface crack in a bonded semi-infinite plane (reference problem A) and (b) edge interface crack (unknown Problem B).

$$\sigma_y + \mathrm{i}\,\tau_{xy} = \frac{K_I + \mathrm{i}\,K_{II}}{(2\pi r)^{\frac{1}{2}}}\left(\frac{r}{2a}\right)^{\mathrm{i}\varepsilon}, \quad r \to 0 \tag{1}$$

where, σ_y, τ_{xy} are the tensile and shear stresses around the crack tip, respectively. r is the radial distance away from the crack tip, a denotes the half length of an center interface crack or the length of an edge interface crack, furthermore, ε is the bi-elastic constant which is defined in equation (2):

$$\varepsilon = \frac{1}{2\pi}\ln\left[\left(\frac{\kappa_1}{\mu_1} + \frac{1}{\mu_2}\right)\left(\frac{\kappa_2}{\mu_2} + \frac{1}{\mu_1}\right)^{-1}\right] \tag{2}$$

$$\kappa_m = \begin{cases} 3 - 4\nu_m & (plane\ strain) \\ 3 - \nu_m(1+\nu_m)^{-1} & (plane\ stress) \end{cases} \tag{3}$$

where μ_m and ν_m are the shear moduli and Poisson's ratios of material m $(m = 1, 2)$. The associated relative COD $\delta_d (d = x, y)$ at a distance r away from the crack tip are defined by

$$\delta_y + \mathrm{i}\,\delta_x = \frac{K_I + \mathrm{i}\,K_{II}}{2(1 + 2\mathrm{i}\,\varepsilon)\cosh(\varepsilon\pi)}\left[\frac{\kappa_1 + 1}{\mu_1} + \frac{\kappa_2 + 1}{\mu_2}\right]\left(\frac{r}{2\pi}\right)^{\frac{1}{2}}\left(\frac{r}{l}\right)^{\mathrm{i}\varepsilon} \tag{4}$$

l is usually selected as an arbitrary length which scales with specimen size or crack length. Here, we have $l = 2a$ to coincide with the definition in equation (1). Equation (4) is rearranged and then it gives the explicit expression of K_I, K_{II} as:

$$K_I = Z\left\{\left(\delta_y - 2\varepsilon\delta_x\right)\cos\left[\varepsilon\ln\left(\frac{r}{l}\right)\right] + \left(\delta_x + 2\varepsilon\delta_y\right)\sin\left[\varepsilon\ln\left(\frac{r}{l}\right)\right]\right\} \tag{5}$$

$$K_{II} = Z\left\{\left(\delta_x + 2\varepsilon\delta_y\right)\cos\left[\varepsilon\ln\left(\frac{r}{l}\right)\right] - \left(\delta_y - 2\varepsilon\delta_x\right)\sin\left[\varepsilon\ln\left(\frac{r}{l}\right)\right]\right\} \tag{6}$$

where the value of S is defined as

$$Z = \frac{2\cosh(\varepsilon\pi)\left[r(2\pi)^{-1}\right]^{-1/2}}{\left(\frac{\kappa_1+1}{\mu_1} + \frac{\kappa_2+1}{\mu_2}\right)} \tag{7}$$

Equations (5) and (6) are rewritten as

$$\frac{K_I}{\delta_y} = Z\left\{\left(\cos Q + 2\varepsilon\sin Q\right) + \left(\sin Q - 2\varepsilon\cos Q\right)\frac{\delta_x}{\delta_y}\right\} \tag{8}$$

$$\frac{K_{II}}{\delta_x} = Z\left\{\left(\cos Q + 2\varepsilon\sin Q\right) - \left(\sin Q - 2\varepsilon\cos Q\right)\frac{\delta_y}{\delta_x}\right\} \tag{9}$$

and

$$Q = \varepsilon\ln\left(r(l)^{-1}\right) \tag{10}$$

In equations (8) and (9), assuming $Q, \varepsilon, \delta_y(\delta_x)^{-1}$ are the same for the reference and unknown problems, then a relationship is given as equation (11).

$$K_I(\delta_y)^{-1} = const,\ K_{II}(\delta_x)^{-1} = const \tag{11}$$

Examining the two interface crack problems A and B shown in figure 1(a) and (b) respectively, when equations (12) and (13) are satisfied, then K_I, K_{II} are proportional to δ_y, δ_x as given in equation (14). Here, the relative COD δ_y, δ_x can be calculated using FE analysis. The SIFs of the reference problem A are analytically solved in advance, therefore, the SIFs of the given unknown problem B can be obtained from equation (14).

$$\begin{pmatrix} Q_A = Q_B \\ \varepsilon_A = \varepsilon_B \end{pmatrix} \rightarrow \begin{pmatrix} \left[\varepsilon\ln\left(r(l)^{-1}\right)\right]_A = \left[\varepsilon\ln\left(r(l)^{-1}\right)\right]_B \\ \varepsilon_A = \varepsilon_B \end{pmatrix} \tag{12}$$

$$\left[\delta_y(\delta_x)^{-1}\right]_A = \left[\delta_y(\delta_x)^{-1}\right]_B \tag{13}$$

$$\left[K_I(\delta_y)^{-1}\right]_A = \left[K_I(\delta_y)^{-1}\right]_B,\ \left[K_{II}(\delta_x)^{-1}\right]_A = \left[K_{II}(\delta_x)^{-1}\right]_B \tag{14}$$

In this research, the bi-material bonded half-planes with a central interface crack subjected to uniform tension and shear as shown in figure 1(a) is employed as the reference problem A of the current numerical procedure. Therefore, the SIFs of the given unknown problem B can be computed using equation (14). The detailed information of the application of the current method as well as its robustness discussions could be found in [4].

3. Results and discussions

3.1. Solutions of the bonded strip for different component heights

The longitudinal asymmetrically bonded strip shown in figure 1(b) is computed for various heights of component 1. The relative crack size is fixed $aW^{-1}=0.1$, the height of component 1 changes from $L_1=0.333W\sim 2W$, and the height of the lower component 2 is fixed to $L_2=1W$. Here, given space limitations, only the SIFs for $L_1=0.2W, 0.1W$ are plotted in figures 2(a) and 2(b), respectively. Similar varying tendencies as in figure 2(a) can also be obtained for $L_1\geq 0.2W$, and then they gradually transfer to the pattern shown in figure 2(b) with the continuous reduction of L_1. As can be seen from figure 2(a), the Maximum value of K_I is always located in the upper left corner $(\alpha=-1,\beta=0)$,when material 1 is extremely stiff compared to material 2, $\alpha\rightarrow 1$, K_I reaches its minimum value, and K_{II} comes closer to its minimum one. However, when material 1 is extremely compliant comparing with material 2, $\alpha\rightarrow -1$, K_I and K_{II} reach their maximum values.

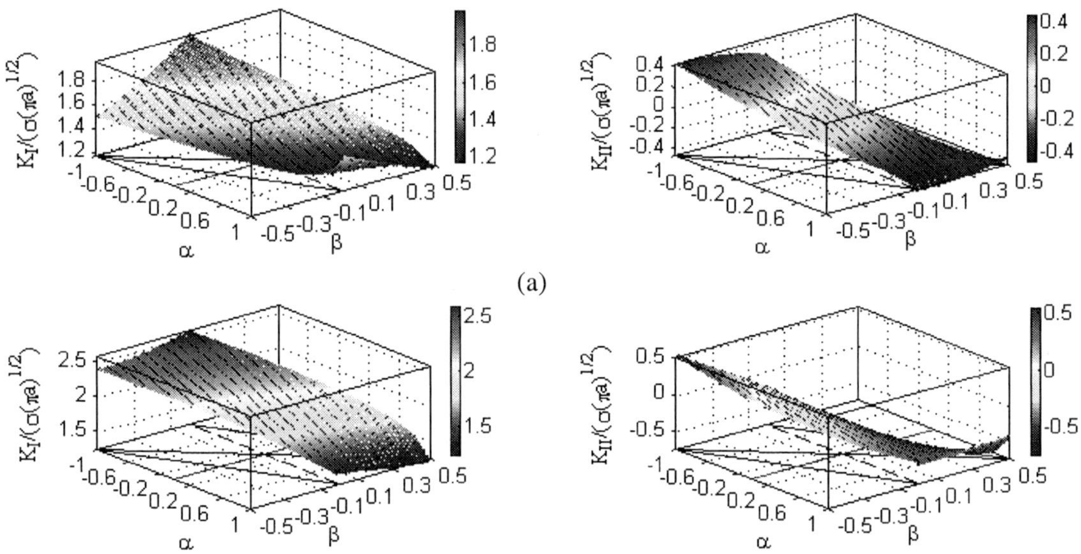

(a)

(b)

Figure 2. 3-D normalized SIFs of (a) $L_1=0.2W$ and (b) $L_1=0.1W$.

3.2. Effect of the bonded component heights

The SIFs are examined and compared for the whole physical admissible typical engineering material combinations distributed in the vicinity of $\beta=0.25\alpha$ by varying $L_1=0.333\sim 2W$. Figure 3(a) shows the values of $K_{I,nor}=K_I\left(K_I^{L_1=W}\right)^{-1}$, all the values are normalized using those of $L_1=W$. It is found that $K_{I,nor}>1$ when $L_1W^{-1}\leq 0.1$, however, for the case of $L_1W^{-1}\geq 0.2$, $K_{I,nor}<1$ can also be observed when $\alpha\rightarrow 1$, and the region of $K_{I,nor}\leq 1$ reduces with decreasing L_1. It means that the SIFs are bigger than those of $L_1=W$ for most the cases of decreasing L_1, and they can also be smaller when material 1 is extremely stiff comparing with material 2. The absolute values of K_{II} are plotted in figure 3(b). The curve of $L_1=W$ are axial symmetry around $\alpha=0$ with the minimum point of $K_{II}=0$ at $\alpha=0$. When the height of the bonded component 1 decreases, the minimum $abs(K_{II})=0$ moves to the left lower part of $\alpha-\beta$ space ($\alpha \to -1$), say, the minimum $abs(K_{II})=0$ is located around $\alpha \to -0.68$ for $L_1W^{-1}=0.1$. However, $abs(K_{II})$ grows monotonically with decreasing L_1 when component 1 is extremely thinner than component 2 ($L_1W^{-1} \le 0.06667$).

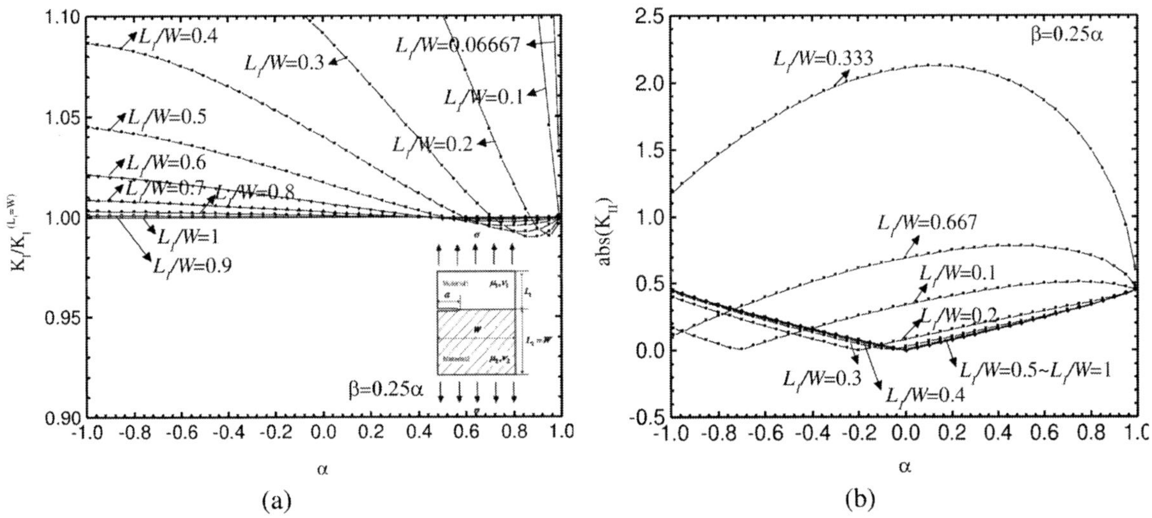

Figure 3. (a) $K_I\left(K_I^{L_1=W}\right)^{-1}$ and (b) $abs(K_{II})$ for $\beta=0.25\alpha$.

4. Conclusions

In this research, the SIFs of an edge interface crack in the bi-material bonded strip subjected to remote tensile loads were calculated with varying various heights of the bonded components. It was found that K_I generally increase with the decrease of L_1/W, and K_{II} monotonically decrease with decreasing L_1/W. However, some exceptions of the SIFs which are smaller than those of $L_1=W$ for decreasing L_1 also exists when material 1 is extremely stiff compared to material 2 ($\alpha \to 1$).

Acknowledgments

This work was supported by grants from National Natural Science Foundation of China (NSFC) (No. 11702160), Natural Science Foundation of Shandong Province, China (No. ZR201702170418) and the Fundamental Research Funds of Shandong University (No. 2016TB009)).

References

[1] Lan X, Noda N A, Mithinaka K and Zhang Y 2011 The effect of material combinations and relative crack size to the stress intensity factors at the crack tip of a bi-material bonded strip *Eng Fract Mech* **78** 2572-84

[2] Noda N A and Lan X 2012 Stress intensity factors for an edge interface crack in a bonded semi-infinite plate for arbitrary material combination *Int J Solids Struct* **49** 1241-51

[3] Lan X, Ji S B, Noda N-A and Cheng Y 2017 The effect of geometric configurations on the elastic behavior of an edge-cracked bonded strip *Adv Mater Sci Eng* **2017** 2380740

[4] Lan X, Ji S B, Noda N A and Cheng Y 2017 Stress intensity factor solutions for several crack problems using the proportional crack opening displacements *Eng Fract Mech* **171** 35-49

[5] Malysev B M and Salganik R L 1965 The strength of adhesive joints using the theory of cracks *Int J Fract Mech* **1** 114-27

IOP Conf. Series: Materials Science and Engineering **369** (2018) 012024 doi:10.1088/1757-899X/369/1/012024

Gap features of layered iron-selenium-tellurium compound below and above the superconducting transition temperature by break-junction spectroscopy combined with STS

T Ekino[1,3], A Sugimoto[1] and A M Gabovich[2]

[1]Hiroshima University, Graduate School of Integrated Arts and Sciences, Higashi-Hiroshima 7398521, Japan
[2]Institute of Physics, National Academy of Sciences of Ukraine, Kyiv 03680, Ukraine

E-mail: ekino@hiroshima-u.ac.jp

Abstract. We studied correlations between the superconducting gap features of Te-substituted FeSe observed by scanning tunnelling spectroscopy (STS) and break-junction tunnelling spectroscopy (BJTS). At bias voltages outside the superconducting gap-energy range, the broad gap structure exists, which becomes the normal-state gap above the critical temperature, T_c. Such behaviour is consistent with the model of the partially gapped density-wave superconductor involving both superconducting gaps and pseudogaps, which has been applied by us earlier to high-T_c cuprates. The similarity suggests that the parent electronic spectrum features should have much in common for these classes of materials.

1. Introduction

From the early stages of our break-junction tunnel spectroscopy (BJTS) measurements on the superconducting $FeSe_xTe_{1-x}$ single crystals, the normal-state gaps were manifested in addition to the Bardeen-Cooper-Schrieffer (BCS)-type superconducting gap features [1,2]. The observed unusual electronic properties of this superconductor including the gap distributions extending to several tenth of meV are in fact intriguing in view of the possible manifestation of the local anomalously high (about 100 K [3]) superconducting critical temperature (T_c). On the other hand, it is reasonable to consider that the normal-state gap is a consequence of the charge- or spin-density-wave formation below the structural/magnetic phase transition, as stems from the phase diagram [4]. In particular, there has been suggested that the lower symmetry C_2 nematic/smectic charge ordering emerges in the iron-based superconductors [5,6]. In such states, the normal-state gap structures can be readily observed by tunnelling spectroscopy, which was proposed theoretically [7]. However, there were not enough direct tunnel measurements of such a normal-state gap.

In this paper, we report the observations of both superconducting and the normal-state gaps for the simple iron-based superconductor, $FeSe_xTe_{1-x}$, using the BJTS techniques as well as the scanning tunnel spectroscopy (STS). We emphasize that the gap of ± 10-20 mV observed by STS in the superconducting state below T_c = 15 K is larger than that revealed in the BJTS measurements [1]. The attention should be paid to the fact that these measurements constitute the most direct and precise tool among the gap-probing experimental techniques. This probe dealing with the conduction electrons themselves immediately leads to the well energy-resolved data needing no further assumptions and relatively easy in interpretation [8].

IOP Conf. Series: Materials Science and Engineering **369** (2018) 012024 doi:10.1088/1757-899X/369/1/012024

2. Experimental procedures

Single crystals of $FeSe_xTe_{1-x}$ were synthesized by a standard procedure. The pristine samples thus obtained were annealed at 673 K for 100 hours [9]. The electron probe micro analyzer (EPMA) was employed to determine the crystal composition. The tunnelling measurements were done using the BJTS and the scanning tunnelling microscopy/spectroscopy (STM/STS). The surface Se/Te composition ratio checked by STM is consistent with the bulk EPMA [2]. In the BJTS technique, clean unaffected superconductor – insulator – superconductor (SIS) junction interface is obtained *in situ* along the crack of the thin single-crystal platelet at temperature as low as $T = 4$ K [10]. In this case, the peak-to-peak separation, i.e., the distance between the gap edges, is $V_{\text{p-p}} = 4\Delta/e$, twice as much as $2\Delta/e$ for the SIN (N=Normal metal) junction, where e (> 0) is the elementary charge. The tunnelling conductance $dI/dV= G(V)$ of the current (I)-voltage (V) characteristics was measured using a low-frequency ac-modulation. The break-junction configuration is usually mechanically unstable, but we can obtain a stable junction, who can survive a long-time bias sweep.The STS apparatus was an Omicron LT-UHV-STM, which was upgraded to suppress external disturbances. The samples were cleaved along the layer *in situ* at the temperature $T = 4$K in an ultra-high vacuum (UHV) chamber with the vapor pressure of ~ 10^{-8} Pa to avoid any contamination of the crystal surface. The tunnelling Pt/Ir tip was cleaned by the high-voltage field emission with the Au target prior to the measurements. The scanning measurements of $G(V)$ were carried out at T= 4.9 – 77 K under the UHV condition of ~ 10^{-8} Pa.

3. Results and discussion

At first, we show the results of the STS measurements in the superconducting state at 4.9 K. Figure 1 shows the distribution map of the gap energy on the cleaved surface of FeSe0.3Te0.7 consisting of the Fe(Se,Te)4 tetrahedral network. The gap energy is determined by the bending edge in G(V) shown for several local regions. In the conductance map, the substantial inhomogeneity of the gapped regions occurs including the clusters of sizes ~ 1 – 2 nm or more, corresponding to regions with differing gap values from almost zero to about 40 meV. As is shown in the representative G(V) examples, the observed gap edge is substantially broadened and asymmetric with respect to zero bias. The STS gap structures of ~ ± 10 – 20 mV are predominantly observed in the vicinity of the excess Fe-atom regions identified by the STM topography, which are locally clustered on the regular atomic sites of surface arrangements [11]. The gapped areas at the excess Fe regions tend to locate at the higher local-barrier-height (LBH) regions with the correlation coefficient between the LBH and the gap being ~ 0.3.

In the crystal locations where almost no gap feature exists, the tunnelling conductance exhibits the broadened tiny hump at zero bias showing the V shaped background conductance concomitant with the minima at ± 10 mV. Such a zero-bias enhancement could be tentatively explained by the Andreev reflection from the superconductor, but it might be observed only when there is a direct point-contact between electrodes showing a low (<10 ohms) junction resistance [12]. In the STS spectra, however, the existence of the high enough junction resistance ~ 10 G ohms is crucial for the vacuum tunnelling from the tip to occur. Therefore, this phenomenon should have another origin, which is discussed below. Since the conductance map presented here expresses the representative feature of the Te-substituted FeSe superconductors at least for our measurements, it means that nanoscale inhomogeneity of the gap energy is inherent to iron-based superconductors similar to what is observed in cuprates [6,13,14].

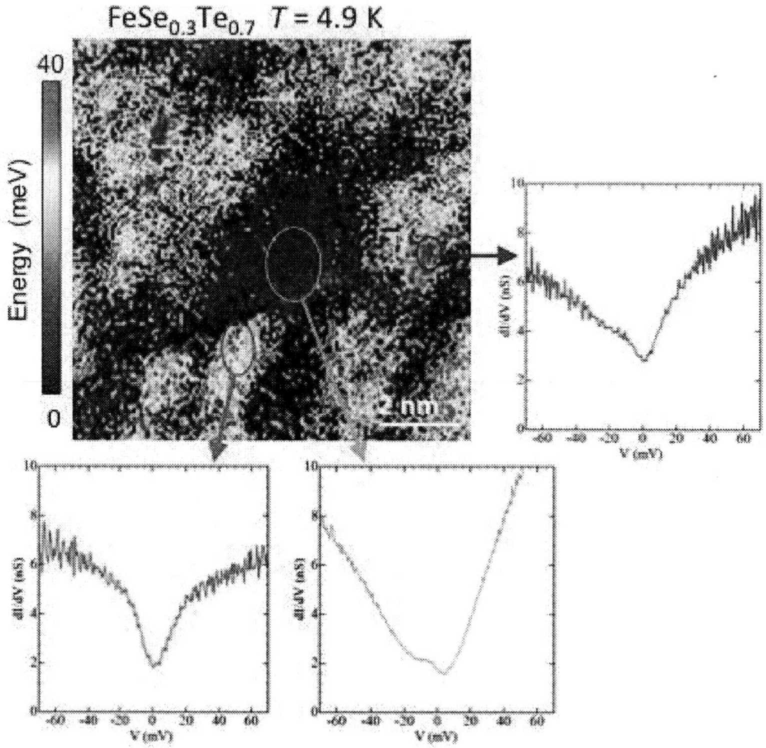

Figure 1. Gap distribution map on the *ab*-plane surface of $FeSe_{0.3}Te_{0.7}$ at 4.9 K. The STS data acquisition area of 10 nm × 10 nm is divided into 128 × 128 segments. STS conductance $G(V)$ is averaged within the indicated area.

The representative STS conductance spectra $G\ (V)$ in the gap regions below T_c are shown in figure 2, together with the corresponding BJTS ones. The STS spectra indicated by the top three curves are taken from the single bias-sweep and are in fact largely broadened, showing the depressions near zero bias correlating with the gap edge energies. The gap energy estimated by the broad-peak positions around ± 10 – 20 mV is surprisingly large as compared with the bulk T_c = 15 K and suggesting the weak-coupling BCS relationship between T_c and the gap (~ 1.3 meV). The fitting of the conductance gives the smaller gap value together with the large broadened energy feature located at the voltage corresponding to the gap itself [11]. The huge gaps found by the present STS were also observed in the superconducting monolayer FeSe film as distinct edge peaks, which disappeared at 60 – 70 K [15]. In the BJTS studies, the obtained BCS-like superconducting gap region is depicted in the left inset [1], demonstrating distinct peaks and well-depressed intra-gap region. These features completely disappear at local T_c ~ 10 K of the junction, which is considerably lower than the bulk T_c = 15 K. The BJTS also reveals the apparent gap scatter, as shown in the right inset for the larger superconducting gap. The pattern is consistent with our STS data. Since the bulk T_c depends on the Se-Te composition ratio, the observed narrow STS gap distribution can be naturally attributed to the random substitution of the Se and Te atoms. From the comparison of the STM topography elucidating the atomic composition ratios and the BJTS distributed gap values, the probed area of BJ is suggested to be less than ~ 50 nm^2.

The bottom curve in figure 2 is the BJTS conductance obtained as the largest gap-like feature of the multi-gap curve with a number of peculiarities. The apparent agreement of the gap-edge positions at ~ ± 10 mV in STS and BJTS data is observed, but it is misleading because the factor of two needed to adjust SIS and SIN spectra should be taken into account. Actually, the outer humps of the BJTS curve at ± 20 mV are twice as large as the broad main gap peaks at ± 10 mV in the STS conductance spectra. At the same time, the most inner peaks of the break-junction spectra agree with the proper scaling of the BJTS and STS results. The inner BJTS double peaks testify the proximity character of the gap at the junction region, which approximately fits the gap-peak positions of the STS curves. Since this BJTS dependence is conspicuously broadened, the inhomogeneity is suggested to be significant for the junction.

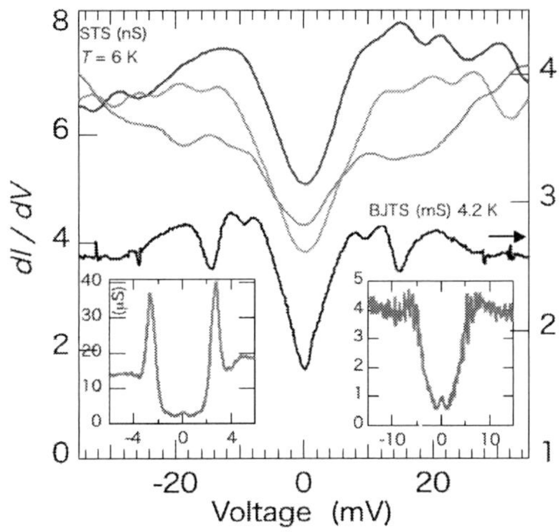

Figure 2. Tunnellingc onductances G (V) of STS (upper three curves) and BJTS (the bottom curve) for $FeSe_{0.5}Te_{0.5}$.The insets show low-bias BJTS G (V) at 4.2 K with representative gap sizes of $4\Delta/e = 5.5$ mV (left) and 12 mV (right) measured under the different junction geometries. The BJTS measurements were carried out for the same crystal piece.

In our previous BJTS measurements, the representative superconducting gap values in $FeSe_{0.5}Te_{0.5}$ were obtained to be less than $2\Delta \sim 10$ meV appropriate to the junction with $T_c = 10 - 11$ K, the latter being lower than the bulk $T_c = 15$ K [1]. This is most likely due to the inevitable surface degradation of the superconductivity involving the crack of the possible non-stoichiometric area in the BJ forming process. On the other hand, when the bulk T_c superconductivity region is located at the junction, the gap value showing the factor of 1.5 or more will be observed because of the optimal electronic structure. One can also suppose that the observed gap-like structures at ± 20 mV in figure 2 characterize the really higher T_c local break-junction sections, as was suggested previously [3,16].

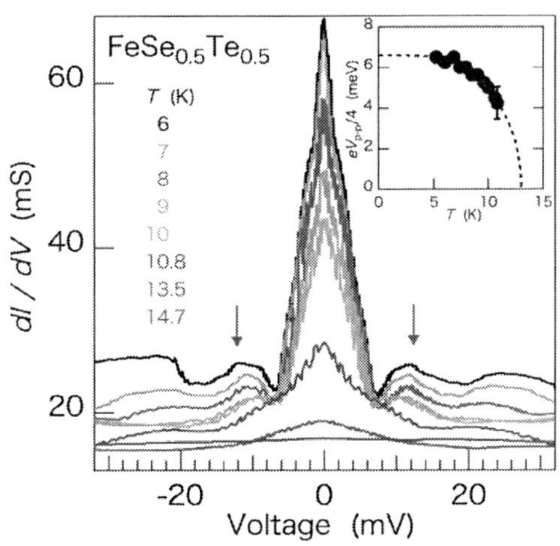

Figure 3. BJTS conductance G(V) for FeSe0.5Te0.5 at different temperatures up to 15 K. The inset shows the T dependence of the hump position compared with the scaled BCS curve.

Figure 3 shows the T-evolution of the break-junction conductance. At low temperatures the BJ spectra demonstrate the giant hump at zero-bias along with the broad peaks at ± 13 mV and the outer weak humps at ± 22 – 25 mV. We attribute the central zero-bias peak to the well-known characteristic weak-link feature of the SIS junction. The break-junction configuration in figure 3 was almost stable mechanically against the temperature changes. The locations of the sideband peaks at ± 13 mV and the broader ones at ± 22 – 25 mV correlate with the characteristic gap-edge energies of the break-junction data in figure 2. The positions of the inner ± 13 mV humps also almost coincide with those observed by the STS as comes about from figures 1 and 2. The spectra gradually smear with T. The sideband gap-related peaks are suppressed to merge into the background at about 12 – 13 K, which is lower than the bulk T_c but it is the highest T, where the superconducting gap remnants are observed by our BJTS in this compound. At the same time, the zero-bias hump, also originating from the superconducting gapping, survives up to at about 15 K, suggesting that the junction area includes a small patch with T_c close to the bulk one. The further T increase above T_c leads to the disappearance of all conductance peculiarities.

The T dependence of the sideband gap-peak energy is plotted in the inset of figure 3. It is readily seen that the T evolution is consistent with the scaled BCS temperature dependence (reduced s-wave and d-wave dependences are very similar), demonstrating that the order parameter, detected by the gapping concerned, corresponds to the conventional mean-field second-order phase transition. Therefore, the features observed at ± 13 mV in the break-junction conductance should be attributed to the superconducting gaps manifesting themselves in the SIS junction configuration.

Note, that the obtained gap values are extremely large as compared to the bulk T_c = 15 K if we adopt the BCS-like theory (even making allowance for the strong-coupling corrections), although the reduced T-dependences, as was shown above, do demonstrate the BCS behavior. Such large gap-to-T_c ratios are well known, e.g., for copper oxides.

One should also indicate that the iron-based high-T_c superconductor NdFeAsOF with T_c= 48 K possesses the gap-to-T_c ratio 2Δ (4 K)/$k_B T_c$~ 5, where k_B is Boltzmann constant [17]. If we suggest that the same ratio is valid in our case, then our gap value of 13 meV should correspond to a local T_c~ 30 K. Superconductivity with T_c as high as this was actually found under the pressure in the FeSe superconductor [16]. Since this gap disappears quickly at lower T ~ 12 – 13 K, it might be a surface rather than a bulk phenomenon.

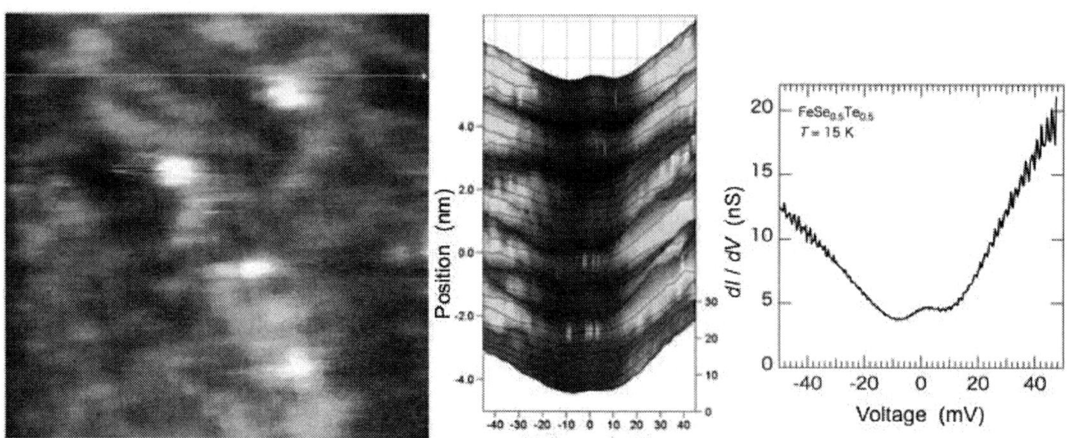

Figure 4. STM topography (left, V = - 50 mV, I = 0.4 nA) and the line profile of G (V) (middle) along the line in the STM topography for $FeSe_{0.5}Te_{0.5}$ at T = 15 K. The right panel is the all-averaged G (V) within the whole area of the STM topography.

Let us examine the STS spectra around T_c in order to clarify the nature of the local gap features from figures 1 and 2. Figure 4 shows the STM/STS of $FeSe_{0.5}Te_{0.5}$ taken at 15 K, which is higher than the local surface T_c. The left panel indicates the STM topography, while the STS conductance line spectra and the all-averaged $G(V)$ are shown in the middle and the right panels, respectively. The STM topographic resolution is worse than that at 5 K. However, the regular atomic lattice grid arrangement is still visible at 15 K with inhomogeneous background electronic modulations [2]. The STS data indicate that there are no quasiparticle gap-like structures. Instead, the zero-bias humps are observed in the whole scanned region with minima in $G(V)$ at ~ ± 10 mV. Since these humps are similar to their counterparts in the superconducting state at 4.9 K in figure 1, they are probably not related to superconductivity, e.g, to Andreev reflection phenomenon. They might have been due to the fluctuating orbital order [18]. However, in our BJTS studies of those samples the hump structures were never observed in the normal state.

The Kondo scattering during the tunneling by the excess Fe atoms on the Pt-Ir STM tip might be another possible origin of the effect. This is supported by our STM observations of the double-periodic superlattice structure in the antiferromagnetic state of the non-superconducting FeTe, which is believed to be possible only when the tip is partially coated by the Fe excess atoms [19]. The more detailed studies of the T dependence of the zero-bias hump can be of help to find its origin, because in the Kondo scenario the hump or peak should decrease logarithmically with T [20]. Assuming this to be the case, the characteristic temperature in terms of the Kondo singlet formation can be estimated as ~ 40 – 50 K from the half-width of the hump at the half maximum.

In our previous report, the high-energy gap surviving in the normal state up to ~ 70 K was observed, but we did not manage to elucidate its interplay with the superconducting gap [1]. The origin of this feature is not fully clear, but it is reasonable to associate it with the density-wave condensation arising from the local nematic electron spectrum distortion [3,4]. In figure 5, we can see the gradual smearing of the central superconducting gap structure, while the outer high-energy gap remains unchanged with the increase of T through the superconducting transition. The independent thermal behavior of the both gaps means that their origin is different. The broadened normal-state gap along with the cusp-like zero-bias dip, existing even when the superconducting gap disappears, argues for the partial Fermi-surface nesting of the parent quasiparticle spectrum. Such broadened gap is seen in the normal state at ~ 20 K, but there is no trace of it at 80 K. Such a behavior can be met in the layered conductors including high-T_c cuprates and dichalcogenides. In that situation, the intrinsic disorder may play the substantial role [6].

From figure 5, the multiple gap structures revealed at 4.2 K and consisting of ± 10 – 20 mV and ± 50 mV peculiarities can be interpreted as being due to the Cooper pairing and the electron-hole one, respectively. These data seems to be similar to the intertwining of the superconducting and pseudogap features in the tunneling data of cuprate superconductors, whatever the pseudo-gap origin [14].

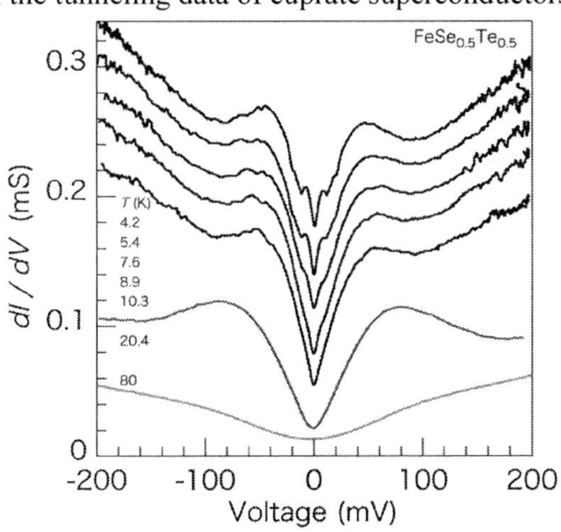

Figure 5. G(V) of BJTS for FeSe0.5Te0.5 below and above Tc. All the curves except that taken at 4.2 K are offset for clarity.

IOP Conf. Series: Materials Science and Engineering **369** (2018) 012024 doi:10.1088/1757-899X/369/1/012024

4. Conclusion

We presented and discussed our scanning-tunneling and break-junction-tunneling data for the single crystal $FeSe_xTe_{1-x}$ (x = 0.3–0.5) superconductors. The characteristic large-gap of ~ 20 – 25 meV observed by both STS and BJTS disappears at temperatures close to but below the bulk T_c =15 K for this compound. The gap-to-T_c ratio, which is the characteristic of the electron pairing mechanism and its strength, appears to be similar to that for copper-oxide superconductors. Both the conductance hump around zero-bias conspicuous in the range between – 10 mV and + 10 mV in STS measurements and the distinct gap-like structures at ± 50 – 70 mV seen in the BJTS above T_c suggest the occurrence of an additional normal-state Fermi-surface instability different from that induced by the Cooper pairing. The difference between two phenomena seems reasonable because the observed features are of different scale, the normal-state-gap low temperature values being substantially larger than T_c and the corresponding superconducting gaps. Moreover, the larger gaps survive far above local T_c. Further experiments are now in progress to elucidate the interplay between two phenomena in more detail.

References

[1] Ekino T, Nagasaka K, Sugimoto A and Gabovich A M 2015 *Physics Procedia* **65** 65-8
[2] Ekino T, Sugimoto A and Gabovich A M 2014 *Low Temp. Phys.* **40** 1182-6
[3] Ge J-F, Liu A-L, Liu C, Gao C-L, Qian D, Xue Q-K, Liu Y and Jia J-F 2014 *Nat. Mater.* **24** 285-9
[4] Cvetkovic V and Tesanovic Z 2009 *Phys. Rev. B* **80** 024512
[5] Fernandes R M, Chubukov A V and Schmalian J 2014 *Nat. Phys.* **10** 97-104
[6] Böhmer A E and Kreisel A 2018 *J. Phys.: Condens. Matter* **30** 023001
[7] Gabovich A M, Voitenko A I, Ekino T, Li M S, Szymczak H and Pekala M 2010 *Adv. Condens. Matter Phys.* **2010** 681070
[8] Wolf E L 2012 *Principles of Electron Tunneling Specroscopy* 2nd ed. (New York: Oxford University Press)
[9] Ukita R, Sugimoto A and Ekino T 2011 *Physica C* **471** 622-4
[10] Ekino T, Takabatake T, Tanaka H and Fujii H 1995 *Phys. Rev. Lett.* **75** 4262-5
[11] Sugimoto A, Sakai Y, Nagasaka K and Ekino T 2015 *Physica C* **518** 23-7
[12] Yanson I K 1974 *J. Exp. Theor. Phys.* **39** 506-13
[13] Hu H, *et al* 2011 *New J. Phys.* **13** 053031
[14] Gabovich A M and Voitenko A I 2016 *Low Temp. Phys.* **42** 1103-14
[15] Wang Q-Y, Li Z, Zhang W-H, Zhang Z-C, *et al* 2012 *Chin. Phys. Lett.* **29** 037402
[16] Medvedev S, *et al* 2009 *Nat. Mater.* **8** 630-3
[17] Ekino T, Sugimoto A, Okabe H, Shohara K, Ukita R, Akimitsu J and Gabovich A M 2010 *Physica C* **470** S358-9
[18] Arham H Z, *et al* 2012 *Phys. Rev. B* **85** 214515
[19] Sugimoto A, Ekino T and Gabovich A M 2014 *Phys. Rev. B* **90** 224503
[20] Appelbaum J A 1967 *Phys. Rev.* **154** 633-43

IOP Conf. Series: Materials Science and Engineering **369** (2018) 012025 doi:10.1088/1757-899X/369/1/012025

Finite element simulation and experimental verification of steel cord extraction of steel cord conveyor belt splice

X G Li[1,2,4], X Y Long[1,2,4], H Q Jiang[1,2] and H B Long[3]

[1]School of Electronic and Information Engineering, Tianjin Polytechnic University, Tianjin, China

[2]Tianjin Key Laboratory of Optoelectronic Detection Technology and System, Tianjin, China

[3]National Energy Investment Group Co., Hami Dananhu Mine, China

E-mail: lixianguo@tjpu.edu.cn/ 1186831688@qq.com

Abstract. The splice is the weakest part of the entire steel cord conveyor belt. And it occurs steel cord twitch fault frequently. If this fault cannot be dealt with timely and accurately, broken belt accidents would be occurred that affecting the safety of production seriously. In this paper, we investigate the steel cord pullout of the steel cord conveyor belt splice by using ABAQUS software. We selected the strength of steel cord conveyor belt ST630, the same as experiment sample in type specification. The finite element model consists of rubber, steel cord and failure unit. And the failure unit is used to simulate the bonding relationship between the steel cord and the rubber. Mooney-Rivlin hyper-elastic model for rubber was employed in the numerical simulations. The pullout force of length 50.0 mm single steel cord,on both sides of a single steel cord and on both sides of the double steel cords each impacted at steel cord conveyor belt splice were numerically computer and typical results obtained have been validated by experimental result. It shows that the relative error between simulation results and experimental results is within 10% and can be considered that the simulation model is reliable. A new method is provided for studying the steel cord twitch fault of the steel cord conveyor belt splice.

1. Introduction

Conveyor belts are the key components of the belt conveyor traction and carrying, which account for more than 40% of the total cost of the belt conveyor [1]. The steel cord twitch fault is the main reason for broken belt and it is needed to be prevented and controlled. According to the statistics, 93.75% broken belt occurred at the splice [2,3]. As the steel cord twitch fault cannot be timely and accurate detection and treatment, broken belt accidents occur frequently, seriously affect the safety of production. Our team earlier developed a non-destructive detecting system for X-ray steel cord conveyor belts. However, it still needs to do a more in-depth study in the quantitative analysis of steel cord twitch. Especially, it is not clear in quantitative relationship between steel core twitch and splice strength. Due to the lack of quantitative theoretical analysis of steel cord twitch, after detecting the steel cord twitch fault, the users of the steel cord conveyor belts only by experience and connector lap margin, select the local processing, redo the connector or replace a section or the entire conveyor belt. The over correction will affect the production and cause a huge waste of resources [4,5].

Finite element simulation technology is an important numerical simulation technology in engineering application. It has been used in the steel cord conveyor belt splice analysis for many years. Conveyor Dynamics Inc. of the United States (CDI) developed a 3-dimensional finite element model of steel cord and isotropic hyper elastic nonlinear rubber. The numerical simulation was carried out by ANSYS software, and the formula for selecting these parameters is also given to guide the design of splice [6]. Adamas uses rubber as an incompressible material and analyzes the splice using finite element simulation techniques under nonlinear large deformation conditions, the simulation results obtained and the experimental results are similar [7]. In 2012, Mazurkiewicz investigated a finite element method that was used to simulate the behavior of a rubber conveyor splice, analyzed the influence of the rubber layer thickness on the strength of the splice, and evaluated the possibility of making predictions the fatigue and the strength of splice, and it is valuable of guiding the structural design of splice by using this model. However, the method and model are aimed at ordinary rubber conveyor belt, not the steel cord conveyor belt [8]. In 2016, Li Yuan *et al* studied finite element analysis method of Aramid fiber rubber belt splice strength, and established the finger splice finite element model. The influence of the width of the finger root, the thickness of the glue and the enhance layer placement direction on the strength of the splice were analyzed [9]. In 2017, Du Wenzheng *et al* found a method of steel cord stress analysis and a study of fatigue life. And the Von-Mises equivalent stress at the contact between the core cord and the inner steel cord is the maximum and the fatigue life is the minimum [10]. The above results show the feasibility and effectiveness of finite element simulation in the analysis of steel core twitch at the steel cord conveyor splice. However, the quantitative relationship between steel core twitch and splice strength based on this technology needs to be further studied.

In this paper, we investigate the steel cord pullout of the steel cord conveyor belt by using ABAQUS software. The pullout force of length 50.0 mm single steel cord, on both sides of a single steel cord and on both sides of the double steel cords each impacted at steel cord conveyor belt were numerically computer. Numerical simulations have been performed on ABAQUS / Explicit finite element code and typical results obtained have been validated by experimental result. Close correlation of typical experimental and numerical results validated the numerical procedure and the model.

2. Constitutive model

2.1. Finite element model establishment

The establishment of the model in the simulation and experiment are selected strength specifications for the ST630 steel cord conveyor belt. When establishing the model of the conveyor belt steel, the steel is simplified and replaced by a cylinder which in order to reduce the difficulty of modeling. The geometric model of the steel cord conveyor belt with length of 50.0 mm single steel cord, on both sides of a single steel cord and on both sides of the double steel cords was made in ABAQUS/CAE shown in figures 1(a)-1(c).

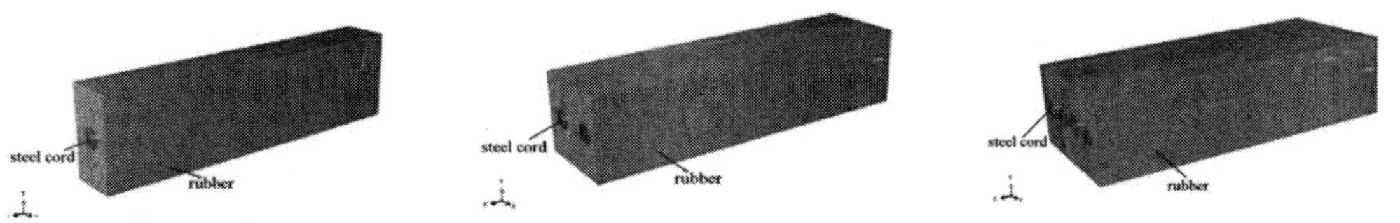

(a) Single steel cord (b) On both sides of a single steel cord (c) On both sides of double steel cords

Figure 1. Finite element steel cord conveyor belt splice model.

2.2. The definition of material constitutive relationship

The numerical simulations were carried out using ABAQUS/ Explicit finite element code. The material model of the rubber, steel cord and failure unit was made in ABAQUS shows in figure 2. The rubber belongs to the hyper elastic material behavior, while the material of the steel cord was linear elastic, the failed unit are set to be the same as the steel cord and adding the material breaking behavior based on the physical criteria.

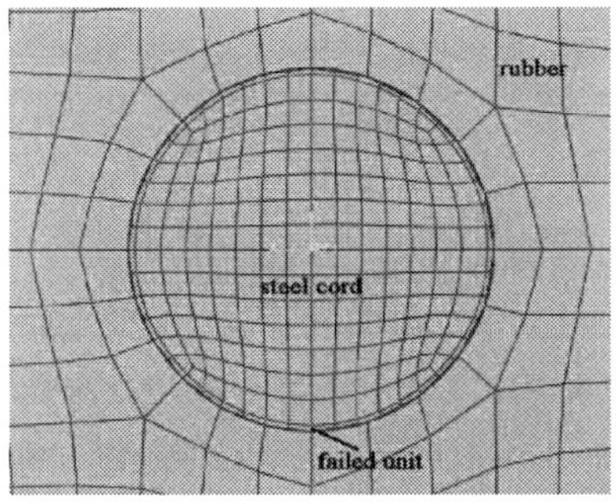

Figure 2. Finite element model material definition.

2.2.1. Rubber materials. In the numerical simulation, the upper and lower coverings of the steel cord conveyor belt are treated as rubber, and the rubber is a kind of polymer, belonging to the polymer material. Its stress-strain behavior belongs to the hyper elastic material behavior and uses the strain potential (U) (strain energy potential) to describe in ABAQUS. Strain potential has many forms, the polynomial strain potential is more commonly used [11]. And the Mooney-Rivlin model is used to describe the stress behavior of rubber in the numerical simulation. Mooney-Rivlin formal model is defined as:

$$U = C_{10}(I_1 - 3) + C_{01}(I_2 - 3) + \frac{1}{D}(J - 1)^2 \quad (1)$$

Where U is the strain potential, J is the elastic volume fraction, I_1 and I_2 are the torsional measures in the material, C_{10} and C_{01} are the shear properties of the material and D is the compressibility (coefficient).

2.2.2. Steel cord material. The steel cord is defined as an ordinary linear elastic material and the material parameters are set as follows: elastic modulus E=210 GPa, Poisson's ratio μ=0.28 and density $\rho=7.85\times10^{-9}$ Ton/mm^3. In the present study, the amount of deformation of the steel cord relative to the rubber during the taping process is very small and in order to save calculation time and improve the efficiency, the steel cord was neglected since it is taken as rigid body.

2.2.3. Failure unit material. Failure unit plays a key role in the whole numerical simulation which is used to simulate the bonding between the steel cord and rubber, in order to get more accurate simulation results, the failure unit should be as small as possible in unit size. The failure of the basic parameters of the unit is set to be consistent with the steel cord, as well as add material plasticity behavior and material damage behavior based on physical criteria.

2.3. Definition of unit failure

The material fracture criteria of failure unit were carried out using shear damage in ABAQUS / Explicit. According to physical criteria, when the node's strain value is greater or equal to a given value, the failure unit will be deleted, thus the steel cord will be separated from the rubber. The detach between steel cord and rubber is based on the establishment of shear failure criteria, and the basis of shear failure criteria is based on the equivalent plastic strain ε. The breaking value of each unit can be determined by equation (2).

$$D = \frac{\varepsilon}{\varepsilon^d} = 1 \quad (2)$$

Where ε is the equivalent strain in the integral step, and ε^d is the failure strain under the current conditions. When D=1, the material fails and deletes the relevant unit.

3. Experiment

3.1. Universal testing machine pullout experiment

Figure 3 shows that carried out experiments in the universal testing machine pullout experiment platform. Figure 4 show the experimental sample with length of 50.0 mm single steel cord ,on both sides of a single steel cord and on both sides of the double steel cords were made in steel cord conveyor belt with strength specification ST630 [12].

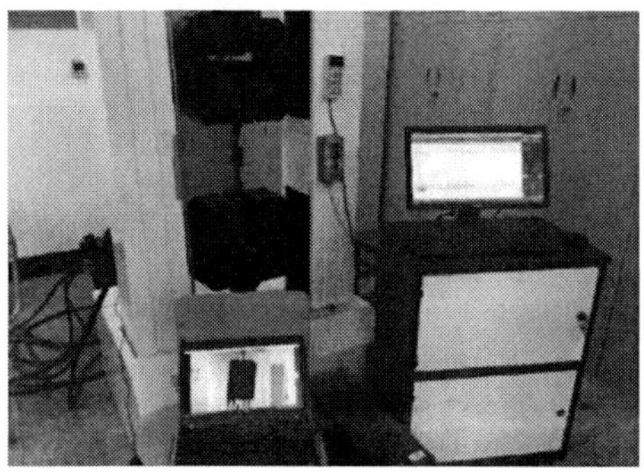

Figure 3. Universal testing machine pullout experiment platform.

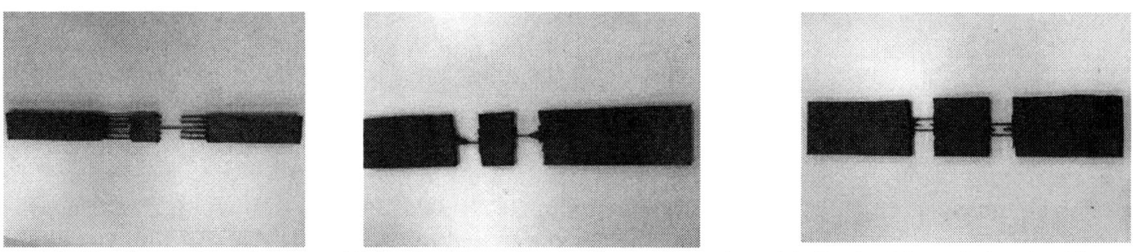

(a) Single steel cord (b) Both sides of a single steel cord (c) Both sides of the double steel cords

Figure 4. Experimental steel cord conveyor belt splice sample.

3.2. Numerical simulation

The numerical simulation with length of 50.0 mm single steel cord pullout of process were carried out using ABAQUS/Explicit in figures 5(a)-5(c). It shows the change of rubber, steel cord and failure unit in the steel cord pullout process.

During the initial stage of the simulation, both the steel cord and the rubber adhered closely and no detach occurred. With the increase of the force exerted on the steel cord, rubber will be pulled out a distance along with the steel cord. And the steel cord extracted constantly, the deformation of rubber is more and more prominent. Until the extraction force reaches the maximum, it happened some degree of de-bonding in the bonding interface between the steel cord and rubber which led to failure, and the steel cord and rubber will be separated. After the extraction force reaches the maximum value, the steel cord is separated from the rubber and is in a slippery state, then the extraction force rapidly drops to a minimum.

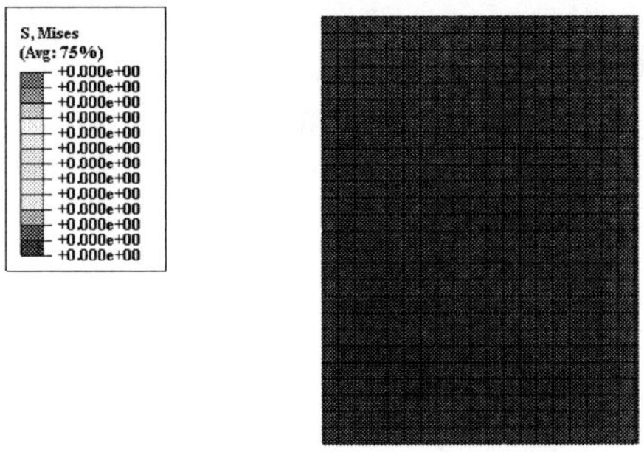

(a) Steel cord is not stressed

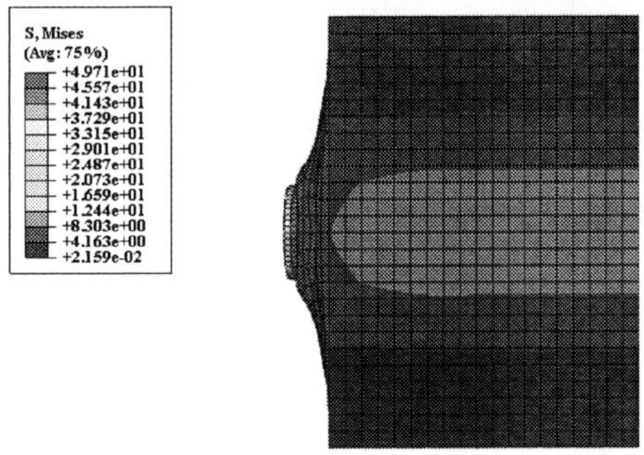

(b) Steel cord begins to be stressed

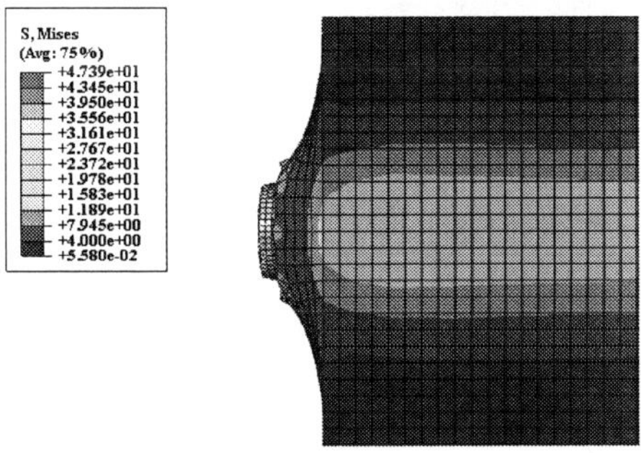

(c) Steel cord is stressed to the maximum

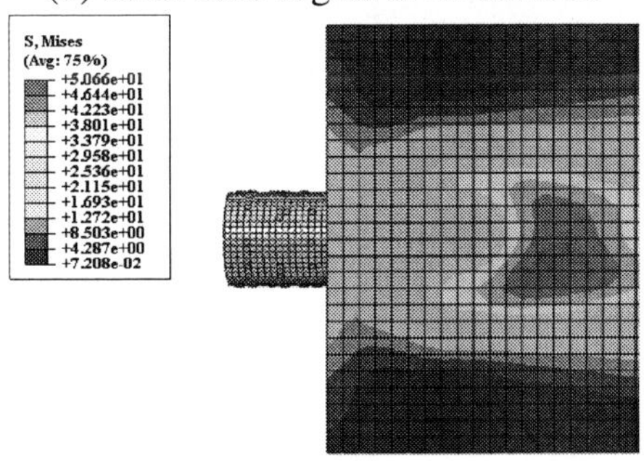

(d) Steel cord and rubber are in a slippery state

Figure 5. Process of steel cord pulled out of rubber.

4. Results and discussion

Simulations were performed on 50.0 mm long models of single steel cord, on both sides of a single steel cord and on both sides of the double steel cords, and the varieties of pullout force with time of the numerical and experimental were obtained in figures 6(a) and 6(b), respectively. It can be seen that the trend of the pullout force with time almost the same by comparing (a) and (b). At the beginning, the extraction force is linearly increased to the maximum, and then quickly dropped to zero. For different numbers of the steel cords, the extraction force is also different.

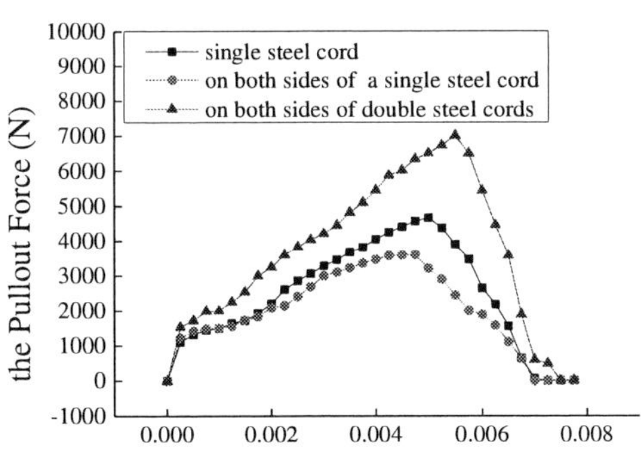

(a) Numerical results

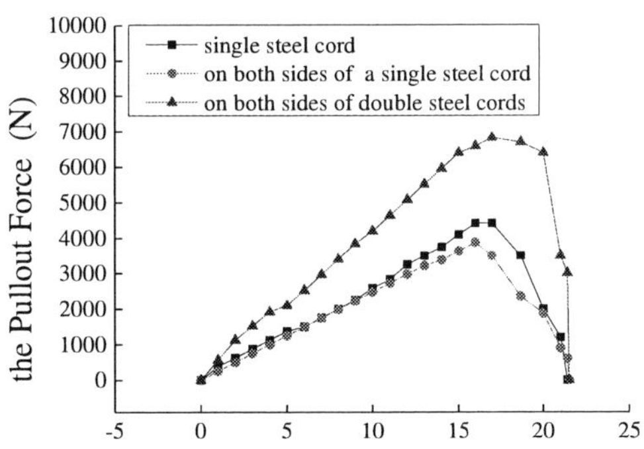

(b) Experimental results

Figure 6. Pullout force - Time graph of different steel cord numbers.

Experimental and numerical results for 50.0 mm long models of single steel cord, on both sides of a single steel cord and on both sides of the double steel cords are shown in table 1. It can be seen that the relative error between simulation results and experimental results is within 10%. As the experiment is not only affected by external environmental factors, such as: surrounding temperature, humidity and damage in sample preparation etc., but the conveyor belt internal factors, for example, there are different degrees of aging or deformation because of the different time of the sample placement in the sample preparation process. Therefore, the error within 10% is acceptable, and it can be considered that the simulation model is reliable.

Table 1. The comparison of the numerical results with the experimental results.

steel cord	Numerical value (N)	Experimental value (N)	Relative error
Single steel cord	4669	4422	5.59%
On both sides of a single steel cord	3660	3995	8.39%
On both sides of double steel cords	7007	6827	2.64%

It shows that the relative error between simulation results and experimental results is within 10% and it can be considered that the simulation model is reliable. It provides a new method for studying the steel cord twitch fault of the steel cord conveyor belt splice.

5. Conclusion

Numerical simulations have been performed on ABAQUS / Explicit finite element code and typical results obtained have been validated by experimental results, close correlation of typical experimental and numerical results validated the numerical procedure and the model.

In this paper, it provides the value of the steel cord pullout force at different steel cord numbers, and the results provide a useful basis and reference for the performance of steel cord conveyor belt splice with theoretical research and application.

Acknowledgments

This work was supported by the National Natural Science Foundation of China (grant number 51504164) and the Tianjin City High School Science & Technology Fund Planning Project (grant number 20130708). The authors also acknowledge the anonymous reviewers for their helpful comments on the manuscript.

References

[1] Zhang M M 2014 Mine conveyor belt longitudinal tear detection system using machine vision technique (Beijing: China University of Mining & Technology Beijing) (In Chinese)
[2] Huang M and Wei R Z 2005 Real time monitoring techniques and fault diagnosis of mining steel cord belt conveyors *Jo China Coal Soc* **30** 245-250 (In Chinese)
[3] Li X X 2013 Research of fault diagnosis method of steel cord conveyor belt based on wavelet singularity and neural network (Taiyuan: Taiyuan University of Technology) (In Chinese)
[4] Li X G, Miao C Y, Zhang Y and Wang W 2012 Automatic fault detection for steel cord conveyor belt based on statistical features *J China Coal Soc* **37** 1233-8 (In Chinese)
[5] Li X G, Miao C Y and Fu Y 2015 Fault detection method and joint elongation calculating for coal mining conveyor belt *Coal Sci Techno* **43** 82-6 (In Chinese)
[6] Nordell L, Qiu X and Sethi V 1991 Belt conveyor steel cord splice analysis using finite element methods *Bulk Solid Handling* **11** 863-8
[7] Adams A 1994 FEA of steel cable conveyor belts splices *Rubber World* **41** 524-9
[8] Mazurkiewicz D 2012 A knowledge base of the functional properties of the conveyor belt adhesive joint for FEM simulation of its stress and strain state *J Adhes Sci Technol* **26** 1429-42
[9] Li Y 2016 Study on splice strength of aramid rubber conveyor belts based on finite element

methods and experimentations (Beijing: Beijing University of Chemical Technology) (In Chinese)

[10] Du W Z, Ma B Z, Cao D Z and Xie Z 2017 Research on stress and fatigue life of steel wire rope *Coal Eng* **49** 134-7 (In Chinese)

[11] Zhuang Z and You X C 2009 *Finite Element Analysis and Application Based on ABAQUS* (Beijing: Tsinghua University Press) (In Chinese)

[12] Li X G, Jiang H Q, Miao C Y, Yang Y L and Long X Y. A test device for tensile test of steel cord conveyor splice (Chinese patent: CN106829388A, 2017.06.13)

IOP Conf. Series: Materials Science and Engineering **369** (2018) 012026 doi:10.1088/1757-899X/369/1/012026

Material conformity and bactericidal properties of high-frequency-pulse-modulated and low-frequency-pulse-excited plasmas

H Okawa[1,3] and T Akitsu[2]

[1]Faculty of Future Industry, Happy Science University, Hitotsumatsuhei 4427–1, Chouseimura, Chiba, Japan
[2]Professor Emeritus of the National University of Yamanashi, Takeda 4-3-11, Kofu, Yamanashi, Japan

E-mail: hiroshi-okawa@happy-science.university

Abstract. Plasma sterilization attracts an increasing attention as an alternative method for chemical sterilization. In this study, we investigate plasma sterilization for practical applications, particularly in dentistry and oral surgery [1]. Helium-diluted oxygen was excited by a dielectric barrier electrode at normal atmospheric pressure. Control of the neutral gas temperature was performed under the plasma sterilization. The relation between the intensity of the spectral emission from the excited oxygen atoms and bactericidal effect was investigated using *Bacillus stearothermophilus* and opportunistic infection bacterium. A comparison is performed with a low-frequency wide-gap discharge. Degradation and material conformity were investigated using the Tyvek unwoven fabric for the sterile package and soft-silicone resin, methyl-methacrylate powder filler used in the dental surgery.

1. Introduction

Plasma sterilization attracts an increasing attention for the prevention of infections in hospitals. Conventional schemes of sterilization include gas sterilization using ethylene oxide and autoclave, pressurized steam, and cleaning with a liquid sterilizing agent. Although ethylene oxide gas sterilization is widely used in the sterilization of medical equipment, in the 1980s, concerns for carcinogenic properties and DNA damage were reported. A strict protocol is followed in the application of this scheme [2]. The application of a high-temperature pressurized steam is limited to heat-resistive materials. Liquid agents, such as 2–20% glutaraldehyde, provide not only a disinfection of pathogenic bacteria but also coagulation of proteins. The residual liquid agent may be a strong irritant to the skin and esophageal surface.

In plasma sterilization, the main role in the bactericidal effect is performed by short-life radicals, using the oxidation and hydrogenation activities of the free-bond and ultra-violet (UV) emissions; no harmful chemical compounds are required. In this study, we present a comparative study of sterilization using helium mixed with oxygen or nitrogen controlled separately, and material conformity to the sterile package, soft silicone forming material, poly-siloxane, and granular bone cement, methyl-methacrylate.

In this experiment, we developed a low-cost power sources to be used in a simple structure of dielectric barrier electrodes, reported for the first time in 1987 by Kanazawa *et al* [3]. The sterilization characteristics of this original atmospheric-pressure glow device were validated by the authors' group in 2001–2003. Since then, numerous applications to surface modification and film synthesis have been reported [4].

The first contribution to the sterilization of pathogenic bacteria was reported in 1996 by Laroussi. Inactivation of various pathogenic microorganisms including viruses and pathogenic proteins was observed in influenza and bovine spongiform encephalopathy [5,6]. Although a model for the effective cause of sterilization has been proposed, such as UV radiation, and oxygen and nitrogen oxide radicals, various experimental schemes have been reported involving the electrode design and frequency, excitation wave-form, and power supply [7–10]. Plasma sterilization needs active oxygen or nitrogen. These agents react with chemical compounds; therefore, it is important to evaluate effects on materials by plasma as well as the bactericidal effect [11]. Only a few studies reported the material conformity and sterilization effect under a controlled power reduction of radical diffusion by the medical equipment in the sterilization process [12].

2. Experimental methods

2.1. Equipment

Helium-diluted oxygen and nitrogen were excited between dielectric barrier electrodes (planar parallel electrodes), excited with a high-frequency power supply or low-frequency pulse inverter at normal atmospheric pressure. Spectral emission of the reactive gas was observed. The relation between the intensity of oxygen radicals and sterilization was confirmed. A diffused glow discharge is difficult to occur when oxygen is used; helium carrier gas with a large electron diffusion leads to a stable glow discharge. The experimental results show that helium dilution is important for the production of dissociated atoms and excitation to the higher level. In the collisional excitation process, Penning excitation, energy transfers from excited helium atoms to excite oxygen atoms or hydroxyl radicals. This process is generally used in a discharge lamp and gaseous discharge laser, in order to decrease the breakdown voltage.

2.1.1. High-frequency discharge device. Figure 1 shows a schematic of the employed system. The helium-diluted mixture is excited between the dielectric barrier electrodes by a modulated radio-frequency power supply at 27.12 MHz (maximum 1 kW). Water cooled the electrode block, a rectangular aluminum parallelepiped with a width, depth, and thickness of 150, 50, and 20 mm, respectively, with a complex perforation, maintained at 20°C. The surface of the electrode was covered with a transparent quartz plate (230 mm × 115 mm, thickness: 3 mm). The main role of these dielectric barrier plates is to prevent the transfer of arc discharge and excitation of the radicals, referred to as non-thermal excitation. In this experiment, the discharge volume is only 3 mm, surrounded by the glass plate and polymer sheet.

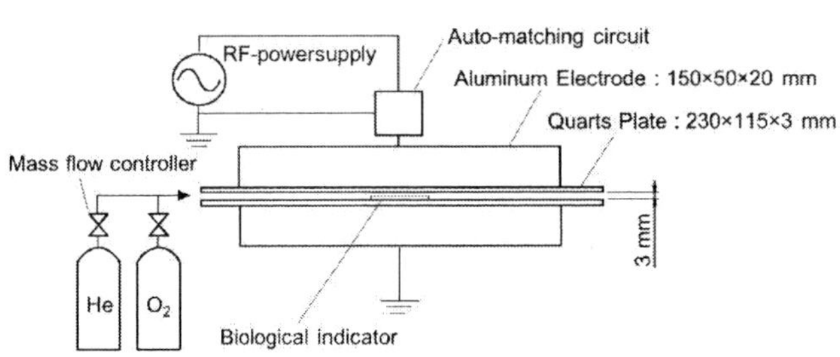

Figure 1. Pulse-modulated high-frequency discharge at normal atmospheric pressure, and schematic.

The design of the diffuser is very important, as the homogeneity of the excitation is determined by the gas flow. The helium-diluted oxygen gas was introduced by a glass diffuser with 10 equally spaced

IOP Conf. Series: Materials Science and Engineering **369** (2018) 012026 doi:10.1088/1757-899X/369/1/012026

perforations, with a diameter of 1 mm, attached at the back of the discharge volume. A similar gas leak plate with equally spaced perforations was placed on the front end of the discharge volume. The gas flow was controlled with a piezo-activated mass flow controller, from 1 mL/min to 5 L/min.

2.1.2. Low-frequency-pulse-inverter plasma device. A pulse-inverter power supply was developed for the low-frequency excitation (50 kHz, maximum 150 W). The electrode had a width, depth, and thickness of 150, 36, and 12 mm, respectively, with a complex perforation for water cooling. The maximum discharge volume had a square shape with dimensions of 150 mm × 36 mm, and gap of 20 mm.

The true input power was estimated by the digitized current waveform multiplied by the voltage, as shown in figure 2. In this experiment, the electrode gap was 15 mm, and the gas flow rate was 3 L/min; no significant change was observed by the change of the gas flow rate and mixture.

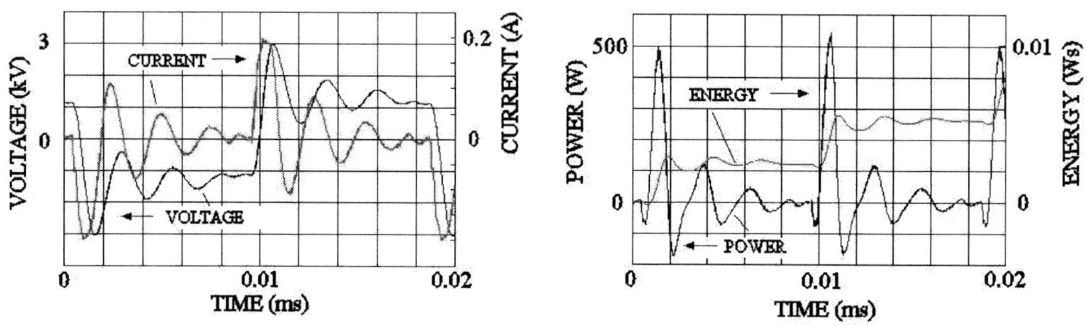

(a) (b)

Figure 2. Waveforms for the estimation of power deposition. (a) Current–voltage waveform. (b) Electric power and accumulated energy.

From the origin, a zero-cross point of the voltage wave form, the integration over one period, from approximately -0.00000920 to 0.00000910 per second gives the average power:

$$P = \frac{1}{T}\int_{-0.0000092}^{0.0000091}[I(t)\cdot V(t)]dt$$
$$= \frac{1}{0.0000183}\cdot 0.00052192 \fallingdotseq 28.5$$

This calculation revealed that most of the input electric power is consumed as reactive electric power, attributed to the charging current to the capacitance of the dielectric barrier.

2.1.3. High-frequency plasma jet device. Figure 3 shows the plasma jet device. The helium-diluted gas mixture is excited with a high-frequency power source at 13.56 MHz, 400 W. The high-frequency power source is an auto-matching servo-device without reflection. The plasma jet device consists of two electrode blocks with the dielectric barrier forming a nozzle at the outlet of the flowing gas. Two aluminum blocks were placed at the opposite sides. The temperature was controlled at 20°C by the circulation of water in the channel. The electrode block had a depth, length (parallel to the gas flow), and thickness of 50, 70 and 20 mm, respectively. The surface of the electrode was covered by a quartz plate with a thickness of 1 mm to prevent the transfer of arc discharge. The discharge gap was 1.5 mm. The gap was narrowed to 0.5 mm at the opening, in order to increase the flow velocity. Figure 3 shows the electrode of the plasma jet device.

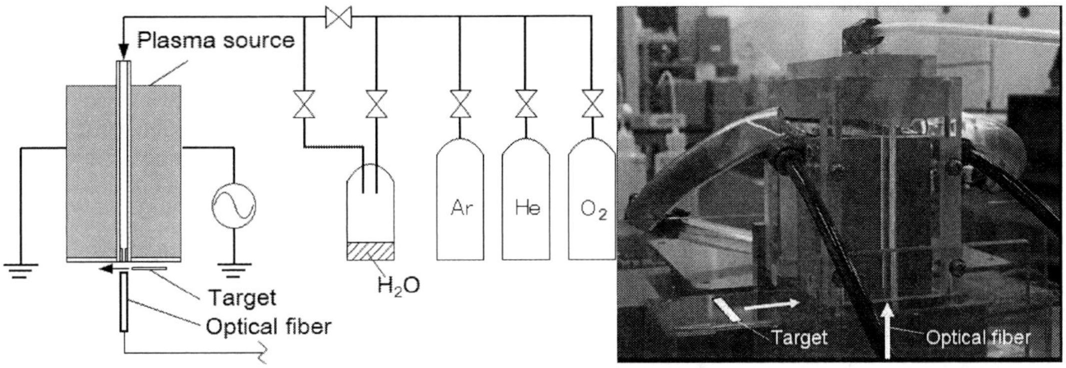

Figure 3. Lateral view of the plasma jet device and schematic of the experimental setup.

The discharge plasma can be observed through the lateral barrier. A biological indicator was passed under the opening of the electrode, at a distance of 2.4 mm, moving at a speed of 3 mm/min. The spectral emission at 309 nm was observed from the opening.

The gas flow was controlled with a piezo-activated mass-flow controller, from 50 mL/min to 5 L/min, and a moisture of distilled water was loaded diverting a part of the gas mixture, approximately 3 mol%.

2.2. Diagnostics

The spectral emission was observed with a Raman spectrometer, MonoSpec18, Thermo Vision Colorado, equipped with a thermoelectrically cooled charge-coupled device (CCD) camera system ST-6V, SBIG (focal length: 156 mm). The optical probe was placed at the end of the outlet of the plasma, aimed to observe a band peak in the vicinity of 309 nm proportional to the excited state of the OH radicals. The gas temperature was measured with an optical sensor (FL-2000, ANRITSU, Japan) inserted into the discharge volume, which was not influenced by the high-frequency electric field.

2.3. Biological indicator

Spore forming bacteria are used as a biological indicator for the validation of the autoclave, *Geobacillus stearothermophilus*, ATCC 7953, sampled from Attest TM 1262 (3M Co., figure 4(a)) inoculated on a filter paper (8.6×10^5 colony forming unit (CFU)). The carrier was dismounted from a sterile cylindrical plastic package in a sterile environment; only the carrier was inserted into the plasma. After the experiment, the carrier was mixed with a recovery medium with a pH indicator, and incubated in a sterile cylindrical package for 48 h, conforming to the protocol prescribed by 3 M. The temperature of the incubator was 56°C for *Geobacillus stearothermophilus.* One biological indicator was set at the test.

An opportunistic infection refers to a disease caused by weak infectious microbes, which is normally prevented by the natural immunity. Various microbes can cause opportunistic infection. In this experiment, the biological indicator of selected microbes was prepared in the following manner. After the culturing in a soybean casein digest (SCD, Becton Dickinson), 0.01 mL was cultured on a solid agar culture medium. One scoop of a platinum inoculation loop was diluted in distilled water. The density was adjusted to 10^6 CFU; the inoculated water was then placed on a glass plate with dimensions of $18 \times 18 \times 0.15$ mm^3, dried in a clean room, and covered with a sterile package of Tyvek, non-woven polyethylene terephthalate sheet with a thickness of 0.15 mm covered with a transparent film (figure 4(b)). The selected samples are Gram negative: *Escherichia coli* ATCC8739, *Salmonella enteritidis*, Gram positive: *Staphylococcus aureus* ATCC6538, and yeast fungus: *Candida albicans* ATCC10231. After the exposure, the biological indicators were immersed in a recovery medium, SCD, 100 mL, and incubated for seven days at 30–35°C (25°C in a *Candida albicans* case). The sterility after the plasma treatment was judged by the turbidity of the recovery medium. The influence of experimental conditions, such as air humidity in the drying process and conservation time, is investigated with an untreated control. Although the turbidity and pH detection show only whether the sterilization process was successful, the colony counting assessment could provide better information. The colony counting was performed with *Geobacillus stearothermophilus* and *Bacillus atrophaeus*, discussed in other reports.

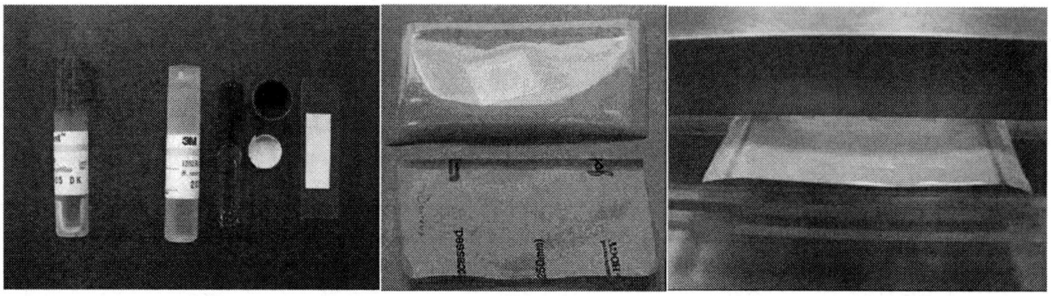

(a) (b) (c)

Figure 4. Biological indicator. Commercially available biological indicators (a) *Attest*™ 1262 and (b) *Tyvek*: non-woven fabric of polyethylene. (c) Exposure to the atmospheric-pressure plasma discharge.

3. Materials, methods, and experimental results

3.1. High-frequency plasma

In the atmospheric pressure discharge, the thermal conductance between the discharge volume and cooled surface is limited. The neutral gas temperature at high frequencies, 13.56 MHz and 27.12 MHz, increases in the wide gaps, which may be desirable from the viewpoint of the treatment. A control scheme of the temperature was developed. Using pulse modulation, the neutral gas temperature of 400°C became as low as 95–100°C. This operation was realized with a gap of 3 mm, electric power of 670 W, pulse width of 10 μs, and duty cycle of 50%. The flow rate of the gas mixture was 1.5 L/min, while that of oxygen was 1 mL/min. The results are shown in table 1. Negative means that the successful sterilization of 3M-type Attest TM 1262, 10^6 CFU, was successfully inactivated. Positive corresponds to the observation of the propagation. One biological indicator was set at the test. In the pulse modulated high-frequency discharges, the discharge is cut-off before the electron energy is conducted or thermalized to heavy ions and neutrals. The results indicate that the best pulse width was 10 μs.

Table 1. Dependence on the pulse width at 27.12 MHz, direct exposure.

Pulse width (μs)		Temperature	Processing time (one pass, minutes)						
ON	OFF	(°C)	0.5	1.0	1.5	2.0	2.5	3.0	3.5
30	30	134	+	-	-	-	-	-	
20	20	125	+	+	-	-	-	-	
15	15	115		+	-	-	-	-	
10	10	90						+	-
7	7	65						+	+

The neutral gas temperature was fixed at 90°C, and the biological indicator was exposed through a sterile package. The right-hand side of table 2 shows the control in the hot-gas environment without the exposure to the plasma.

Table 2. Opportunistic infection bacteria, exposure through the sterile package.

Species	Plasma treatment (min)					Treatment at 90°C (min)				
	1	3	5	10	15	1	3	5	10	15
Escherichia coli ATCC8739 (1.6×10^7)	-	-	-	-	-	+	-	-	-	-
Salmonella enteritidis (3.5×10^7)	-	-	-	-	-	+	+	+	-	-
Staphylococcus aureus ATCC6538 (4.7×10^7)	+	+	-	-	-	+	+	+	+	+
Candida albicans ATCC10231 (5.1×10^6)	-	-	-	-	-	+	-	-	-	-

Power source: 670 W, ON: 10 μs, OFF: 10 μs, He: 1.5 L/min, O_2: 1 mL/min

3.2. Low-frequency discharge

3.2.1. Sterilization experiment. Table 3(a)(b) shows the results of the sterilization experiment. The gas mixture was He (3 L/min) and O_2 or N_2 (3 mL/min). In this experiment, the gap was 15 mm. Although the gap is wider than that of the high-frequency discharge, the neutral gas can be maintained at a lower temperature of approximately 60°C. The number of the trial is indicated by the symbols. Although the nitrogen case shows an unknown opportunistic dependence, it should be noted that a certain number of the experiment indicated a grey area, such as " - + - ", corresponding to a partially complete sterilization of the 10^{-6}-th order in 5 min.

Table 3. Antibacterial effect at a low frequency; comparison between oxygen and nitrogen mixture.

(a) Sterilization experiment (direct exposure)

Flow rate (L/min)			Time (min)		
He	O_2	N_2	5	10	15
3	0.003	0	+ + +	+ + +	- - -
3	0	0.003	- + -	+ + +	+ - -

(b) Sterilization experiment (exposure inside a small syringe)

Flow rate (L/min)			Time (min)				
He	O_2	N_2	30	35	40	45	60
3	0.003	0	+++++	++	- - - - -	- - - -	-
3	0	0.003	+	-	+ - +	- + - +	- -

In the low-frequency device, the discharge was excited by a non-sinusoidal pulse voltage. The electron energy exceeds the dissociative ionization energy of the working gas, oxygen; the so-called advanced oxidation is triggered with a lower thermal heating effect.

The band spectrum in the vicinity of 309 nm was observed for a 0.003/3-mixture with oxygen. This peak was emitted by the OH radicals; no peak was observed with nitrogen.

3.2.2. Surface treatment of polymer materials in dentistry. A low-frequency discharge was excited between the dielectric barrier electrodes, at a distance of 20 mm. Soft-silicone impressions of a back teeth, Exafine^R and Exahigh-flex ^{R}, GC Japan, were analyzed.

The power input was operated at 150 W. The neutral gas temperature increased to approximately 80°C. After 10 min of exposure, the surface of the impression material was studied with a reflective Fourier-transform infrared (FTIR) photometer. A typical example of the change in the Exafine^R sample is shown in figure 5. The transparencies of the surfaces, in the Exafine^R and Exahigh-flex^R samples, exhibited significant decreases at 1,500 cm^{-1} and 2,800 cm^{-1}. These changes were localized at the surface; no significant change was observed in the cut-out samples, and no significant change in the form was induced by the plasma exposure.

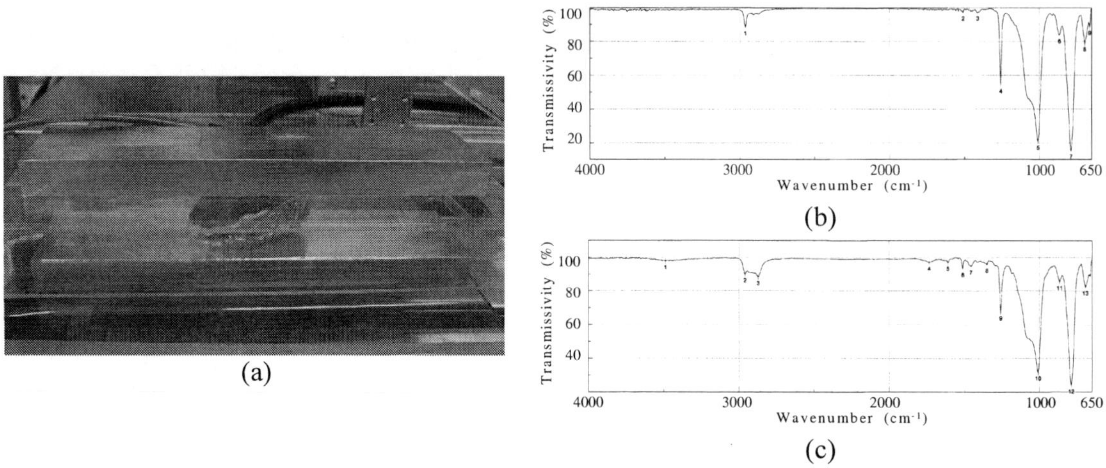

Figure 5. Plasma treatment of an impression material *Exafine*[R] (GC). (a) Impression material. (b) FTIR diagrams; control without a plasma treatment, and (c) after a plasma treatment, at 150 W and 50 kHz, for 10 min.

The results of the sterilization experiment suggest that the helium-diluted oxygen low-frequency plasma can be a candidate for the precaution of infection by viral hepatitis and human immunodeficiency virus (HIV), through the impression material.

Further, the effect of the plasma exposure to the granular bone cement resin was investigated. In general, these powders are stored or kept open for a subsequent use without a sterilization process, as a chemical sterilization may reduce the biological conformity. Two 0.16-g scoops of bone cement included in Super-Bond® were measured and expanded in a sterile petri dish (figure 6(a)). Super-Bond® is a self-curing adhesive resin cement including 4-META, MMA/TBB: 4-methacryloyloxyethyl trimellitate anhydride; Methyl methacrylate; Tri-n-butyl borane polymerization agent. Immediately after the exposure to the helium-diluted oxygen plasma, MMA monomer was mixed to form poly (methyl methacrylate) (PMMA) solid (figure 6(b)). The surface hardness was measured using a Vickers hardness tester. The average hardness of the control is 17.2 HV (Vickers Hardness), whereas the granular bone cement after the 10 min of plasma exposure formed a solid with a decreased hardness of 15.4 HV. This change may be in a permissible range as a tooth filling material, corresponding to an increase in the flexibility. This result reveals that surface phenomena, such as oxidation or addition of carbonyl groups, occurred on the filler.

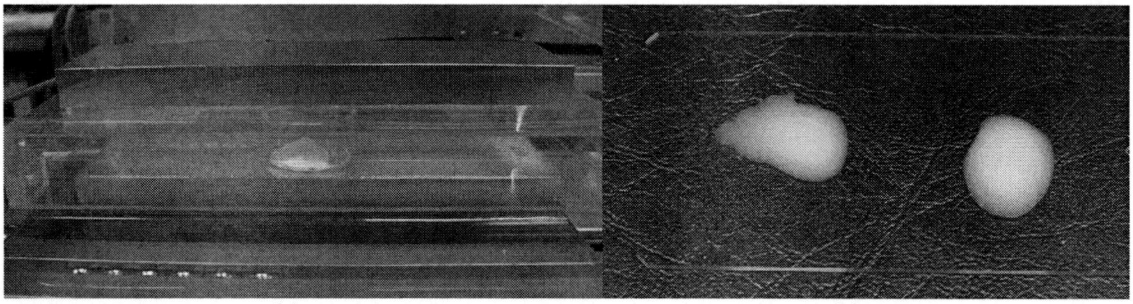

(a) (b)

Figure 6. Plasma treatment of a granular acrylic resin. (a) Plasma treatment of a granular acrylic resin under atmospheric pressure. (b) Polymerized powder mixture (left) without treatment and (right) plasma-treated.

3.2.3. High-frequency plasma jet device sterilization experiment. We performed a sterilization using the plasma jet device and sterility assessment with the biological indicator, *Geobacillus stearothermophilus*, and pH detection. Table 4 shows the results of the sterilization experiment. The sterilization was performed in a helium carrier without a gaseous mixture, but with the water vapor with a density at the saturated water vapor pressure at 20°C, approximately 3 mol%. The detection of the oxygen-containing functional groups was performed using X-ray photoelectron spectroscopy (XPS).

Table 4. Antibacterial effect at a high-frequency plasma jet.

Discharge medium	O_2	No mixture	50 mL/min	No mixture	50 mL/min
	H_2O	No mixture	No mixture	Vapor	Vapor
He (5 L/min)		-	+	+	+
He (2.5 L/min) + Ar (2.5/min)		-	+	+	+

The band spectrum in the vicinity of 309 nm was observed for a mixture with oxygen. This peak was emitted by the OH radicals generated from the migrated water molecules, without mixture; no peak was observed with oxygen or water vapor mixture.

4. Conclusion

A plasma sterilization was realized at the normal atmospheric pressure. The performances were compared for three different types. In each experiment, a $6Log_{10}$-level sterilization of the biological indicators, *Geobacillus stearothermophilus* ATCC7953, was achieved in 3–15 min, which may be a reasonable waiting time in practical applications. In the high-frequency discharge, the biological indicator was sterilized. The thickness of the object was limited by the discharge gap of 3 mm. This thickness was determined by the control of the neutral gas temperature. The low-frequency discharges, where a larger object can be inserted into the discharge gap of 20 mm, require a prolonged processing time. The low-frequency-pulse-excited device and high-frequency plasma jet device can overcome the discharge volume limit, as shown with the high-frequency discharge device.

The plasma sterilization process introduces oxidation or addition of carbonyl groups on the surface of polymer materials. The evaluation of the damage, or benefit of this effect, require further studies. Soft-silicone impression resin and granular PMMA were treated using a low-frequency discharge; an evaluation of the material conformity was performed. The sterility of such medical product is guaranteed by the factory. In general, one unit of the bone cement can be used several times. It should not be discarded; the remaining part is stored, and could be used in a subsequent treatment. A sterilizer for such a small amount of medical material may find a place in the market. In dentistry or oral surgery, such sterilizer could be employed as a precaution against the risk of droplet infection, owing to the infectious contaminations of blood or saliva. The bactericidal effect and material conformity require further studies, for a practical use.

A gas-plasma-based approach is useful for the sterilization of fragile medical devices, in particular, in the reprocessing of endoscopes. The sterilization characteristics need further evaluations with different types of microorganisms [13].

References

[1] Cha S and Park Y-S 2014 Plasma in dentistry *Clinical Plasma Medicine* **2** 4-10
[2] ATSDR (Agency for Toxic Substances and Disease Registry) 2003 Medical Management Guidelines (MMGs) for Ethylene Oxide ($[CH_2]_2O$) CAS 75-21-8, UN#1040
[3] Kanazawa S, Kogoma M, Moriwaki T and Okazaki S 1987 Carbon film formation by cold plasma at atmospheric pressure ISPC8, Tokyo, Japan, August 31 - September 4, Edited by K Akashi and A Kinbara *Proceedings of 8th International Symposium on Plasma Chemistry* pp 1839–44
[4] Kogoma M, Prat R, Suwa T, Okazaki S and Inomata T 1997 Plasma processing of polymers,

NATO ASI series E *Applied Science* **346** 379
[5] Laroussi M 1996 Sterilization of contaminated matter with an atmospheric pressure plasma *IEEE Trans Plasma Sci* **24** 1188-91
[6] Laroussi M, Alexeff I and Kang W L 2000 Biological decontamination by non-thermal plasmas *IEEE Trans Plasma Sci* **28** 184-8
[7] Ohkawa H, Akitsu T, Tsuji M, Kimura H and Fukushima K 2005 Initiation and microbial disinfection characteristics of wide gap atmospheric-pressure glow discharge using soft-x-ray ionization *Plasma Process Polym* **2** 120-6
[8] Ohkawa H, Akitsu T, Tsuji M, Kimura H, Kogoma M and Fukushima K 2006 Pulse-modulated, high-frequency plasma sterilization at atmospheric pressure *Surf Coat Technol* **2005** 829-35
[9] Siti Khadijah binti Zaaba, Akitsu T, Ohkawa H, Onishi K, Katayama-Hirayama K, Tsuji M, Shimizu N and Imanishi Y 2010 Plasma disinfection and deterioration of surgical tools at atmospheric pressure plasma *IEE J Trans Fund Mater* **130** 355-61
[10] Akitsu T, Khadijah binti Zaaba S, Ban K, Onishi K, Ohkawa H, Katayama-Hirayama K, Tsuji M, Shimizu N, Imanishi Y and Kogoma M 2011 Plasma sterilization at normal atmospheric pressure—comparison of working gas and organic loading effect in medical applications (Hauppauge, NY: Nova Science) pp 185–213
[11] Masao Y, Kenichi M and Takashi I 2011 Surface modification by cold-plasma technique for dental implants-Bio-functionalization with binding pharmaceuticals *Japanese Dental Science Review* **47** 89-101
[12] Hiroshi O and Tetsuya A 2015 Bactericidal characteristics and material conformity of atmospheric-pressure glow discharge *Plasma Medicine* **4** 37-47
[13] Sakudo A, *et al* 2011 Inactivation of Viruses *Sterilization and Disinfection by Plasma, Nove Scientific* Chapter **2.4** 49-56

Electrochemical one-step synthesis of hybrid nanocomposites Au/polymer

L G Kolzunova[1,3], E V Shchitovskaya[1,2,3] and M A Karpenko[1]

[1]Russian Academy of Sciences. Far Eastern Branch. Institute of Chemistry, Vladivostok 690022, Russian Federation
[2]Far Eastern Federal University, Vladivostok 690091, Russian Federation

E-mail: kolzunova@ich.dvo.ru/shchitovskaya@gmail.com

Abstract. Possibility of one-step formation of composites – polymeric film/Au nanoparticles by electropolymerization of monomers (acrylamide and its derivatives, formaldehyde, chitosan) at presence of a tetrachloroauric acid in the aqua medium is shown. It is confirmed by X-ray diffraction and Transmission electron microscopy that gold particles exist in nonconductive polymeric matrix as Au^0. The distribution of AuNPs by size was revealed by small-angle X-ray scattering method. It was found that the radius of gold particles is affected by the adding of chitosan into the original monomer composition. The maximum radius of the AuNPs is 35 Å in the absence of chitosan, whereas after its addition, gold particles have been aggregated and their sizes increased to 140 Å. The masses of polymer films and composites based on them rise with electropolymerisation time increasing and reach maximum after 600 seconds of electrolysis. The masses of hybrid films with AuNPs are 26-40% higher than for blank polymethylolacrylamide film.

1. Introduction

Polymers are a very important class of compounds but the increasing request for superior performance needs the improvement of their physical and mechanical properties. It drives researchers towards the synthesis of composite materials, such as hybrid materials obtained by coupling organic polymers with inorganic fillers [1,2]. The incorporation of polymers with other materials can produce new hybrids with distinct properties that are not observed in the individual components. Usually, conductive polymers (polyaniline, polypyrrole, polythiophene and their derivatives [3]) or nonconductive polymers (polymethylolacrylamide [4]) are used as polymer matrix. Noble metal nanoparticles (Au, Ag, Pt and Pd) [3-7], graphene [6] and metal oxides (SnO_2, Fe_3O_4) [8,9] can be used as inorganic fillers.

Among numerous hybrids, polymer/noble metal nanoparticle materials have attracted the most intensive attention in the past few years. Recently, gold nanoparticles (AuNPs) are considered as an effective catalyst for chemical, electrochemical, and photochemical reactions [10-12]. In this case, the AuNPs can be present in the reaction medium either by themselves or in the volume of a solid inert carrier which can be a polymer film or other polymeric material. In this case, the introduction of an AuNPs into an inert matrix is more preferable, since an active reaction medium (for example, electrically conductive polymers) can have a destabilizing effect on nanoparticles, cause the aggregation of AuNPs, and, as a consequence, the loss of initial catalytic properties inherent in exactly non-aggregated AuNPs.

There are two main ways to implement this concept. In the first case, a polymer film is first synthesized, which is then impregnated with an Au^0 precursor ($HAuCl_4$) and then conditions are created under which the precursor is converted to an AuNPs [12,13]. Such a technology, at least, is a three-stage process. In the second case, the precursors Au^0 and the polymer film (monomers, initiators) are together in one electrolyte and during the course of the corresponding reactions simultaneously AuNPs and a polymer film are formed [14,15]. Such a one-step approach allows to significantly reducing the time of obtaining this type of catalyst.

The electropolymerization reaction allows one to combine the formation of a polymer film and AuNPs in one process. At the moment, an electropolymerization system based on polypyrrole is known, which makes it possible to produce polymer (polymer film)/AuNPs composites [14]. This composition has the feature that the polymer obtained on its basis belongs to the class of electro conductive, which contributes to the electrochemical formation of the AuNPs. However, our studies have shown that nanoparticles of a noble metal can also be introduced into a non-conductive polymer matrix. So, it was shown [4] that the electropolymerization composition developed by us on the basis of acrylamide and formaldehyde makes it possible to obtain a polymethylolacrylamide film/Pt. It was found that carrying out electropolymerization in a monomer composition with the addition of H_2PtCl_6 leads to the simultaneous occurrence of several processes: the formation of a polymer matrix, the electro reduction of $PtCl_6^{2-}$ ions to Pt^0, and the introduction of platinum nanoparticles into the volume of the polymer film. Moreover, it was found that the introduction of such an additive as chitosan into the electropolymerization composition intensifies the process of formation of the polymethylolacrylamide/Pt composite. Based on these data, we investigated the possibility of expanding the range of metal-polymer compositions by introducing an AuNPs into a non-conductive polymer film (carrier) during the electropolymerization process.

2. Materials and methods

The following substances were used for the formation of metal-polymer composites: Acrylamide (AA) 2 crystallized (Paneco), Formaldehyde (F) GOST 1625-89 (Neva Reactive), N, N'-methylene-bis-acrylamide (MBAA) 98% (Paneco), Zinc chloride 97-100.5% Extra Pure (Scharlau), $HAuCl_4{\cdot}4H_2O$ (Voikov plant), chitosan hydrochloride (Cht) (Russia), hydrochloric acid (Neva Reactive).

The compositions of electrolytes for electropolymerization are given in table 1.

Table 1. Electropolymerization compositions.

Component	Concentration	Polymer (Film)			
		1	2	3	4
$CH_2{=}CHC(O)NH_2$	3 mol/l	+	+	+	+
HCHO	3 mol/l	+	+	+	+
$CH_2(NHCOCH{=}CH_2)_2$	0.05 mol/l	+	+	+	+
$ZnCl_2$ [a]	0.2 mol/l	+	+	+	+
Chitosan	0.1 %	-	+	-	+
$HAuCl_4$	1 mmol/l	-	-	+	+

[a] Zinc chloride is an indirect initiator of the polymerization of acrylamide [16,17].

Synthesis and electrochemical analysis of composites was carried out using the potentiostat/galvanostat IPC-Pro 8.0 ("Cronas". Russia). Electropolymerization was carried out in glass cells with a volume of 10 cm^3 at a cathode potential of -1.2 V for 3 minutes (unless otherwise specified). Separation of the anode and cathode space was not used. The working electrode (cathode) was a stainless steel AISI 304 rod (S = 1.0-1.5 cm^2). We used a platinum plate as an auxiliary electrode (anode) (S = 7.5 cm^2). The reference electrode is a saturated silver chloride electrode EVL-1M3 with a Luggin capillary.

Titanium electrodes were preliminarily subjected to electrochemical polishing for 2-3 seconds in a hot mixture (3:1) of HNO_3 and HF acids. Stainless steel rods were ground and degreased with a paste of finely dispersed magnesium oxide. All electrodes were thoroughly rinsed with deionized water. To separate the composite polymethylolacrylamide film/AuNPs from the electrode surface, the latter was immersed in 0.1 mol/l HCl for ~ 90-100s, after which the film lost its adhesion to the electrode due to dissolution of a thin metallic zinc sub-layer [17].

X-ray diffraction was performed on Advance D-8 diffractometer (Bruker-AXS, Germany).

Transmission electron microscopy (TEM) was performed on Libra-200FE (Carl Zeiss, Germany).

The particle size distribution of gold nanoparticles was assessed by small angle X-ray scattering (SAXS) using a HECUSS3-MICRO-PIX diffractometer (Austria).

3. Results and discussion

In the process of electropolymerization for all electrolytes, as the polymer layer forms on the electrode surface, the current density is reduced by overlapping the cathode with a layer of a new weakly conducting phase. Under identical conditions (420 s), the smallest residual current (i_r) is observed on a polymer coating that does not contain metallic particles (figure 1, curves 1, 2). While the inclusion of gold nanoparticles in the polymer leads to an increasing i_r (figure 1, curve 3).

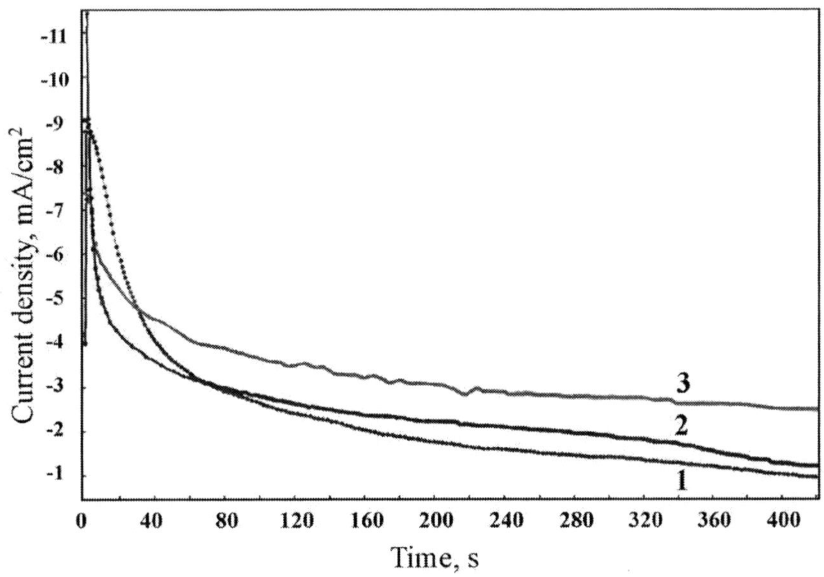

Figure 1. Dependence of current density on the electropolymerization time: polymer 1 (1), polymer 2 (2), polymer 4 (3).

Studies have shown that the mass of different types of polymer films depends on the time of synthesis (figure 2).

It should be noted, that the films of the initial polymer with chitosan (figure 2, curve 2) have a lower mass than without chitosan (figure 2, curve 1). This effect is due to the less moisture-absorbing capacity of films containing a water-insoluble fraction of chitosan, their greater density and smaller thickness. On the other hand, films with inclusion of gold particles (figure 2, curves 3, 4) have a large mass in comparison with films without AuNPs (figure 2, curves 1, 2). And this difference reaches 10-15 mg at a synthesis time of more than 60 s, which is 26-40% of the total mass of the composite.

Polymer films obtained by electropolymerization from an aqueous solution based on acrylamide are an elastic, colorless and transparent material in the swollen state in water. With electropolymerization in the presence of $HAuCl_4$, the color of the films becomes a pink-lilac, which visually confirms the inclusion of gold particles in the polymer. The intensity of coloration depends on

IOP Conf. Series: Materials Science and Engineering **369** (2018) 012027 doi:10.1088/1757-899X/369/1/012027

the amount of AuNPs in the film. The color of the composite becomes intensive lilac in the presence of chitosan in the electrolyte solution.

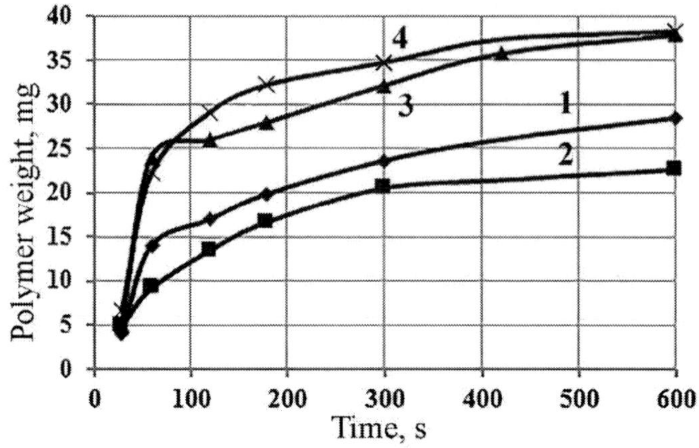

Figure 2. Dependence of the polymer mass on the polymerization time: polymer 1 (1), polymer 2 (2), polymer 3 (3), polymer 4 (4).

It should be noted, that the initial electropolymerization solution acquires a yellow color after the addition of $HAuCl_4$. This color does not change for at least 20 minutes, despite the fact that the solution contains such a strong reducing agent as formaldehyde (table 1). X-ray small-angle scattering (SAXS) study of the solutions 3 and 4, which contain $HAuCl_4$, showed the absence of any nanoparticles in electrolytes (the solutions were held for 15 minutes). This is evidenced by the fact that the intensity curves for both solutions are identical (merged) and represent a practically straight line (figure 3), which does not allow us to construct a particle size distribution spectrum. The result obtained allows us to conclude that formaldehyde does not reduce gold in this water-monomer composition.

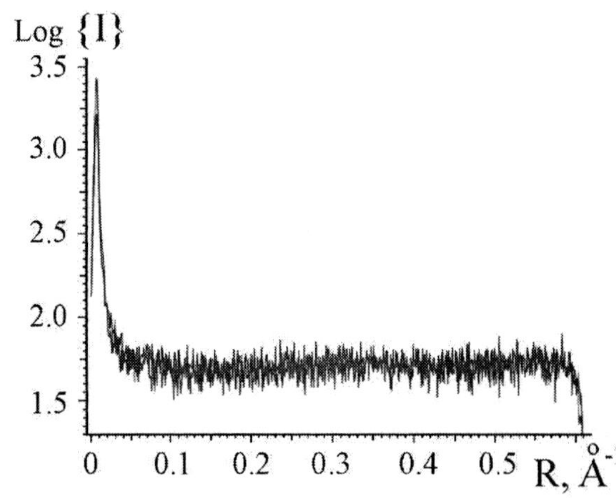

Figure 3. Spectrum of small-angle X-ray scattering of solutions 3 and 4.

Further studies have shown that gold nanoparticles are formed and embedded in the film during the electropolymerization process, i.e. simultaneously with the formation of the polymer film itself. The presence of metallic gold in polymer films is confirmed by the results of X-ray diffraction analysis (figure 4) and Transmission electron microscopy (TEM) (figure 5).

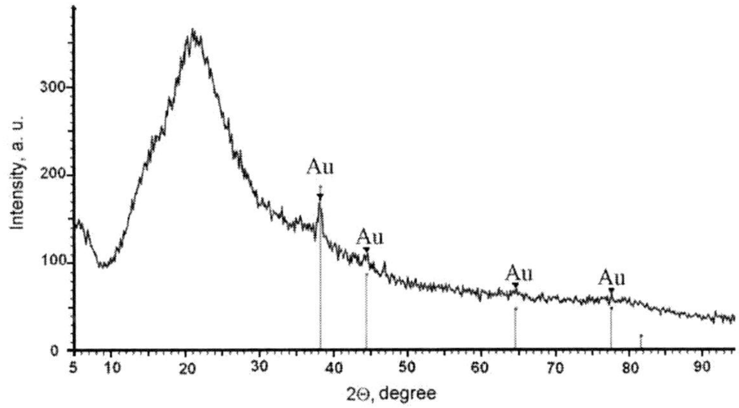

Figure 4. X-ray diffraction patterns of polymethylolacrylamide/AuNPs film.

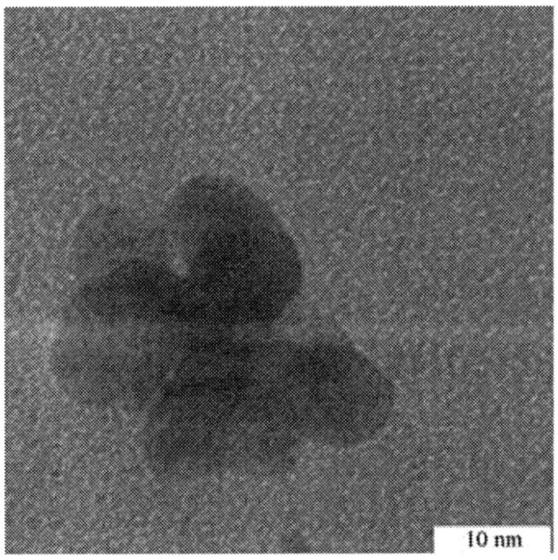

Figure 5. TEM image of polymethylolacrylamide/AuNPs film (polymer 3).

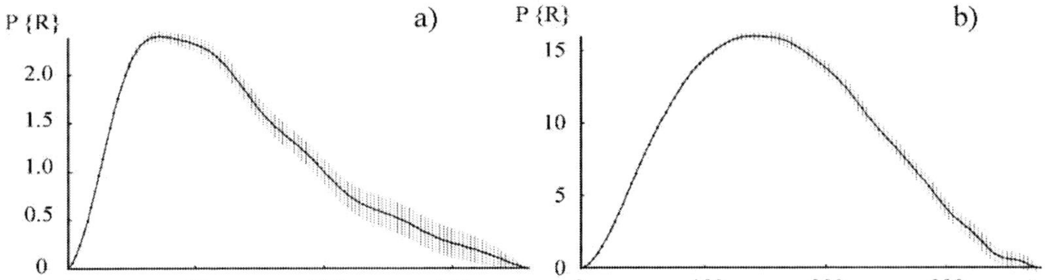

Figure 6. SAXS spectra of the distribution of AuNPs particles in size for polymer 3 (a) and polymer 4 (b).

As follow from figure 5, there are small Au^0 crystalline particles (~ 10 nm) aggregated to ~ 40 nm in the film. These TEM data and SAXS results are in good agreement. SAXS method revealed the presence of AuNPs in electrosynthesized polymer films (figure 6). The AuNPs distribution by size was established. It was found that the particle size of gold is affected by the adding of chitosan into the original monomer composition. As it follows from figure 6, the maximum size (radius), of the AuNPs is 35-40 Å in the absence of chitosan, whereas after its addition, gold particles have been aggregated to 150 Å.

The chitosan addition effect is in good agreement with the data [18-20]. It follow from [18,19] chitosan and $AuCl_4^-$ are formed a complex. Authors [20] found that the UV initiated synthesis of gold nanoparticles in the chitosan presence, which simultaneously performs stabilizing and restoring functions, results in the formation of larger gold nanoparticles.

4. Conclusions

The possibility of electrochemical formation of metal/polymer composites was studied. It is shown that the method of electropolymerization of aqueous solutions based on acrylamide and formaldehyde in the presence of $HAuCl_4$ makes it possible to obtain particles of nano-sized gold stabilized in a polymer matrix (polymer film). The peculiarity of this approach is that this process is carried out in one-stage, when simultaneously non-conductive (dielectric) polymer matrix and gold particles are formed and Au^0 is captured by the polymer film. The inclusion of AuNPs in polymethylolacrylamide was confirmed visually, as well as by X-ray diffraction analysis, small-angle X-ray scattering and TEM data.

Acknowledgments

The work was supported by Russian Government State Order (0265-2014-0001). AAAA-A17-117-03-09-100-84-4.

References

[1] Blanco I and Bottino F A 2012 *AIP Conference Proceedings* **1459(1)** 247-9
[2] Blanco I, Bottino F A, Cicala G and Latteri A 2013 *Polym. Degrad. Stabil.* **98(12)** 2564-70
[3] Hana J, Wanga M, Hua Y, Zhoub Ch and Guo R 2017 *Prog. Polym. Sci.* **70** 52-91
[4] Shchitovskaya E V, Kolzunova L G, Kuryavyi V G and Slobodyuk A B 2015 *Russ. J. Electrochem.* **51** 1097-107
[5] Zheng Y, Wang H and Ma Z 2017 *Microchim. Acta* **184** 4269-77
[6] Ma Y, Shen X L, Zeng Q and Wang L S 2017 *Microchim. Acta* **184** 4469-76
[7] Jin S A, Heo Y, Lin L K, Deering A J, Chiu G T C, Allebach J P and Stanciu L A 2017 *Microchim. Acta* **184** 4879-86
[8] Kafi A K M, Wali Q, Jose R, Biswas T K and Yusoff M M 2017 *Microchim. Acta* **184** 4443-50
[9] Ahmad H, Ahmad A and Islam S S 2017 *Microchim. Acta* **184** 2007-14
[10] Haghshenas E, Madrakian T and Afkhami A 2015 *Mater. Sci. and Engineer* **57** 205-14
[11] Priecela P, Salamia H A, Padillaa R H, Zhongb Z and Lopez-Sancheza J A 2016 *Chin. J. Chem.* **10** 1619-50
[12] Zinchenko A, Miwa Y, Lopatina L I, Sergeyev V G and Murata S 2014 *Appl. Mater. Interface* **6** 3226-32
[13] Clukaya C J, Grabill C N, Hettinger M A, Dutta A, Freppona D J, Robledo A, Heinrich H, Bhattacharya A and Kuebler S M 2014 *Appl. Surf. Sci.* **292** 128-36
[14] García-Hernández C, García-Cabezón C, Medina-Plazal C, Martín-Pedrosa F, Blanco Y, Antonio de Saja J and Rodríguez-Méndez M L 2015 *Beilstein J. Nanotechnol.* **6** 2052-61
[15] Vinu Mohan A M, Aswini K K and Biju V M 2014 *Sensors and Actuator* **196** 406-12
[16] Collins G L and Thomas N W 1977 *J. Polymer Science: Polymer Chemistry Edition* **15** 1819-31
[17] Kolzunova L G 2012 Polymer Films: Properties, Performance and Applications *Electropolymerization as the Method of Producing of Functional Polymer Films and Coatings. Series: Materials Science and Technologies* (USA: Nova Science Publishers, Inc.

NY) pp 1-108

[18] Wang Q Z, Chen X G, Liu N, Wang S X, Liu C S, Meng X H and Liu C G 2006 *Carbohydr. Polym.* **65** 194-201

[19] Cataldo S, Crea F, Gianguzza A, Pettignano A and Piazzese D 2009 *J. Mol. Liq.* **148** 120-6

[20] Yakimovich N O, Smirnova L A, Gracheva T A, Klychkov K S, Bityurin N M and Aleksandrov A P 2008 *Polym. Sci. Ser. B* **50** 238-42

Thermoelectric properties of topological insulators based on π-stacked molecular and graphene-belt systems with hydrogen bonds

M Wierzbowska

Institute of High Pressure Physics, Polish Academy of Sciences, Sokołowska 29/37, 01-142 Warsaw, Poland

E-mail: wierzbowska@unipress.waw.pl

Abstract. Recently, new two-dimensional topological crystalline insulators were theoretically predicted in the family of molecular networks and graphene nanobelts. The planar structures are vertically stacked in columns formed via the π-type interactions between benzene rings. Moreover, the molecular or graphene-belt parts interact with each other via the hydrogen bonds, due to their termination with O and H. These organic materials possess flat band structures, which are characteristic for weakly interacting systems. As a consequence, the thermolectric power is high and excess that of graphene nearly four times.

1. Introduction

The thermoelectric power of materials should be high for technological applications in "green" electronics. The Seebeck effect is the electric voltage generation in a conductor under a temperature gradient, and it is used in the heat sink devices. This parameter is dependent on a slope of the energy bands. It grows for the more flat parts of the electronic structures. It is natural, thus, that the organic or nanostructure systems with weak interactions are often the high thermopower materials. The type of the chemical bonds between molecules in the organic crystals might influence the thermoelectrics. The same mechanism holds when a pressure is applied.

Recently, the molecular and graphene-belt based systems were theoretically studied for their interesting electronic structure properties, namely the topological edge states [1, 2]. Their 2D structures were built via vertical stacking, i.e. the π-type interactions, of the 1D parts. Additionally, the 1D molecular structures were obtained via hydrogen bonds which form due to the oxygen and hydrogen termination of the opposite ends of the molecules. On the other hand, the same type of O- and H-termination of the zigzag edges of the graphene nanobelts introduces the hydrogen bonds in the vertical direction, in contrast to the horizontal bonds in the molecular systems. Both types of interactions, the π-stacking and the O...H weak bonds, are very easy to tailor their strength with the applied pressure. Therefore, the band structures and thermopower are easily tunable with pressure. Another flexibility in the design is obtained via a change of an order of the carbon-rings stacking (namely AA, AB or ABC type) and an order of the vertical overlap of the O- and H-terminated edges.

Here, a report is given on the thermopower in this interesting new family of the topological crystalline insulators. The next section describes the methods of the calculations, and the following section presents the detailed atomic structure of the systems and the calculated Seebeck constant at various modifications of the conformations.

2. Methods and computational details

Firstly, the density-functional theory (DFT) [3] calculations were done with the use of the plane-wave pseudopotential code Quantum Espresso [4]. The gradient-corrected funtional of the Perdew-Burke-Ernzerhof parametrization [5] was chosen. The ultrasoft [6] and normconserving [7] pseudopotentials were used for the atomic cores of the graphene belts and the molecular system, respectively. The plane-wave cutoffs of 30 Ry and 60 Ry were sufficient for the convergence with the ultrasoft and normconserving pseudopotentials, respectively. The uniform k-mesh of Monkhorst and Pack type [8] was used, and checked for the convergence in each direction of the calculated systems (around 8 to 12 points).

The self-consistent DFT electronic density was used for the calculations of the thermoelectric data. In this step, the maximally localized Wannier functions [9, 10] were obtained with the wannier90 code [11]. Then, the energy bands were accurately interpolated using the Hamiltonian represented in the Wannier functions. The thermoelectric parameters are defined as gradients of the energy bands, as described below. Thus, the Boltzwann code [12], which is the wannier90 postprocessing code and treats the problem in a semiclassical way, was used.

The transport distribution function (TDF) is a 3D tensor $\Sigma_{ij}(E)$ defined as a function of the energy E as follows:

$$\begin{aligned} \Sigma_{ij}(E) &= \frac{1}{V}\sum_{n,\mathbf{k}} v_i(n,\mathbf{k})v_j(n,\mathbf{k})\ \tau_{n\mathbf{k}}\ \delta(E-E_{n,\mathbf{k}}), \\ v_i(n,\mathbf{k}) &= \frac{1}{\bar{h}}\frac{\partial E_{n,\mathbf{k}}}{\partial k_i}, \end{aligned} \quad (1)$$

where V is the system volume, $v_i(n,\mathbf{k})$ is the band velocity, $E_{n,\mathbf{k}}$ is the band dispersion, and $\tau_{n\mathbf{k}}$ is the relaxation time dependent on the band index n and the reciprocal space $\mathbf{k}$. The TDF enters an expression for the 3D tensor of the Seebeck coefficients S_{ij}, which depends on the chemical potential μ and temperature T, and is defined as:

$$S_{ij}(\mu,T) = \frac{1}{eT}\ K_{ij,1}(\mu,T)\ K_{ij,2}^{-1}(\mu,T), \quad (2)$$

where

$$\begin{aligned} K_{ij,1}(\mu,T) &= \int_{-\infty}^{\infty} dE\ \left(-\frac{\partial f(E,\mu,T)}{\partial E}\right)\ (E-\mu)\ \Sigma_{ij}(E), \\ K_{ij,2}(\mu,T) &= \int_{-\infty}^{\infty} dE\ \left(-\frac{\partial f(E,\mu,T)}{\partial E}\right)\ \Sigma_{ij}(E). \end{aligned} \quad (3)$$

In the above formulae, the function $f(E,\mu,T)$ is the Fermi-Dirac distribution and e is the electron charge. It is important to note that the Seebeck coefficients are independent on the relaxation time, which was fixed for 10 fs for the TDF calculations. The k-mesh for the bands interpolation and gradients was 1000 points for each periodic direction of the systems.

3. Results

3.1. Molecular topological insulators

The molecular system built on phenalene derivatives has been shown to possess the edge states which enable us to classify this material as a weak topological crystalline insulator [1].

IOP Conf. Series: Materials Science and Engineering **369** (2018) 012028 doi:10.1088/1757-899X/369/1/012028

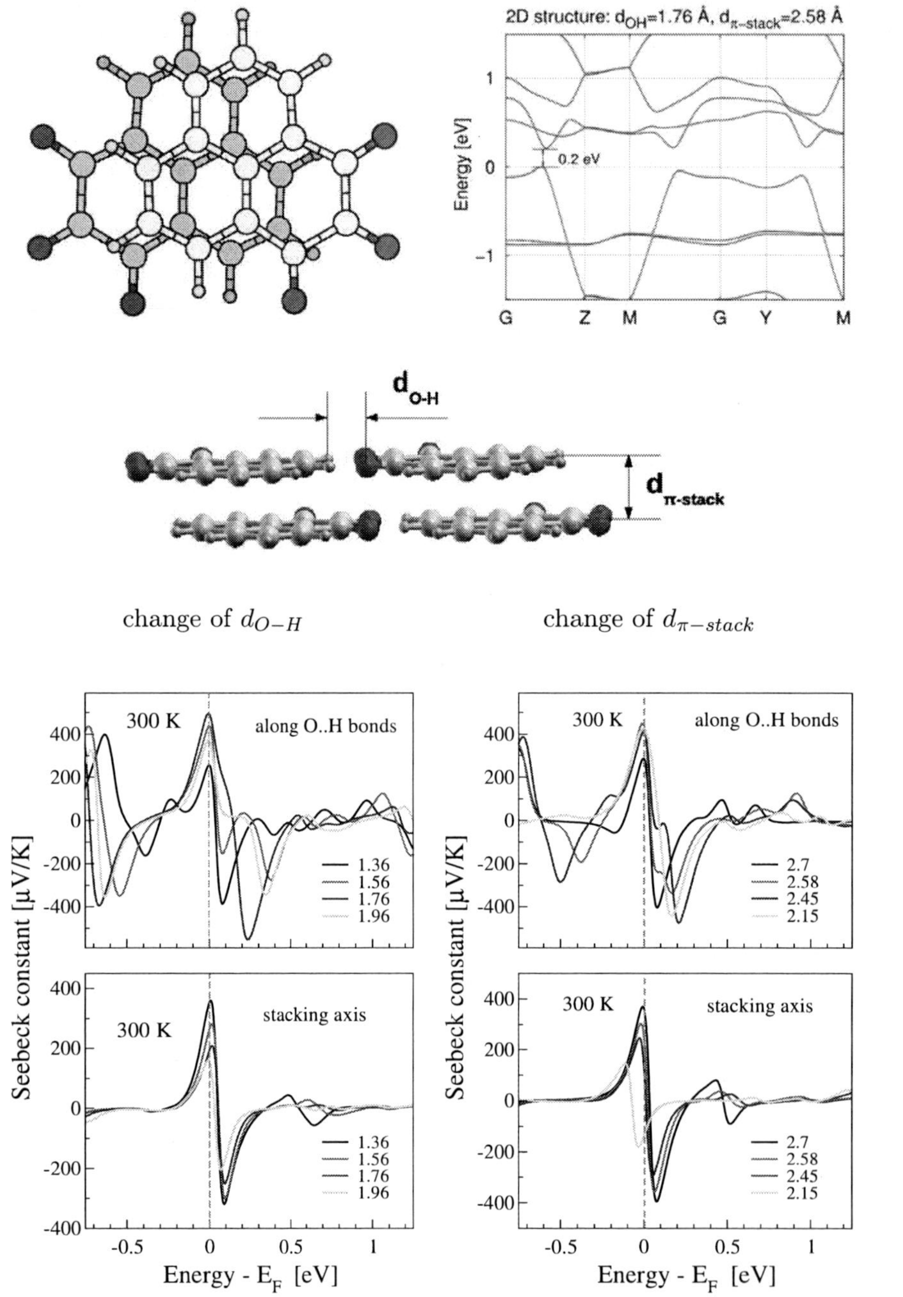

Figure 1. The atomic structure and bandlines of molecular topological system. The Seebeck constant is calculated for the molecular structures at various intermolecular distances in $\AA$ units.

The atomic structure and the energy bands calculated for chosen intermolecular distances are presented in Figure 1. The Seebeck constant as a function of the chemical potential is also presented in Figure 1 for two periodic directions of the system: the stacking axis (Z cartesian coordinate) and the O..H-bonds axis (Y coordinate). The temperature was chosen for 300 K and the intermolecular distances were varied in the stacking and planar directions. The maxima and minima of the Seebeck constant as a function of the chemical potential move near the Fermi level with a change of the $d_{\pi-stack}$ and $d_{O..H}$ parameters. This is due to a change of the band gap. The band structures of all geometric cases were published in the supporting information of Ref. [1]. The thermopower at the top of the valence band excess for some intermolecular distances even 400 μV/K. This is a value four times larger than this parameter calculated for graphene [13], and also larger than the Seebeck constant obtained by other authors for the τ-type organic superconductors [14].

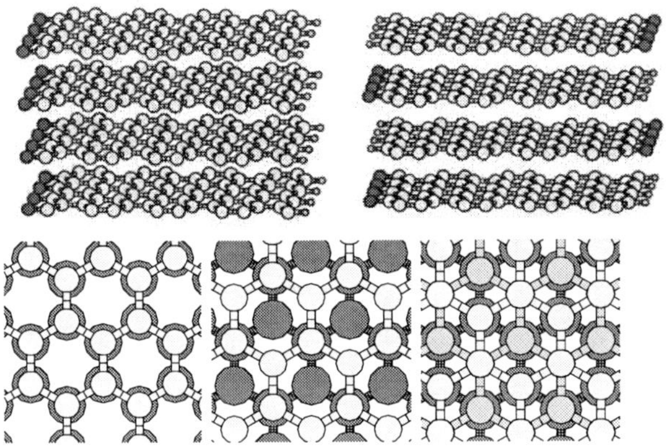

Figure 2. The atomic structure of stacked graphene belts - the order of stakicking of the zigzag O-edge (on-top or reversed) and the order of the carbon-rings stacking (AA, AB and ABC type).

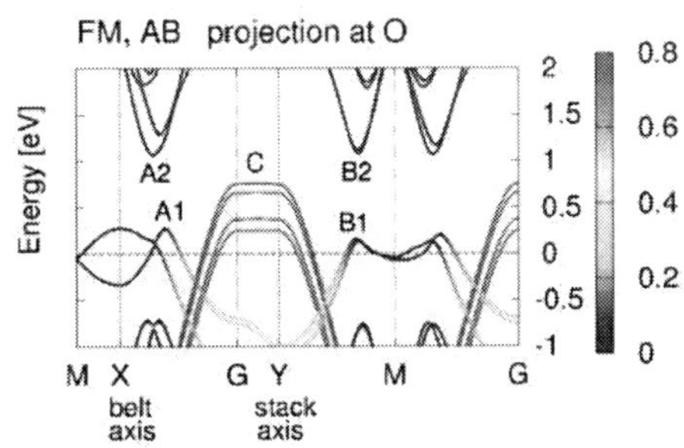

Figure 3. The schematic band structure of the stacked graphene belts. Color projection means the contribution of O-localized Wannier functions in the Bloch function.

3.2. Graphene-belt based topological insulators

Graphene has too small bands splitting due to the spin-orbit coupling to be a "classical" topological insulator. However, some graphene based systems were theoretically predicted to be the crystalline topological insulators [15, 16, 17]. Recently, it has been found that the graphene nanobelts, which are O- and H-terminated at the zigzag edge and stacked in the AB-type order (similar to graphite), are 2D quantum Hall materials [2].

The atomic structure and stacking orders are presented in Figure 2. The carbon rings can be stacked in AA, AB and ABC type. While the zigzag edges are O- and H-terminated on the opposite sides of the belt. These belts are stacked with O-edge on-top of the O-edge of the neighbor, or in the reversed order with H-edge between two O-edges. Moreover, these systems are magnetic with the spin moments localized mainly at oxygens. The coupling between these local moments might be ferro- (FM) or antiferromagnetic (AF). The Fermi level lays around 0.8 eV below the top of the valence band, because these belts are naturally p-type semiconductors due to the asymmetric termination of the edges.

The interactions along the stacking axis are very weak, and these systems have a flat band structure in this direction. Moreover, the mentioned bands flattening is energetically located near the valence band top. The schematic picture of the band structure is presented in Figure 3. The Wannier-functions analysis shows that the Bloch functions of the flat regions are mainly built with the oxygen-localized contributions. The flattening of the bandlines gives a mathematical reason for the large Seebeck constant. This observation was made for doped graphene and generalized for all systems with the similar bands [18]. Therefore, we can expect a

IOP Conf. Series: Materials Science and Engineering **369** (2018) 012028 doi:10.1088/1757-899X/369/1/012028

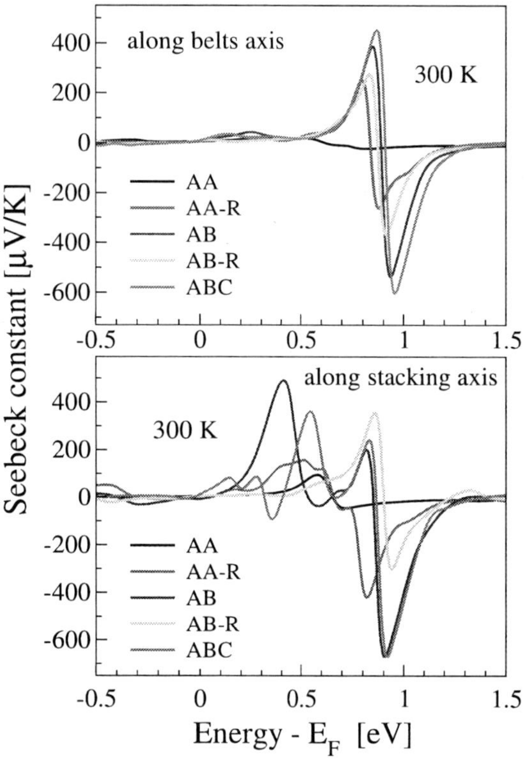

Figure 4. The Seebeck constant calculated for the ferromagnetic phase of various stacks.

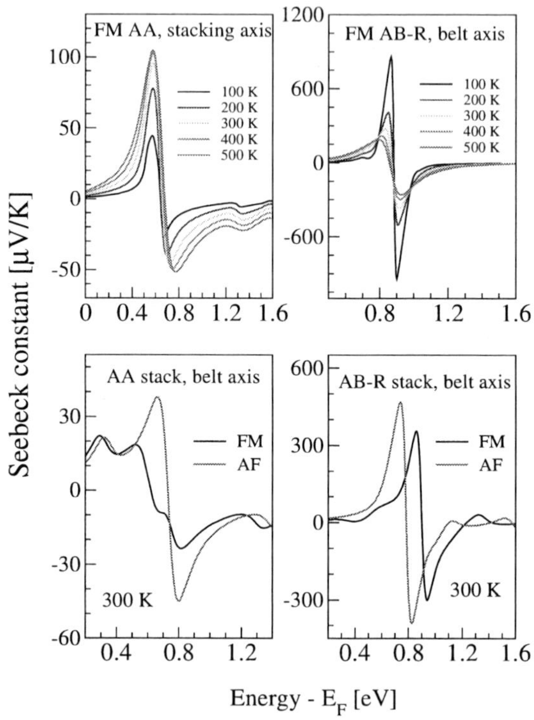

Figure 5. The temperature dependence and magnetic-phase impact on the thermoelectrics of the stacked graphene belts.

high thermopower in the described systems too. The band structures of all belts configurations are presented in the supporting information. The Seebeck constant as a function of the chemical potential is presented for the studied systems in Figure 4. The dependence of thermopower on temperature and magnetic coupling is presented for two chosen stackings in Figure 5.

For some configurations, the top of the valence band is dominated by the carbon-rings contributions to the Bloch functions. For the other cases, the flat oxygen-dominated bandlines move towards the top of the valence manifold. Some systems, namely the AA-stacked and AA-stacked reversed belts in both magnetic states, are metallic either by a "classical" bandgap closure or via the indirect negative band gap. They possess much lower Seebeck constant in comparison to the rest of the systems. Additionally, the temperature dependence of the thermopower is increasing with temperature for the metallic cases, while it is standard (decreasing with T) for the semiconducting (gapped) systems. The AB-stacked and ABC-stacked systems possess more than one flat region in their band structures. This fact has a fingerprint in the multiple peaks curvature of the thermopower function. The graphene nanobelts, similarly to the molecular topological crystalline insulators, are high thermopower materials with the maximal values of the Seebeck constant also excessing 400 μV/K at room temperature.

4. Conclusions

The topological insulators are potential materials for future electronic devices. The newly proposed topological crystalline insulators, which were found by theoretical modeling with the DFT and tight-binding approaches, are built of light atoms only (C, O, H). They contain molecules or graphene nanobelts interacting via the π-stacking and hydrogen bonds [1, 2]. Their thermoelectric power is high - four times larger than that of graphene [13]. This is due to the weak intermolecular interactions, and - as a consequence - the flat electronic bands [18].

Acknowledgments
Calculations have been performed in the Cyfronet Computer Centre using Prometheus computer which is a part of the PL-Grid Infrastructure. This work has been supported by The National Science Centre of Poland via the Project No. 2016/23/B/ST8/03480.

Reference

[1] Wierzbowska M 2017 *Sci. Rep.* **7** 9888–10
[2] Wierzbowska M *submitted*
[3] Kohn W and Sham L J 1965 *Phys. Rev.*A **140** 1133–6
[4] Giannozzi P et al. 2009 *J. Phys.: Condens. Matter.* **21** 395502–19
[5] Perdew J P, Burke, K and Ernzerhof M 1996 *Phys. Rev. Lett.* **77** 3865–4
[6] Vanderbilt D 1990 *Phys. Rev.* B **41** R7892–4
[7] Kleinman L and Bylander D M 1982 *Phys. Rev. Lett.* **48** 1425–4
[8] Monkhorst H D and Pack J D 1976 *Phys. Rev.* B **13** 5188–5
[9] Marzari N and Vanderbilt D 1997 *Phys. Rev.* B **56** 12847–9
[10] Marzari N, Mostofi A A, Yates J R, Souza I and Vanderbilt D 2012 *Rev. Mod. Phys.* **84** 1419–57
[11] Mostofi A A et al. 2008 *Comput. Phys. Commun.* **178** 685–15
[12] Pizzi G, Volja D, Kozinsky B, Fornari M and Marzari N 2014 *Comput. Phys. Commun.* **185** 422–8
[13] Wierzbowska M, Dominiak A, Pizzi G 2014 *2D Materials* **1** 035002–11
[14] Aizawa H, Kuroki K, Yoshino H, Mousdis G A, Papavassiliou G C, Murata K 2014 *J. Phys. Soc. Jap.* **83** 104705–5
[15] Kindermann M 2015 *Phys. Rev. Lett.* **114** 226802–5
[16] Martin I, Blanter Y and Morpurgo A 2008 *Phys. Rev. Lett.* **100** 036804–4
[17] Vaezi A, Liang Y, Ngai D H, Yang L and Kim E-A 2013 *Phys. Rev.* X **3** 021018–9
[18] Wierzbowska M and Dominiak A 2014 *Carbon* **80** 255–14

Deformation behaviour of Cu-Al clad composites produced by rotary swaging

L Kunčická[1,2] and R Kocich[1]

[1]Regional Materials Science and Technology Centre, VŠB TU Ostrava 17. listopadu 15, 70833 Ostrava 8, CZ

E-mail: lenka.kuncicka@vsb.cz

Abstract. Al/Cu composites are an advantageous perspective material applicable in various industrial branches, from electrotechnics to transportation industry. This study focused on the investigation of Al/Cu clad composites produced by rotary swaging at two different temperatures, 20°C and 250°C. The composites were swaged from the original 30 mm down to 5 mm with the total swaging degree of 3.58, however, samples were acquired after multiple steps. The influences of the processing conditions on the structure were studied via scanning electron microscopy; the analyses mainly focused on the deformation behaviour of the component metals and the possible development of intermetallic phases on their interfaces, as well as on the grains orientation. During processing, the radial swaging forces were recorded with our own developed KOMAFU S600 system for dynamic detection of swaging forces. According to the results of the analyses, the swaging temperature influenced significantly the behaviour of the composites, as did also the total imposed strain. The composite swaged at 250°C was affected more notably, the cross-sections of the Al wires in the composite were deformed due to the influence of the radial swaging dies movement more significantly than in the composite swaged at 20°C. This effect was evident for all the investigated swaging steps and increased with increasing total imposed strain. The higher swaging temperature also decreased the plastic flow of the material; the deformation work was 730.3 kJ for 250°C composite and 650.7 kJ for the 20°C one. Tensile testing revealed similar effect; while the UTS for both the composites was slightly higher than 280 MPa, the plasticity of 250°C composite was evidently higher.

1. Introduction

Advances in production and manufacturing technologies introduce increasing demands on the innovative materials, one of which are composite materials advantageously providing the ability to create various combinations of mechanical, as well as physical properties. This makes them suitable for various applications e.g. in elecrotechnics, aerospace and marine industry, as well as in biomedicine [1–3]. Among the favourable forms of composites are clad composites, sometimes denoted as hybrid materials. Various clad composite systems, such as Ti-TiBw [4], Ti/Cu [5], Ni/WC [6], Al/Cu/steel [7] and Mg/Al/steel [8], have already been studied. Nevertheless, the majority of the studies deal with the Al/Cu system [9–12]. This combination of materials is favourable for its beneficial thermal and electrical conductivity and low density, as well as the advantageous weight and cost when compared to Cu. Cu/Al clad composites can for example be applied for air-cooling fans, and armoured cables [13], and as alternatives to copper in aerospace and automotive industry and electrotechnics.

Among the conventionally applied processes for composites manufacturing are various types of welding [14–16]. However, they feature the disadvantage of high processing temperatures introducing local structure modifications, which can result in the development of internal stresses or formation of intermetallic compounds at the interfaces (although these can also be introduced by a heat treatment) [17]. Nevertheless, composites can also successfully be manufactured at ambient temperatures (under cold conditions) by application of large plastic (shear) strains via conventional methods (rolling [8], extrusion [18] etc.), as well as non-conventional ones (severe plastic deformation – SPD – processes [18–21]).

One of the most favourable methods to manufacture long composite work-pieces is rotary swaging (RS), which can advantageously be used to manufacture copper-based-clad-composite wires to be used as conductors in the automotive. RS also ensures dimensional accuracy and high quality surface and therefore eliminates subsequent machining. As recently reported by Kocich *et al*, RS is suitable for fabrication of clad composites in forms of wires, tubes and long rods [17]. Zhang *et al* [22] used RS for production of bimetallic tubes.

The aim of this study was to investigate the influence of the swaging technology, which was performed at two different temperatures, on the structural development of the processed Al/Cu clad composites. Among the aims was to observe the influence of the selected processing temperature on the structure development and possible occurrence of intermetallic compounds at the Al-Cu interfaces. The radial swaging forces, affecting primarily the radial flow of the processed material, were recorded during swaging and the properties of the swaged pieces were assessed via tensile testing. Scanning electron microscopy was further used for supplemental analyses of texture development.

2. Experimental methods

The Cu/Al clad composites were produced using the rotary swaging (RS) technology. The original materials were electro-conductive commercial purity (CP) Cu (0.002% O, 0.015% P, 0.002% Zn, bal. Cu) and electro-conductive CP Al (0.20% Si, 0.25% Fe, 0.05% Cu, balance Al). To observe the influence of the preparation conditions, RS was performed at two different temperatures (room temperature for composite *A* and the temperature of 250°C for composite *B*), the selection of which ensued from our previous research [13,17].

For both the samples, the original outer diameter of the Cu shell was 30 mm and the diameters of the 19 Al wires were 3 mm (figure 1). Both the composites were processed to 5 mm and, during swaging, 10 mm, 6 mm and 5 mm samples were acquired. The individual reduction ratios summarized in table 1 were computed using the $\varphi = \ln (S_0/S_n)$ relation, where S_0, S_n were the entering and final cross-sectional areas, respectively. Our own KOMAFU S600 system for dynamic detection of swaging forces was used to record the radial swaging forces during processing [23].

The investigations of the swaged composites mechanical properties were performed via tensile testing. The parameters for the tensile tests were the following: specimens with length of 150 mm; strain rate of $1.47\text{x}10^{-3}$ s^{-1}. Detailed analyses of the structures of the swaged composites and the interfaces between the individual components were performed via scanning electron microscopy (SEM) and using the energy dispersive spectroscopy (EDX) and electron backscatter diffraction (EBSD) analyses. The samples were ground on SiC papers and electrolytically polished. EBSD analyses were processed on transversal cuts using a JEOL FESEM 7000F device with an EDAX DigiView High Resolution Camera.

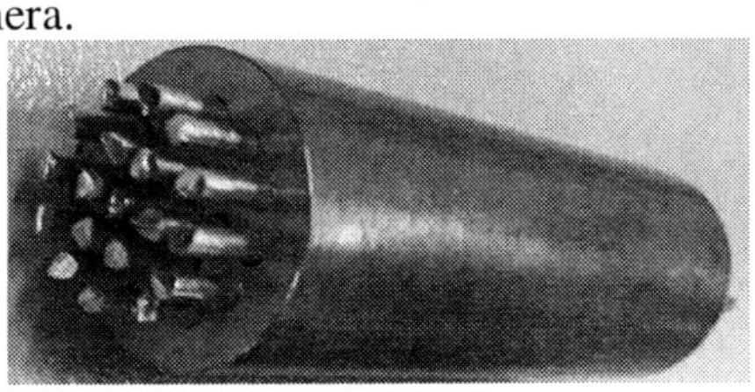

Figure 1. Original material before swaging.

Table 1. Swaging degrees for individual reductions.

RS pass number	total deformation degree φ (-)
1	0.36
2	0.81
3	1.08
4	1.39
5	1.75
6	2.20
7	2.77
8	3.22
9	3.58

3. Results and discussion

3.1. Deformation behaviour

The wires within the composites were categorized as the axial wire, the inter-layer wires and the outer-layer wires. Wires shape changes after swaging were obvious, however, the plastic flow was different for composites *A* and *B*.

For all the *A* and *B* composites, the shape of the axial wire cross-section changed the lest; its diameter was comparable for both the 10 mm samples (the diameters of circumscribed circles were 943 μm for composite *A* and 935 μm for composite *B*). However, the axial wire diameters were larger for composite *A* than for composite *B* with continuing swaging.

The shape changes of the inter-layer wires were more notable. Their original shapes were preserved the most in 10 mm composite *A*, however, 10 mm composite *B* already exhibited evident changes. To evaluate the deformation of the wires cross-sections, the ratios of two perpendicular dimensions were calculated (the largest dimension related to the largest perpendicular dimension expressed in percent) and summarized in table 2. For composite *A*, the ratio of the two dimensions was approximately 40% for the 6 mm and 5 mm samples, while for composite *B* it was approximately 30%. The reduction steps thus had comparably homogenous effects on the composites, which behaved more or less as single-component materials. However, the radial flow was more intense for composite *B*, the wires cross-sections of which got flatter and more elongated, due to the elevated swaging temperature, which decreased the composite flow stress, supported the development of restoration processes, and thus facilitated the plastic flow of both the composite components [17].

The mentioned phenomena affected the most the behaviour of the outer-layer wires. Whereas the cross-sections of the outer-layer wires in 10 mm composite *A* changed comparably to the inter-layer ones, the 10 mm composite *B* (figure 2(a)) outer-layer wires cross-sections exhibited significant deformation – their elongation was more evident than for the inter-layer wires. The most notable cross-section deformation exhibited the wires in 5 mm composite *B* (figure 2(b)).

Table 2. Differences in two largest perpendicular wires dimensions.

Diameter (mm)	10	6	5
Inter-layer wires			
Composite A	98%	40%	43%
Composite B	95%	33%	29%
Outer-layer wires			
Composite A	94%	24%	34%
Composite B	73%	17%	10%

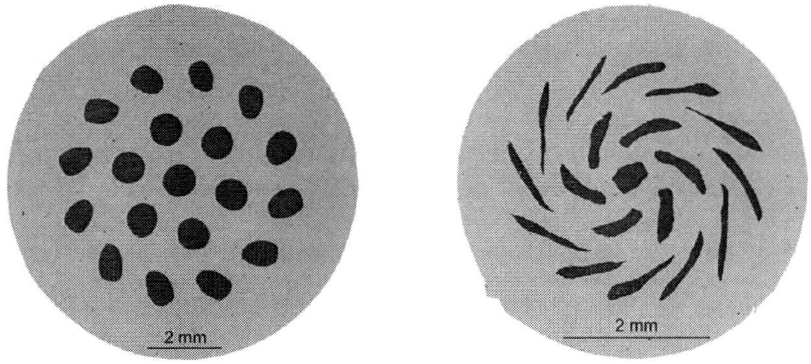

(a) (b)

Figure 2. SEM image of: cross section of 10 mm composite *B* (a); cross section of 5 mm composite *B* (b).

The specific shapes of the Al wires in the swaged composites were caused primarily by the radial shear component of the imposed strain introduced by the simultaneous effect of rotation and centripetal movement of the hammering dies [23]. The swaging force vector consists of the normal and radial components, the normal component of which influences the axial plastic flow, while the radial component influences the radial plastic flow of the swaged material.

3.2. Swaging parameters

The swaging force developments in time for the 10 mm samples are shown in figures 3(a) and 3(b). The figures clearly show that lower forces were recorded for composite *B*, which corresponds to its lower flow stress introduced by the elevated swaging temperature, as written above. The average swaging force was 80 kN for this composite, while for composite *A* it was almost 90 kN. The occurring restoration processes influencing the flow stress and facilitating the plastic flow thus affected also the swaging forces and the deformation work, which was 750.3 kJ for composite A and 717.7 kJ for composite B.

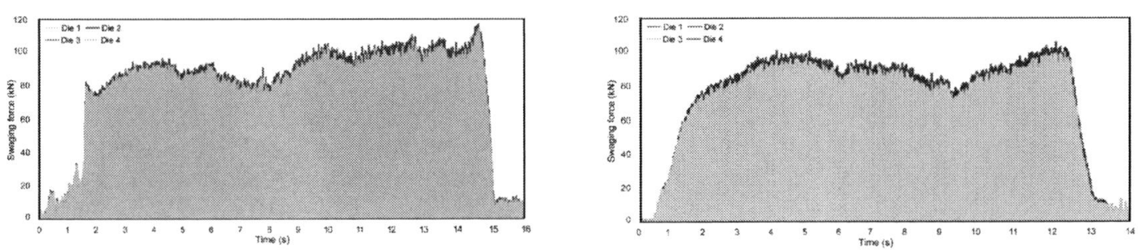

(a) (b)

Figure 3. Development of swaging forces: 10 mm composite *A*, (a); 10 mm composite *B* (b).

The development of the swaging force during the single pass was also different for both the composites. For composite *A*, the force increased steeply in the moment when the swaged material got to the reduction zone of the dies and further increased slightly with continuing swaging. This phenomenon was related to the effect of the free end [17]. At the beginning of swaging, the material is not constrained in the axial direction and its flow is easier than in the subsequent swaging stages, in which the radial movement of the material is constrained by the dies and the axial movement by the swaged end. The load to the machine then increases with progressing swaging pass [1]. On the other hand, the force for composite *B* exhibited more or less steady development during the whole pass, which can be attributed to the elevated swaging temperature and decreased flow stress.

A more detailed study of the relation of structure phenomena and the swaging forces development for multiple steps is being prepared for a subsequent publication.

3.3. Microstructure development

The evolution of (sub)structure and possible development of intermetallic phases were investigated via scanning electron microscopy (SEM). The presence of intermetallics was observed via SEM-EDX analyses. For the 10 mm samples, the development of the intermetallic phases on the interfaces of both the composites *A* and *B* was neglectable. However, the intermetallics started to be notable especially for composite *B* with continuing swaging. The presence of mixed phases was observed for all the 6 mm and 5 mm samples, but the extent and form of their development was different.

As can be seen in figure 4(a), 6 mm composite *A* exhibited localized thin bands of intermetallics on the very interfaces of the two metals; the detected phases were Al_2Cu_3 and Al_4Cu_9. The intermetallics were cracked in some locations, but mostly adjoined the wires and did not crumble away. On the other hand, the intermetallics within composite *B* occurred in a greater extent (figure 4(b)); the detected phases were AlCu, Al_2Cu_3 and Al_4Cu_9. The fact that recent studies have shown even the temperature of 300°C to be safe from the viewpoint of intermetallics development [13,17] points to the fact that the original swaging temperature of 250°C increased notably during swaging due to the development of deformation heat [24]. Similarly to the behaviour of the Al wires, the behaviour of intermetallics was also affected by the radial movement of the dies causing local accumulation of intermetallics in preferential locations. Although intermetallic phases are known to be brittle [2], the elevated temperature and high imposed strain imparted them higher plasticity and, although mostly crashed, they surrendered to the behaviour of the surrounding material and exhibited the tendency for radial plastic flow, as observed also for the Al wires (documented in Section 3.1).

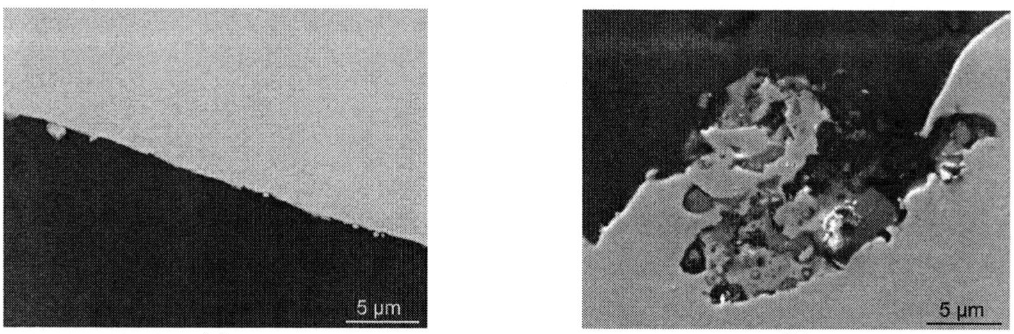

(a) (b)

Figure 4. Development of intermetallic phases for 6 mm composite A (a); 6 mm composite B (b).

3.4. Texture

Due to the limited length of this contribution, only a short insight into the texture development via pole figures (PFs) acquired perpendicular to the swaging axis is presented herein, while a deeper study is being prepared for a subsequent publication.

Figures 5(a) and 5(b) depict the texture of the central Al wire of composite *A* after swaging to 10 mm and 5 mm, respectively, whereas figures 5(c) and 5(d) show the texture of the Cu sheet of composite *A* (middle diameter) after swaging to 10 mm and 5 mm, respectively. The grains within the Al wires and also the Cu sheet exhibited a strong tendency to form the <111> fibre (Copper texture) [25], however, especially the Cu sheet featured the presence of the <001> fibre (Goss texture), too. Similar texture development was observed in FCC metals after imposing shear strain (deformation via severe plastic deformation processes) [26]. The 10 mm sample featured larger deviations from the ideal texture orientations, the intensities of which increased with continuing swaging. The texture intensities were higher for the 5 mm sample due to the high imposed shear strain, although restoration processes can be expected to have occurred during swaging from 10 mm to 5 mm. Nevertheless, these would most probably primarily influence the grain size, which will more in detail be analysed in a subsequent publication. These findings document the influence of the radial (shear) component of the swaging force.

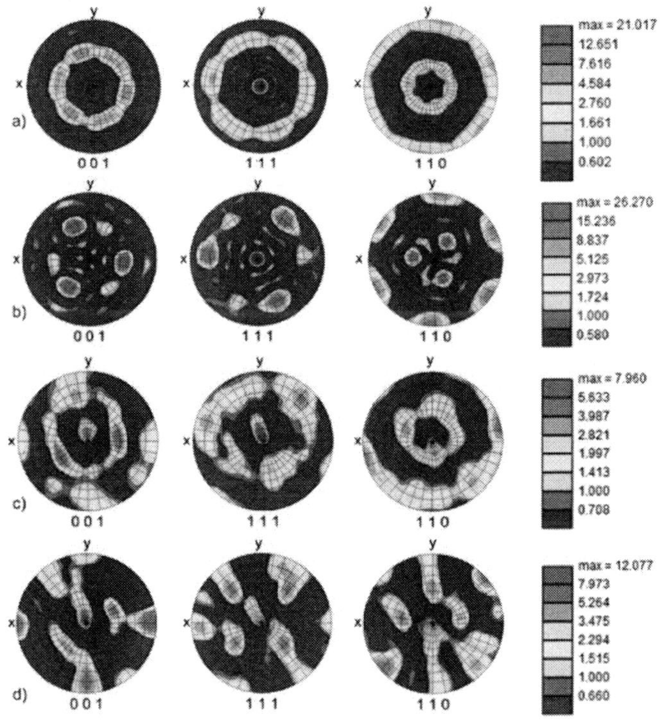

Figure 5. Textures in composite A: central Al wire in 10 mm sample (a); central Al wire in 5 mm sample (b); Cu sheet (middle diameter) in 10 mm sample (c); Cu sheet (middle diameter) in 5 mm sample (d).

3.5. Tensile tests

In order to correspond with the measurements of the swaging parameters, the herein presented tensile tests results were acquired for the 10 mm samples. The final stress-strain curves are depicted in figure 6. The Figure shows that the ultimate tensile strength (UTS) was comparable for both the composites (UTS of 285 MPa and 282 MPa for composites *A* and *B*, respectively). However, their plasticity and deformation behaviour, indicated by the curve slope, were quite different.

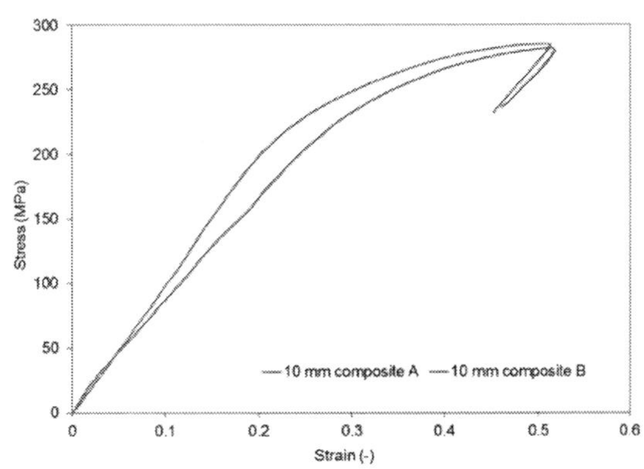

Figure 6. Stress-strain curves for 10 mm composites *A* and *B*.

The curve for composite *A* exhibited steeper initial increase, which points to higher deformation strengthening. The elongation before reaching the yield strength and the elongation to failure were also lower for composite *A* than for composite *B*. This proves the structure of composite *A* to have a higher accumulation of defects (precipitates) and dislocations than the structure of composite *B* [27]. These findings correspond to the above drawn partial conclusions; the elevated swaging temperature provided composite *B* with better plasticity, and facilitated the plastic flow when compared to composite *A* featuring more obstacles for dislocations movement.

4. Conclusions

This study was focused on the deformation behaviour of Al/Cu clad composites produced using rotary swaging at two different swaging temperatures (25°C and 250°C). The temperature of 250°C advantageously decreased the flow stress of the composite, which resulted in decreased swaging forces and deformation work (lower load on the machine), as well as increased plasticity, while the 20°C one exhibited significant work hardening. However, for 6 mm and 5 mm samples, 250°C together with the high imposed (shear) strain induced local development of intermetallic phases on the interfaces of the two metals, which exhibited tendencies to crack. The texture analyses revealed a strong formation of <111> fibre in both the components. A deeper study of mechanical, as well as electrical properties is being prepared for a subsequent publication. According to the results, the rotary swaging technology was successfully proven as practically applicable for fabrication of Al/Cu clad composites in the industrial scale.

Acknowledgments

This paper was created in the Project no.LO1203 "Regional Materials Science and Technology Centre – Feasibility Program" funded by Ministry of Education, Youth and Sports of the Czech Republic.

References

[1] Kunčická L, Lowe T C, Davis C F, Kocich R and Pohludka M 2015 *Mater. Sci. Eng. A* **646** 234–41
[2] Rajan T P D and Pai B C 2014 *Acta Metall. Sin. (English Lett.)* **27** 825–38
[3] Tjong S C and Mai Y-W 2008 *Compos. Sci. Technol.* **68** 583–601
[4] Liu B X, *et al* 2014 *J. Alloys Compd.* **602** 187–92
[5] Hosseini M and Danesh Manesh H 2015 *Mater. Des.* **81** 122–32
[6] Lohit R B and Bhovi P M 2017 *Mater. Today Proc.* **4** 2975–80
[7] Lee J E, *et al* 2007 *J. Mater. Process. Technol.* **187–188** 546–9
[8] Kim I K and Hong S I 2013 *Mater. Des.* **49** 935–44
[9] Kim I K and Hong S I 2013 *Mater. Des.* **47** 590–8
[10] Tolaminejad B, Hoseini Athar M M, Arabi H and Karimi Taheri A 2015 *Eng. Sci. Technol. An Int. J.* **19** 254–9
[11] Kim I K and Hong S I 2014 *Mater. Des.* **57** 625–31
[12] Kocich R, Kunčická L, Macháčková A and Šofer M 2017 *Mater. Des.* **123** 137-46
[13] Kocich R, Macháčková A, Kunčická L and Fojtík F 2015 *Mater. Des.* **71** 36–47
[14] Zhang Y, Sun D Q, Gu X Y, Duan Z Z and Li H M 2018 *Mater. Lett.* **212** 54–7
[15] Hokamoto K, *et al* 2009 *J. Alloys Compd.* **472** 507–11
[16] Deqing W, Ziyuan S and Ruobin Q 2007 *Scr. Mater.* **56** 369–72
[17] Kocich R, *et al* 2016 *Mater. Des.* **90** 379–88
[18] Wu C W and Hsu R Q 2006 *J. Mater. Process. Technol.* **174** 312–7
[19] Schneider J, Cobb J, Carpenter J S and Mara N A 2018 *J. Mater. Sci. Technol.* **1** 92-101
[20] Wu K, *et al* 2010 *Mater. Sci. Eng. A* **527** 3073–8
[21] Kocich R, Greger M and Macháčková A 2010 *Proc. 19th Int. Metall. Mater. Conf. Metal 2010 (Roznov Pod Radhostem)* (Ostrava: Tanger Ltd.) pp 166-71
[22] Zhang Q, *et al* 2017 *Int. J. Adv. Manuf. Technol.* **89** 163–73

[23] Kocich R, Kunčická L, Dohnalík D, Macháčková A and Šofer M 2016 *Int. J. Refract. Met. Hard Mater.* **61** 264–72
[24] Humphreys F J and Hetherly M 2004 *Recrystallization and Related Annealing Phenomena* (Elsevier Ltd.)
[25] Engler O and Randle V 2010 *Introduction to Texture Analysis, Macrotexture, Microtexture, and Orientation Mapping* (Taylor and Francis Group)
[26] Beyerlein I J and Tóth L S 2009 *Prog. Mater. Sci.* **54** 427-510
[27] Kocich R, Kunčická L, Král P and Macháčková A 2016 *Mater. Charact.* **119** 75–83

Energy use in repairs by cover concrete replacement or silane treatment for extending service life of chloride-exposed concrete structures

A Petcherdchoo

Department of Civil Engineering, Faculty of Engineering, King Mongkut's University of Technology North Bangkok, Thailand

E-mail: aruzp@kmutnb.ac.th

Abstract. In this study, the service life of repaired concrete structures under chloride environment is predicted. This prediction is performed by considering the mechanism of chloride ion diffusion using the partial differential equation (PDE) of the Fick's second law. The one-dimensional PDE cannot simply be solved, when concrete structures are cyclically repaired with cover concrete replacement or silane treatment. The difficulty is encountered in solving position-dependent chloride profile and diffusion coefficient after repairs. In order to remedy the difficulty, the finite difference method is used. By virtue of numerical computation, the position-dependent chloride profile can be treated position by position. And, based on the Crank-Nicolson scheme, a proper formulation embedded with position-dependent diffusion coefficient can be derived. By using the aforementioned idea, position- and time-dependent chloride ion concentration profiles for concrete structures with repairs can be calculated and shown, and their service life can be predicted. Moreover, the use of energy in different repair actions is also considered for comparison. From the study, it is found that repairs can control rebar corrosion and/or concrete cracking depending on repair actions.

1. Introduction

The attack of chlorides (Cl^-) is one of the significant phenomenon in the mechanism of concrete structure deterioration. When the quantity of the chlorides (Cl^-) near the circumstance of rebars is equal to its precarious value in addition to plenty amount of O_2 and humidity, rebar corrosion probably occurs causing concrete cracking and reducing the bond strength between concrete and rebar. Later, the debonding may decrease structural concrete strength. So, the corrosion of rebars detrimentally influences the safe and serviceable state, and also shortens the service life of structural concrete.

To prevent the deterioration, a proper repair action is necessary. Here, the action to slow down the rate of deterioration by implementing cyclic repairs; like replacing cover concrete or treating by silanes, is considered in order to extend the deterioration time. To predict the extension of the deterioration time or the service life of structural concrete with repairs, the evaluation of Cl^- penetration by a quantitative approach is recommended [1-3].

For the quantitative assessment, this study shades the light on the Cl^- diffusion determined by the Fick's second law. Two major processes are considered, i.e., chloride penetration through concrete without and with repairs. First, the surface chloride concentration (boundary condition) and the diffusion coefficient (material property) were supposed to be constant. Accordingly, solving one-dimensional partial differential equation (1-D PDE) is straightforward. However, when repairing by

IOP Conf. Series: Materials Science and Engineering **369** (2018) 012030 doi:10.1088/1757-899X/369/1/012030

cover concrete replacement or silane treatment for service life extension in the second process is implemented, the assumption cannot be kept. To eliminate this problem, the Crank-Nicolson-based finite difference method (FDM) is used. In addition, the environmental impact in terms of energy use in different repairs is compared.

2. Chloride ion penetration in concrete

2.1. Cover concrete replacement

REHABCON [1] defined cover concrete replacement to be an action which aimed to remove original cover concrete and replace it by new materials, for example, concrete, repair material etc. To cope with this, there are two main stages to be considered in chloride ion penetration over exposure time:

2.1.1. Penetration of chlorides (Cl^-) from outer surface toward old (original) concrete. The basic 1-D PDE for explaining the diffusion of chlorides [4] is expressed as

$$\frac{\partial C}{\partial t} = \frac{\partial}{\partial x} D \frac{\partial C}{\partial x} \tag{1}$$

in which C is the quantity of chlorides in terms of location x and time t, and D is the concrete diffusion coefficient. According to (1), a Crank-Nicholson scheme for (1) is in the following form [5]

$$\frac{c_{i,j+1} - c_{i,j}}{\Delta t} = \frac{D}{2}\left[\frac{\left(c_{i+1,j+1} - 2c_{i,j+1} + c_{i-1,j+1}\right)}{(\Delta x)^2} + \frac{\left(c_{i+1,j} - 2c_{i,j} + c_{i-1,j}\right)}{(\Delta x)^2}\right] \tag{2}$$

where $c_{x,t}$ is the quantity of chlorides at a mesh point x and time t. And, Δx is the mesh size (2 mm) and Δt is the time step (1 week).

2.1.2. Penetration and re-penetration of chlorides (Cl^-) after cover replacement. From figure 1(a), cover concrete is removed up to the cover replacement depth of x_p (or concrete cover depth), so the chlorides within the cover concrete are removed as well. Then, a repair material is replaced up to removed concretes (shaded area). The amount of unremoved chlorides (Cl^-) inside old concrete is shown by the Cl^- profile at time t_i. Then, three major steps occur as shown in figure 1(b). In the first step at time t_i, the chlorides inside old concrete are ready for re-penetrating toward the repair material, the problem involving solving the PDE with position-dependent Cl^- profile, or C(x,t), is met.

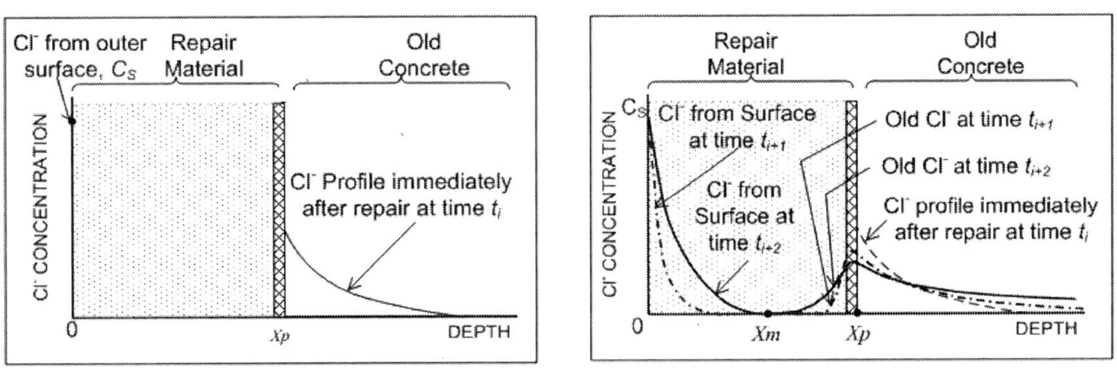

(a) Instantly subsequent to replacement (b) Subsequent to replacement

Figure 1. Cl^- profile subsequent to cover replacement [1].

In the second step at time t_{i+1}, the re-penetrating chlorides diffuse from old concrete toward the repair material, the problem involves position-dependent diffusion coefficient, or D(x), because the diffusion coefficients of old concrete is different from that of the repair material. Therefore, the partial differential equation in (1) can mathematically be updated as

$$\frac{\partial C}{\partial t} = \frac{\partial}{\partial x} D(x) \frac{\partial C}{\partial x} \tag{3}$$

In the third step at time t_{i+2}, when the newly diffused Cl^- from concrete surface meet with the re-penetrating chlorides (Cl^-) at the location x_m, the problem in solving the partial differential equation will again be met. If the number of repairs is more than one, it is even more complex. To avoid these problems, the Crank-Nicolson numerical scheme is introduced in calculation. By considering the three steps simultaneously, the problem solely deals with the transport of chlorides (Cl^-) toward concretes comprising dissimilar diffusion coefficients. Using [6], the numerical solution for (3) is written as

$$\frac{c_{i,j+1} - c_{i,j}}{\Delta t} = \frac{1}{2}\left[\frac{\left[D_{i+1/2}(c_{i+1} - c_i)_{,j+1} - D_{i-1/2}(c_i - c_{i-1})_{,j+1}\right]}{(\Delta x)^2} + \frac{\left[D_{i+1/2}(c_{i+1} - c_i)_{,j} - D_{i-1/2}(c_i - c_{i-1})_{,j}\right]}{(\Delta x)^2}\right] \tag{4}$$

where $D_{i+1/2}$ and $D_{i-1/2}$ is equal to $(D_i + D_{i+1})/2$ and $(D_{i-1} + D_i)/2$, respectively. In calculation, whenever the cover is replaced within the location of x, its diffusion coefficient is updated, e.g., $(D_x)_0$ is $(D_x)_{CR}$. Moreover, $(D_x)_0$ is the diffusion coefficient of old concrete at the location of x, and $(D_x)_{CR}$ is the diffusion coefficient of repair material at x. Taking advantage of the FDM, the position-dependent Cl^- profile is considered location by location, hence this problem can simply be solved. Moreover, if the diffusion coefficients become constant, (3) and (4) reduces to (1) and (2), respectively.

2.2. Silane treatment (SL)

Silane treatment could be classified to be a type of penetrating sealers that reacted with hardened concrete within concrete pore to create a hydrophobic or non-wettable surface [7]. Hence, two stages occurs over the exposure time of concrete structures with silane treatments as:

2.2.1. Penetration of chlorides (Cl^-) from outer surface toward old concrete. To consider this, (1) and (2) are applicable as explained in Section *2.1.1.*

2.2.2. Penetration of chlorides (Cl^-) toward silane-treated and old concrete. At time t_i in figure 2, silanes are treated at concrete surface. By this treatment, silanes react with the hardened concrete within concrete pore up to the treating depth of x_p.

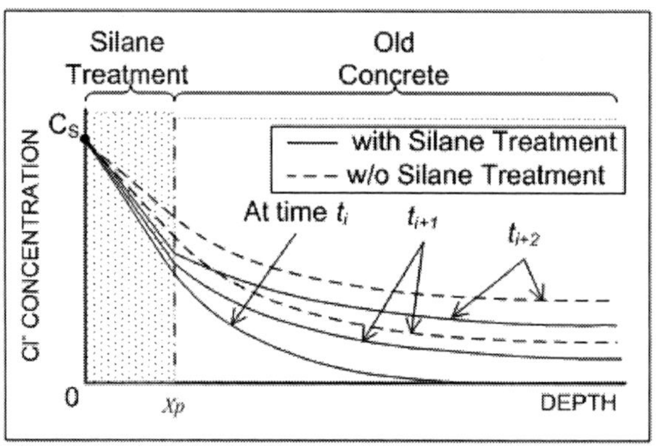

Figure 2. Chloride profile for concretes without and with treating by silanes.

Silane treatment such as that shown in figure 2 will result in the problem involving position-dependent diffusion coefficient, or D(x), because the diffusion coefficient of silane-treating concretes is dissimilar to the diffusion coefficient of old concrete. For this, the partial differential equation is identical with (3), so (4) is appropriate. In computation, the diffusion coefficient of concretes with silane treatment is updated after treatment. But, the unremoved chlorides in concrete with treatment are not set as zero, because they are not taken off by treatment (see figure 2).

2.3. Surface chlorides

In computation, the surface chloride could be treated as a boundary condition. Its value relies upon several factors, for instance, concrete mix or properties, the condition of exposure, and the distance of structural concretes from the source of chloride ions [8]. Due to these factors, the surface chloride functions were proposed in several forms [9]. Here, the constant surface chloride is given as 13 kg/m^3 for evaluating structural concretes situated in rigorous zone of the sea [10].

2.4. Diffusion coefficient

The diffusion coefficient of concrete could be treated as a material property. Its value relies on material types, for instance, normal concrete, repair concrete, silane-treating concretes and so on. Here, the diffusion coefficient of old concretes and repair materials is selected to be 1×10^{-8} cm^2/s [11] and 90% of the original concrete (0.9×10^{-8} cm^2/s), respectively, whereas the diffusion coefficient of silane-treating concretes is given as 0.29×10^{-8} cm^2/s [11]. In comparing, the diffusion coefficient for silane-treating concretes according to [11] is compatible a literature [12].

2.5. Time of repair implementation and lifetime for repairing

Cover replacement is utilized when the quantity of chlorides (Cl^-) on a monitoring (threshold) depth is equal to its precarious value (threshold-based). JSCE [13] specified the precarious value of 1.2 kg/m^3 for rebar corrosion initiation, whereas a researcher [14] specified the precarious value of 2.0 kg/m^3 for concrete cracking initiation. Here, the precarious value is given to be 1.2 kg/m^3 so as to control rebar corrosion (CR1) and 2.0 kg/m^3 so as to control concrete cracking (CR2). These two types of repair action are implemented whenever its precarious value is reached as shown in table 1.

Table 1. Time of repair implementation and lifetime for repairing.

Repairs	Notation	Time of repair implementation	Lifetime
Cover replacement	CR1	whenever $Cl_{TH,T} = 1.2$ kg/m^3	whenever $Cl_{TH,T} = 1.2$ kg/m^3 again
	CR2	whenever $Cl_{TH,T} = 2.0$ kg/m^3	whenever $Cl_{TH,T} = 2.0$ kg/m^3 again
Silane Treatment	SL1	every 7 yrs	5 yrs after treatment
	SL2	every 5 yrs	5 yrs after treatment

Silane treatment is utilized according to a defined cyclic time (time-based). NCHRP-558 [7] claimed that the lifetime of silane treatment relies on the condition of exposures, for instance, ultraviolet, moisture, and etc. Moreover, its lifetime felt between 5 and 7 years prior to re-implementation. According to these, the time of treating by silanes is selected at every 7 years or 5 year (SL1 or SL2, respectively), whereas its lifetime is determined as 5 years after treatment (see table 1). SL2 is utilized every 5 years or after the end of lifetime of the former treatment in order for restoring the quality of treatment continuously.

3. Illustrative case studies

Illustrative case studies are given here; structural concretes with replacing cover and that with treating by silanes. In computation, the monitoring depth to evaluate the critical time is found to be 80 mm for structural concretes situated within seawater [13].

3.1. Cover concrete replacement (CR)

Using the aforementioned data, the diffusion of chlorides (Cl^-) toward the depth of structural concretes is determined in figures 3(a) to 3(d). In figure 3(a), the outer chlorides (Cl^-) transport toward the old (original) concretes with the surface chloride of 13 kg/m^3, when the time passes by. At the 42nd week after 35 years, the Cl^- profile hits the precarious value of 1.2 kg/m^3 (CR1) at the monitoring depth as presented by the profile at year 35B (B: Prior to repair), so-called critical time. With cover replacement (shown by the shaded area within cover concrete in figure 3(b)), the Cl^- profile converts the profile at year 35A (A: Subsequent to repair). At year 36, chlorides (Cl^-) from the surface transport to the concretes (shown by the profile near concrete surface), and the unremoved chlorides inside old concretes both re-penetrate to the repair material and penetrate to old concretes (shown by the profile near the monitoring depth). The diffusion process of chlorides (Cl^-) still lasts. At year 73 in figure 3(c), the Cl^- profile at the monitoring depth again becomes equal to the precarious value, the repair identical with the former one is re-implemented, so the behavior of chloride diffusion is similar. At year 75, the design service life [15] is reached, the computation is terminated as shown in figure 3(d).

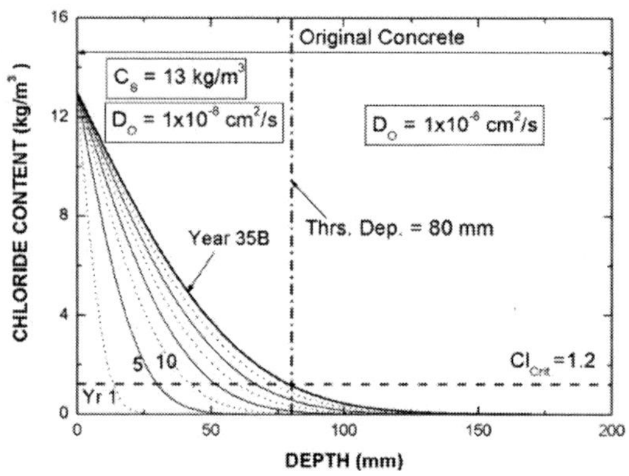

(a) Yrs. 1 - 35B

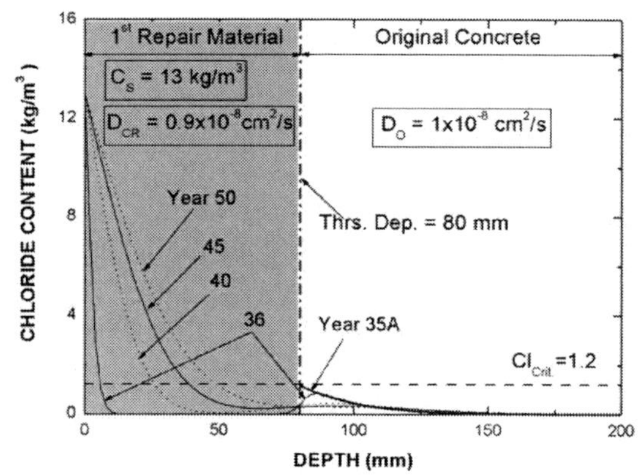

(b) Yrs. 35A - 50

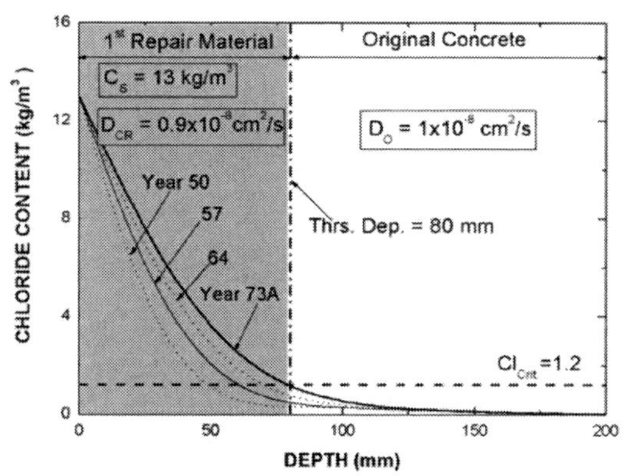

(c) Yrs. 50 - 73B

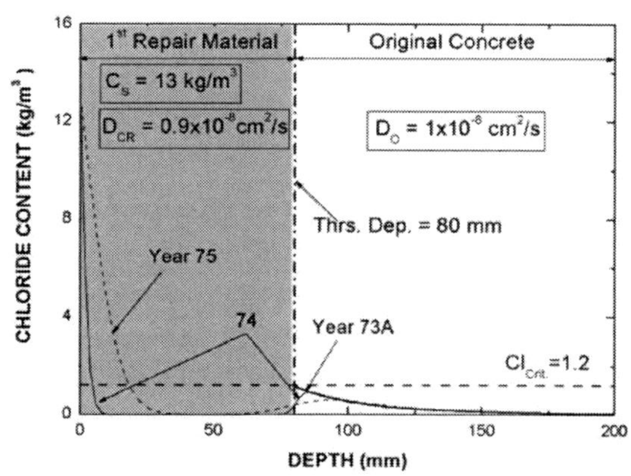

(d) Yrs. 73A - 75

Figure 3. Space-dependent chloride profile with replacement over 80 mm concrete cover.

3.2. Silane treatment (SL)

In the second case study, treating by silanes is used instead of cover replacement. The position-dependent Cl^- profiles could be presented in figures 4(a) and 4(b). Figure 4(a) shows the profiles for concretes with silane treatment utilized every 7 years (SL1). It is noted that the shaded area which covers over 30 mm from concrete surface shows the silane-treated zone (x_p). By comparing at year 34 in figure 4(b), the quantity of chlorides (Cl^-) at the monitoring depth for concretes with treating by SL1 is lower than that without. This shows the benefit of SL1 for extending the critical time. It is also noted that there is no silane-treated zone in this figure, because the quality of silanes implemented at year 28 vanishes at year 33 (5 years after implementation). The next silane treatment occurs at year 35.

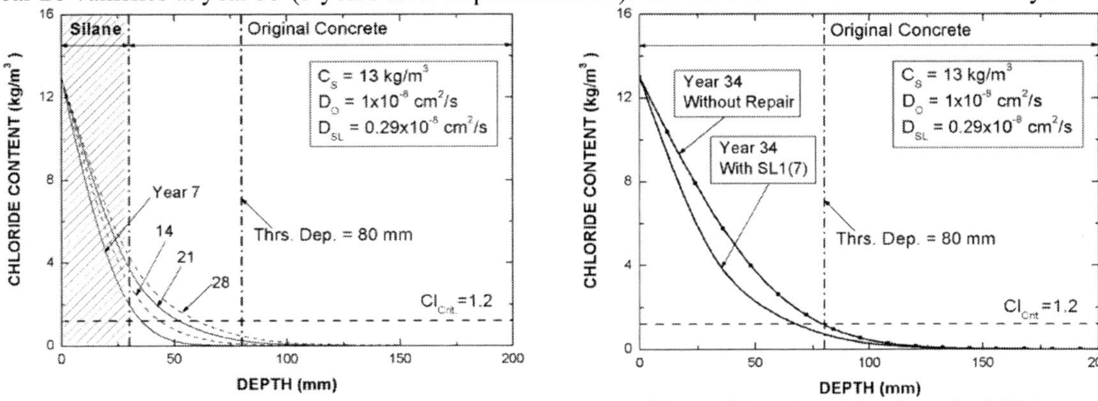

(a) Profiles: Yrs. 7 to 28 w/ SL1 (b) Profiles: Yrs. 35 w/o and w/ SL1

Figure 4. Space-dependent chloride profile with and without silane treatment.

3.3. Comparison

The Cl^- profiles for concretes with replacing cover at 80-mm monitoring depth is depicted with time as presented in figure 5(a). With no replacement, the chlorides unceasingly rise possibly causing rebar corrosion. When the service life of structural concretes is specified to be the time that the Cl^- at the 80-mm monitoring depth hits 1.2 kg/m^3 (CR1) or 2.0 kg/m^3 (CR2), the service life is about 35 years for corrosion control or 50 years for crack control, respectively.

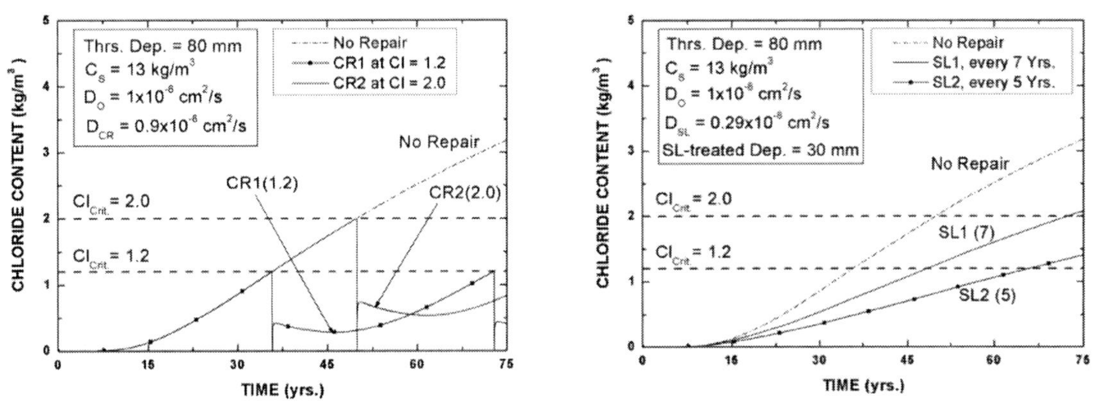

(a) Cover concrete replacement (CR1 or CR2) (b) Silane treatment (SL1 or SL2)

Figure 5. Time-dependent chloride profile with repairs.

If cover replacement is utilized for controlling rebar corrosion at the time that the Cl^- profile hits 1.2 kg/m^3 in year 35, the service life becomes extended for 38 years before hitting the precarious value again at year 73 in figure 5(a). At the repair time (year 35), the quantity of Cl^- reduces to zero because of removing the chlorides with concretes, but the quantity of chlorides (Cl^-) instantly rises owing to abrupt re-penetration of Cl^- from old concretes. Then, the diffusion process of chlorides still lasts. At year 73, the Cl^- profile hits the precarious value one more time, the repair identical with the former one is re-implemented (see figures 3(c) and 3(d)). Hence, the cover replacements at years 35 and 73 extends the service life of structural concretes over 75 years. If CR2 is instead implemented for controlling crack, there is only one repair within 75 years due to the precarious value of 2.0 kg/m^3. It should be notices that the transport of chlorides in concretes with corrosion control (CR1) is different from that with crack control (CR2) in spite of the same repair. This implies that the concrete structure with crack control needs fewer repairs but allows higher risk of deterioration due to rebar corrosion.

With treating concrete by SL1 or SL2 as presented in figure 5(b)), the service life of corrosion-control structural concretes could be extended to year 48 or 66, respectively. With SL1 or SL2, the service life of crack-control structural concretes can however be extended to year 73 or more than year 75, respectively. Thus, concrete structures with SL2 are free of crack for 75 years.

Cover replacement result in eliminating the chloride ions in concrete cover, while silane treatment retards the penetration of chlorides as shown in figures 5(a) and 5(b), respectively. The question is which repair is more environment-friendly. Based on literatures [16,17], the energy use is found to be 1194 and 76 MJ per m^2 for cover replacement and silane treatment, respectively. For 75 years, the energy use for repairs by CR1 and CR2 is equal to 2388 (twice) and 1194 (once) MJ/m^2, respectively, while that by SL1 and SL2 is equal to 760 (10 times) and 1140 (15 times) MJ/m^2, respectively. So, if the energy use and corrosion control for 75 years is desirable, an optimum repair is CR1. But if the energy use and crack control is desirable, an optimum one becomes SL2.

4. Conclusions

Concrete cover replacement is used to control rebar corrosion or concrete cracking by controlling the quantity of chlorides (Cl^-) at the monitoring depth, while treating by silanes is used to extend the time that the Cl^- at the monitoring depth hits the precarious value. Moreover, the optimum repair action relies on what to control (corrosion or crack). And, concrete structures with crack control need fewer repairs leading to lower energy use, but allow more rebar corrosion leading to higher risk of failure.

Acknowledgments
This research was funded by King Mongkut's University of Technology North Bangkok. Contract no. KMUTNB-61-GOV-B-16.

References

[1] REHABCON 2004 Final report on the evaluation of alternative repair and upgrading options: Strategy for maintenance and rehabilitation in concrete structures (LIT, Sweden)
[2] Petcherdchoo A 2015 *Journal of Advanced Concrete Technology* **13(3)** 205-13
[3] Petcherdchoo A 2016 *Cement and Concrete Composites* **74** 88-99
[4] Crank J 1975 *The Mathematics of Diffusion* (Oxford: The Clarendon Press)
[5] von Rosenberg D U 1969 *Methods for the Numerical Solution of Partial Differential Equations* (Elsevier Publishing Co.)
[6] Press W H, Teukolsky S A, Vetterling W T and Flannery B P 1996 *Numerical Recipes in C: The Art of Scientific Computing* (Cambridge University Press, Second Edition)
[7] NCHRP-558 2006 *Manual on Service Life of Corrosion-Damaged Reinforced Concrete Bridge Superstructure Elements* (Transportaiton Research Board of the National Academies)
[8] Costa A and Appleton J 1999 *Material and Structures* **32(5)** 354-9
[9] Petcherdchoo A 2017 *Cement and Concrete Research* **102** 136-48
[10] DPT 2012 *Standdard for considering durability and service life 1332-55* (Dept. of Public Work and Town & Country Planning, Ministry of Interior, Thailand) (in Thai)
[11] Shimomura T 2005 *Proc. Int. Workshop on Durability of Reinforced Concrete under Combined Mechanical and Climatic Loads* (Qingdao Technological University, Qingdao, China)
[12] Medeiros M H F and Helene P 2009 *Construction and Building Materials* **23(3)** 1476-84
[13] JSCE 2002 *Standard specification for durability of concrete* (Concr. Lib.) (in Japanese)
[14] Moriwake A 1996 Study on durability and maintenance of reinforced concrete jetty deck against chloride induced deterioration Ph.D. thesis (Tokyo Inst. of Tech., Tokyo, Japan)
[15] Tikalsky P J, Tepke D and Schokker A 2003 *2nd Rilem Workshop on Life Prediction and Aging Management of Concrete Structures* (Paris 5-6 May 2003)
[16] Arskog V, Fossdal S and Gjorv O E 2004 *International Workshop on Sustainable Development*

and Concrete Techchnology (Beijing)
[17] Petcherdchoo A 2015 *RILEM Bookseries* **10** 35-41

Investigation of crystalline structure of plasticized poly (lactic acid)/ Banana nanofibers composites

T Farid[1,2,3], V N Herrera[2] and O Kristiina[2]

[1]Chemical Engineering Department, Faculty of Science and Technology, University of Ghardaia, 47000 Ghardaia, Algeria
[2]Composite Centre Sweden, Division of Materials Science, Lulea University of Technology, 97187 Lulea, Sweden

E-mail: ftouaiti@gmail.com

Abstract. Polylactic acid (PLA) is a promising biodegradable candidate to replace synthetic commodity plastics in many applications. However, this polymer shows high brittleness, slow rate and lower degree of crystallization. The addition of plasticizing agents can enhance the toughness, but its effects on the crystallization behavior remain inconclusive. Therefore, this research is aiming to cast light on this area. Using differential scanning calorimetry (DSC) at a 2°C/min cooling rate, extruded neat PLA samples showed lower degree of crystallinity and thermal stability. This material shows cold crystallization upon heating and does recrystallize prior melting. These results indicate a clear instability in the crystalline state are confirmed by the crystallographic results by the X-ray diffractions (XRD) pattern and atomic force microscopic imagery. The addition of around 20 wt% of glycerol triacetate (GTA) with 1wt% of banana nanofibers (BNF) almost doubled the crystallinity. This modification is believed to occur through a dilution mechanism in order to increase crystallization rate yielding a more stable crystalline structure as shown by the XRD. However, the dynamic mechanical thermal analysis (DMTA) showed a 30 to 50% reduction in the room temperature storage modulus (stiffness) is in plasticized samples when compared to neat 100% PLA. Although these results shows the possibility to enhance the crystallization through a combination of plasticizing and nanoreinforcing effects, further studies is still needed to optimize the material formulation in order to find the best ratios to secure both a good crystallization and mechanical properties. This will definitively result in a new material that can be used for current and futuristic applications.

1. Introduction

The negative environmental impact of oil based plastics motivates scientists to search for eco-friendly substitutes. The biodegradable polymers like Agro-polymers and Biopolyesters seemed a suitable candidate since they possess many attractive characteristics like renewability, sustainability and compostability [1-11]. Intensive investigations of bio-polyesters revealed the potentials of polylactic acid (PLA) as a replacement for commodity plastics and as an interesting material in many fields like biomedical [12,13], pharmaceutical [14] and packaging [15,16]. Similar to commodity plastics, PLA shows appealing characteristics such as good mechanical properties [17,18], high clarity [19] and simple processability [20,21]. However, its intrinsic brittleness [18], low melt strength [22,23], low thermal stability [24] and slow crystallization [25] prevent its domination over oil based plastics in many applications [26-30].

During the last decade, many studies on the morphology and structure of PLA tried to improve its fragility and poor crystallinity. Different strategies have been adopted for this purposes such as blending with softer polymers [31,32], chemical modification and copolymerisation [33,34] and heterogeneous crystallization through using sterocomplex and other polyesters [35-44].

Currently, PLA is intensively used with nanocellulose reinforcement in order to make biodegradable nanocomposites. The results showed an overall improvement in the mechanical, optical and thermal properties of these materials [45-47]. Unexpectedly, the Triacetin plasticizer which was primarily intended to ensure good dispersion of nanoreinforcements, showed the potential of improving PLA crystallinity. This surprising result needs clarification through understanding the origin of this behavior as well as the mechanism by which it works. The aim of this work is to answer these questions by a thorough experimental investigation.

2. Experimental

2.1. Materials

Polylactic acid made by FUTERRO (Escanaffles, Belgium) is used in this work. The data sheet showed that this high molecular weight L-PLA which is primarily used for extrusion has a glass transition temperature (Tg) of 56 ± 4°C and a crystallisation temperature (Tc) of 160±15°C. The glycerol triacetate (GTA) is purchased from Sigma-Aldrich (Stockholm). This high purity liquid plasticizer has a molecular weight of 218 g/mol. The Banana nanofibers (BNF) were prepared from bleached banana waste pulp using mechanical fibrillation method using Masuko super mass collider [47]. These nanofibers have diameter that varies from 10 to 60 nm.

The samples were prepared using a liquid feeding extrusion process in similar manner as reported by Oksman et al 2006 [48].

The formulation of each material is given in table 1. Neat PLA is used as a reference, whereas plasticized and reinforced PLA samples are used to study the impact of the plasticizer and the reinforcing agents on the crystallinity.

Table 1. Composition of different materials.

Material	Polylactic acid (wt%)	GTA plasticizer (wt%)	Banananofibre (wt%)
Neat PLA	100	0	0
Plasticized PLA	80	20	0
Reinforced PLA	79	20	1

2.2. Thermal transitions (DSC)

A Mettler Toledo DSC822e (Schwerzenbach, Switzerland) is used to study different thermal transitions within the materials. Samples weighting 6 mg were cut from each extruded material and placed in an aluminum pan to be thermally scanned under N_2 atmosphere from 30 to 200°C at a heating rate of 10°C/min. In order to thoroughly investigate the crystallization process, two cooling rates of 20°C/min and 2°C/min are used.

2.3. Crystal structure (XRD)

A Siemens D5000 Diffractometer (Berlin) with a CU-Kα radiation source (λ=0.154 nm) was used to study the internal crystalline structure in these materials. The tests were performed on thin films made from extruded samples. The equipment was operated at 40 KV and 45 mA.

2.4. Thermomechanical properties (DMTA)

The thermomechanical properties were investigated using a QA 800 analyzer from TA instruments (USA). The analysis was conducted in tensile mode under 3°C/min heating rate and 1Hz frequency. The temperature scan ranges from 30°C to 200°C in air.

2.5. Atomic force microscopy (AFM)

The morphology and structure of PLA samples was investigated using a Nanoscope V AFM. From extruded materials, rectangular samples (2 mm×10 mm) were cut and fixed on an AFM specimen holder (Insert 16702448). The samples was placed on the Leica EM UC7 ultramicrotome to prepare a rectangular flat surface (400 µm × 600 µm) for Tapping mode™ AFM imaging. In order to visualize the spherulitic structure, the scanning was performed on large area (120 µm × 120 µm). The morphology was characterized by measuring the average spherulites size which was calculated from the diameter (cross-section) of at least 9 deferent spherulites.

2.6. Optical microscopy

Samples weighting around 5 mg from neat, plasticized and reinforced PLA were cut and placed on rectangular (75 mm x 25 mm) glass microscope slide (Vista Vision™). The slide was heated at 190°C for 10 min before placing another slide on the top to cover the molten matter. In order to create a very thin film, the sandwich was pressed gently then left to cool down in air. A second heating cycle at 190°C for 5 min was used to release any stress that might be induced during the compression and to prepare the thin film for morphology investigation.

Using Nikon's inverted optical microscope (Eclipse MA 200™), thin films were observed under polarized light to visualize the spherulitic morphology. In addition, dark field mode was also used to clearly observe the spherulites since a good contrast between the high refractive crystalline spherulites and the amorphous region is expected in this mode [49,50].

3. Results and discussion

3.1. Thermal analysis

Thermal transition curves of neat, plasticized and reinforced PLA measured using differential scanning calorimetry are shown in figure 1. Neat PLA shows a clear second order transition (Tg) at 64°C, after which it goes through two cold crystallization stages at T_{cc1}= 100°C and T_{cc2}=161°C before melting (T_m) at 178°C. Plasticized and reinforced PLA materials showed the absence of the glass transition and cold crystallization peaks. Thermal properties such as the melting enthalpy, thermal transitions and degree of crystallinity estimated from the DSC thermograph are presented in table 2.

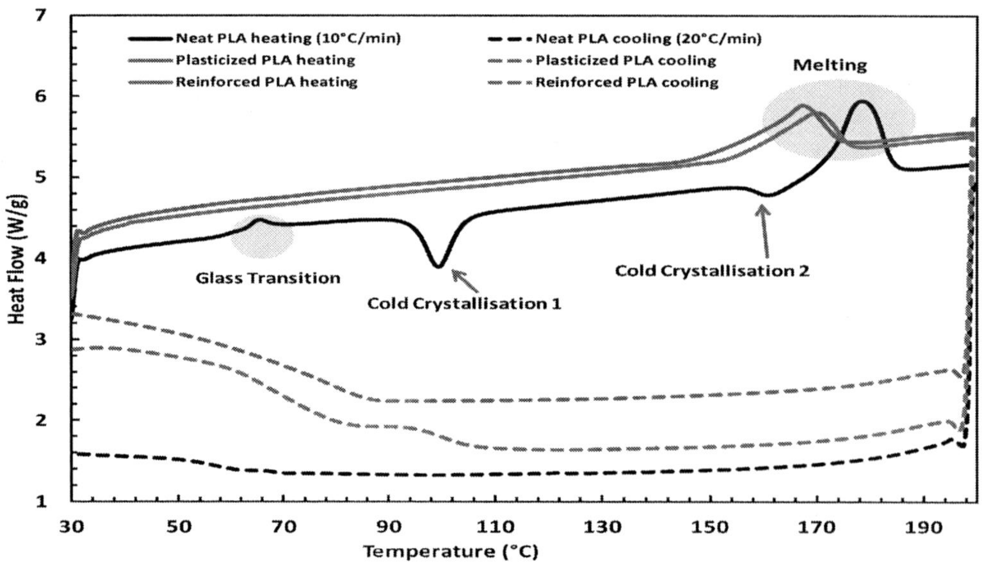

Figure 1. DSC thermograph of different materials.

Table 2. DSC thermal properties of Neat, Plasticized and reinforced PLA.

Material	Cold crystalization				Melting		Glass transition	Crystallinity
	T_{CC1} (°C)	ΔH_{cc1} (J/g)	T_{CC2} (°C)	ΔH_{cc2} (J/g)	T_m (°C)	ΔH_m (J/g)	Tg (°C)	Xc (%)
Neat PLA	100	22.7	161	5	178	48	64	22
Plastcized PLA	-	-	-	-	167	39	-	41
Reinforced PLA	-	-	-	-	170	35	-	37

The absence of the glass transition ramp in plasticized and reinforced PLA can be attributed to the lake of sensitivity in the DSC. Previous studies showed a 20°C decrease in the glass transition temperature when adding GTA to neat PLA which is a clear indication of the plasticizing ability of GTA [51-53].

At 100°C, neat PLA chains gain enough mobility to disentangle from the amorphous phase and rearrange in an ordered 3-D crystallites. The exothermic peak at 161°C is due to melt-recrystallization of early formed crystallites since they possess low thermal stability therefore they recrystallize at the vicinity of the melting temperature. The recrystallization peak indicate that previously formed crystallites did not assume the stable orthorhombic α-crystal form [54,55]. The thermographs of plasticized and reinforced PLA (in contrary to neat PLA) showed no signs of cold crystallization which proves that their crystallites are made of thermally stable α- form.

The crystallites size distribution can be also estimated since it is directly proportional to the width of the melting peak [56]. Plasticized PLA with a melting peak span ΔT_m=29°C has the largest crystallites size distribution whereas, neat PLA with ΔT_m =20°C has the narrowest size distribution. Reinforced PLA has a melting temperature of 169°C and a peak span of 26°C indicating a slightly narrower crystallite size distribution compared to plasticized PLA.

The highest melting temperature in these three samples is observed for the neat PLA (T_m=178°C). This indicates that Neat sample has the highest crystallite average size, whereas plasticized PLA has the lowest crystallites average size since it has the lowest melting temperature T_m= 167°C.

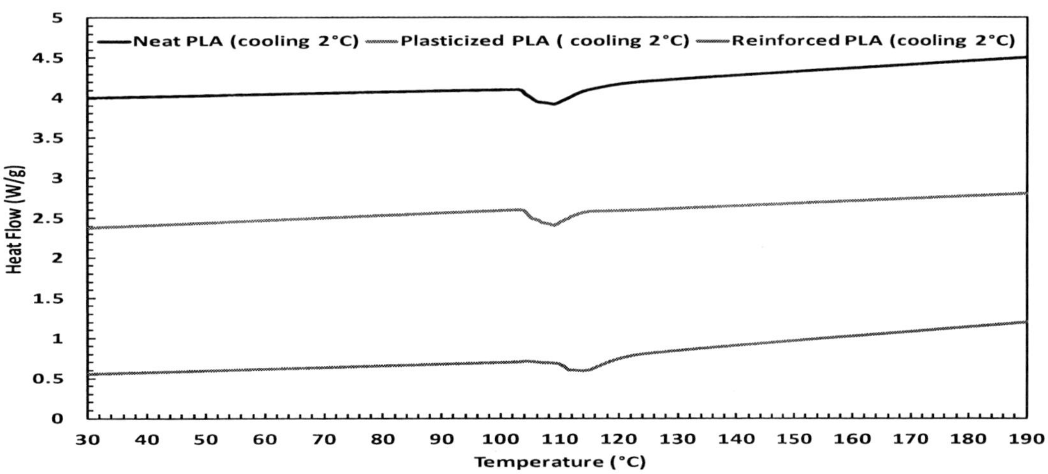

Figure 2. DSC thermograph for cooling cycle at 2°C/min.

During the cooling cycle at 20°C/min (figure 1), these materials showed no crystallization peaks which would indicate the formation of amorphous materials. The PLA crystallization process is known to be slow and at this cooling rate, the crystal growth will decrease drastically since PLA molecules do not possess enough mobility to diffuse to the crystallites. Instead, they freeze in their high entropic state resulting in minimal crystallinity. As this PLA is intended to be extruded, lower cooling rate would be adequate for realistic study. Additionally, the sensitivity of DSC would improve at low rates. Therefore, a slow cooling rate of 2°C/min scan is performed and the thermograph is presented in figure 2.

Neat, plasticized and reinforced PLA materials showed exothermic crystallization peaks at 108°C, 109°C and 112°C respectively. The slightly higher crystallization temperature of reinforced PLA compared to other samples can be explained as heterogeneous crystallization resulting from the fiber surface which can act as nucleating agent. Since lower crystallization temperature in PLA leads to deformed crystals [57,58], neat PLA is believed to have unstable α´-crystal form

Using Equation1, The degree of crystallinity (Xc) can be calculated from the enthalpy of fusion (ΔHm), the enthalpy of crystallization (ΔHc) and the theoretical enthalpy of fusion (ΔHm^0= 93 J/g) [59]. The results are presented in table 2.

$$Xc= ((\Delta Hm - \Delta Hc) / \Delta Hm^{\circ}) \times 100\% \qquad (1)$$

The addition of GTA plasticizer enhanced the crystallinity through increasing the PLA chains mobility and allowing them to easily disentangle from the amorphous molten state and rearrange in an order manner to form crystallites. However, the lower crystallization temperature (109°C) did not offer enough thermal energy for these chains to diffuse toward the crystallite surface which eventually reduced the overall degree of crystallinity.

Among the three samples, reinforced PLA showed the highest crystallization temperature (112°C), it is therefore expected to possess the highest degree of crystallinity. Surprisingly, plasticized PLA showed higher level than reinforced PLA. This strange result calls for further investigation using proper technique that cannot be dependent on the heating /cooling rates assessment. X-ray diffraction technique was used for deeper insight in the structure of these materials.

3.2. Crystallinity analysis

Using the WAXD spectrum ranging from 5° to 25°, it was possibe to identify the crystalline structure of neat, plasticized and reinforced PLA samples. As shown in figure 3, these materials showed typical semi-crystalline spectra containing both amorphous "halo" and crystalline peaks [60].

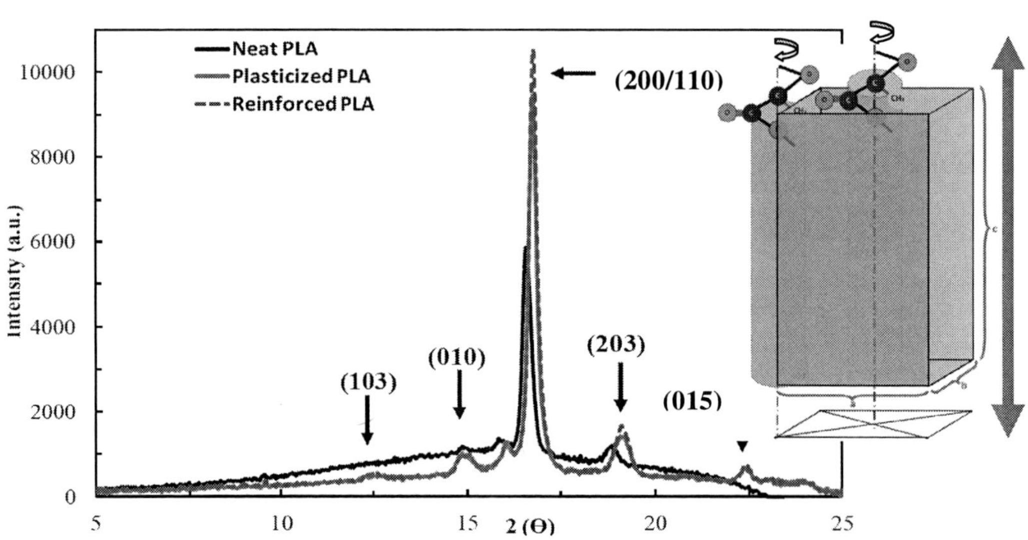

Figure 3. XRD spectra of different samples with an orthorhombic cell (upper corner).

Neat PLA showed fewer diffraction peaks compared to plasticized and reinforced PLA. Based on the diffraction patterns presented in figure 3, PLA chains in plasticized and reinforced samples are crystallized in the α-form orthorhombic crystals [11]. In addition to the absence of (103) and (015) peaks, neat PLA pattern shows a slight shift of (200/110) and (203) peaks positions toward lower angles. These observations can be attributed to low degree of order and deformation in the crystalline structure [61-64]. It is therefore clear that neat PLA crystallites adopted the distorted α′-form orthorhombic crystal

The orthorhombic cell parameters were calculated by combining the Bragg's law with equation (2). The geometrical characteristics of the (010), (200) and the (203) plans were used to obtain three mathematical equations with which the parameters can be obtained.

$$(d_{(hkl)})^{-2}=(h/a)^2+(k/b)^2+(l/c)^2 \quad (2)$$

The crystal parameters for plasticized and reinforced PLA were a=1.06 nm, b= 0.59 nm, c=2.78 nm. Whereas those of neat PLA where found to be a′=1.08 nm, b′=0.59 nm, c′=2.87 nm. These values are in good agreement with literature [65,66]. The large cell dimensions in neat sample are due to the stiff nature of PLA chains [67,68]. The oxygen and methyl side groups in the PLA chains (figure 3) result in low degree of symmetry therefore, PLA molecules folds in large cells compared to polymers with flexible symmetrical chains.

Under appropriate crystallization conditions (T, t, P), PLA adopt the left-handed 10_3 helix conformation (10Å rise for every 3 monomer units) in α-form crystal [69-71]. while, under abnormal conditions (low crystallization temperature, fast cooling) the chains keep the 10_3 helix conformation with a perturbation in the rotational and the longitudinal ordering resulting from the intermolecular dipole-dipole interaction between methyl and oxygen side groups of adjacent chains [54]. The 4% (a′xb′xc′/(axbxc)) increase in the α′-crystal volume resulted from the loosely packed chains in this crystal. Therefore the plasticizer enhances the firm packing of PLA chains in their crystals.

The peak intensities observed in reinforced and plasticized PLA patterns are higher than neat PLA which reveals the higher level of crystallinity. The areas under crystalline and amorphous peaks can be used to calculate the degree of crystallinity [72,73]. Furthermore, the lamella L_θ which characterize the packing ability of polymer chains can be estimated by Scherrer equation (equation (3)). Using the main peak characteristics (peak angle (θ) and Full width at half maximum height in radians (β)) for (200) and (203) plans, the thickness (L_θ) can be calculated using wave length λ=0.154 nm and Scherre constant K=0.9 [70,74].

$$L_\theta = (k\lambda)/(\beta \cos\theta) \quad (3)$$

In order to calculate the exact value of β, an open source curve fitting software (*curve expert professional 2.4.0*) was used on the (200) and (203) peaks. Gaussian function was found to result in the best fitting with an error less than 9%. Table 3 summarizes the results including the inter-plannar distance d, β and L_θ.

Table 3. Detailed analysis of crystalline region in neat, plasticized and reinforced PLA.

	ΔT^{-1}	(200/110) plan			(203) plan			Xc (%)
	(10^{-3})	d_{200} (nm)	β_{200} (rad)	L_{200} (nm)	d_{203} (nm)	β_{203} (rad)	L_{203} (nm)	
Neat PLA	14.2	0.54	0.0055	25	0.47	0.0134	10	28
Plastcized PLA	17.2	0.53	0.0045	31	0.46	0.0128	11	47
Reinforced PLA	17.5	0.53	0.0043	32	0.46	0.0117	12	50

A 20% increase in crystallinity is observed upon plasticization of PLA. Similarly, additions of 1% of BNF resulted in almost double the level of crystallinity observed in neat PLA. The mechanisms by which GTA and BNF affect crystallinity are different and will be discussed later. The distortion in the unit cell of neat PLA can be clearly seen by the increase in the inter-plannar distance in these two main crystalline peaks. Besides Scherrer equation, the lamellae thickness (L_0) can be calculated based on Gibbs-Thomson equation using the undercooling ($\Delta T=T_m-T_c$) and the enthalpy of fusion (ΔH_m) from the DSC thermograph [75]. This equation shows that the lamella thickness is inversely proportional to the degree of supercooling (ΔT). This relationship is fulfilled according to values presented in table 3. Thus, neat PLA has thinner lamellae compared to plasticized and reinforced samples.

In figure 1, the high melting temperature of neat PLA indicates the formation of thicker lamellae in contradiction with the results in table 3. The first peak (100°C) in figure 1 shows the formation of new crystallites (α′-form), whereas the second peak (161°C) revealed the recrystallisation of these crystals into the stable α-form. This process will result in thickening the lamellae in neat PLA [76-79]. The X-ray experiment was conducted at room temperature before these thermal transitions occur therefore, the thickening due to this process cannot be observed. Nevertheless both techniques agreed on the fact that the GTA plasticizer improved the level of crystallinity in the polylactic acid polymer.

The XRD results in table 3 showed that reinforced PLA has the highest crystallinity in contradiction to what DSC results showed in table 2. Since reinforced PLA has higher crystallization temperature than plasticized PLA, the crystallinity should be also high. Therefore, the theoretical enthalpy of fusion estimated by Fischer [59] based on Flory theory for polymers and copolymers is quite inaccurate when dealing with trans-crystallization in heterogeneous nucleation.

3.3. Thermomechanical analysis

GTA plasticizer and BNF reinforcement do clearly affect Thermomechanical properties of PLA as shown in figure 4. The modulus of neat PLA (~3 GPa) is higher than plasticized and reinforced samples. Around Tg (64°C), the modulus falls sharply below the moduli of other samples. This can be attributed to the fast rearrangement in the amorphous region resulted in large internal macromolecular motion suggesting that neat PLA molecules were frozen in stressful conformation during cooling stage. Around 100°C ±10, neat PLA gained certain elasticity resulting in a slight increase in the modulus. This is clearly due to cold crystallization as showed in the DSC thermograph.

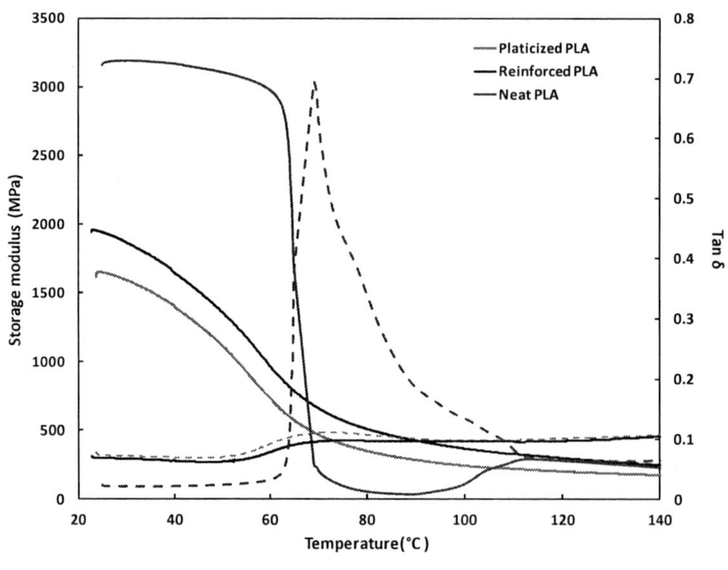

Figure 4. Storage modulus and loss factor of different samples.

Plasticized PLA sample showed a 50% reduction in elasticity compared to neat sample. This is expected since GTA plasticizer allows for higher elongations at low level of applied stress. However, the transition from glassy to rubber was smoother compared to that of neat PLA as it occurred along large temperature range (50°C to 80°C). Therefore, plasticizer induced high thermal stability and negligible level of residual stress. The enhanced chain mobility in plasticized PLA is clearly manifested by the low elasticity at the rubbery state. Loss factor peaks (dashed lines in figure 4) for plasticized and reinforced PLA samples are broader and shorter compared to neat PLA which is believed to be due to high level of crystallinity [80].

Adding 1% BNF to plasticized PLA resulted in 10 % increase in the glassy state modulus. The introduction of stiffer phase within the viscoelastic soft matrix will increase the elasticity. At the rubbery plateau, certain molecular motion is observed in the material which can be attributed to weak interfacial interaction between the plasticized matrix and BNF surface. The general trends shows that GTA improves the thermal stability but it deteriorate the PLA elasticity. The addition of 1% BNF showed negligible impact on the modulus.

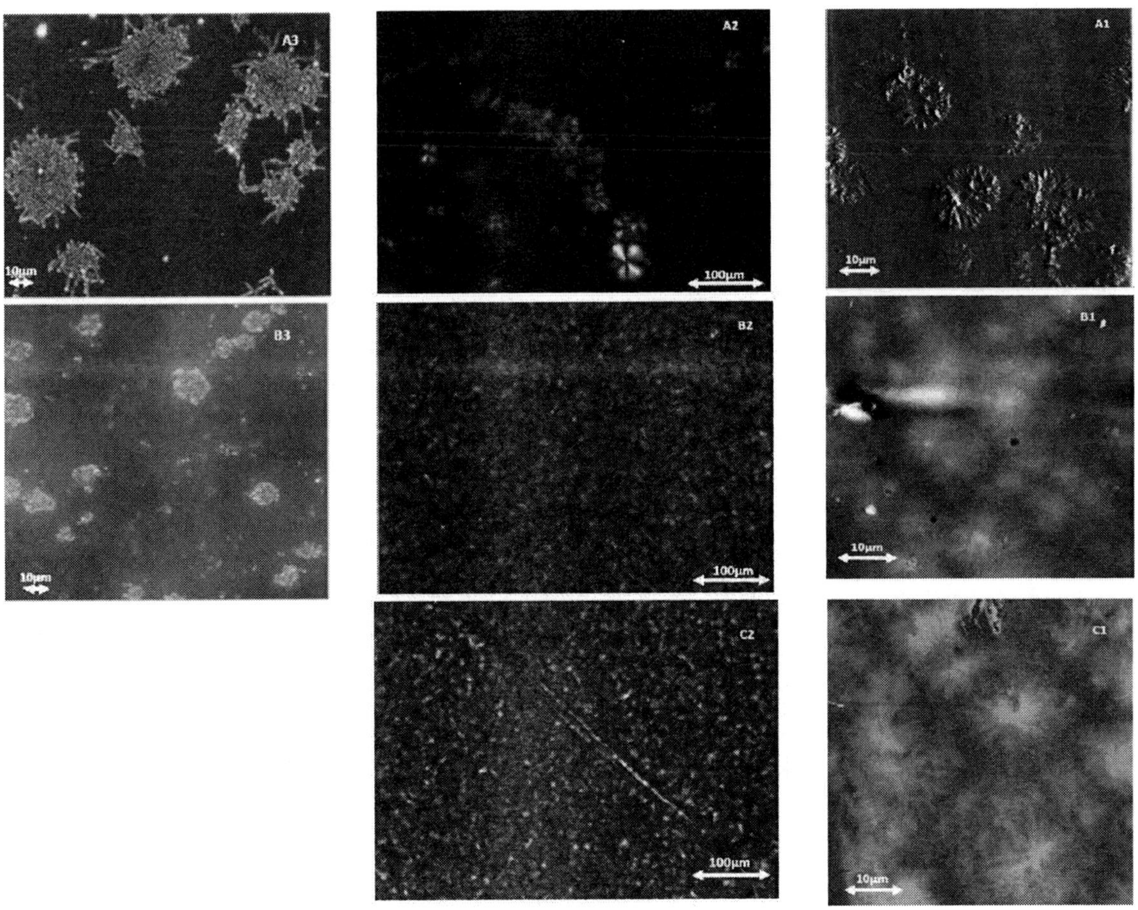

Figure 5. Crystalline morphology of neat (A), plasticized (B) and reinforced PLA (C) using AFM (1), Polarized (2) and dark field (3) imaging.

3.4. Microscopy analysis

Atomic force microscopy and optical microscopes are used to investigate the crystalline morphology of PLA samples. Images in figure 5 show the spherulitic microstructure of neat, plasticized and reinforced PLA. In (A1) image, neat PLA showed isolated cauliflower-like spherulites [81], whereas

IOP Conf. Series: Materials Science and Engineering **369** (2018) 012031 doi:10.1088/1757-899X/369/1/012031

plasticized and reinforced PLA showed a coalescent spherulitic structure (B1, C1). The average size of PLA spherulites measured using AFM was 22 μm ± 7. This large size is expected to result in a brittle behavior [68,82]. The few scattered spherulites in neat PLA (A2) revealed the formation of few nucleation sites during crystallization. Furthermore, the isolated spherulites is a clear indication of low crystallization rates since at high rate they would grow to form a spherulitic polyhedral structure [83]. In the dark field image (A3) few lamellae bundles are present indicating an early stage of spherulites formation [84].

In plasticized PLA, high nuclei density is observed (B1, B2, B3) as expected [85]. Using AFM, the average spherulites size was 9 μm ± 3, which would show enhancement in the plasticized PLA toughness. In reinforced PLA, the nucleation occurs at two different locations (C2); the plasticized matrix and on the fiber surface which is suitable location for Tran-scrystallization [86,87]. The spherulites average size (11 μm ± 5) is smaller than neat PLA which is due to the enhancement in nucleation process.

In the literature a 30°C increase in the crystallization temperature (Tc) were obtained when 1% Poly-D-lactic acid was added to PLLA sample as heterogeneous crystallization agent [88]. Whereas, adding 1% of BNF resulted in 4°C increase in Tc (figure 2). The spherulites average size in reinforced PLA was higher than that of plasticized PLA. These results highlight the poor nucleation efficiency of BNF. This poor nucleation efficiency may result from surface contaminations occurred during pretreatment process prior to extrusion. The fiber was first subjected to solvent exchange (water/aceton) after fibrillation, then dispersed in GTA plasticizer before mixing it with molten PLA matrix. Therefore, the true contact between the real BNF surface and PLA matrix is minor. Nevertheless, the enhancement in the degree of crystallinity at 1% BNF shows its potential as a heterogeneous nucleating agent.

3.5. The mechanism by which GTA enhanced crystallinity

In theory, the enhancement in chain mobility is suggested to improve polymer crystallinity. The depression of glass transition temperature will expand the temperature range where macromolecules are able to crystallize in optimum conformation. This logic idea does not explain why in many studies GTA plasticizer decreased Tg but it did not enhance crystallinity [89,90]. According to literature, the plasticizer has also a thermodynamic impact since it increases the lamellae lateral surface energy at the growing front of spherulites [91]. It is therefore clear that the plasticizer impact is beyond a simple depression of Tg, and a new mechanism should be proposed taking into account the literature and the current results

The GTA plasticizer has similar chemical nature as PLA polymer since they both possess methyl and carbonyl groups and develops similar intermolecular forces. This explains why they both have high degree of miscibility (similar solubility parameter (δ=20.1 $MPa^{0.5}$) and compatibility (interaction factor=0.34) [89]. As a consequence, the plasticizer's small molecules can easily mixed and diffuse through PLA chains. When PLA and GTA are mixed together in melt, GTA molecules will easily diffuse into PLA chains to lubricate them and diminish the internal friction. Consequently, the entanglements density within this plasticized melt is expected to be far below that of neat PLA resulting in reduction in the entropic barrier for nuclei formation. This can explains the large number of spherulites in plasticized and reinforced PLA (figure 5). Meanwhile, the PLA macromolecules have the opportunity to adopt easily the optimum crystallographic conformation (α-form) and form many stable nuclei (figure 3).

After the formation of stable nuclei, more macromolecular segments are packed together forming lamellae as the early stage of spherulites formation (image B3 in figure 5). The GTA molecules are excluded from the crystals as proven by XRD Spectrum (figure 3). These small impurities gather at the growing lamellae interface (front) with the super-cooled melt [83]. Upon cooling, the crystal growth is directly dependent on the chains diffusion ability through the melt into the lamellae at the growing spherulites front as shown in figure 6. The continuously accumulating GTA molecules will decrease the local melt viscosity at the vicinity of the growing lamellae front (figure 6). Additionally, they increase the lamellae lateral surface energy leading to faster growth which explains the fully developed polyhedral spherulitic morphology in figure 5 (images B1, B2, C1, C2). As temperature drops, the diffusion activation energy decreases and the structure freezes. Finally, the morphology will show crystalline regions where PLA chains are concentrated in the spherulites and an amorphous region at the spherulites edges (figure 6) mostly rich of GTA molecules.

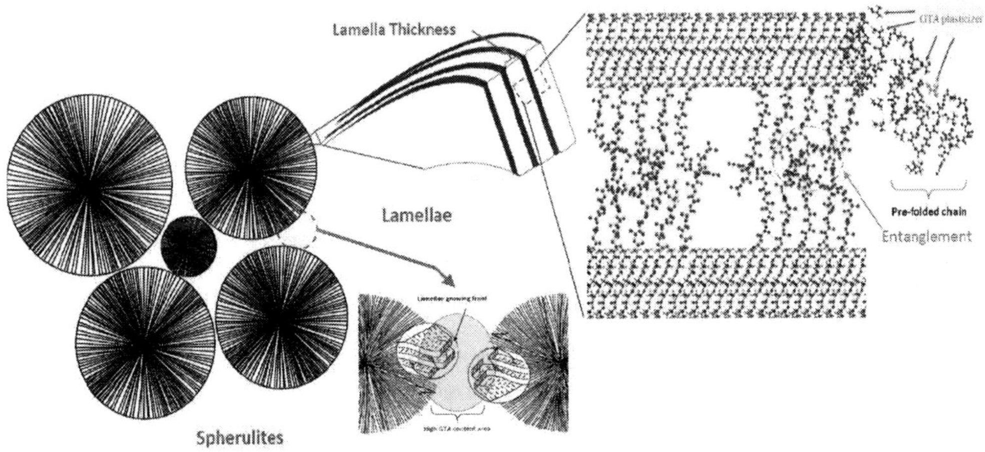

Figure 6. Illustration of the plasticization enhanced crystallinity mechanism.

The dilution mechanism proposed here agrees with the work of Tsuji and his co-workers when they found that dilution effect assists the nucleation and increases the nuclei density [92]. In order to efficiently enhance crystallinity, the optimum GTA plasticizer concentration has to be reached to trigger the dilution effect. This concentration is believed to dependent on many factors such as the polymer molecular weight, cooling rate and chemical interactions and compatibility.

The GTA concentration should take into account the thermomechanical properties; the loss of 50% of storage modulus (figure 4) must be addressed when looking for suitable formulation. Therefore, the optimum GTA content should assure the maximum crystallinity with the lowest possible loss in elastic modulus.

4. Conclusions

This study shows that at a cooling rate of around 2°C/min, extruded neat PLA samples showed lower degree of crystallinity and thermal stability. This material shows cold crystallization upon heating with a recrystallization before melting. These indications of instability in the crystalline state are confirmed by the crystallographic (XRD) and microscopic studies (AFM and optical microscopy).

The addition of 20 wt% of glycerol triacetate (GTA) with 1wt% of banana nanofibers (BNF) almost doubled the crystallinity. This modification believed to act through a dilution mechanism in order to increase crystallization rate yielding a more stable crystalline structure. However, 30% to 50% reduction in the room temperature storage modulus (stiffness) is observed. Thus further optimization studies of PLA/GTA formulation is needed. This is believed to result in a new material with the ability to replace commodity plastics.

5. Outlook

Future contribution to this work can be summarized in two objectives:

- Find the minimum GTA concentration to enhance crystallinity.
- Find the best GTA content for concentrated BNF suspension (3-5% BNF).

Acknowledgments
The author would like to thank the *Kempe Stiftelserna* Sweden for the financial support. This work was also part of the activities of Wallenberg Wood Science Center (WWSC).

References
[1] Averous L and Pollet E 2012 *Environmental Silicate Nano-BioComposites* (London: Springer)
[2] Della valle G, Boche Y, Colonna P and Vergnes B 1995 *Carbohyd. Polym.* **28(3)** 255-64
[3] Oksman K and Mathew A P 2014, *Bionanocomposites: Processing, Characterization and Properties* (Hackensack: World Scientific Publishing Co.)
[4] Segar A D and Merrill E W 1995 *Polymer* **36(9)** 1883-6
[5] Olsen B D 2013 *AIChE J* **59** 3558-68
[6] Chung H and Washburn N R 2013 *Green Mat.* **1** 137-60
[7] Auhorn J W 2006, *Handbook of Paper and Board* (Ravensburg: WILEY-VCH verlag GmbH&Co. KGaA)
[8] Yu L, *et al* 2006 *Prog. Polym. Sci* **31(6)** 576-602
[9] Kellerhals M B, *et al* 2000 *Macromolecules* **33** 4690-8
[10] Lo R G, *et al* 2014 *Eur. Polym. J.* **54** 138-50
[11] El-Hadi A, *et al* 2002 *Polym. Test.* **21** 665-74
[12] Mano J F, *et al* 2004 *Compos. Sci. Technol.* **64** 789-817
[13] Kohn J, *et al* 1996 *Biomaterials Science: Introduction to Materials in Medicine* (London: Academic Press)
[14] Williams S F and Martin D P 2005 *Biopolymers Online* **4** 91-103
[15] Philip S, Kashavarz T and Roy I 2007 *J. Chem. Technol. Biotechnology* **82(3)** 233-47
[16] Van T R, *et al* 2000 *Biobased Packaging Materials for the Food Industry-Status and Perspective* (KVL: Frederiksberg) 8-33
[17] Ikada Y and Tsuji H 2000 *Macromol. Rapid Commun.* **21** 117
[18] Shumigin D, *et al* 2011 *Mater. Sci+ (Medziagotyra)* **17** 32-7
[19] Kim S H, *et al* 1991 *Macromol. Chem.* **193** 1623
[20] Jorda-vilaplana A, *et al* 2014 *Eur. Polym. J.* **58** 23-33
[21] Jamshidian M, *et al* 2010 *Compr. Rev. Food Sci. Food Saf.* **9** 552-71
[22] Xu H J, *et al* 2014 *Ind. Eng. Chem. Res.* **53** 1150-9
[23] Di Y W Innance, *et al* 2005 *Macromol. Mater. Eng.* **290** 1083-90
[24] Fan Y *et al* 2004 *Polm. Degrad. Stab.* **86** 197-208
[25] Pei A H, *et al* 2010 *Compos. Sci. Technol.* **70** 815-21
[26] Rayn C M, *et al* 1997 *TAPPI Proc. Polym. Laminations coat Conf.* 139-44
[27] Dorgan.JR. *et al* 1999 *J. Rheoloy.* **43** 1141-55
[28] Sungsanit K, *et al* 2010 *Korea-Australia Rheol J.* **22** 187-95
[29] Ahmed J, *et al* 2010 *J. Food Sci.* **75** 17-24
[30] Taubner V and Shishoo R 2001 *J. Appl. Polym. Sci.* **79** 2128-35
[31] Ruedee J, *et al* 2011 *J. Appl. Polym. Sci* **124** 5027-36
[32] Anderson K S, *et al* 2003 *J. Appl. Polym. Sci* **89** 3757
[33] Qi R, Luo M and Huang M 2011 *J. Appl. Polym. Sci* **120** 2699
[34] Li Y and Shimizu H 2009 *Eur. Polym. J.* **45** 738-46
[35] Huda M S, *et al* 2005 *J. Mater. Sci.* **40** 4221-9
[36] Tsuji H and Ikada Y 1992 *Macromolecules* **25** 5719-23
[37] Tsuji H, *et al* 1992 *Macromolecules* **25** 4114-8
[38] Tsuji H and Ikada Y 1993 *Macromolecules* **26** 6918
[39] Serizawa T, *et al* 2001 *Macromolecules* **34** 1996-2001
[40] Amass W, *et al* 1998 *Polym. Int.* **47** 89-144
[41] De S P and Kovacs A J 1968 *Biopolymers* **6** 299-306
[42] Hoogsteen W, *et al* 1990 *Macromolecules* **23** 634-42

[43] Okihara T, *et al* 1988 *Bull. Inst. Chem. Res.* **66** 271
[44] Araque-Monrós M C, *et al* 2013 *Polym. Degrad. Stab.* **98** 1563-70
[45] Herrera N, *et al* 2016 *Mater. Des.* **92** 846-52
[46] Herrera N, *et al* 2016 *Compos Part A: Appl. Sci. Manuf.* **83** 89-97
[47] Herrera N, *et al* 2015 *Compos. Sci. Technol.* **106** 149-55
[48] Oksman K, *et al* 2006 *Compos. Sci. Technol.* **66** 2776-84
[49] Price F P 1959 *J. Polym. Sci.* **37** 71-89
[50] Bassett D C 2007 *Morphology Encyclopedia of Polymer Science and Technology* **1** 1-31
[51] Murariu, M *et al* 2008 *Polym. Adv. Technol.* **19(6)** 636-46
[52] Labrecque L V, *et al* 1997 *J. Appl. Polym. Sci* **66(8)** 1507-13
[53] Arias V, *et al* 2013 *J. Appl. Polym. Sci* **130(4)** 2962-70
[54] Cho T-Y and Gert S 2006 *Polymer* **47(4)** 1036-43
[55] Iannace S and Nicolais L 1997 *J. Appl. Polym. Sci* **64(5)** 911-9
[56] Kanig G, *et al* 1982 *Colloid. Polym. Sci.* **260(4)** 356-77
[57] Kawai T, *et al* 2007 *Macromolecules* **40(26)** 9463-9
[58] Auriemma F, *et al* 2005 *Adv. Polym. Sci.* **181** 1-74
[59] Fischer E W, *et al* 1973 *Colloid. Polym. Sci.* **251** 980-90
[60] Sperling L H 2005 *Introduction to Physical Polymer Science* (Hoboken: John Wiley & Sons, Inc)
[61] Zhang J, *et al* 2008 *Macromolecules* **41(4)** 1352-7
[62] Pan P J, *et al* 2007 *Macromolecules* **40(19)** 6898-905
[63] Zhang J, *et al* 2005 *Macromolecules* **38(19)** 8012-21
[64] Zhang J, *et al* 2008 *Macromolecular Symposia* **242(1)** 274-8
[65] Pan P J and Yoshio I 2009 *Prog. Polym. Sci.* **34(7)** 605-40
[66] Brizzolara D, *et al* 1996 *Macromolecules* **29(1)** 191-7
[67] Cornelis A P, *et al* 1996 *Macromol. Chem. Phys.* **197** 2219-29
[68] Grijpma D W, *et al* 1994 *Macromol. Chem. Phys.* **195(5)** 1649-63
[69] Di Lorenzo M L, *et al* 2011 *Thermochim. Acta* **522** 110-7
[70] Marega C, *et al* 1992 *Die Makromolekulare Chemie* **193** 1599-606
[71] De Santis P, *et al* 1968 *Biopolymers* **6** 299-306
[72] Reddy C S, *et al* 2006 *Polym. Plast. Technol. Eng.* **45(7)** 815-20
[73] Alexander L, 1971 *J. Mater.Sci.* **6(1)** 93
[74] Oliveira J E, *et al* 2013 *Adv. Mater.Sci.Eng.* **2013** 1-14
[75] Muthukumar M 2004 *Adv. Chem. Phys.* **128** 1-64
[76] Weeks J J 1963 *J. Res. Natl Bur. Stand. A* **67** 441-51
[77] Chiang R, *et al* 1961 *J. Am. Chem. Soc* **83(13)** 2857-62
[78] Mandelkern L, *et al* 1952 *J. Am. Chem. Soc* **74(15)** 3949-51
[79] Quinn F A, *et al* 1958 *J. Am. Chem. Soc* **80(13)** 3178-82
[80] Picciochi R, *et al* 2007 *Colloid. Polym. Sci.* **285(5)** 575-80
[81] Liu R, *et al* 2014 *J. Polym. Sc: Polym. Phy.* **52** 1476-89
[82] Raff R A V, *et al* 1970 *Northwest Science* **44** 184-204
[83] Magill J H 2001 *J. Mater. Sci.* **36** 3143-64
[84] Leia Y-G, *et al* 2003 *Polymer* **44** 4673-9
[85] Wong S, *et al* 2003 *Polym. Eng. Sci.* **43** 1566-75
[86] Mathew A P, *et al* 2006 *J. Appl. Polym. Sci* **101** 300-10
[87] Al-Hadi A 2014 *Polimeros* **24** 9-16
[88] Anderson K S, *et al* 2006 *Polymer* **47** 2030-5
[89] Murariu M, *et al* 2008 *Eur. Polym. J.* **44** 3842-52
[90] Tanrattanakul V, *et al* 2014 *Express Polym. Lett.* **8** 387-96
[91] Xiao H, *et al* 2009 *J. Appl. Polym. Sci.* **113** 112-21
[92] Tsuji H, *et al* 2009 *ACS Applied Materials & Interfaces* **1** 1719-30

IOP Conf. Series: Materials Science and Engineering **369** (2018) 012032 doi:10.1088/1757-899X/369/1/012032

Numerical modelling in friction lap joining of aluminium alloy and carbon-fiber-reinforced-plastic sheets

A Das[1], H S Bang[1] and H S Bang[1,2]

[1]Department of Welding and Joining Science Engineering, Chosun University, Korea

E-mail: banghs@chosun.ac.kr

Abstract. Multi-material combinations of aluminium alloy and carbon-fiber-reinforced-plastics (CFRP) have gained attention in automotive and aerospace industries to enhance fuel efficiency and strength-to-weight ratio of components. Various limitations of laser beam welding, adhesive bonding and mechanical fasteners make these processes inefficient to join metal and CFRP sheets. Friction lap joining is an alternative choice for the same. Comprehensive studies in friction lap joining of aluminium to CFRP sheets are essential and scare in the literature. The present work reports a combined theoretical and experimental study in joining of AA5052 and CFRP sheets using friction lap joining process. A three-dimensional finite element based heat transfer model is developed to compute the temperature fields and thermal cycles. The computed results are validated extensively with the corresponding experimentally measured results.

1. Introduction

Lightweight materials of aluminium alloy and carbon-fiber-reinforced-plastic (CFRP) can significantly reduce the weight of the aerospace and automotive structures [1]. Various processes are attempted for joining aluminium and plastic with techniques such as adhesive bonding, mechanical fasteners, additive manufacturing, laser beam joining, ultrasonic spot welding and friction stir joining [1-6]. Adhesive bonding is not suitable due to its long time. Mechanical fasteners are used with relative success, however, high stress concentration close to the holes and additional weight of fasteners have been a problem. Another alternative is the laser beam welding process that has high operational cost [3]. Thus friction stir joining (FSW) has become a prominent choice due to its capability to join materials with low distortion.

Nagatsuka *et al* developed a method called friction lap joining process to join 3 mm thick AA5052 alloy and 2 mm thick polyamide (PA6) CFRP sheets [5]. Unlike FSW, a probe less cylindrical rotating tool is used in friction lap joining (FLJ) process to apply pressure to the metal surface and to generate heat at the tool/metal interface. The tool in FLJ process does not assist material flow like the probe tool performs in FSW process [5]. The generated heat at the tool/aluminium interface in FLJ process is conducted through aluminium alloy to the joint interface with an aim to partially melt the CFRP sheets and subsequently make a bond between metal and plastic after solidification of molten CFRP. The authors reported grinding on aluminium surface enhanced joint strength up to 60 MPa as the rough aluminium surface made a mechanical joint with CFRP sheets through anchoring effect [5]. In another study, authors found increasing Mg content in aluminium alloys improved the tensile strength of Al to CFRP joint [7]. In contrast to FSW process, Zhang *et al* reported maximum joint strength of around 39 MPa in laser beam joining of AA7050 alloy and PA6 CFRP sheets while Jung *et al* achieved joint strength of approximately 30 MPa [1]. Although several studies examined the joint characteristic of aluminium to CFRP sheets in friction and laser beam joining processes. However, a quantitative characteristics understanding of the influence of welding conditions on the peak temperature, weld thermal cycle especially along the weld joint interface are rarely studied in friction lap joining of aluminium to CFRP.

In the present work, a combined theoretical and experimental study is reported in joining of AA5052 alloy and polyamide 66 (PA6.6) CFRP sheets in lap joint configuration using cylindrical tool assisted friction joining process. A three-dimensional finite element based heat transfer model is developed to compute the temperature fields and thermal cycles. The computed results are validated extensively with the corresponding experimentally measured results.

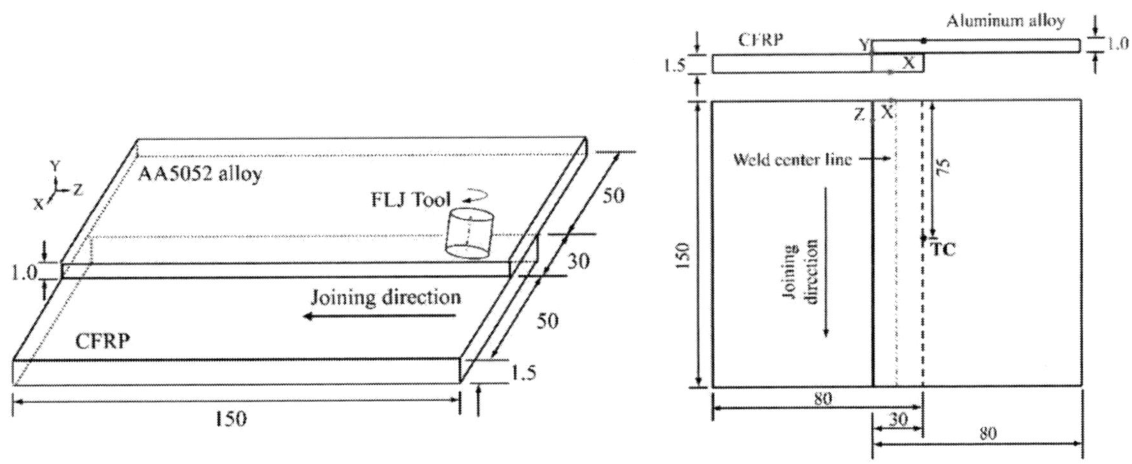

Figure 1. Schematic representations of specimen for friction lap joining of AA5052 alloy and CFRP sheets.

Table 1. Chemical composition of tool and workpiece materials in Wt% and joining conditions.

	Mg	Mn	Zn	Fe	Si	Cr	Cu	Ti	Al	Thickness
AA5052	2.5	0.1	0.1	0.4	0.25	0.15	0.1	0.15	Bal.	1 mm
	Polymer		Fiber		Fabric		Melting Temperature		Thickness	
CFRP	PA6.6		Carbon		Twill 2/2		533 K		1.5 mm	
Joining conditions	Rotational speed		Plunge depth		Joining speed (mm/s)		Surface roughness			
	400 rpm		0.2 mm		0.8 and 1.0		0.3 μm			

2. Experimental investigation

Figure 1 shows the schematic diagram of the dimensions of specimen. AA5052 alloy sheets of 1 mm thickness are joined with polyamide 66 (PA6.6) CFRP sheets in lap configuration by friction lap joining process. A gantry type computer numerical control operated friction stir welding machine (WINXEN made) is used in the present work. The tool is confirmed to WC–12% CO made with shoulder diameter of 18 mm and tilted at an angle of 2° with the vertical axis along the direction of joining. The aluminium surface, which is in direct contact with CFRP sheet, is ground to achieve surface roughness around 3 um. Table 1 shows the details of work pieces and joining conditions. The joining conditions are chosen after performing several trial experiments to achieve continuous bead profile with no visible crack and distortion. The thermal cycles are measured during joining by K–type

IOP Conf. Series: Materials Science and Engineering **369** (2018) 012032 doi:10.1088/1757-899X/369/1/012032

thermocouple located at the top of aluminium sheets approximately at a distance of 30 mm from the aluminium edge as shown in figure 1. The weld dimensions are measured in transverse cross-section direction after polishing.

3. Theoretical formulation

A transient three-dimensional heat conduction analysis is developed to simulate friction lap joining process using the governing differential equation as [8]

$$\frac{\partial}{\partial x}\left(k\frac{\partial T}{\partial x}\right)+\frac{\partial}{\partial y}\left(k\frac{\partial T}{\partial y}\right)+\frac{\partial}{\partial z}\left(k\frac{\partial T}{\partial z}\right)+\dot{Q}=\rho C\frac{\partial T}{\partial t} \tag{1}$$

where ρ, k and C refer to the density, thermal conductivity and specific heat, respectively; T and t refer to the temperature and time variables, respectively. The term $\dot{Q}$ depicts internal heat generation per unit volume. The boundary condition is represented as [9]

$$k\left(\frac{\partial T}{\partial n}\right)-q_s+h(T-T_0)+\sigma\varepsilon\left(T^4-T_0^4\right)=0 \tag{2}$$

where n represents normal direction; h, σ and ε represent convective heat transfer coefficient, emissivity and Stephen-Boltzmann constant, respectively. The first term indicates heat loss due to conduction from the surface, the second term indicates applied heat flux on to the surface, and third and fourth terms indicate heat loss due to convection and radiation, respectively. The heat loss due to radiation is negligible in friction lap joining as the maximum temperature is below the melting temperature of the aluminium. A lumped heat transfer coefficient $h_0(T-T_0)^{0.25}$, where h_0=70 W/m^{2}K, is used at the bottom of workpiece surfaces to account for the heat losses and a constant heat transfer coefficient of 20 W/m^{2}K is applied rest of the surfaces.

The frictional heat (q_s) generated along the tool shoulder–workpiece contact area is computed analytically using equation (3) [9]

$$q_S=\eta_h\left[\eta_m(1-\delta)\tau_y+\delta\mu_f P\right]\times(\omega r-U\mathrm{Sin}\,\theta) \tag{3}$$

where η_m and η_h refer to fractional heat generated due to mechanical work of sticking friction and fraction of total heat goes in the workpiece, respectively. It is reported that the value of η_m for aluminium alloys is varying within the range of 0.2 – 0.6; in the present work, it is considered as 0.4 [9]. r depicts the radial distance from tool axis, and P, ω and U represent axial pressure, rotational and linear speeds of the tool, respectively. τ_y refers to temperature dependent shear yield strength of workpiece that is represented as $\tau_y=\sigma_y/3^{1/2}$, where σ_y is the yield strength based on von Mises yield criteria. η_h is estimated considering one-dimensional steady state heat flow as [9]

$$\eta_h=\frac{\sqrt{(k\rho C)_{workpiece}}}{\sqrt{(k\rho C)_{workpiece}}+\sqrt{(k\rho C)_{tool}}} \tag{4}$$

The thermo-physical properties of the materials are shown in table 2. The terms, δ and μ_f represent the local variations in frictional sliding and coefficient of friction along the tool/ workpiece interface and are estimated as [9,10]

$$\delta=-0.026+0.5\times exp(r\omega/1.87) \tag{5}$$

$$\mu_f=0.51\times exp(-\delta r\omega) \tag{6}$$

Table 2. Thermophysical properties of aluminium alloy and steel sheets.

Density, kg/m^3	AA5052: 2696.8 CFRP: 2136.7
Liquidus temperature, K	AA5052:925 K CFRP: 533 K
Specific heat, J/ kg K	AA5052: $929.0-0.627T+1.5\times10^{-3}T^2+4\times10^{-8}T^3$ CFRP: 1200
Thermal conductivity, W/ m K	AA5052: $25.2+0.398T+7\times10^{-6}T^2-3\times10^{-7}T^3$ CFRP: 0.27278
Yield Strength, MPa	AA5052: $13.52+263.25\times[1+\exp\{(T-456.5)/29\}]^{-1}$

4. Result and discussion

4.1. Experimental bead profile
Figures 2(a) and 2(b) show top and bottom surface of the joint bead profile at a joining speed of 1 mm/s for a constant rotational speed of 400 rpm. Figure 2(a) shows smooth and continuous bead is formed with low distortion and little amount of flash on the aluminium surface. At the bottom side, shown in figure 2(b), small area of PA6.6 is melted (indicate by white block) by the frictional heat from aluminium surface and finally joined with aluminium alloy. A Similar type of bead profile is reported by Nagatsuka *et al* [7] and Liu *et al* [11] in friction lap joining of aluminium and CFRP.

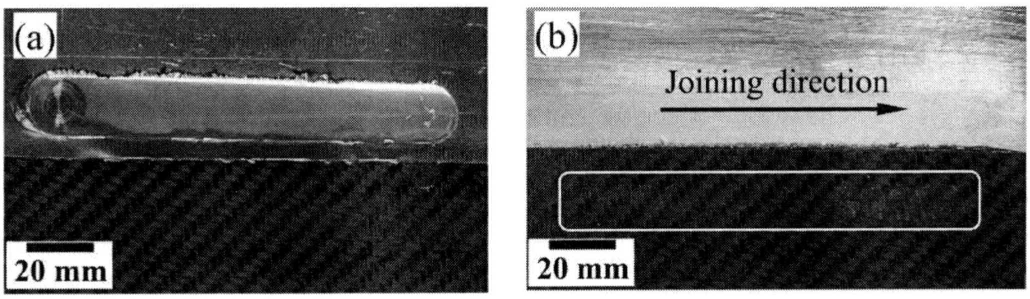

Figure 2. (a) Top and (b) the bottom surface of joint bead profiles prepared at a constant rotational speed of 400 rpm for joining speed of 1 mm/s. Melted portion of CFRP is represented by white block.

4.2. Thermal cycles
Figure 3 shows a comparison between the computed and the corresponding experimentally measured thermal cycles at the thermocouple monitoring location (figure 1) for two different joining speeds of 0.8 and 1.0 mm/s. The computed and the corresponding experimentally measured peak temperature are 474.2 and 486 K in figure 3(a), and 463.5 and 470.4 K in figure 3(b), respectively. A Decrease in peak temperature in figures 3(a) and 3(b) is attributed to the reduction in heat generation per unit length of the joint with an increase in joining speed. Overall, a fair agreement between the computed and the corresponding measured thermal cycles is depicted in figure 3 that provides a premise to utilise the numerical model for a quantitative understanding of friction lap joining process to join aluminium alloys and PA6.6 CFRP sheets.

4.3. Computed temperature profiles
Figure 4 shows the computed temperature profile of transverse section in friction lap joining of aluminium to CFRP sheets for speeds of 0.8 and 1 mm/s. The position of the computed temperature profile is taken across the central axis of the tool. It can be noted that the workpieces undergo

maximum temperature region of 550 – 650 K, which is above the decomposition temperature of base matrix plastic and below the solidus temperature of the AA5052 alloy. The maximum temperature region in figures 4 indicates frictional heat generated at the tool/aluminium interface is sufficient to melt the CFRP sheet along the joint interface. A comparison of figures 4(a) and 4(b) shows the size of the maximum temperature region is decreased with increase in joining speed that can be attributed to the lower heat generation per unit length of the joint.

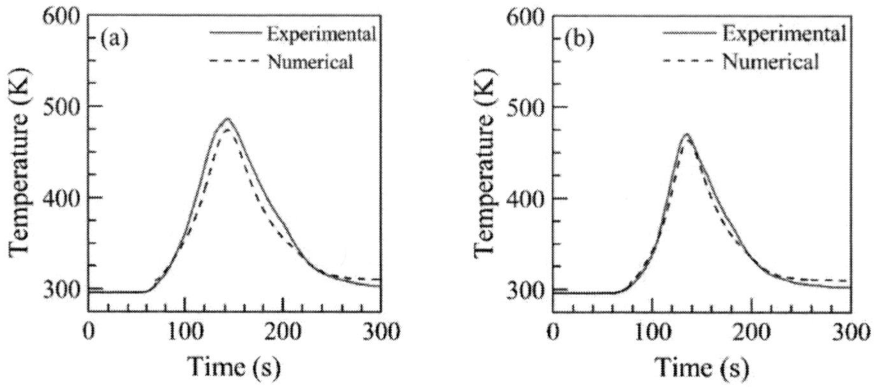

Figure 3. Comparison of numerically computed and corresponding experimentally measured thermal cycles at thermocouple monitoring locations at a constant rotational speed of 400 rpm for different joining speeds of (a) 0.8 and (b) 1.0 mm/s. The computed and measured peak temperatures (K) are: (a) (474.2, 486), (b) (463.5, 470.4).

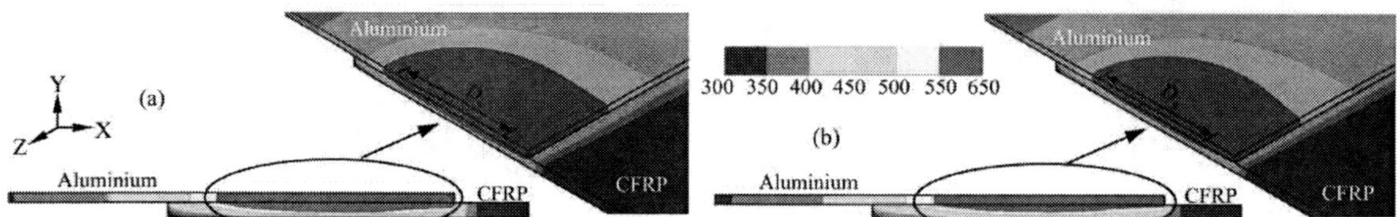

Figure 4. Computed temperature fields in friction lap joining of aluminium to CFRP sheets at a rotational speed of 400 rpm and at different joining speeds of (a) 0.8 and (b) 1 mm/s. D_s represents shoulder diameter and temperature in K is represented with colour band.

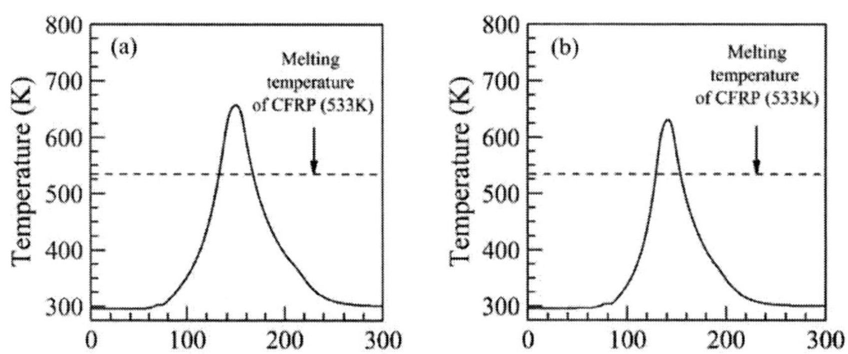

Figure 5. Computed thermal cycles at the joint interface in friction lap joining of aluminium to CFRP sheets at a rotational speed of 400 rpm and at joining speeds of (mm/s) (a) 0.8 (b) 1. The computed peak temperatures are (a) 657.7 K and (b) 625.5 K.

Further, numerically computed thermal cycles are evaluated at the joint interface where peak temperature is maximum. Figure 5(a) and 5(b) show the computed thermal cycles at the joint interface for joining speeds of 0.8 and 1 mm/s, respectively. The computed peak temperatures are 657.7 and 625.5 K, respectively in figure 5(a) and 5(b) which are above the decomposition temperature CFRP.

As a result, the friction lap joining process proves to be a useful technique where frictional heat generated at the tool/workpiece interface is transferred easily to decompose the CFRP sheets for making a bond with aluminium alloy. It is, however, necessary to examine further the microstructural characteristics especially extent of the formation of bubble along the joint interface, flow of molten plastic into the microgrooves of aluminium alloy and formation of oxide layer along the interface to understand the bonding mechanism between aluminium and CFRP.

5. Conclusion

An integrated theoretical and experimental study is reported on friction lap joining of automotive aluminium alloys and carbon-fiber-reinforced plastic sheets. A three-dimensional conduction based heat transfer model is developed considering heat generation due to friction and plastic deformation at the tool/workpiece interface. A fair agreement is achieved between the numerically computed and the corresponding experimentally measured weld thermal cycles. For the overall joining conditions considered in this study, the samples prepared are within a specific range of rotational and joining speeds have exhibited sound bead profiles without any visible discontinuity.

Acknowledgment

The authors acknowledge the research fund of Chosun University' 2018 to accomplish this study.

References

[1] Zhang Z, Shan J, Tan X and Zhang J 2017 *Int. J. Adv. Manuf. Technol.* **90** 3465-72
[2] Matsuzaki R, Shibata M and Todoroki A 2008 *Compos. Part A – Appl. S.* **39** 786-95
[3] Jung K W, Kawahito Y, Takahashi M and Katayama S 2013 *J. Laser Appl.* **25** 032003-1-6
[4] Balle F, Wagner G and Eifler D 2007 *Mat.-wiss. u. Werkstofftech.* **38** 934-8
[5] Nagatsuka K, Yoshida S, Tsuchiya A and Nakata K 2015 *Compos. Part B-Eng.* **73** 82-8
[6] Sames W J, List F A, Pannala S, Dehoff R R and Babu S S 216 *Int. Mater. Rev.* **61** 315-60
[7] Nagatsuka K, Onoda T, Okada T and Nakata K 2017 *Weld. Int.* **31** 9-16
[8] Arora A, Nandan R, Reynolds A P and DebRoy T 2009 *Scr. Mater.* **60** 13-6
[9] Mehta M, Arora A, De A and DebRoy T 2011 *Metall. Mater. Trans. A* **42A** 2716-22
[10] Mehta M, Chatterjee K and De A 2013 *Sci. Technol. Weld. Join.* **18** 191-7
[11] Liu F C, Liao J and Nakata K 2014 *Mater. Des.* **54** 236-44

IOP Conf. Series: Materials Science and Engineering **369** (2018) 012033 doi:10.1088/1757-899X/369/1/012033

Friction stir lap joining of automotive aluminium alloy and carbon-fiber-reinforced plastic

H S Bang[1], A Das[1,3], S Lee[2] and H S Bang[1]

[1]Department of Welding and Joining Science Engineering, Chosun University, Korea
[2]Graduate School of Welding and Joining Science Engineering, Chosun University, Korea

E-mail: atd.kgec@gmail.com

Abstract. Multi-material combination such as aluminium alloys and carbon-fiber-reinforced plastics (CFRP) are increasingly used in the aircraft and automobile industries to enhance strength-to-weight ratio of the respective parts and components. Various processes such as adhesive bonding, mechanical fasteners and laser beam joining were employed to join metal alloy and CFRP sheets. However, long processing time of adhesive bonding, extra weight induced by mechanical fasteners and high operating cost of the laser is major limitations of these processes. Therefore, friction stir welding is an alternative choice to overcome those limitations in joining of CFRP and aluminium alloys. In the present work, an attempt is undertaken to join AA5052 alloy and polyamide 66 CFRP sheets by friction stir lap joining technique using pinned and pin-less tools. The joint qualities are investigated extensively at different joining conditions using two different types of tools and surface ground aluminium sheets. The results show that pin-less tool and surface ground aluminium alloy can provide the suitable joint with maximum joint strength around 8 MPa.

1. Introduction

Carbon-fiber-reinforced plastics (CFRP) have characteristic of lightweight, high strength-to-weight ratio, excellent corrosion resistance and outstanding chemical and fatigue resistance [1-3]. As a result, multi-material joining of lightweight aluminium alloy to CFRP sheets has gained tremendous attention in automotive and aerospace industries to reduce the weight of the components and improve fuel efficiency [4,5]. However, hybrid joining of aluminium and CFRP is a bit difficult due to their widely different thermo-physical and chemical properties [5-7].

Various processes are attempted for dissimilar joining of plastic to aluminium alloy such as adhesive bonding [8], mechanical fasteners e.g. bolt and rivets [9], ultrasonic welding [10], laser beam based joining [3,4] and friction based joining [5,11-13]. Adhesive bonding needs long processing time for joining, hence this process is not suitable for mass production considering the present demand of high production [7]. Mechanical fasteners increase overall weight of the components and deteriorate mechanical strength of the joint owing to the development of stress concentration at the hole [8]. Another alternative is laser beam based joining process that can join various kinds of plastic and metals at high speed. However, high cost of laser unit is the main limitation of this process [4,5]. In friction stir joining process, generation of heat at the tool – metal interface subsequently melts plastic to join with the metal alloy. In addition to a small amount of deformation of the base metal is one of the leading advantages of this process. Therefore, friction stir welding (FSW) has become a prominent choice to join CFRPs and metal alloys [5,7,10].

Ngatsuka *et al* [5,13] used friction lap joining process for joining AA5052 alloy and polyamide 6 (PA6) carbon-fiber-reinforced plastic by pin-less tool. In this process, generated frictional heat at the tool-aluminium interface conducts from Al to the plastic and melts small region of plastic close to the interface. The joint is finally completed when melted plastic, in contact with Al under pressure, solidifies. The authors suggested that aluminium and CFRP sheets were joined together through the formation of the oxide layer at the interface. The authors further reported ground AA5052 alloy surface increased joint strength from 1.0 to 2.9 kN [5]. Liu *et al* [7] joined AA6061 alloy to MC Nylon-6 plastic using friction stir lap welding and found maximum joint strength around 8 MPa. Esteves *et al* [12] could improve joint strength in friction spot joining of AA6181 to CFRP sheets by increasing tool rotational speed as higher speed formed larger joining area. In contrast to friction spot joining processes, Jung *et al* [3] achieved maximum joint strength around 3 kN in joining of AA5052 alloy to PA6 CFRP sheets using continuous wave diode laser beam. Authors suggested that presence of oxide layer at the joint interface could form molecular level chemical bonding in between aluminium and melted CFRP sheets. The authors witnessed similar observation in laser beam joining of PA6 CFRP with zinc coated steel and stainless steel sheets [6,14]. In another study, Zhang *et al* [4] reported protrusion on aluminium surface could improve joint strength in laser beam joining of AA7050 alloy to CFRP sheets. The authors suggested protrusions acted as a locking material that produced tight bonding at the interface and increased joint strength to 39 MPa. The above studies provide significant insights in joining of aluminium to CFRP sheets by different processes. However, quantitative understanding of the influence of welding conditions on joint properties in friction stir joining of aluminium to CFRP sheets is scarce in the open literature.

Present work reports an experimental study in joining of AA5052 alloy to CFRP (polyamide 66) sheets using friction lap joining (FLJ) process. The primary emphasis is quantitative understanding of joint characteristics using pinned and pin-less tools. Further, the effect of ground aluminium surface on joint quality is investigated. The joint bead profiles and joint strength are examined and correlated with the processing conditions.

2. Experimental Investigation

AA5052 aluminium alloy and polyamide 66 (PA6.6) carbon-fiber-reinforced plastic (CFRP) sheets were joined in lap configuration with Al sheets placed on top of the CFRP sheets (figure 1). The thickness of the CFRP sheets was 1.5 mm, and aluminium sheets were 2.5 and 1.0 mm, respectively. A typical gantry type FSW machine (made by WINXEN) was employed to join the sample sheets using WC-12% CO made rotating pinned and pin-less tools. Table 1 shows the chemical composition and ultimate tensile strength (UTS) of the sheets. The shoulder diameter of both pinned and pin-less tools was 18 mm, and pin height and the root diameter of the pinned tool were 1.8 and 6 mm, respectively. Both tools were tilted at an angle of 2° with the vertical axis in the direction of linear travel. Friction stir tools were employed to generate frictional heat and apply pressure at the tool-metal interface simultaneously. The pressure induced by the tools was measured using load cell coupled with the FSW machine. The axial pressure induced by the tool was around 300 kgf. Further surface of aluminium alloy sheets was wet ground up to surface roughness (R_a) of around 3 μm to compare the joint properties with un-grind aluminium alloy sheets. First, 2.5 mm thick AA5052 alloy sheets were joined with 1.5 mm thick CFRP sheets using pinned tool. Next, pin-less tool was used to join 1 mm thick AA5052 alloy sheets and 1 mm thick Al alloy to 1.5 mm thick CFRP sheets, respectively. Aluminium sheets of 1 mm thicknesses were joined in lap configuration to obtain suitable joining conditions. The processing conditions of aluminium to CFRP joint, shown in table 2, were set after several trials on Al to Al sheets and Al to CFRP sheets to achieve continuous bead appearance under visual inspection.

The joint bead profiles and the respective dimensions were characterised along the transverse cross-section after polishing. The interface of the joint was examined using a Hitachi S-4800 scanning electron microscope (SEM) and energy dispersive spectroscopy (EDS). The tensile strength of the joint was measured on a Shimadzu EHF-EG200KN-40L universal tensile machine at a cross-head speed of 0.5 mm/min. Three tensile specimens were tested for each process condition to examine the variability in measured joint strengths.

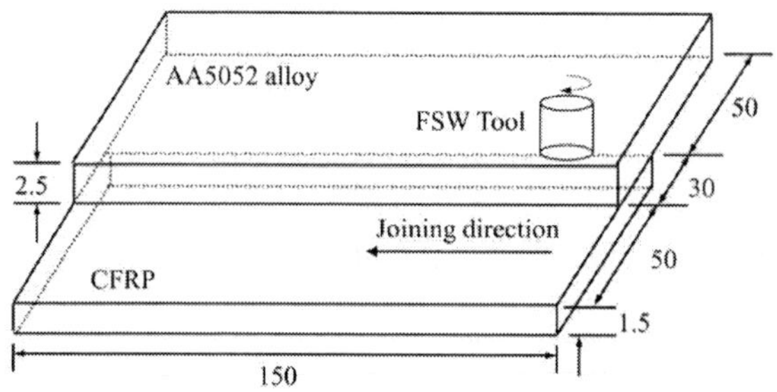

Figure 1. Schematic diagram of experimental set-up for joining of AA55052 alloy and PA6.6 CFRP sheets.

Table 1. Chemical composition (in wt%) and ultimate tensile strength of workpiece materials.

	Mg	Mn	Zn	Fe	Si	Cr	Cu	Ti	Al	UTS (MPa)	Thickness (mm)
AA5052	2.5	0.1	0.1	0.4	0.25	0.15	0.1	0.15	Bal.	252	1 and 2.5
	Polymer		Fiber		Fabric	Melting Temp		UTS (MPa)		Thickness (mm)	
CFRP	PA6.6		Carbon		Twill 2/2	260°C		785		1.5	

Table 2. Process conditions used for joining of AA5052 alloy to CFRP sheets.

	Rotational speed (RPM)	Plunge depth (mm)	Travel speed (mm/s)	Surface roughness (μm)	Sheet thickness (mm)	
					Al	CFRP
Pinned tool	400	0.8, 0.9, 1.0	1.0	0	2.5	1.5
Pin-less tool	400	0.2	0.6, 0.8, 1.0	0 and 0.3	1.0	1.5

3. Results and discussion

3.1. Experimental bead profile

Figure 2 shows the bead profile in dissimilar joining of 2.5 mm thick AA5052 alloy to 1.5 mm thick CFRP sheets using pinned tool at different plunge depth of 0.8, 0.9 and 1.0 mm for constant rotational and travelling speed of 400 rpm and 1 mm/s, respectively. Figure 2 depicts continuous joint; however, tunnel defect appears on the upper surface of aluminium alloy. The width of the tunnel was increased with an increase of the plunge depth. All joints exhibited low strength and failed under the application of force by hand. This can be attributed as the lower CFRP material is hard, it is not easy for the tool pin to be inserted into the CFRP, which causes insufficient stirring and frictional heat generation between the upper Al alloy and the lower CFRP material. In order to solve this problem, it was aimed to compare and evaluate joint properties using pin-less tool which could increase the generation of frictional heat between upper and lower material.

In friction stir lap joining by pin-less tool, the aluminium sheet thickness was reduced from 2.5 to 1 mm to conduct sufficient amount of generated frictional heat from the tool-aluminium interface to the joint interface for making a suitable joint. Figure 3 shows bead profiles and corresponding bead cross-

IOP Conf. Series: Materials Science and Engineering **369** (2018) 012033 doi:10.1088/1757-899X/369/1/012033

sections of Al to Al joint at different joining speeds of 0.2 and 0.3 mm/s for constant tool rotational speed and plunge depth of 500 rpm and 0.3 mm, respectively. Figures 3(a) and 3(b) show smooth and continuous bead with improper plunging at the starting point. This happens because of inappropriate clamping. Figure 3(c) depicts the formation of a continuous bead at a travel speed of 0.3 mm/s; however, figure 3(d) shows a presence of small gap at the interface, indicating insufficient joint. This could be attributed to an increase in travel speed decreased heat input per unit length resulting in formation of gap at the interface. Therefore, travel speed of 0.2 mm/s is suitable for joining of Al to CFRP.

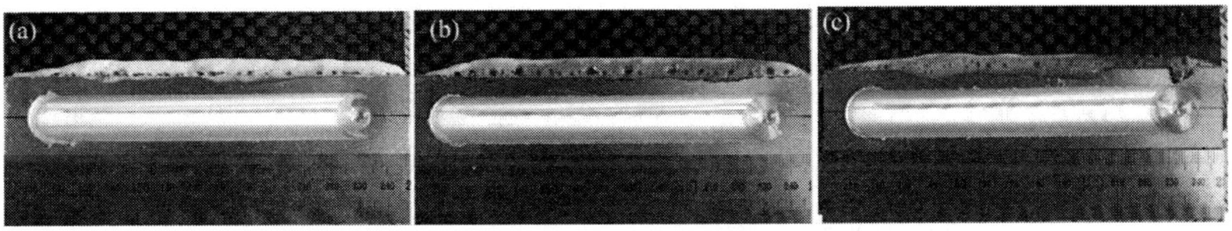

Figure 2. Bead profiles of Al and CFRP joint using pinned tool at different plunge depths of (a) 0.8, (b) 0.9 and (c) 1.0 mm for constant rotational and travel speed of 400 rpm and 1.0 mm/s, respectively.

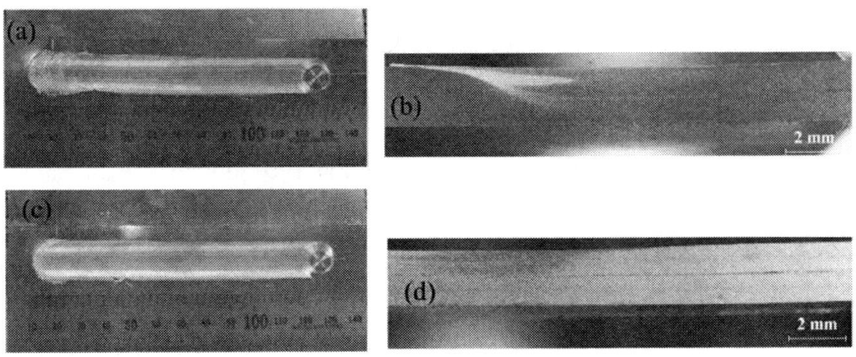

Figure 3. Bead profiles and corresponding macrographs of Al to Al joint using pin-less tool at different travel speeds of (a, b) 0.2 and (c, d) 0.3 mm/s for constant rotational speed and plunge depth of 500 rpm and 0.3 mm, respectively.

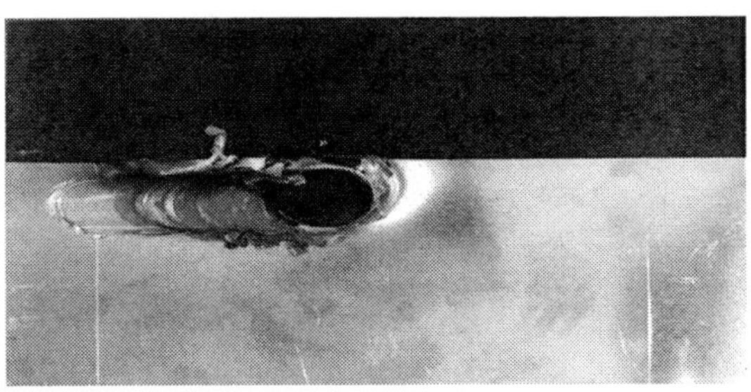

Figure 4. Al to CFRP joint bead profile at a joining speed of 0.2 mm/s for rotational speed of 500 rpm and plunge depth of 0.3 mm.

Figure 4 shows bead profile of 1 mm thick AA5052 alloy to 1.5 mm thick CFRP sheets at a plunge depth of 0.3 mm for rotational and travel speed of 500 rpm and 0.2 mm/s, respectively. Figure 4 depicts rotational speed of 500 rpm causes deformation and warppage in the upper Al alloy sheet that can be attributed to the low thermal conductivity of CFRP sheet, not conducting heat easily. As a result generated heat at the tool-workpiece interface is trapped in the aluminium surface and consequently distorts the aluminium sheet.

In order to control the heat input as mentioned above, the joints were made at higher travel speeds of 0.6 to 1.0 mm/s with lower rotational speed and plunge depth of 400 rpm and 0.2 mm, respectively. The bead profiles and cross-section of the bead for different travel speeds are shown in figure 5. Continuous joints were formed at high travel speeds with decrease in deformation of aluminium alloy sheets. Cross-section of the bead profiles for travel speeds of 0.6 and 0.8 mm/s could not be measured due to their breakage during cutting of the specimens. Cross-sectional macrograph of bead profile at joining speed of 1.0 mm/s, shown in figure 5(d), depicts bonding between Al alloy and CFRP sheets to some extent at the interface. This could be attributed to heat conducted from tool-aluminium interface partially melt CFRP sheet and consequently made a bond with aluminium alloy.

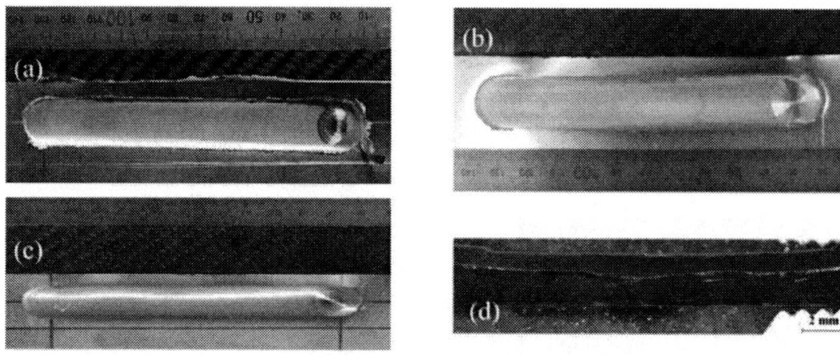

Figure 5. Bead profiles and macrographs of Al to CFRP joint at a rotational speed of 400 rpm, plunge depth of 0.2 mm and different travel speeds of (a) 0.6, (b) 0.8 and (c,d) 1 mm/s.

Further, surface roughness of R_a 3 μm was prepared on the joining side of AA5052 sheet by wet grinding with emery paper to realise the influence of roughness on joint properties. Figure 6 shows bead profile and cross-section of joint at a rotational speed of 400 rpm, plunge depth of 0.2 mm and travel speed of 1 mm/s. Figure 6(a) shows continuous and smooth bead is formed with low distortion of aluminium sheets. Cross-sectional profile in figure 6(b) depicts frictional heat from top AA5052 surface partially melts CFRP near the joint interface and makes a continuous bond in between Al and CFRP sheet.

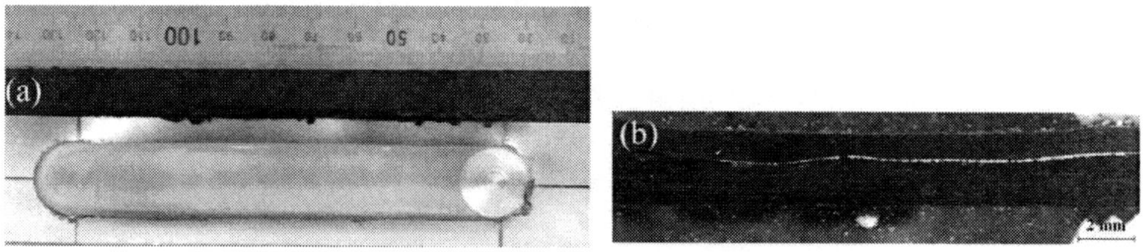

Figure 6. Bead profile and cross-sectional macrograph of wet ground Al alloy and CFRP sheets at rotational speed of 400 rpm, plunge depth of 0.2 mm and travel speed of 1 mm/s.

3.2. Characterization of joint interface

Figure 7 shows the scanning electron microscope image with results of EDS analysis at the interface of ground aluminium to CFRP joint for the rotational speed of 400 rpm and travelling speed of 1.0 mm/s. Figure 7(a) depicts aluminium alloy makes a strong bond with polyamide 66 near the joint interface. This could be attributed to microgrooves in the ground aluminium surface filled with molten CFRP matrix led to form a tight bond in between Al alloy and CFRP sheets. Similar phenomena were reported earlier by several researchers for the joints of aluminium alloy to CFRP sheets [4,5]. The elemental concentrations across joint interface were examined further using energy dispersive X-Ray spectrometer (EDS) based point analysis. Figure 7 shows positions of EDS analysis at the joint interface and concentrations of Al, Mg and C elements at the interface. The EDS analysis was carried out at four points with two across the interface (points 5and 6), one on the aluminium (point 3) and one on the CFRP surface (point 4). The concentration of different elements in Wt% at different points is shown in figure 7(b). The elemental concentrations of Al and C at points 5 and 6 confirm chemical bonding between aluminium and molten CFRP. It is, however, necessary to examine further the microstructural characteristics at the joint interface using TEM to get a deeper understanding of the bond between aluminium and CFRP.

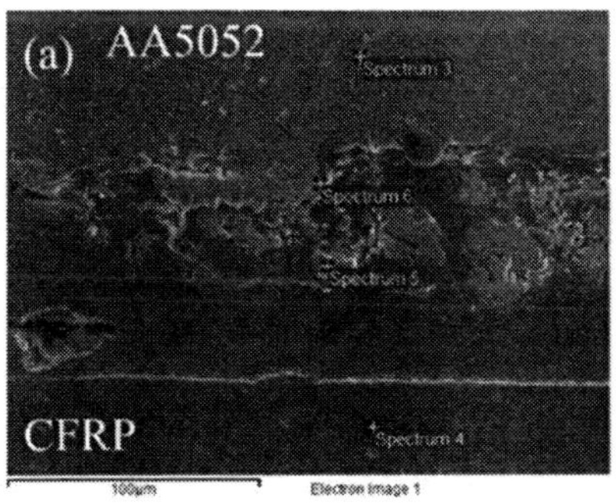

(b)

Point No	Al (Wt%)	Mg (Wt%)	C (Wt%)
Point 3	97.39	2.61	0
Point 4	0	0	100
Point 5	76.73	2.61	20.66
Point 6	86.57	4.61	8.82

Figure 7. (a) Cross-sectional SEM image and locations of EDS analysis of ground Al alloy to CFRP joint, (b) elemental concentration of EDS analysis.

3.3. Tensile shear strength
Figures 8(a) and 8(b) show the load-displacement and stress-strain curve of the joint between ground AA5052 alloy and CFRP sheets corresponding to the rotational and traverse speed of 400 rpm and 1.0 mm/s, respectively. The load-bearing capacity of the joint was measured in kN (figure 8(a)) and joint strength was estimated in MPa (figure 8(b)) as the ratio of the load and the original joint cross-section area. Figures 8(a) and 8(b) depict failure load and maximum joint strength respectively are around 2 kN and 7.6 MPa. Although all specimens were fractured at the joint interface, however, CFRP stick to the AA5052 was present on the fracture surface of the AA5052 sheet. The similar joint strength was achieved by Liu *et al* [7] in friction stir lap joining of AA6061 alloy to MC Nylon-6 plastic.

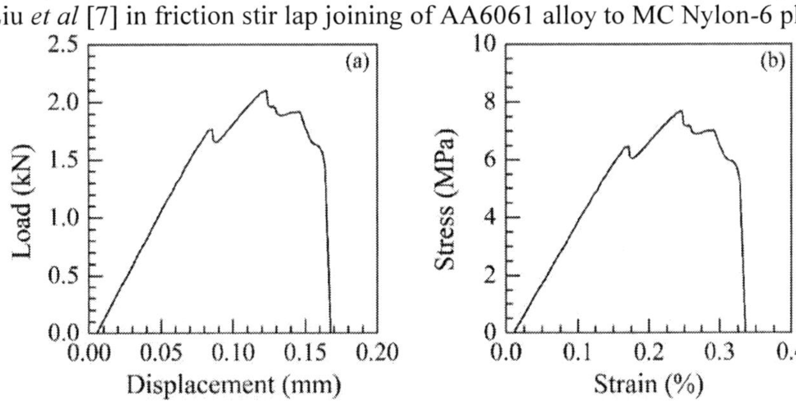

Figure 8. (a) Load-displacement and (b) stress-strain curve of Al and CFRP joint at rotational and travel speed 400 rpm and 1.0 mm/s, respectively.

Figures 9(a) and 9(b) show fracture sample and scanning electron image of fracture aluminium surface. Presence of residual CFRP adhere to aluminium surface can be seen in figure 9(a). EDS area analysis was carried out further, and results showed the existence of significant elements were Al and C with a concentration of 36.18 and 63.82 Wt%, respectively. It could be confirmed that the fracture occurred along the joint surface but partially happened in between Al alloy and the CFRP base material. It is, however, necessary to improve the joint strength of aluminium to CFRP sheets. Alternative joining process such as friction spot joining (FSJ) could be employed further to investigate the joint properties of aluminium to CFRP sheets.

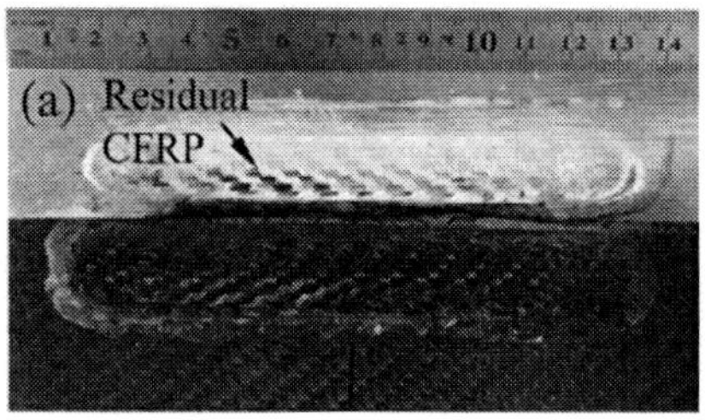

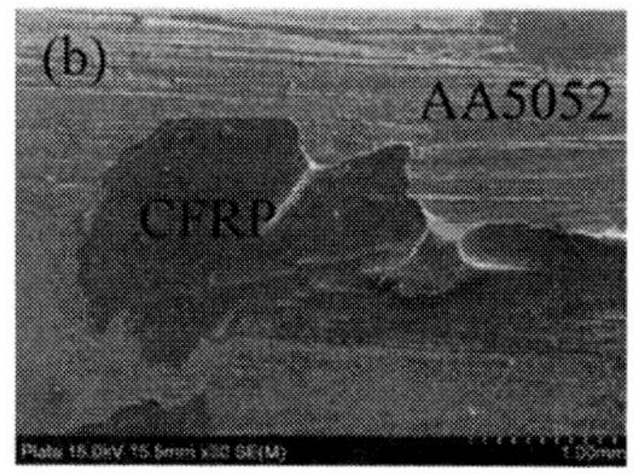

Figure 9. (a) Fracture sample and (b) SEM image of fracture sample at rotational speed of 400 rpm and travelling speed of 1.0 mm/s.

4. Conclusion

Following conclusions are derived from the present study:

- A detailed experimental study in friction lap joining of AA5052 to carbon-fiber-reinforced plastic sheets with pinned and pin-less tools are presented in the present work. Although reasonable joint strength is achieved, however, joining process parameters to get good joint is very restricted.
- Pinned tool exhibits tunnel defect and lower joint strength. That is because CFRP material is hard, it is not easy for the tool pin to be inserted into the CFRP, which causes insufficient stirring and frictional heat generation between the upper Al alloy and the lower CFRP sheet. On the other hand, pin-less tool produces continuous joints with reduction of deformation of Al alloy sheets.
- Maximum joint strength around 8 MPa is achieved in joining of ground Al to CFRP sheets using pin-less tool at a constant plunge depth of 0.2 mm for a rotational speed of 400 rpm and travel speed of 1 mm/s. Ground Al surface acts as locking material to interlock molten CFRP into Al alloy and leads to improve joint strength.

Acknowledgment

The authors acknowledge National Research Foundation of Korea (NRF) funded by the Ministry of Education (2016R1d1A3B03935978) to carry out this study.

References

[1] Pramanik A, Basak A K, Dong Y, Sarkar P K, Uddin M S, Littlefair G, Dixit A R and Chattopadhyaya S 2017 *Compos. Part A* **101** 1-29
[2] Moeller F, Thomy C, Vollertsen F, Schiebel P, Hoffmeister C and Herrmann A S 2010 *Phys. Procedia* **5** 37-45
[3] Jung K W, Kawahito Y, Takahashi M and Katayama S 2013 *J. Laser Appl.* **25** 032003-1-6
[4] Zhang Z, Shan J, Tan X and Zhang J 2017 *Int. J. Adv. Manuf. Technol.* **90** 3465-72
[5] Nagatsuka K, Yoshida S, Tsuchiya A and Nakata K 2015 *Compos. Part B-Eng.* **73** 82-8
[6] Jung K W, Kawahito Y, Takahashi M and Katayama S 2013 *Mater. Des.* **47** 179-88
[7] Liu F C, Liao J and Nakata K 2014 *Mater. Des.* **54** 236-44

[8] Arenas J M, Alía C, Narbón J J, Ocana R and Gonzalez C 2013 *Compos. Part B-Eng.* **44** 417-23
[9] Kolesnikov B, Herbeck L and Fink A 2008 *Compos. Struct.* **83** 368-80
[10] Balle F, Wagner G and Eifler D 2007 *Mat.-wiss. u. Werkstofftech.* **38** 934-8
[11] Goushegir S M 2016 *Weld. World.* **60** 1073-93
[12] Esteves J V, Goushegir S M, Dos Santos J F, Canto L B, Hage E and Amancio-Filho S T 2015 *Mater. Des.* **66** 437-45
[13] Nagatsuka K, Onoda T, Okada T and Nakata K 2017 *Weld. Int.* **31** 9-16
[14] Jung K W, Kawahito M and Katayama S 2013 *Sci. Technol. Weld. Joi.* **16** 676-80

IOP Conf. Series: Materials Science and Engineering **369** (2018) 012034 doi:10.1088/1757-899X/369/1/012034

Effects of different overlap lengths and composite adherend thicknesses on the performance of adhesively-bonded joints under tensile and bending loadings

F Kadioglu[1,3], E Avil[2], M E Ercan[2] and T Aydogan[2]

[1]Ankara Yildirim Beyazit University, Ankara, 06050, Turkey
[2]Ferka Aviation, Ankara, 06830, Turkey

E-mail: ferkadioglu@gmail.com

Abstract. Fiber-reinforced polymer composites are being used in an increasingly wide range of products. They are particularly popular in automotive and aerospace sectors because they offer an attractive combination of stiffness, strength and low mass. Adhesively-bonded joints of such materials are preferred by many designers due to their assembling advantages over other traditional mechanical joining systems, such as bolted and riveted joints. In this study, some experimental works have been carried out on adhesively-bonded adherends manufactured from a woven carbon fiber-reinforced polymer matrix composite (Hexply 8552S/A280-5H, produced by Hexcel), by using a film adhesive (AF163-2K produced by 3 M). The bonded specimens were prepared in the Single Lap Joint (SLJ) configuration, and tested in tensile and also in four-point bending loading. In order to assess the joint performance, three different overlap lengths, 15 mm, 25 mm and 40 mm, and two different thicknesses of the composite adherends, 2 mm and 3 mm, were used. The results shown that the parameters are controlled by the loading modes; while the overlap length increases the joint performance significantly in tensile loading, the opposite was the case for those in bending loading, which was affected mainly by the adherend thicknesses. The results were related to the mechanisms of joint failures; while the joints in the tensile failed in the adhesive layer with some exceptions, those in the bending mainly failed in the plies adjacent to the layer. The current study indicates that one of the important factors affecting the joint strength of the adherends manufactured from the laminated composites is the local failure of the plies. It is thought more focused-studies would be needed to lessen such problems, which would be possible via in-depth numerical analysis.

1. Introduction

In today's modern automotive and aerospace applications, the materials offering an attractive combination of stiffness, strength and low mass are the main point of interest for many engineers and scientists. In general, the materials with high strength-to-weight ratio, design flexibility, high impact and fatigue resistance are demanded by these applications, and these requirements are mostly met by fiber-reinforced polymer matrix composite materials [1,2]. Assembling such materials is a challenging subject for the designers of the relevant sectors. Some traditional joining methods such as bolted or riveted joints have been used for many years [3,4], however, it has been revealed that the hole existence in the composite adherends is the cause of the main stress concentrations leading to a poor joint performance. On the other hand, adhesively-bonded joints of such materials are found to have some advantages over the traditional joining methods, as they provide relatively uniform stress distributions and continuity in the bonded region, resulting in relatively higher joint performance [5].

It is known that there are many parameters such as composite bonding methods, surface preparation, mechanical properties of adherends and adhesives, geometrical parameters (adhesive and adherend thickness, overlap length, stacking sequence, ply angle, fillet etc.) affecting the strength of the bonded joints, which have been reviewed in detail [6]. For example, the effects of the bonding methods between the co-cured joint and the secondary bonded joint method under mode I (opening) and mixed-mode loading were investigated by Mohan *et al* [7,8], and it was found that the co-cured bonded joints showed lower strength than the secondary bonded joints in both loading conditions. Similar results were supported by another work [9]. To protect the bonded region from all contaminants such as lubricants, dusts, and micro-organisms, the surface preparation of the adherends is vital for a good joint performance. There are many works on this subject recommending different surface preparations including chemical and physical treatments for a good surface wettability, surface energy, good activation of material surfaces to be bonded [10,11]. It is found that peel ply is one of the techniques being able to protect the surface from the contamination and also to maintain the specific surface texture [12,13]. Related to the effects of the adhesive layer, some experimental works [14,15] have been conducted on the bonded joints, and the results have shown that the joint strength decreases as the adhesive thickness increases, which is supported by some numerical works as well [16,17]. In designing bonded joints, a uniform stress distribution is desired, however, this is not the case for many reasons; for example, some stress concentrations occur due to applied load eccentricity, different ratio of adherend/adhesive elastic constants and geometric discontinuity etc. Remedies to these problems have been proposed by some researchers using different techniques [18-21]. It is important to note that a laminate with different angles of fiber orientations has also a significant effect on the joint performance. It was found that a ply with relatively small fiber angles results in relatively higher load bearing capacity [22]. Another important parameter affecting the joint performance is different overlap lengths. Works carried out have shown that increase in overlap length results in increase of joint performance, in general, depending on mechanical behavior of adhesive used and fiber angles of plies [23,24]. One of the most important enemies of the adhesive joints is the degradation effects of environmental conditions such as moisture, temperature and ultra violet radiation, which have been studied by some researchers [25-28].

As mentioned above, many efforts have made to investigate shear performance of adhesives in the SLJ configuration in tensile loading. However, it is believed that the adhesive joints are subjected to tensile as well as bending loading during their service life, and so it is worth investigating these two loading modes simultaneously. The aim of this works is to investigate the performance of adhesively-bonded joints using the adherends manufactured from a woven carbon fiber-reinforced polymer matrix prepreg, and an adhesive film. The joints in single lap configuration were subjected to tensile loading for their shear performance, and to four-point bending loading for their pure bending performance. While the thickness of the adhesive layer was kept constant, 0.2 mm, two different thicknesses of adherends, 2 mm and 3 mm, and three different overlap lengths, 15 mm, 25 mm and 40 mm, were used as parameters to assess their effects on the joint performance. For the current experimental work, a special consideration was also given to failure modes and also failure mechanisms to gain an in-depth insight into the joint behavior, which will help to model the joint configuration by conducting a numerical analysis, to be conducted later on.

2. Experimental works

The materials used in this study are an adhesive film, AF163-2K (TM) produced by 3M, and adherends manufactured from a prepreg of woven carbon fiber-reinforced epoxy matrix composite, Hexply 8552S/A280-5H, produced by Hexcel. While the adhesive used was cured at 125°C for 60 minutes under a pressure of 2 bars, the cure procedure for the prepreg of the composite adherends was 120°C for 120 minutes, after an initial heating-up procedure of 80°C for 90 minutes under a pressure of 5 bars. The adherends were manufactured from the plates of the cured prepregs, and machined to the required dimensions. The joints were bonded in a specially designed mould in which the specimens were allowed to be manufactured with different adherend thicknesses and different overlap lengths. For the study, while the thickness of the adhesive layer was kept constant, 0.2 mm, two different thicknesses of adherends were used, 2 mm and 3 mm, to be able to assess the effects of different adherend thicknesses on the joint performance in tensile and bending loadings. Three different overlap lengths, 15 mm, 25 mm and 40 mm, were also tested for the same purpose. Geometrical details of the single lap joint (SLJ) specimens tested are shown in figure 1, prepared according to the ASTM1002 standard specifications. Surface preparation of the adherends to be bonded was made using peel-ply removal as the sole surface preparation technique, and all the specimens were tested at 23°C room temperature and 50% relative humidity, to avoid effects of different environmental conditions. Four different samples of joints were tested for each type to see if the results are repeatable.

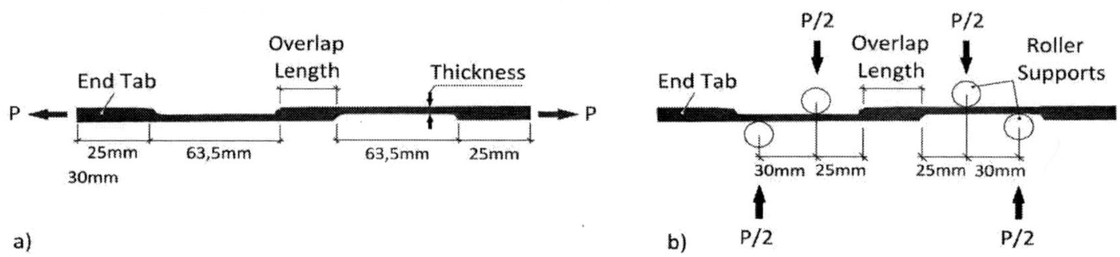

Figure 1. Dimensions of the SLJ configuration in (a) tensile and (b) bending loading.

Some of the mechanical properties of the adherend and the adhesive film are shown in table 1, where σ and E and υ are the tensile stress, Young's modulus and Piossion's ratio of the materials used, and where subscripts 11 and 22 are related to the coordinate system in x and y directions, respectively.

Table 1. Mechanical properties of the materials used for the adhesively-bonded joints.

Adherend	σ_{11} (MPa)	σ_{22} (MPa)	E_{11}(MPa)	E_{22}(MPa)
	850	850	65000	65000
Adhesive	σ (MPa)	E (MPa)	υ	
	48	1800	0.3	

3. Results and discussions

Figures 2(a) and 2(b) show the typical curves of load against crosshead displacement of the joints with the 2 mm and 3 mm adherend thicknesses, respectively, for different overlap lengths under the tensile loading. The curves show that increasing the overlap lengths results in an increase in the displacements, which is an indication of the high energy stored within the samples. The failure load against the overlap length is shown in figure 3(a) for the both adherend thicknesses. The figure shows that there is generally a linear relationship between the failure load and the overlap length, and that the effect of the different adherend thicknesses on the failure load is not important. However, samples with 40 mm of 3 mm adherend thickness gave less failure load compared to those with the 2 mm thickness. While the minimum failure load is about 10 kN for the joints with 15 mm overlap length, the maximum value is about 24 kN for those with the 40 mm, and the failure load for the 25 mm overlap length is about 15 kN. The plot of average shear stress against the overlap length is presented in figure 3(b). The joints with adherend thickness of 2 mm and 3 mm showed a stress value of 20 MPa and 30 MPa respectively. It is seen that the joints with 2 mm adherend thickness shows generally higher shear stress values compared to those with the 3 mm, which is thought to be because of the load eccentricity being large for the thicker adherends. For the tensile shear test, all the specimens with 15 mm and 25 mm overlap lengths showed a cohesive surface failure (the failure within the adhesive layer (see figure 4(a)), whereas those with 40 mm overlap lengths, under high failure loads, experienced a mixed failure, failure of plies adjacent to the adhesive layer as well as failure of adhesive layer itself (see figure 4(b)). In this situation, more damage of the plies was the case for the joints with the 3 mm, compared to those with the 2 mm. It was clear that the joints with the higher adherend thickness experienced more peel effects at the ends of the overlap length, due to applied load eccentricity, which could be the cause of the more plies damage. The stress concentrations (peel effects) at the end of the overlap have been studied in a previous numerical work [22], which shows the stress distributions in adhesive layer in SLJ configuration is not uniform.

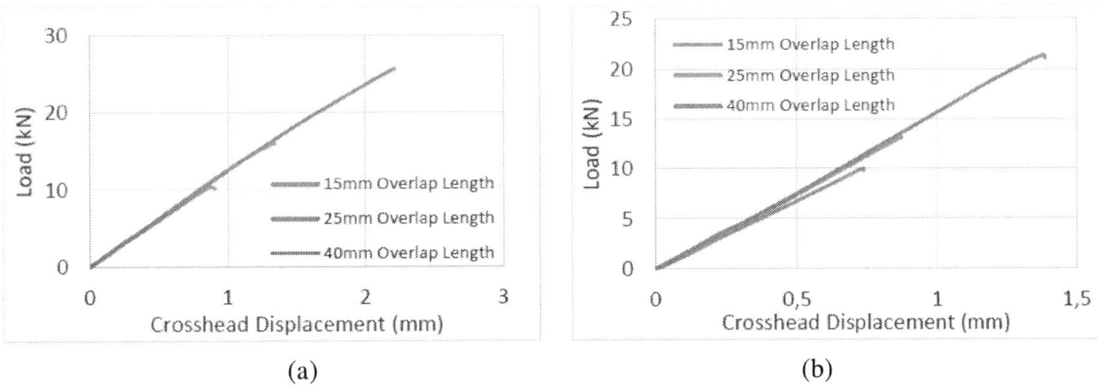

Figure 2. Plots of load against crosshead displacement for the SLJ specimens in tensile loading, for the adherends with (a) 2 mm and (b) 3 mm thicknesses.

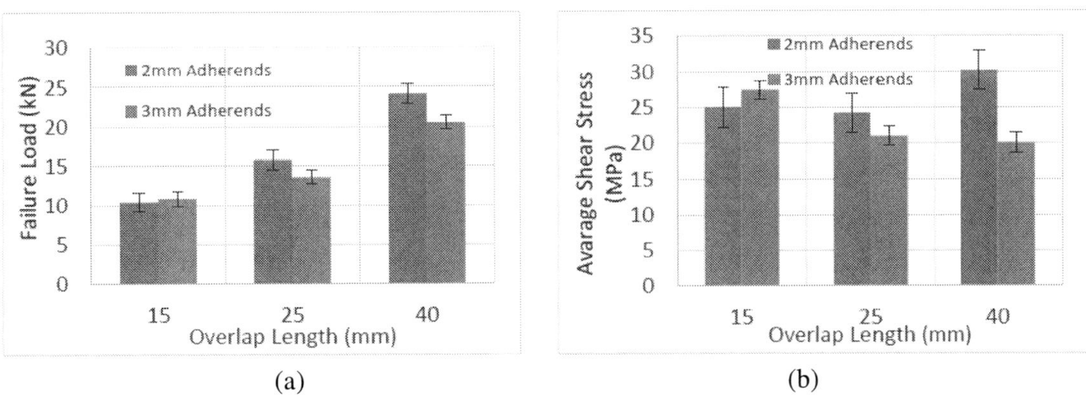

Figure 3. (a) Failure load against overlap length and, (b) Average shear stress against overlap length for the SLJ in tensile loading.

For all the joints with 15 mm and 25 mm overlap lengths, the surface showed nearly the same failure as presented in figure 4(a). In this case, no initial failure of the adherends was observed and the failure mechanism was reasonably symmetric. Stress concentrations due to the peel effects were clear at the ends of the overlap, which was before a catastrophic failure, coincided in the middle of the joint. For the joints with 40 mm overlap length, the joint failure was not only controlled by the adhesive but also accompanied by the failure of plies adjacent to the layer (see figure 4(b)).

Some representative curves of bending load against crosshead displacement for the joints with 2 mm and 3 mm adherend thicknesses are shown in figures 5(a) and 5(b), respectively. From the curves, it is seen that increasing overlap length does not results in a significance increase in the failure load and crosshead displacement; while all the joints with 2 mm adherend thicknesses fail about 1000 N, those with the 3 mm fail around 2000 N, implying an increase of 100% for the thicker adherends. This is considered to be due to the relatively small deflection of the thick adherends experiencing under the bending, compared to the thin ones, which is believed to have important effects on the stress concentrations at the edges of the overlap length, the main cause of failure initiation. Figures 6(a) and 6(b) show the plot of the bending failure load and the bending moment against the overlap length, respectively, which indicate the values from the thick adherends double compared to the thin ones, which is again related to the adherend deflections under the bending.

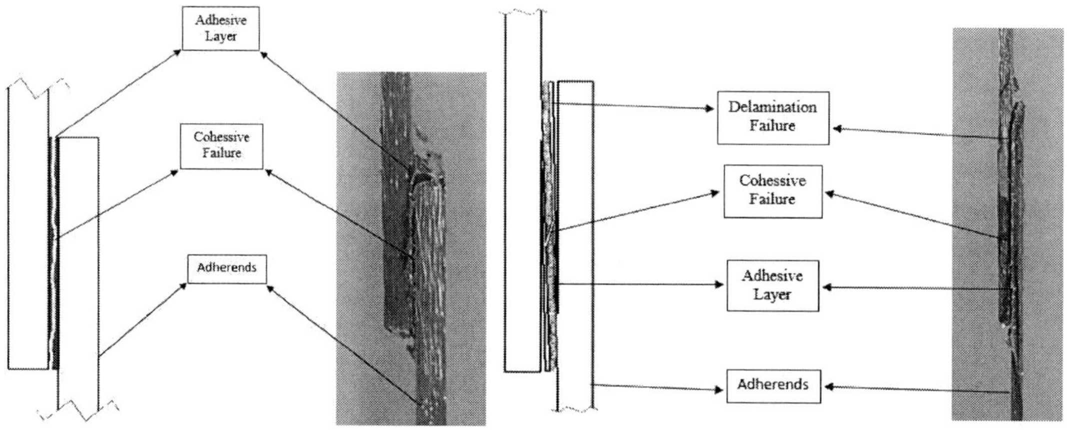

Figure 4. Failure mechanisms for the SLJ in tensile loading, (a) cohesive failure and (b) mix failure.

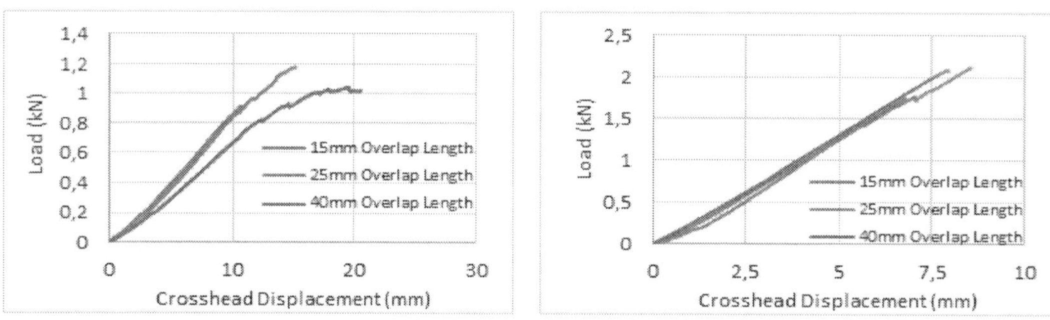

(a) (b)

Figure 5. Plots of load against crosshead displacement for the SLJ specimens in bending loading, for the adherends with (a) 2 mm and (b) 3 mm thicknesses.

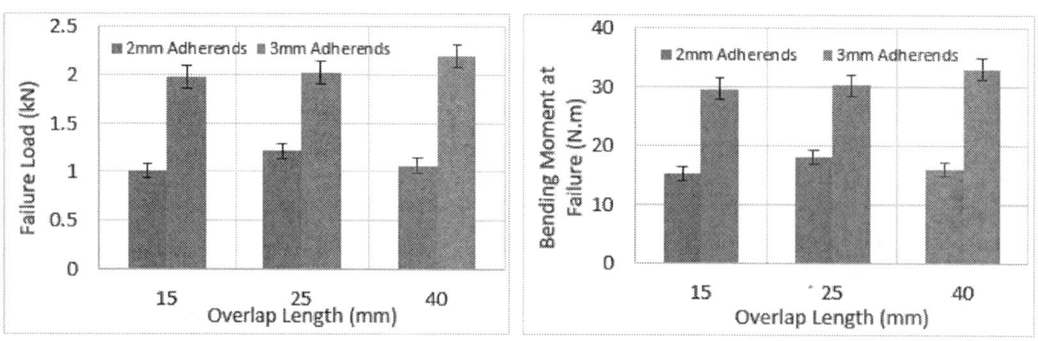

Figure 6. (a) Failure load against overlap length and (b) Bending moment against overlap length for the SLJ in bending loading.

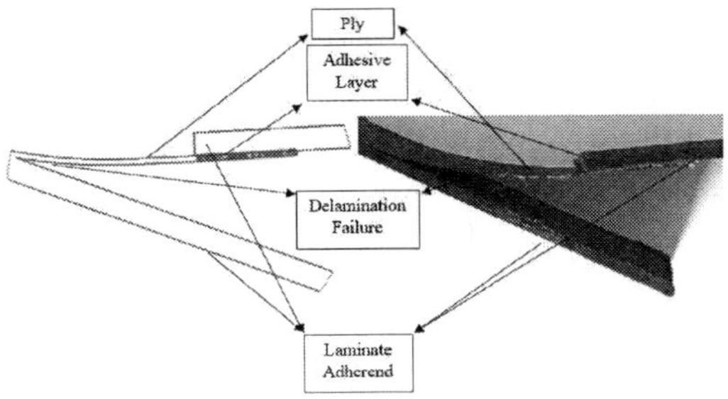

Figure 7. Failure mechanisms for the SLJ in bending loading.

In all cases, the adherends used in bending loading experienced plies failures adjacent to the adhesive layer, and no failure was witnessed in the adhesive layer during the test. As can be seen from figure 1(b), one end of the overlap region is in tension (right side) while the other end is in compression (left side), which are the possible points where the failure may initiate. As far as the critical points at the overlap edges are concerned, the failure should be expected at the side in tension since polymers are always stronger in compression rather than in tension [5], which was exactly the case for all the joints in bending loading; it was clear that initial failure started at the edge in tension and then propagated towards the edge in compression, result in in a delamination failure of plies of the composite laminates (adherends) (see figure 7). It should be note that the peel stress in the laminated composite is mainly controlled by the matrix (polymer) part of the material, rather than the fibers.

4. Conclusions

It has been shown that the joints with three different overlap lengths, and two different thicknesses of adherends manufactured from a woven carbon fiber-reinforced polymer matrix composite had different mechanical performance depending on the loading modes, tensile and bending; for the tensile loading case, increase in the overlap length resulted in increase in the joint performance almost linearly, except those with 40 mm overlap length of 3 mm adherend thickness, which had low strength. The decisive parameter was the change in the overlap lengths, in general, other than adherend thickness for this loading mode. However, for the joints under four-point bending, the decisive parameter was the adherend thickness rather than the overlap length, and the thick adherends contributed to the performance around 100% increase, compared to the thin ones. There are mainly two different failure mechanisms for the tensile loading case; a failure in the adhesive layer resulting in cohesive failure, and a mix failure, partly the adhesive layer and also the plies for those with 40 mm overlap length. For the bending loading case, all the failure occurred in the piles adjacent to the adhesive layer. A detailed numerical analysis would help explain this phenomenon.

Acknowledgments
We would like to thank to the Turkish Aerospace Industry for supporting the current work.

References

[1] Baker A A and Scott M L 2016 *Composite Materials for Aircraft Structures* 3rd Ed. (American Institute of Aeronautics and Astronautics)
[2] Sano T and Srivatsan T S 2015 *Advanced Composites for Aerospace, Marine and Land Applications II* (John Wiley & Sons)
[3] Giannopoulos I K, Doroni-Dawes D, Kourousis K I and Yasaee M 2017 Effects of bolt torque tightening on the strength and fatigue life of airframe FRP laminate bolted joints Composites *Part B* **125** 19-26
[4] Yoo S Y, Kim C H, Kweon J H and Choi J H 2016 The structural analysis and strength evaluation of rivet nut joint for composite repair *Compos. Struct.* **136** 662-8
[5] Adams R D, Comyn J and Wake W C 1997 *Structural Adhesive Joints in Engineering* 2nd Ed.

(Elsevier Applied Science)

[6] Budhe S, Banea M D, de Barros S and da Silva L F M 2017 An updated review of adhesively bonded joints in composite materials *Int. J. Adhes. Adhes.* **72** 30-42

[7] Mohan J, Ivanković A and Murphy N 2014 Mode I fracture toughness of co-cured and secondary bonded composite joints *Int. J. Adhes. Adhes.* **51** 13-22

[8] Mohan J, Ivanković A and Murphy N 2015 Mixed-mode fracture toughness of co-cured and secondary bonded composite joints *Eng. Fract. Mech.* **134** 148-67

[9] Song M G, Kweon J H, Choi J H, Byun J H, Song M H, Shin S J and Lee J 2010 Effect of manufacturing methods on the shear strength of composite single-lap bonded joints *Compos. Struct.* **92** 2194-202

[10] Iqbal H S M, Bhowmik S and Benedictus R 2010 Surface modification of high performance polymers by atmospheric pressure plasma and failure mechanism of adhesive bonded joints *Int. J. Adhes. Adhes.* **30** 418-24

[11] Encinas N, Oakley B R, Belcher M A, Blohowiak K Y, Dillingham R G, Abenojar J and Martinez M A 2014 Surface modification of aircraft used composites for adhesive bonding *Int. J. Adhes. Adhes.* **50** 157-63

[12] Kanerva M and Saarela O 2013 The peel ply surface treatment for adhesive bonding of composites: A review *Int. J. Adhes. Adhes.* **43** 60-9

[13] Buchmann C, Langer S, Filsinger J and Drechsler K 2016 Analysis of the removal of peel ply from CFRP surfaces *Composites Part B* **89** 352-61

[14] Anyfantis K N and Tsouvalis N G 2013 Loading and fracture response of CFRP-to-steel adhesively bonded joints with thick adherents – Part I: experiments *Compos. Struct.* **96** 850-7

[15] Marzi S, Biel A and Stigh U 2011 On experimental methods to investigate the effect of layer thickness on the fracture behavior of adhesively bonded joints *Int. J. Adhes. Adhes.* **31** 840-50

[16] Ji G, Ouyang Z, Li G, Ibekwe S and Pang S S 2010 Effects of adhesive thickness on global and local Mode-I interfacial fracture of bonded joints *Int. J. Solids Struct.* **47** 2445-58

[17] Xu W and Wei Y 2012 Strength and interface failure mechanism of adhesive joints *Int. J. Adhes. Adhes.* **34** 80-92

[18] Mahi B E, Benrahou K H, Belakhdar K, Tounsi A and Bedia E A A 2014 Effect of the tapered end of a FRP plate on the interfacial stresses in a strengthened beam used in civil engineering applications *Mech. Compos. Mater.* **50** 467-76

[19] Li J, Yan Y, Zhang T and Liang Z 2015 Experimental study of adhesively bonded CFRP joints subjected to tensile loads *Int. J. Adhes. Adhes.* **57** 95-104

[20] Campilho R D S G, de Mouro M F S F and Domingues J J M S 2009 Numerical prediction on the tensile residual strength of repaired CFRP under different geometric changes *Int. J. Adhes. Adhes.* **29** 195-205

[21] Marques E A S, Campilho R D S G and da Silva L F M 2016 Geometrical study of mixed adhesive joints for high temperature applications *J. Adhes. Sci. Technol.* **30** 691-707

[22] Kadioglu F, Demiral M, Avil E, Ercan M E and Aydogan T 2018 Performance of adhesively-bonded joints of laminated composite materials under different loading modes *2018 AIAA/ASCE/AHS/ASC Structures, Structural Dynamics, and Materials Conference* AIAA SciTech Forum (AIAA 2018-0222)

[23] Meneghetti G, Quaresimin M and Ricotta M 2010 Influence of the interface ply orientation on the fatigue behaviour of bonded joints in composite materials *Int. J. Fatigue* **32** 82-93

[24] Kadioglu F and Puskul H 2016 Effects of different fiber orientations on the shear strength performance of composite adhesive joints *World Academy of Science, Engineering and Technology Int. J. Mater. Metall. Eng.* **10** 65-8

[25] Aniskevich K, Aniskevich A, Arnautov A and Jansons J 2012 Mechanical properties of pultruded glass fiber-reinforced plastic after moistening *Compos. Struct.* **94** 2914-9

[26] Banea M D, da Silva L F M and Campilho R D S G 2012 Effect of temperature on tensile strength and mode I fracture toughness of a high temperature epoxy adhesive *J. Adhes. Sci. Technol.* **26** 939-53
[27] Meng M, Rizvi M J, Grove S M and Le H R 2015 Effects of hygrothermal stress on the failure of CFRP composites *Compos. Struct.* **133** 1024-35
[28] Liu S, Cheng X, Zhang Q, Zhang J, Bao J and Guo X 2016 An investigation of hygrothermal effects on adhesive materials and double lap shear joints of CFRP composite laminates *Composites Part B* **91** 431-40

IOP Conf. Series: Materials Science and Engineering **369** (2018) 012035 doi:10.1088/1757-899X/369/1/012035

Modelling the failure of near-edge impacted carbon fibre-reinforced composite subjected to shear loading

J Šedek

Czech Aerospace Research Centre - VZLÚ, Beranových 130, Prague – Letňany, 199 05, Czech Republic

E-mail: sedek@vzlu.cz

Abstract. The failure of an impacted carbon fibre-reinforced composite is analysed, where a double impact near the edge of a large centre hole and subsequent shear loading are simulated. Three levels of impact with energies of 15 J, 25 J and 50 J are investigated, along with an impact-free panel used as control. The numerical finite-element (FE) method together with cohesive zone modelling is employed. The composite material is represented by the three-dimensional ply-by-ply FE model. The numerical results of the impacts are compared with respect to the damaged areas as described in the FE model and it is found that only the 50 J impact causes the impactor to penetrate the composite. Subsequent shear loading shows two failures during loading. The first is caused by a local loss of stability in the impacted region whereas the second is a failure in the tension diagonal of the panel. The load carrying capacity, when compared to the impact-free panel, shows on average, a reduction of 37% at the first failure. The FE modelling technique utilizing the ply-by-ply 3D model and cohesive zone representing the ply interactions is successfully adopted to simulate the shear test of the near-edge impacted composite. However, the behaviour of the compression region including the impact zone does not entirely agree with that experimentally observed during the shear test. In future studies, the compressive strength and local loss of stability should be more precisely addressed.

1. Introduction

In view of their increasing application in transportation, civil industry, power industry, military and sport equipment, and more generally, in all types of light weight applications, the impact damage of composite materials has been extensively investigated in the past decades. For composite materials, prior to manufacturing, the design must take into account not only the loading conditions but also the reliability and serviceability of the composite part. Numerical methods have been used in the design process.

Simulating the impact damage in a composite material is not an easy task, since several impact characteristics such as the impact energy expressed in terms of the impactor mass and velocity, the impactor shape, and how these factors are related to the vibrational response of the composite must be taken into account, With respect to the velocities used in impact testing, there are three classifications: low velocity impact (considered to be up to 10 m/s) [1], high velocity (in the range ~ 20–50 m/s) [2], and ballistic impact (~ 900 m/s) [3]. Low impact velocities and energies often cause damage that is difficult to detect by visual inspection. The impactor does not penetrate the composite, but bounces back, causing under-surface and often back-surface damage. In this case, the delamination within the structure often develops around the impact zone. Such damages are commonly termed "barely visible impact damages (BVID)". In contrast, ballistic impacts result in penetration of the surface, causing minor local delamination. In this case, it has been observed that the damage is concentrated at the impact and spreads from the entry face towards the exit face, during which the size of the damage increases linearly with distance [4].

Low energy impact damage is a major problem from the point of view of detection. High and medium energy impacts cause surface damages that can be detected relatively easily during inspection, but low energy impact damage mainly induces internal damage. Therefore, non-destructive inspection techniques (NDI) need to be used to detect such damage. For example, ultrasonic C-scan imaging or laser shearography have been successfully used in BVID inspection [5,6].

The investigation of impact damage often involves simulation techniques to predict the size and type of damage. Due to the above-mentioned reasons, a number of studies on BVID have been reported. However, most research groups have studied central impact damage [6-10], while near-edge impact has not been covered. Accidental impacts of various kinds can occur on several locations during manufacture and in-service. Near-edge or on-the-edge impact is severe in those zones of composite structure, where compressive in-plane loading is present [11]. This work deals with two impact damages created at points opposite to each other near a hole edge in a composite panel with a central large hole. The hole represents a lightening cut in the shear-loaded part, which can be the frame rib. The damages created by impacts at different energy levels are studied. The shear response following the impacts is also investigated, and it is generally expected that the load-carrying capacity is reduced following impact.

2. Analysis

The finite element (FE) method, together with cohesive zone modelling technique, was employed for the analysis. A three-dimensional ply-by-ply model is used to represent a 3.4 mm-thick composite specimen comprising T700GC carbon fibres and the RTM6 resin matrix. Unidirectional layers (24 in number) are stacked in the sequence $[45/-45_2/45/90/-45/45_2/-45/90/-45/45]_s$ and a large 40 mm diameter hole is located at the centre of a square panel with a side dimension of 234 mm, as shown in figures 1 and 2. The dimensions were chosen to fit the picture-frame test apparatus used in VZLÚ laboratories and are close to specimens referred in [12]. The impactor-head radius was chosen to be 8 mm. Details of the model parameters are given in a previous report [13], wherein the behaviour of the model with respect to experimental observations is presented. In the model, a frictionless contact has been assumed. In the normal direction, the hard-kinematic contact algorithm was used between the impactor and plies, and among the plies themselves, the same algorithm was used as well.

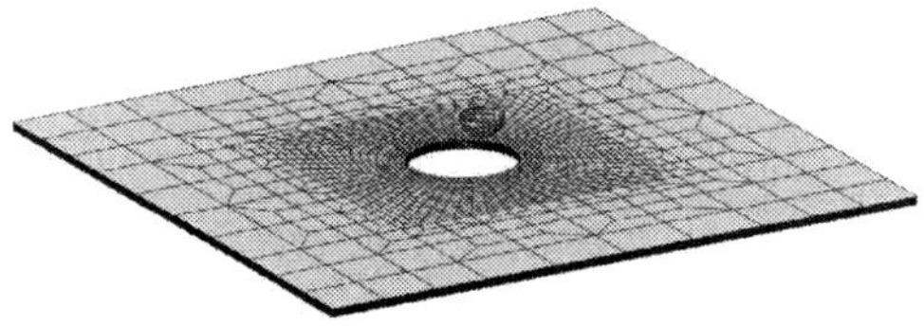

Figure 1. FE model of the composite panel for analysis.

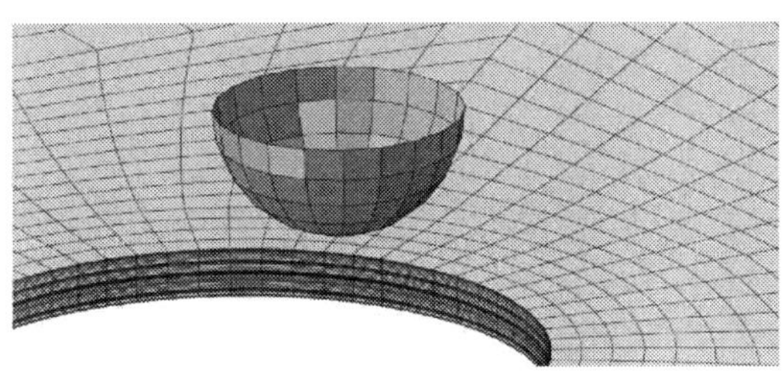

Figure 2. Details of the FE model, composite panel and impactor.

Progressive ply damage in fibre tension, fibre compression, matrix tension, and matrix compression as described by Hashin and Rotem [14] have been incorporated in the model; the material parameters are shown in table 1. E_{ii} and G_{ii} denote the Young's and shear modulus in direction ii, ν_{12} is Poisson's ratio, X^T/X^C is the longitudinal ply strength in tension/compression, Y^T/Y^C is the transverse ply strength in tension/compression and S^L/S^T is the longitudinal/transverse shear ply strength. Damage evolution is described by reducing the stiffness in the elastic matrix C_D and the resulting stress σ is computed from strains ε using equation (1).

$$\sigma = C_D \varepsilon \qquad (1)$$

The progressive cohesive zone damage with bi-linear law was also included; more details are given in a previous report [13].

Table 1. Ply material properties.

E_{11} [MPa]	E_{22} [MPa]	ν_{12} [-]	G_{12} [MPa]	G_{13} [MPa]	G_{23} [MPa]
110000	7400	0.3	4200	4200	4200
X^T [MPa]	X^C [MPa]	Y^T [MPa]	Y^C [MPa]	S^L [MPa]	S^T [MPa]
2300	1500	66	220	93	93

2.1. *Impact testing*

Low velocity impacts at the coupon level were tested using the falling weight apparatus simulating tool/equipment drop and BVID. Simulations of the impact were performed at three energy levels of 15 J, 25 J and 50 J, using an impactor velocity of 10 m/s. The energy of 50 J is recommended for tool/equipment drop by the military standard DEF STAN 00-970 [15]; the energy of 15 J is the experimentally determined value for impact causing BVID [13], and the energy level 25 J was considered so as to include an impact level between the two extremes. The impact was performed twice; the first impact was directed near the edge of the large hole at the diagonal and the second was directed to a location at the opposite side on the same diagonal. The duration of the impact was much longer than the time needed for elastic waves to reach the element boundaries, which means that the deflection and load would have a similar relation as in static loading. Even so, when studying BVID, the explicit integration scheme is most often applied [5,6]. Therefore, the impact stage was solved using the ABAQUS/Explicit solver. During impact, the specimen was supported at the outer squared margin with a coarse mesh to prevent deflection in the normal direction, whereas, the inner part including the hole was free to deflect during impacts on both sides. In-plane displacement was allowed with only the two opposite corners being constrained so as to hold the composite in the initial position during impact. The other analysis details are similar to those described in a previous report [13].

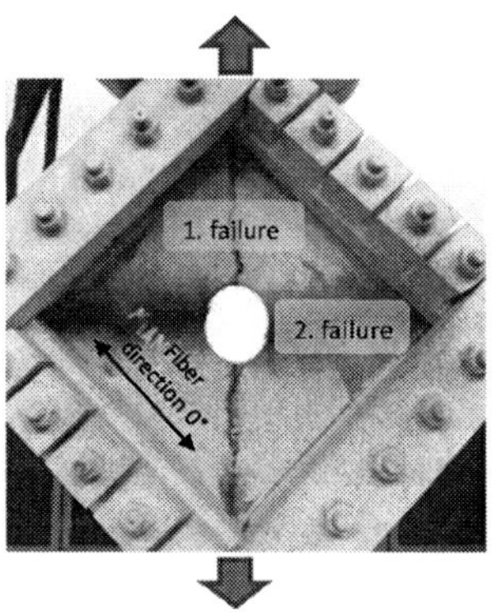

Figure 3. Shear loading of the composite panel with a large hole; experimental arrangement; panel impacted with 15 J energy.

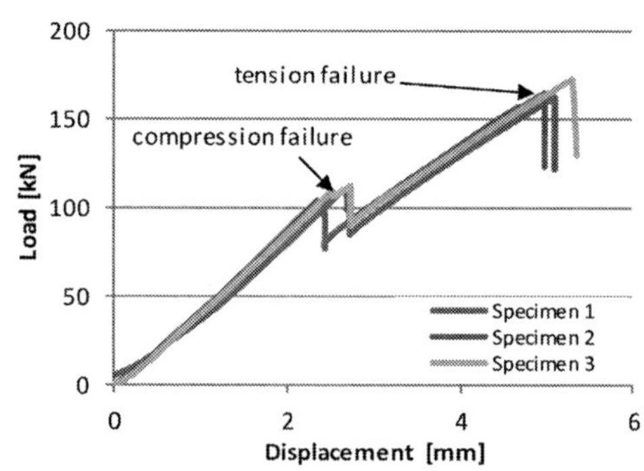

Figure 4. Experimental load-displacement curves for shear loading of impacted panels following impact with an energy of 15 J.

2.2. *Shear loading*

Shear loading was applied following the impacts in such a manner as to have the impacted zones in the compression-loaded diagonal of the panel, as shown in figure 3. The impact-free panel was also included in the simulation to compare the effect of the impacts on the strength. The experimental load carrying capacity shows two rapid drops due to failures in the panel diagonals. The first failure is caused by material compression in the diagonal parallel to the loading force applied at the panel corners and the second failure is caused by a simple tension in the perpendicular diagonal. The experimentally measured load-displacement curves for the panel impacted with an energy of 15 J are shown in figure 4.

Although the shear test is essentially quasi-static, it was also simulated by means of the dynamic FE code ABAQUS/Explicit. The explicit solver was used in view of the fact that if an implicit FE code was used, high material and geometrical nonlinearities were observed, resulting in severe convergence difficulties. Accordingly, a displacement loading was applied at the two opposite corners of the plate that was sufficiently slow and smooth, so as to maintain the kinetic energy lower than the internal energy of the system.

3. Results

3.1. Numerical results of the impact

Impacts were applied individually in succession on opposite sides close to the edge of the hole. It is seen that different energy levels result in different types of damages. The 15 J impacts cause BVID and the impactors bounce back. In addition, the damage caused by both the impacts is almost the same in shape and size, which indicates that there is no influence of the previously applied impact on the second one. That the two 15 J impacts are independent is demonstrated by comparing the two delamination and damaged areas both numerically and experimentally [13]. The 25 J impact is more severe, but no penetration of the impactor is predicted numerically; here too, the impactors bounce back. The damaged and delaminated area is increased compared to the 15 J impact. The layers below the impact point are damaged, in contrast to the 15 J impact, where the damage occurs mainly within the top half of the layers. Increasing the impact energy up to 50 J leads to the penetration of impactors. During 50 J impact, the edge bends in a manner similar to that observed for impact at lower energy levels, but the elastic energy is not sufficient to stop the impactor. At a certain displacement, the back-surface tension is high enough to break the back layers. The resistance of the material against the movement of the impactor disappears, causing the impactor to penetrate and pass through the composite.

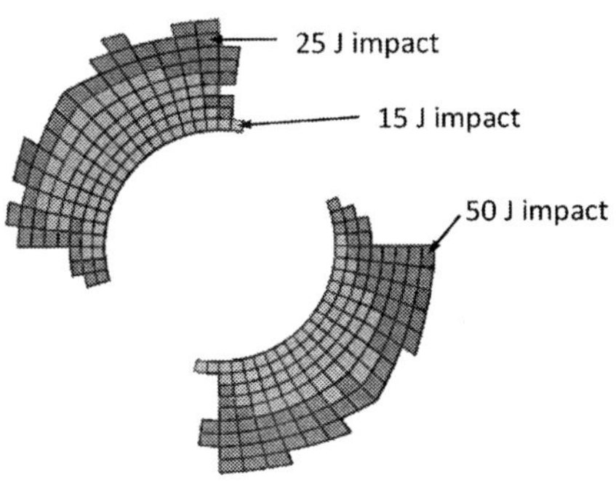

Figure 5. Delamination zone after impacts at energies of 15 J, 25 J and 50 J; Abaqus damage post processing parameter $SDEG_{min} = 0.999$.

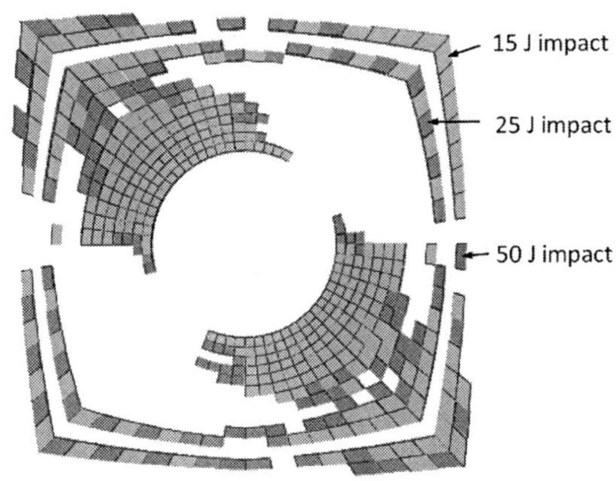

Figure 6. Ply damage zone after impacts with energies of 15 J, 25 J and 50 J; Abaqus damage post processing parameter $DMICRTMAXVAL_{min} = 0.999$.

Viewed from the top, the area of damage in the layers appears to be comparable in size for all the different energies used. However, going through the layers, it is seen that only the upper half of the layers is damaged under 15 J impact, whereas, during impact with 25 J energy, all the layers are broken and the detached back layers are damaged. The 50 J impact causes severe damage in all the layers during the penetrating process of the impactor through the composite. The delamination increases with higher energy impacts, but the 25 J and 50 J impacts cause similar delamination. During post processing, full damage and delamination are assumed at damage parameters of 0.999. The comparison of the damaged area and delamination for the different energy levels of impacts are shown in figures 5 and figure 6. The area of damage caused by the 50 J impact overlaps that caused by the 25 J impact, and the 15 J impact area is shown at the top. In figure 6, the "square" shape shows lamina damage of the back layers caused by the supports. At this location, the damage is through matrix tension failure mode according to Hashin's failure criteria [14]. The clamping applied during the impacts simulates a connection of the part to the surrounding structure. In the absence of clamping, the response of the specimen could be totally different, but for example, the wing rib is constrained around its perimeter as well, which means that the boundary conditions have been applied correctly. The surrounding white colour indicates that in the other elements no exceeding of the damage parameter is observed.

The cross-sections of the impact area after impact for all the energy levels are shown in figure 7. Although the increased displacement of the damaged elements with energy of impact is clearly seen, the damaged area as viewed from the top is similar for the 25 J and 50 J impacts, as mentioned above. During the 25 J impact, the back layers are damaged almost as severely as in the case of 50 J impact, when the impactor penetrates the composite case. The damage also develops in the normal direction through the composite and does not spread in the plane. Thus, it seems that when increasing the impact energy to over 25 J, no new elements are damaged and the damage in the affected elements only progresses further.

In figure 7, the distorted elements under 25 J and 50 J impact can be seen. There seems to be an interpenetration of the elements, but this is the consequence of the contact algorithm, where master nodes can penetrate the slave surface. The effect is seen after leaving the impactor when the nodes of master and slave surfaces do not match each other due to the different displacements of the element caused by damage development. During impact, the corresponding contact nodes are very close and this effect does not appear. However, the stiffness of severely damaged elements is too low, which in reality behaves as ply failure, and penetration of the layers occurs as well. Using smaller elements could help but would increase the cost.

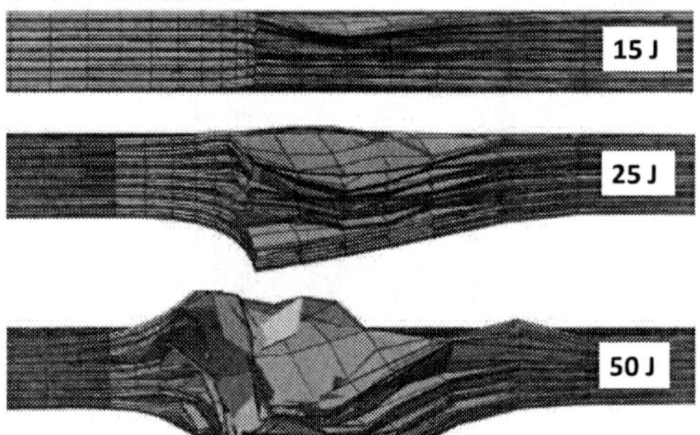

Figure 7. Displacement in cross-section of damaged elements after impact at energy levels of 15 J, 25 J and 50 J; FE simulation.

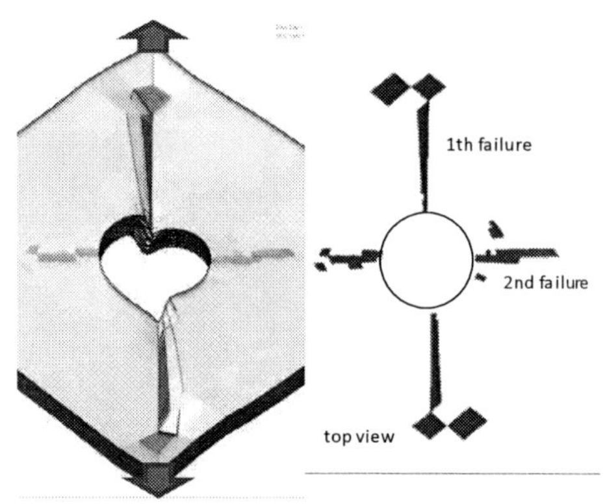

Figure 8. Fibre tension and fibre compression failures due to shear loading – Impact energy is 15 J; FE simulation.

3.2. Numerical results of shear loading

Shear loading following impact is characterised by two failures along the diagonals of the panel, as determined experimentally. The behaviour as determined using the numerical model during the shear loading is also characterised by two failures in the diagonals, as shown in figure 8. When using the 3-D ply-by-ply model, the sensitivity to impact energy levels is well demonstrated, but the behaviour of the load-displacement curve does not fully correspond to that observed experimentally [13]. The rapid loss of stability in the impacted zone loaded in compression, which is observed experimentally, was not simulated precisely by the model. A progressive development of failure was observed in the simulation. Nevertheless, the load at which the development of compressive failure starts agrees well with the corresponding experimentally determined value. Thus, the model is inconsistent in the description of the local loss of stability in the compression region. The reason for this may be in the micro-mechanism of failure. Hashin's model does not adequately take into account, the micro-buckling of the fibres in the compression zone. A more sophisticated model could predict a more realistic behaviour in compression, for example, the LARC failure model developed at the Langley Research Centre [16]. However, certain parameters to be used in the modelling remain very difficult to determine and thus, their implementation into commercial FE codes is not fully satisfactory.

The load-displacement curves following impacts are similar, as shown in figure 9. The first failure in the compressive zone is influenced by the impact, but the second failure in the tension diagonal occurs at approximately the same load level, since it depends particularly on the fibre strength under tension. The initial parts of the load-displacement curves are identical to that observed for the impact-free panel However, starting from approximately half of the maximum load of the impact-free panel, the curves for the impacted panels start to deviate due to the damage developed in the compression zone. The damaged zone and the delaminated area resulting from higher energy impacts, do not differ much, and therefore, their responses to the loading are similar. The load levels at the first and the second failure are presented in figure 10 with respect to the maximum load of the impact-free panel. In this case, the compression and tension failures occur at similar load levels, but the load level at the first failure is slightly higher. Contrary to the impact-free case, the load level at the compressive failure of impacted panels is lower than that at tension failure. Thus, for the impacted panels, compressive and tension failures occur on average at 37% and 15% lower load level, respectively.

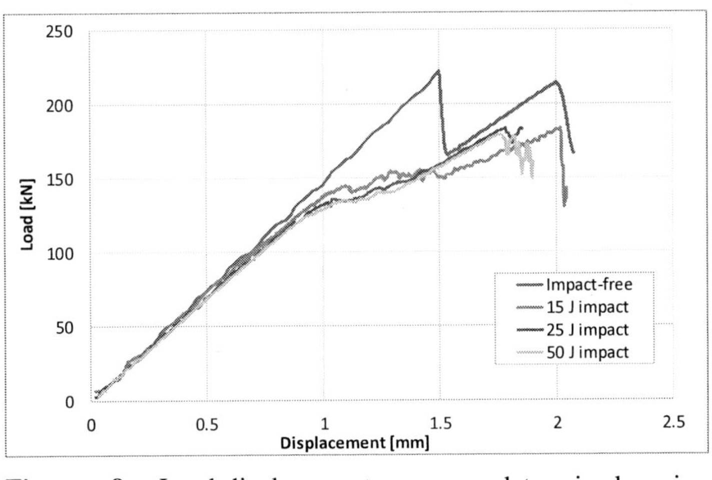

Figure 9. Load-displacement curves determined using the numerical model

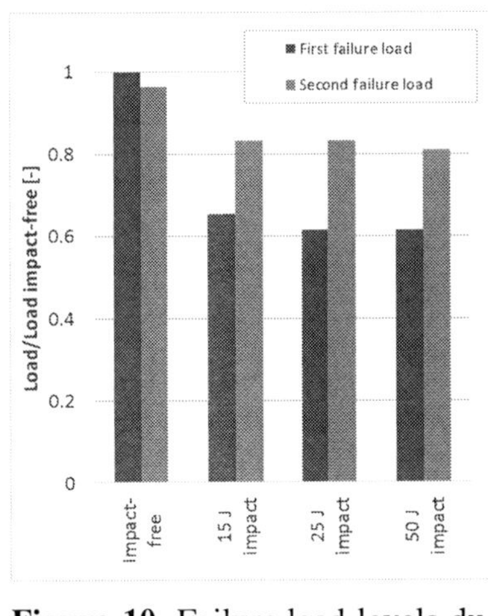

Figure 10. Failure load levels due to shear loading; FE simulation.

4. Conclusion
The impact damage and subsequent resistance to shear loading were investigated numerically and the effect of impact energy was studied using a combination of a 3-D ply-by-ply FE model and the cohesive zone technique. Impacts of three energy levels and an impact-free panel were analysed. An impact energy of 50 J caused the penetration of the impactors, but the damaged zone in the plies was comparable to the impact caused at the energy level of 25 J. The delamination area increased with impact energy, but the delamination areas were similar for impact energy levels of 25 J and 50 J.

Numerical results showed, as expected, that the impact-free panel can carry a higher shear load than the impacted panels. During loading, two failures occurred; compressive failure occurred in the panel diagonal parallel to the applied load and tension failure occurred in the second diagonal. The response on the loading in the load-displacement curve was similar at low loads, but for loads greater than 0.5, the of impact-free panel failure load the curves of impacted panels deviated. The load at the compressive failure was on average 37% lower than that for the impact-free panel. Only the 50 J impact caused the impactor to penetrate the composite, but the reduction in load-carrying capacity was similar to the case with no penetration. The 15 J and 25 J energy impacts caused damage, which can be considered to be BVID.

Numerical investigation of the near-edge impact and subsequent shear loading showed the effect of the impact on the load-carrying capacity of a composite panel with a large hole. Following impact, the load-carrying capacity was dramatically reduced. The danger of BVID near the edge of the central hole was clearly demonstrated. The loss of load-carrying capacity as a result of BVID, was as severe as in the case where penetration by the impactor occurred.

Acknowledgment
This work was funded by the Ministry of Industry and Trade of the Czech Republic in the framework of the Institutional support of Research organizations.

References
[1] Umer R, Alhussein H, Zhou J and Cantwell W 2016 The mechanical properties of 3D woven composites *J Compos Mater* doi:10.1177/0021998316681187
[2] Hao A, Sun B, Qui Y and Gu B 2008 Dynamic properties of 3-D orthogonal woven composite T-beam under transverse impact *Compos Part A Appl Sci Manuf* **39** 1073-82
[3] Walter T R, Subhash G, Sankar B V and Yen C F 2009 Damage modes in 3D glass fiber epoxy woven composites under high rate of impact loading *Compos Part B Eng* **40** 584-9
[4] Davis M J and Jones R 1983 Damage tolerance of fibre composite laminates *Fracture Mechanics Technology Applied Evaluation and Structure Design* eds Sih G C *et al* (Dordrecht: Springer) pp 635-55
[5] Michalcová L and Kadlec M 2016 Carbon/epoxy composite delamination analysis by acoustic emission method under various environmental conditions *Eng. Failure Analysis* **69** 88-96
[6] Kadlec M and Růžek R 2012 A Comparison of laser shearography and C-scan for assessing a glass/epoxy laminate impact damage *Appl Compos Mater* **19** 393-407
[7] Alemi-Ardakani M, Milani A S and Yannacopoulos S 2014 On complexities of impact simulation of fiber reinforced polymer composites: a simplified modeling framework *Scientific World Journal* 382525 http://dx.doi.org/10.1155/2014/382525
[8] Liu H, Falzon B G and Tan W 2018 Experimental and numerical studies on the impact response of damage-tolerant hybrid unidirectional/woven carbon-fibre reinforced composite laminates *Composites* B **136** 101-18
[9] Sun X C and Hallett S 2017 Barely visible impact damage in scaled composite laminates: experiments and numerical simulations *International Journal of Impact Engineering* **109** 178-95
[10] Farooq U, Mzler P and Kandola B K 2009 Prediction of barely visible impact damage in

composite panels subjected to blunt nose impact *Research and Innovation Conference 2009* 4 http://digitalcommons.bolton.ac.uk/ri_2009/4

[11] Malhotra A and Guild F J 2014 Impact damage to composite laminates: effect of impact location *Appl Compos Mater* **21** 165-77

[12] Iwai H, Uemura M and Okubo H 1989 Research on in-plane shear testing methods of advanced composite panels *Journal of the Japan Society for Composite Materials* **15** 109-15

[13] Šedek J and Bělský P 2017 Numerical evaluation of barely visible impact damage in carbon fibre-reinforced composite panel loaded in shear *Material and Contact Characterisation VIII* **116** WIT Press 73-85

[14] Hashin Z and Rotem A 1973 A fatigue criterion for fiber-reinforced materials *Journal of Composite Materials* **7** 448-64

[15] DEF STAN 00-970 2015 Desigh and airworthiness requirements for service aircraft PART 1/12 section 3 Leaflet 44 *UK Ministry of Defence*

[16] Pinho S T, Dávil C G, Camanho P P, Ianucci L and Robinson P 2005 Failure Models and Criteria for FRP Under In-Plane or Three-Dimensional Stress States Including Shear Non-Linearity *NASA* TM-2005-213530

Fluid flow characteristics during polymer flooding

S L Yao [1,2], H E Dou [1], M Wu[1] and H J Zhang [1]

[1] Research Institute of Petroleum Exploration and Development Beijing, China

E-mail: yaoshanglin@petrochina.com.cn

Abstract. At present the main problems of polymer flooding is the high injection pressure which could not guarantee the later injection. In this paper the analyses of polymer's physical properties and its solution's variable movement characteristics in porous media reveal the inevitable trend of decrease in injection capacity and liquid production due to the increase of fluid viscosity and flow rate with more flow resistance. The injection rate makes the primary contribution to the active viscosity of the polymer solution in porous media. The higher injection rate, the greater shearing degradation and the more the viscosity loss. Besides the quantitative variation, the rate also changes qualitatively as that the injection rate demonstrates composite change of injection intensity and density. Due to the different adjustment function of the polymer solution on its injection profile, there should be different adjustment model of rates in such stages. Here in combination of the on-site recognitions, several conclusions and recommendations are made based on the study of the injection pattern adjustment during polymer flooding to improve the pressure distribution system, which would be a meaningful reference for extensive polymer flooding in the petroleum industry.

Key words: fluid flow law; Injection rate.

1. Introduction

Polymer solution is a kind of typical non-newtonian fluid due to not only its non-newtonian viscosity, but also the strong viscoelasticity of flowing in porous media [1]. In the process of field development by water drive, the mixture of oil and water can be described as a Newtonian fluid. But for polymer flooding development, because of shearing degradation, dilution, adsorption and retention etc. the fluid property of polymer flooding is more complex than that of water flooding.

2. Formation fluid characteristics during polymer flooding

2.1. Water solubility of polymers

The polymer molecules have charged groups which is able to react intensively with polar molecules through hydrogen bonds. So that water is a good solvent for polymer while oil is not. After the polymer solution flows into the formation, dipole water molecules form surrounding macromolecule of polymer a solvation layer as bound water by adsorption or hydrogen bond. At the same time due to the electrostatic repulsion of charged groups between the polymer molecules, the coil volume increases to enhance the internal friction of active molecules, then to improve the water viscosity. But herein oil phase flowing is not affected [2].

2.2. Rheological properties of polymer solutions in porous media

The polymer solution is a kind of typical non-Newtonian fluid, of which the flow behavior can describe by the law of power exponent.

$$\mu=\tau/r=Hrn-1$$

(μ-viscosity,τ-shearing stress, r-shearing rate, H-consistency coefficient, n- index of power exponent law.)

The working viscosity of polymer solution in the pay zone is mainly affected by the shearing rate. From several methods of calculation, the shearing rate is mainly related to the injection rate and pore throat size, such as the Pang Cheng equation.

$$r=(3n+1)\,Q/An\sqrt{8k\varphi}$$

(Q-injection volumetric rate; A-infiltration area; K-effective permeability;φ-porosity.)

For the same core, the faster the injection rate, the quicker the degradation of the polymer and the greater the loss of viscosity. For cores with different permeability, under the same injection rate, the viscosity loss of polymer solution in high permeability reservoir is small, and the viscosity loss of polymer in low permeability reservoir is large.

3. Fluid flowing and its dynamic characteristics during polymer flooding

After the polymer solution is injected into the porous medium, the change of fluid flowing is mainly demonstrated in two aspects [3]:

3.1. Influence of fluid flowing rate variation on dynamic characteristics

For non-Newtonian fluids, the modified Darcy's law shows the effect of the change of fluid viscosity on various parameters.

$$\upsilon= Q/F= K\Delta P/\mu L$$

(υ-infiltration rate; Q- volumetric rate; F- cross section area of fluid flowing;ΔP-fluid pressure difference;μ-fluid viscosity; K- permeability; L-length.)

From the equation above, the fluid infiltration rate (V) is directly proportional to injection strength (Q/F), permeability (K) and fluid pressure (P), but inversely proportional to the fluid viscosity (u). With the injection of polymer solution, the viscosity of fluid in the formation is increasing, which will inevitably lead to the decline of fluid infiltration rate, the increase of infiltration resistance and the decrease of injection rate and production rate. This is the main reason why it is difficult to enhance the liquid production in the later stage of polymer flooding, although the pressure of the formation is high. From the dynamic characteristics, after polymer injection the pressure rise in injectors while the flowing pressure drops in producers with significant reduction of both water absorption index and production index (Table 1). For example, in the Block N-Ⅱ by polymer flooding, water absorption index decreased by 22.58%, liquid production the index decreased by 43.15%.

3.2. Effect of fluid flow direction on dynamic characteristics

For water drive development, due to the great heterogeneity between different pay zones, inside different layers and at different surface locations, the infiltration rate of the injection fluid along the high permeability layer (mainstream line) is faster so that the middle and low permeability (non mainstream) oil layer is inhibited. Because of polymer's solubility in water, the polymer solution into the reservoir of high permeability (the main line) firstly increase the fluid viscosity herein to increase flow resistance, then to reduce the infiltration rate, so as to change fluid flow direction to the low permeability reservoir (non mainstream line). However, after the entrance of polymer solution into the middle & low permeability reservoir (non mainstream line), the increase of fluid viscosity will decrease the infiltration rate of low permeability layer (non mainstream line), and then speed up the infiltration rate of high permeability reservoir (mainstream line). Hence in the polymer flooding process, the profile adjustment would affect the development effect of polymer flooding. At different stages of polymer flooding, the reservoirs injection profile changes. The suction capacity of high permeability declines to the highest extent during the low value period and rises continuously along with the decrease of suction capacity of low & middle permeability layers during the recovery period. From injection pressure changes of polymer flooding blocks, the initial injection pressure increased rapidly and in the water cut decline period, due to the great increase of infiltration area, low permeability layer flow more and more rapid while pressure incremental amount becomes small. Therefore, it is necessary to limit the injection pressure rise of the polymer flooding to ensure the later normal injection [4].

Table 1. Variation of absorption and liquid production index during polymer flooding at Sabei Development Zone.

Blocks	Water absorption index			liquid production index			Reduction difference (%)
	Before polymer injection (m^3/d.MPa)	After polymer injection (m^3/d.MPa)	Reduction extent (%)	Before polymer injection (t/d.MPa)	After polymer injection (t/d.MPa)	Reduction extent (%)	
Block N-II WE	32.29	25.00	22.58	63.45	36.07	43.15	20.57
Block N-II WW	26.07	20.50	21.37	42.46	28.49	32.9	11.53
Block N-III WE	19.90	19.69	1.06	37.63	26.22	30.32	29.26
Block N-III WW	21.80	16.36	24.95	28.74	19.42	32.43	7.47

The polymer fluid movement changes firstly surrounding the injection well. Due to the different profile adjustments by polymer at different stages, the injection production method should be adjusted (see Table 2), especially during the initial period when the profile is adjusted at low rate with displacement at high rate to control the decrease amount of both injection and liquid production capacity [5].

And the improved pressure system will guarantee the later normal injection. At the early stage of polymer injection, the main purpose is to reduce the infiltration capacity of high permeability layer (main line) through formation of slugs surrounding injection well. To reduce the degradation of the polymer, it is necessary to reduce injection intensity, to improve concentration to control the pressure rising rate, so as to provide favorable conditions for late injection concentration adjustment. At the declining period of water cut, because of the incremental of infiltration area, the increase of injection intensity and reduction of concentration at this time will not cause significant increase of injection pressure. At the same time due to the increase of amount, the working viscosity of polymer in high permeability layer (main line) will not be affected and the decrease of concentration will accelerate the decline of polymer degradation in middle & low permeability layers (non mainstream line) without plugging the surrounding area of wellbore [6]. When it comes into the low water cut period, in order to reduce water flow rate and improve oil flow rate inside layers, the injection concentration should be enhanced on the basis of injection intensity. At the same time, due to the most obvious profile control effect of the high permeability layer (main line), the relative permeability of the high permeability layer (main line) will decline with the largest amount. When it comes into the water cut recovery period, with the recovery of suction capacity in high permeability layers, deep profile adjustment could be adopted to decrease the injection intensity by improvement of polymer concentration to inhibit liquid breakthrough of fluid along the high permeability layer. For wells with obvious

IOP Conf. Series: Materials Science and Engineering **369** (2018) 012036 doi:10.1088/1757-899X/369/1/012036

contradiction the stratified injection method could be used to tap the low permeability layer (non mainstream line) potential.

Table 2. Adjustment of injection and production methods at different stage of polymer injection.

Stages	Adjustments of Injection & Production	Purposes	Characteristics of Injectors and Producers
Initial	1.Adjust profile fundamentally	Plug pay zones of high permeability (old water channel)	1.Injection pressure rises by great amount and flowing pressure of producer declines 2.Suction capacity declines in high permeability layers
	2.Reduce injection intensity and improve its concentration	Form plugs inside pay zones of high permeability surrounding wellbores	
Declining	1.Improve injection intensity and reduce its concentration	Inhibit the incremental of infiltration resistance inside pay zones of middle & low permeability surrounding wellbores and increase the infiltration rate (non-mainstream line)	1.Injection pressure keep stable and flowing pressure of producer recovers 2.Suction capacity rises with great amount in middle & low permeability layers
Low value	1. Maintain injection intensity and improve its concentration	Reduce water flow rate inside pay zones and increase oil flow rate	1.Injection pressure rises a bit 2.Suction capacity keeps relatively stable in middle & low permeability layers 3.Liquid production volume declines
	2.Stimulate reservoirs	Narrow differences inside, between layers and at horizontal locations	
Recovery	1.Adjust profile fundamentally	Inhibit the breakthrough of injected liquid along layers of high permeability	1.Injection pressure rises by great amount 2.Suction capacity rises in high permeability layers
	2. Subdivide layers	Release inter-layers contradiction	
	3.Reduce injection intensity and improve its concentration	Form plugs inside pay zones of high permeability surrounding wellbores	

Block N-Ⅱ WE is the fifth polymer flooding block with water cut currently in declining period. Through the adjustment of injection pattern after polymer flooding and grasp best occasions of the effective wells, there appears an obvious characteristic as significant improvement of injection condition based on the polymer flooding effect of the Block. After the polymer injection compared to the same period, the injection pressure was significantly lower than that of other blocks (see Table 3). At the beginning time of effect the injection pressure rised by 2.0MPa with current (16 months after injection) injection pressure of 11.2MPa, while the other blocks of the injection pressure rised by 3.0-4.0MPa. If compared with the performance before polymer injection, the injection pressure rised by 3.0MPa, and the injection pressure is close to 13.0MPa at other blocks. At the beginning of effect the apparent water absorption index of Block N-Ⅱ WE decreased by 46.28% while the Block N-III WW decreased by 50.56%. The injection pressure control effectively ensured the injection rate improvement after the polymer flooding effect. At present, the injection rate of the Block increased to 0.150PV/a, and the injection rate of other blocks can only stay at 0.12PV/a after a year [7].

IOP Conf. Series: Materials Science and Engineering **369** (2018) 012036 doi:10.1088/1757-899X/369/1/012036

Table 3. Comparison of flooding injection status in Sabei Development Zone.

Blocks	Fracturing Pressure (MPa)	Before Polymer Flooding			Effect Time			Currently		
		Injection Pressure (MPa)	Daily Injection (m^3)	Apparent Water absorption index (m^3/d.MPa)	Injection Pressure (MPa)	Daily Injection (m^3)	Apparent Water absorption index (m^3/d.MPa)	Injection Pressure (MPa)	Daily Injection (m^3)	Apparent Water absorption index (m^3/d.MPa)
Block N-II WW	13.7	7.3	108	14.79	11.3	202	17.88	12.5	205	16.40
Block N-II WE	13.2	8.5	220	25.88	11.7	204	17.44	12.8	200	15.63
Block N-III WW	14.1	8.3	106	15.70	11.6	113	9.74	12.8	110	8.59
Block N-III WE	14.1	7.1	105	14.71	11.1	137	12.34	12.1	127	10.50
Block N-II EW	13.6	8.2	94	11.44	10.2	85	8.37	11.2	122	10.92

From the development effects in the above table, the effective time is basically the same. After injection of 5 months, the water cut began to fall. The dosage was 48PV.mg/l which is lower than Block N-II WE by 13PV.mg/l and bigger than Block N-III WW by 5PV.mg/l. In the two Blocks the average single well liquid production is 147t, bigger than that before polymer injection by 29t while the number of the other Blocks after polymer injection declines, to ensure the whole block development effect. The production rate of both geological reserves and recoverable reserves reached 4.20% and 8.22%, respectively while the corresponding recovery degrees reaches 4.10% and 7.04%, respectively which is only lower than the number of Block N-II WE (see Table 4).

Table 4. Comparison of the effect of polymer flooding in Sabei Development Zone.

Blocks	Time	Polymer dosage (PV.mg/l)	Daily liquid Production of one well (t)	Daily oil Production of one well (t)	Water Cut (%)	Production rate (%)		Stage recovery degree(%)		Enhanced oil Production by 1 ton of polymer (t)
						geological reserves	recoverable reserves	geological reserves	recoverable reserves	
Block N-II WE	Before polymer injection		162	17	89.60	2.22	4.3			
	Effective time	61	128	16	87.78	2.07	4.00	1.17	2.37	6
	17months	176	150	45	69.13	5.99	11.60	5.32	10.39	87
Block N-II WW	Before polymer injection		136	10	92.53	2.01	3.90			
	Effective time	81	139	12	92.35	2.28	4.43	0.75	1.46	4
	17months	175	118	21	82.63	4.09	7.95	2.82	4.49	29
Block N-III WW	Before polymer injection		122	8	93.55	1.68	3.00			
	Effective time	43	118	9	92.23	1.91	3.41	0.47	0.90	7
	17months	161	126	23	81.73	4.78	8.55	3.54	5.75	72
Block N-II WE	Before polymer injection		160	9	94.49	1.81	4.28			
	Effective time	113	155	11	93.00	2.18	5.17	0.50	0.98	4
	17months	184	136	19	86.52	3.64	8.60	2.08	3.26	33
Block N-II EW	Before polymer injection		118	7	93.81	1.17	3.07			
	Effective time	48	143	10	92.79	2.17	4.26	1.11	2.18	6
	17months	161	147	20	86.45	4.20	8.22	4.10	7.04	48

4. Conclusions

1. The injection rate has a great influence on the rheological properties of polymer solution, the higher injection rate, the larger shearing degradation and the greater viscosity loss. The injection rate is related to not only the quantity variation but also the concentration.

2. After the polymer injection, with the injection of polymer solution, the viscosity of the formation fluid increases to enhance the resistance and reduce flow rate with essential drop of both injection and production capacity. But injection and production adjustment through slow adjustment rate and high displacement rate can control the decline extent of injectivity and liquid productivity during the early stage of polymer injection.

3. In the whole process of polymer flooding, the profile adjustment has direct effect on the development performance. In different stages variable functions of profile adjustment by polymer solution result in different dynamic characteristics.

4. During the whole polymer flooding process, it is necessary to adjust the injection rate considering the injection intensity and concentration. In the initial stage of polymer injection, injection intensity should be reduced and injection concentration should be enhanced to some extent to control the rising injection pressure, so as to ensure the layering profile control and the second profile control in later stage of polymer flooding to improve the integral development effect.

References

[1] Hu B Z 1997 *Oil Production Engineering by Polymer Flooding* (Beijing: Beijing Petroleum Industry Press) p 16
[2] Jin Y S 1985 *Oil Production Geological Engineering* (Beijing: Beijing Petroleum Industry Press) p 42
[3] Wang K L and Wang F L 1997 *Research on Improving Polymer Flooding Technology* (Beijing: Beijing Petroleum Industry Press) p 51
[4] Ge J L 1982 *Oil and Gas Infiltration Mechanics* (Beijing: Beijing Petroleum Industry Press) p 64
[5] Shu C Q, Li H B, Zhou C J and Chen Z H 2005 Study of the coefficient of resistance and residual resistance factor and displacing efficiency of hydrophobically associating polymer *Offshore Oil* **25** 58-60
[6] L W, Zhao X T, Han Y X, Meng F Z and Zhang X H 2010 Research progress on heat and salt resistance polymer flooding *Special Oil & Gas Reservoirs* **17** 11-4, 38
[7] Lan Y B, Yang Q Y and Li B H 2006 Experimental research on sweep efficiency and oil displacement efficiency of polymer flooding *Acta Petrolei. Sinica* **27** 64-8

Investigation of dynamic properties of a polymer matrix composite with different angles of fiber orientations

F Kadioglu[1,3], T Coskun[2] and M Elfarra[1]

[1]Ankara Yildirim Beyazit University, Ankara, 06050, Turkey
[2]Ferka Aviation, Ankara, 06830, Turkey

E-mail: ferkadioglu@gmail.com

Abstract. For the dynamic values of fiber-reinforced polymer matrix composite materials, elastic modulus and damping values are emphasized, and the two are desired to be high as much as possible, as the first is related to load bearing capacity, the latter provides the capability of energy absorption. In the composites, while fibers are usually utilized for reinforcement providing high elastic modulus and so high strength, matrix introduces a medium for high damping. Correct measurement of damping values is a critical step in designing composite materials. The aim of the current study is to measure the dynamic values of a glass fiber-reinforced polymer matrix composite, Hexply 913/33%/UD280, produced by Hexcel, using a vibrating beam technique. The specimens with different angles of fiber orientations (0, ±10°, ±20°, ±35, ±45°, ±55°, ±70, ±80 and 90) were manufactured from the composite prepreg and subjected to the clamped-free boundary conditions. Two different methods, the half power bandwidth and the logarithmic free decay, were used to measure the damping values to be able to compare the results. It has been revealed that the dynamic values are affected by the fiber orientations; for high flexural modulus the specimens with small angles of orientation, but for high damping those with large angles of orientation should be preferred. In general, the results are comparable, and the free decay method gave smaller values compared to the bandwidth method, with a little exception. It is suggested that the results (data) obtained from the test can be used for modal analysis reliably.

1. Introduction

Use of fiber-reinforced polymer matrix composites find many applications especially in aerospace and automotive sectors, due to their strength to weight ratio, damping properties, impact performance, fatigue life and resistance to corrosion etc. Therefore, their dynamical behaviors at low and high stress levels are investigated by many researchers [1-5]. In such materials, high strength capability is provided by the fibers and manipulated by their angles of orientations. On the other hand, matrix as a constituent material introduces high damping into the structures and serves for load transfer [6,7]. It is known that aerospace vehicles and automobiles are always under mechanical vibrations during their service life, which makes the components made out of composite materials to be handled. There are many efforts to increase damping value of the materials using different techniques [6-11]. Their points of interest mainly focus on matrix, interfacial between fiber and matrix, inter-laminar effects and stacking sequences [8,9]. Koratkar *et al* [10] measured more than 1000% increases in the loss modulus but a little reduction in the storage modulus when polycarbonate was enhanced by 2% weight fraction of single-walled carbon nanotubes (SWNTs). It was believed that frictional sliding at the SWNT interfaces was the reason for the enhanced energy dissipation. This was supported by the analysis of Zhou *et al* [11], who developed a model for the frictional sliding damping mechanism based on interfacial frictional motion between the nanotubes and the polymer matrix. Somen *et al* [12] used a novel liquid Methylmethacrylate (MMA) thermoplastic resin to enhance vibration damping of a composite made up of thick and thin plies of carbon fibers. After developing the novel composite system, the liquid MMA was found to be 27% more efficient in improving the structural damping compared to the epoxy resin. In another work [7], the damping behavior of continuous carbon fiber and flax fiber reinforced polymer (CFRP and FFRP) composites was studied by comparing angle-ply laminates. The results showed that about 2-3 times better damping of FFRP compared to CFRP at low frequency and low strain, and that the damping of both materials increases with increasing angle-ply orientation below 300 Hz. In a similar works [4], a multi-scaled methodology by using flax fibers and carbon nanotubes (CNTs) was proposed to enhance the damping property of carbon fiber reinforced epoxy composites. The damping property was measured by free vibration test and the strength and modulus were obtained by tensile and flexural tests. Effects of stacking sequences of flax fibers and the addition of CNTs on both the damping property and the mechanical properties of carbon fiber reinforced composites were investigated. Results clearly showed that the damping property of carbon fiber reinforced composites was improved greatly by laying flax fibers on the outmost layers of the composites. With the addition of CNTs, the damping property was further enhanced.

It is well known that reliable measurements of damping can only be possible via sophisticated measurement systems, and therefore it has been a subject of investigations for many researchers [13-15]. Some alternatives techniques were developed using such systems, which are able to give the results by a non-contact system [16,17].

As mentioned above, many efforts have been made by many researchers so far to increase damping values in the composite structures using different methods. The aim of this work is to obtain a desired damping and vibration capability in the structures by altering the orientation angles of fibers, balanced and symmetric. For this purpose, the dynamic properties of the composite materials were investigated via an experimental test set-up with an impact hammer and a vibration laser head. A non-contact response from the impacted specimens with clamped-free end conditions was obtained, thanks to the laser head. It is believed the data obtained by the set-up could be used for the modal analysis.

2. The experimental works

2.1. The specimens

The material used in this study was a glass fiber reinforced polymer matrix composite, Hexply 913/33%/UD280, produced by Hexcel, and the cure conditions for the prepreg of the composite was 130°C for 120 minutes under a pressure of 5 bars. The specimens with 0, ±10°, ±20°, ±35, ±45°, ±55°, ±70, ±80 and 90 fiber orientations were machined from the cured plates with 10 layers, and some details of them are shown in table 1. The test was carried out under a controlled environment, at room temperature (23°C) and 50% relative humidity, to avoid environmental effects on the specimens, and about four specimens of each type were tested to see if the results are repeatable.

Table 1. The details of the composite specimens used for the experimental work.

Fiber Angle	Width [mm]	Thickness [mm]	Length [mm]	Mass [g]	Density [kg/m^3]
0°	25.44	1.909	202.8	17.70	1797.46
± 10°	24.98	1.873	200.4	16.68	1778.91
± 20°	24.97	1.842	196.5	16.24	1796.58
± 35°	24.89	1.853	200.1	16.52	1789.72
± 45°	25.64	1.937	200.5	17.62	1769.14
± 55°	25.76	1.833	199.9	16.46	1744.20
± 70°	25.62	1.932	201.0	17.61	1770.47
± 80°	25.65	1.958	199.4	17.50	1747.68
90°	25.82	1.881	201.0	16.91	1732.60

2.2. The test set-up

For the dynamic measurements of the specimens, a vibrating beam technique was used, shown in figure 1. The experimental set-up includes a manual impact hammer to generate impulse and a non-contact laser head to pick up the response from the specimen. The impacted specimen is aimed to give its natural frequencies within a required frequency domain, via some distinguished peak values of the amplitude, depending on the mode shapes. A sophisticated data acquisition card was used to collect the digital values of velocity (amplitude) versus time, provided by the laser head which is also able to give displacement values. The set-up is supported by a software to process the data according to the required output format for visualization and analysis. The natural frequency, elastic modulus and damping values of the specimens could be measured easily via the technique. For the damping values, two different methods, the half power bandwidth and the logarithmic free decay, were used, and more details about the methods are explained below.

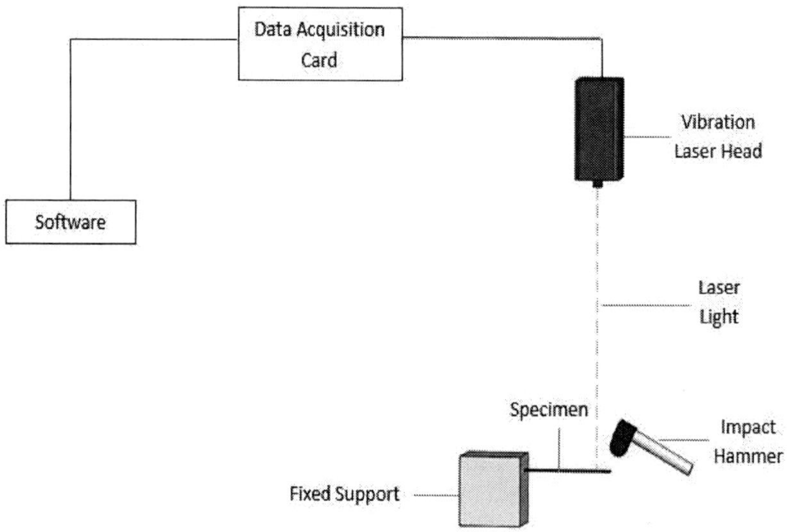

Figure 1. Experimental set-up for the specimens with clamped-free end conditions.

3. Theory

3.1. Measurement of flexural modulus

For every solid obeying Hook's Law, there is a specific natural frequency which is a function of its elastic modulus, geometry, density and the mode number. The natural frequency of a beam, f_n, is found as [18]:

$$f_n = \frac{1}{2\pi}\left(\frac{\lambda_n}{l}\right)^2 \sqrt{\frac{EI}{\rho t b}} \tag{1}$$

where $\lambda_n = 1.875$ is the 1st eigenvalue for clamped-free end conditions, n is the mode number, E is the flexural modulus, I is the second moment of area, l is the length, ρ is the mass density, t is the thickness, and b is the width of the beam. The flexural modulus of the specimen was measured using equation (1), and only the first (fundamental) mode was taken into account for all the measurements.

3.2. Measurement of damping

For an elastic solid structure, damping is defined as the conversion of mechanical energy into thermal energy, and it is defined in a number of different, yet related ways [19]. In this study, the half-power bandwidth method and the free decay method were used for measuring the damping values, and the results were compared to see if they are in agreement.

3.2.1. The half-power bandwidth method. This is determined from the curve of velocity amplitude against frequency, obtained when the specimen is impacted by the hammer. The 'half-power bandwidth' is (f_2-f_1) where f_2, f_1 are the frequencies at which the amplitude falls to $1/\sqrt{2}$ of its maximum value, reached at f_n, the resonant frequency. The loss factor, η is defined as,

$$\eta = \frac{f_2 - f_1}{f_n} \tag{2}$$

A schematic representation of the definition is illustrated in figure 2(a).

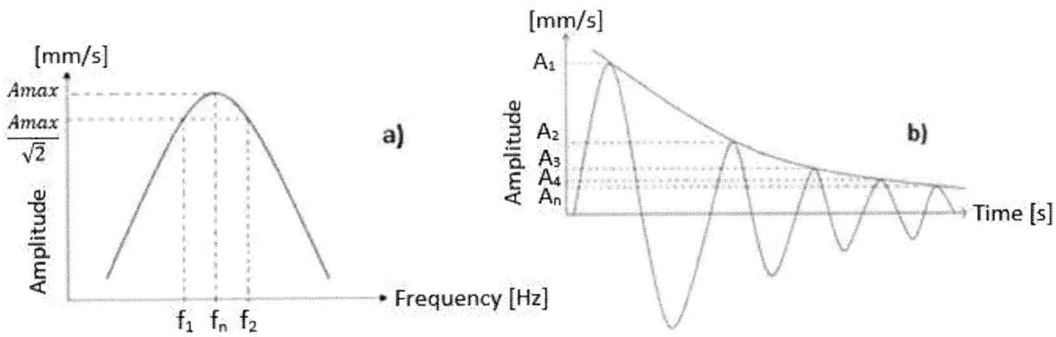

Figure 2. Schematic representations of the definitions of a) the half-power bandwidth method, b) the logarithmic free decay method.

3.2.2. The logarithmic free decay method. This is found by observing the decay of free vibrations (a 'free decay' method). The specimen is vibrated using the hammer, and a trace obtained of vibration amplitude against time, as illustrated in figure 2(b), is obtained. The logarithmic decrement, δ is then given by:

$$\delta = \frac{1}{n}\ln(\frac{A_1}{A_{n+1}}) \tag{3}$$

Where A_1 is the amplitude of the first cycle, and A_{n+1} is the amplitude of the $(n+1)^{th}$ cycle.

For convenience, damping is usually presented in Specific Damping Capacity (SDC), Ψ, which is defined as the ratio of the energy dissipated per cycle to the maximum elastic energy stored per cycle, per unit volume [20]. This ratio is usually expressed as a percentage.

For small damping, it is known there is relationship as follows [19],

$$\eta = \frac{\Psi}{2\pi} = \frac{f_2 - f_1}{f_n} = \frac{\delta}{\pi} \tag{4}$$

4. Results and discussions

The variation of natural frequencies of the specimens with different angles of fiber orientation is shown in figure 3(a). It is seen from the figure that the natural frequency is greatly dependent on the fiber orientations, and that small angles of fibers result in relatively high natural frequencies. From equitation (1), it is clear that the frequency is affected by the dimensions of the specimens under test, to eliminate this, the specimens tested have almost the same dimensions (see table 1), which implies the change in the frequency is only attributed to the different angles of orientation. While the maximum natural frequency is about 47 Hz for the specimens with the 0° orientation, the minimum value is around 25 Hz for those with the 90°. The figure clearly shows that there is a linear decrease in the frequency for the specimens with from the 0° up to the 55° orientations, and then it levels up for those with fiber orientation angles ≥ 55°.

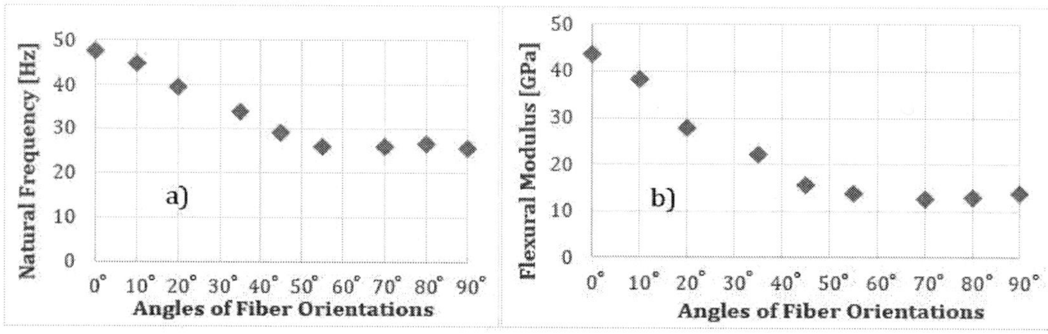

Figure 3. Variation of (a) the natural frequency, and (b) the flexural modulus of the composite specimens with different angles of fiber orientations.

The variation of flexural modulus with the fiber orientations are shown is figure 3(b), the response is similar to that shown in figure 3(a). A maximum value of the modulus, about 44 GPa, is found for the specimens with the 0° orientations, and the minimum value was about 15 GPa for those above the 55° orientations. The potential reason for this occurrence is because the specimens with from the 0° up to the 45° orientations are controlled mainly by the fibers, and those with the 45° orientations are in somewhere controlled by the both, fibers and matrix, and those with above 45° orientations are dominated by the matrix of the material. It is important to note that the specimens with high strength could only be possible for those with small angles, and these are especially recommended for high loading capacity.

As explained above, for the damping measurements of the specimens, two different methods, the logarithmic free decay and half power bandwidth, were used, and the results are presented in figure 4, which is the variation of SDC against the different fiber orientation angles. The figure shows that the results are totally opposite to that in figure 3(b); while the maximum values of the SDC are about 7% for the specimens with the higher angles, ±45°, ±55°, ±70, ±80° and ±90°, the minimum value is about 2% for those with the 0°. These results are directly related to the mechanical behavior of the fiber and the matrix forming the content of the composite material. It is clear that the matrix as a polymer has a high damping value, but a low value of elastic modulus, however, the specimens controlled by the behavior of the fibers should give low damping, but high elastic modulus. The current results are in agreement with the results reported by Adams and Maheri [20].

As shown in figure 4 the damping results obtained from the two methods are comparable, in general, however, those with the ±20° orientations showed a little bit different value; the results from the logarithmic decrement method yields higher than those from the bandwidth method, opposite for the other specimens. Typical responses of the specimens measured by the laser head, after impacted by the hammer, are shown in figures 5(a) and 5(b), the cases for the bandwidth and for the free decay methods, respectively. It should be noted that the first figure (figure 5(a)) indicates the natural frequencies between 1st and 4th modes, and that only the 1st mode was used for the measurements. The damping values of the specimens were obtained by using equations (2) and (3), and then converted into SDC by using equation (4). A previous work showed that especially the specimens with ±45° fiber orientations could provide a high damping value with a reasonable high strength, which could be preferred for structural purposes and for vibratory conditions [16].

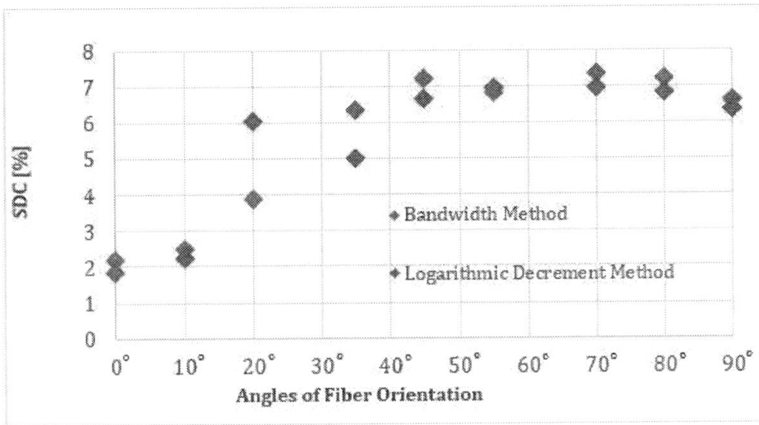

Figure 4. Variation of Specific Damping Capacity (SDC) with angles of fiber orientations from the two methods.

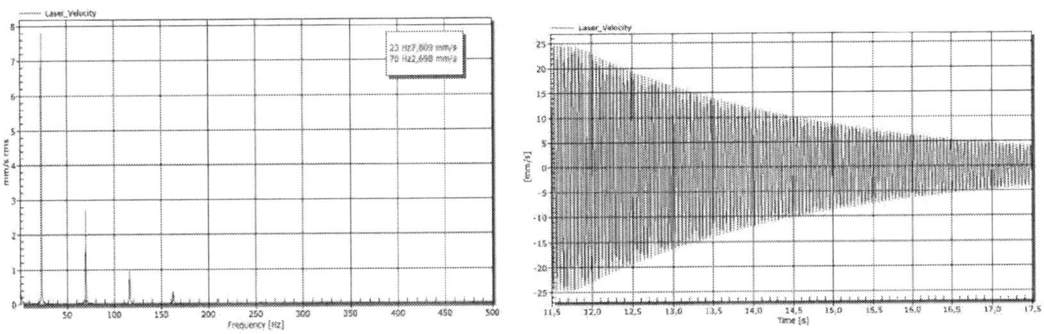

Figure 5. The response from the impacted specimens for the damping measurements of a) the half-power bandwidth method, and b) the logarithmic free decay method.

5. Conclusions

It is recommended that correct measurements of the dynamic properties of the composite materials are important for design purposes. The results obtained from the vibrating beam technique indicated that the dynamic values of the material were controlled by the angles of fiber orientations. While high values of the natural frequency and flexural modulus were provided by the specimens with the small angles, high damping values were obtained from those with the high angles, ±45°, ±55°, ±70, ±80° and ±90°. In general, the results from the two methods, bandwidth and free decay, were in agreement, and the test set-up with the non-contact response measurement gave reliable and repeatable data which could be used for the modal analysis.

Acknowledgments
We would like to thank to the Turkish Aerospace Industry for supporting the current work.

References

[1] Hung P Y, Lau K T, Cheng L K, Leng J and Hui D 2018 Impact response of hybrid carbon/glass fibre reinforced polymer composites designed for engineering applications *Composites Part B* **133** 86-90
[2] Rozylo P, Debski H and Kubiak T 2017 A model of low-velocity impact damage of composite plates subjected to Compression-After-Impact (CAI) testing *Compos. Struct.* **181** 158-70
[3] Chandra R, Singh S P and Gupta K 2003 A study of damping in fiber-reinforced composites *J.*

Sound Vib. **262** 475-96

[4] Li Y, Cai S and Huang X 2017 Multi-scaled enhancement of damping property for carbon fiber reinforced composites *Compos. Sci. Technol.* **146** 1-9

[5] Elias A, Laurin F, Kaminski M and Gornetb L 2017 Experimental and numerical investigations of low energy/velocity impact damage generated in 3D woven composite with polymer matrix *Compos. Struct.* **159** 228-39

[6] Rueppel M, Rion J, Dransfeld C, Fischer C and Masania K 2017 Damping of carbon fibre and flax fibre angle-ply composite laminates *Compos. Sci. Technol.* **146** 1-9

[7] Rueppel M, Rion J, Dransfeld C and Masania K 2016 Damping of carbon fibre and flax fibre reinforced angle ply polymers *ECCM17- 17th European Conference on Composite Materials* (Munich, Germany)

[8] Jen M H R, Kau Y S and Ong C L 1994 Effect of matrix resin on the response in a centrally notched composite laminate *Compos. Struct.* **29** 99-106

[9] Manjunatha C M, Taylor A C and Kincloch A J 2009 The effect of rubber micro- particles and silica nano-particles on the tensile fatigue behavior of a glass fibre epoxy composite *J. Mater. Sci.* **44** 342-5

[10] Koratkar N A, Suhr J, Joshi A, Kane R S, Schadler L S and Ajayan P M 2005 Characterizing energy dissipation in single-walled carbon nanotube polycarbonate composites *Appl. Phys. Lett.* **87** 063102

[11] Zhou X, Shin E, Wang K W and Bakis C E 2004 Interfacial damping characteristics of carbon nanotube-based composites *Compos. Sci. Technol.* **64** 2425-37

[12] Somen K B, Pavel P and Sunil C J 2017 Enhanced vibration damping and dynamic mechanical characteristics of composites with novel pseudo-thermoset matrix system *Compos. Struct.* **179** 502-13

[13] Rao M D and Crocker M J 1990 Vibrations of bonded beams with a single lap adhesive joint *J. Sound Vib.* **92** 299-309

[14] Qian G L, Hoa S V and Xiao X 1997 A vibration method for measuring mechanical properties of composite, theory and experiment *Compos. Struct.* **39** 31-8

[15] Guild F J and Adams R D 1981 A new technique for the measurement of the specific damping capacity of beam in flexure *J. Phys. E: Sci. Instrum.* **14** 355-63

[16] Kadioglu F, Sekerci H U and Coskun T 2018 A novel method to measure dynamic properties of composite materials *AIAA/ASCE/AHS/ASC Structures, Structural Dynamics, and Materials Conference, AIAA SciTech Forum* (AIAA 2018-0226)

[17] Kadioglu F 2009 Measurement of dynamic properties of composites in vibration by means of a non-contact mechanism *J. Reinf. Plast. Compos.* **28** 1459-67

[18] Den Hartog J P 1947 *Mechanical Vibrations* (New York, London: McGraw-Hill Book Company Inc.)

[19] Singh M M 1993 Dynamic properties of fibre reinforced polymers exposed to aqueous conditions (Department of Mechanical Engineering, University of Bristol)

[20] Adams R D and Maheri M R 1994 Dynamic flexural properties of anisotropic fibrous composite beams *Compos. Sci. Technol.* **50** 497-514

A numerical procedure for failure mode detection of masonry arches reinforced with fiber reinforced polymeric materials

S Galassi

Department of Architecture, Section of Materials and Structures, University of Florence, Piazza Brunelleschi 6, 50121 Florence, Italy

E-mail: stefano.galassi@unifi.it

Abstract. In this paper a mechanical model of masonry arches strengthened with fibre-reinforced composite materials and the relevant numerical procedure for the analysis are proposed. The arch is modelled by using an assemblage of rigid blocks that are connected together and, also to the supporting structures, by mortar joints. The presence of the reinforcement, usually a sheet placed at the intrados or the extrados, prevents the occurrence of cracks that could activate possible collapse mechanisms, due to tensile failure of the mortar joints. Therefore, in a reinforced arch failure generally occurs in a different way from the URM arch. The numerical procedure proposed checks, as a function of an external incremental load, the inner stress state in the arch, in the reinforcement and in the adhesive layer. In so doing, it then provides a prediction of failure modes. Results obtained from experimental tests, carried out on four in-scale models performed in a laboratory, have been compared with those provided by the numerical procedure, implemented in *ArchiVAULT,* a software developed by the author. In this regard, the numerical procedure is an extension of previous works. Although additional experimental investigations are necessary, these former results confirm that the proposed numerical procedure is promising.

1. Introduction

In recent years innovative materials and techniques [1-6] for strengthening and repairing architectural and monumental heritage constructions [7,8], as well as ruins in the archaeological sites [9], have been widely used. In particular, masonry arches and vaults have been successfully strengthened using composite materials applied at the intrados or the extrados. For the analysis of these structures, the author has recently presented new numerical tools [10-15].

The intervention of strengthening or consolidation made using composite materials is more targeted and less invasive in comparison to the traditional technologies and materials employed in the recent past. Among the advantages of using these materials, it is worth mentioning the following FRP properties: lightness, high resistance, high elastic modulus and stiffness, reduced costs due to the ease of transport and assembly, minimal invasiveness, rapid installment in the structure, reversibility of the intervention.

Composite materials and masonry are very compatible [16] (such as in the case of concrete reinforced by steel) and each of these two materials offers particular performance and strengths in the structure: the masonry supports compressive forces, while tensile forces, which cannot be supported by the masonry, are diverted to the fibers, which are designed specifically to perform this function. The reinforced masonry is, in turn, configured as an innovative material, thus its design must consider all possible failure modes, that may occur, which are different from the standard masonry failure ones. More specifically, the following five states should be verified:

- Equilibrium of the structure,
- Compressive force in the masonry,
- Shear force in the mortar joints,
- Adhesion between the reinforcement and the structure,
- Tensile force in the reinforcements.

According to the limit analysis laws, the following five limits correspond to the above mentioned states and, in order to guarantee safety, they must not be exceeded:

- Collapse mechanism or tensile cracks,
- Crushing of the masonry,
- Sliding of one brick upon another in correspondence of the mortar joints,
- Delamination, often a "rip-off failure", that is a failure of the union between masonry and the reinforcement, on the side of masonry (removal of part of the brick),
- FRP rupture.

According to the strength hierarchy criterion, the design of the reinforcement should be aimed at predicting, choosing and preferring one particular failure mode over the others. Consequently, one can select which type of composite and adhesive is better to use, the cross section of the reinforcement and its placement on the substrate, always relying on the tensile strength of the fibers and compressive strength of the masonry. For the assessment of the behaviour of a masonry arch reinforced with FRP materials, the few numerical procedures available in the literature take into account only some failure modes [17,18]. Therefore, a targeted mechanical model and numerical procedure are herein proposed. The tool discussed below is an extension of previous works [19,20] that, in the current formulation, allows one to detect the most probable failure mode of a reinforced arch.

2. Mechanical model

The mechanical model of a masonry arch (figure 1) is assumed to be composed of a finite number of rigid-blocks assembled by elastic mortar joints. In the mortar joints, all the elasticity (i.e.: deformability) of the arch is concentrated. Mortar joints are modeled, in a discrete way, by a device composed of four links orthogonal to the mid-surface of the joint, plus an additional link arranged along that surface.

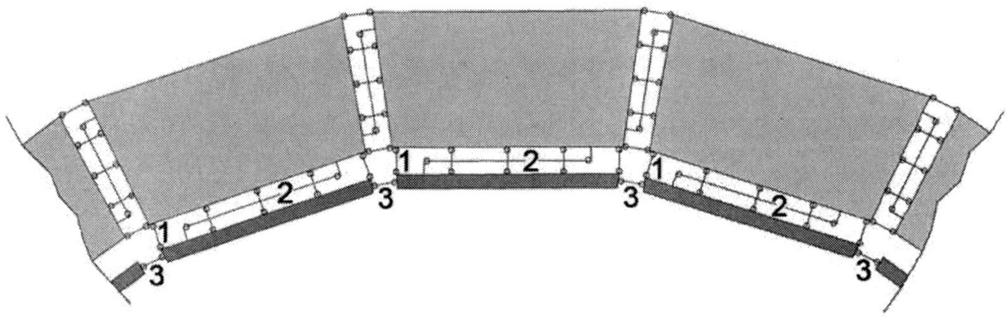

Figure 1. Mechanical model of the reinforced arch: mortar, adhesive and FRP links.

In order to consider the main aspects of the relationship between structure and reinforcement, the latter is assumed to be composed of "frp-type" elements equal to the number of masonry blocks, with their weight, although negligible, linked together and to the structure. The FRP elements are placed at the intrados or extrados of the arch and are also assumed to be rigid.

IOP Conf. Series: Materials Science and Engineering **369** (2018) 012038 doi:10.1088/1757-899X/369/1/012038

The interface which connects fiber with the structure is modeled by a device composed of one set of links, capable of transmitting compressive forces and low tensile forces (type 1 links) and an additional link, capable of transmitting the shear force (type 2 link).

The joint linking one reinforcement element with the next one is composed of one link, capable of transmitting only a tensile force (type 3 link). In type 3 links all the elasticity of the reinforcement is assumed to be concentrated. The first and the last links simulate "external constraints", that is "links with the first structural element unreinforced"; therefore, they represent the so-called "anchorage length".

Type 1 links measure the peeling forces, and they are the tensile forces responsible for the delamination failure which is due to the typical curvilinear shape of the arch (in the case of reinforcements placed at the intrados). However, the reinforcement detachment can also be provoked by the loss of adhesion, which is checked through type 2 links. Indeed, type 3 links, provide, interface by interface, the local tensile force in the reinforcement, and give information about the possibility of its rupture if it exceeds the FRP tensile strength.

3. Numerical procedure

The numerical procedure for the analysis of a reinforced rigid-block arch, modeled in the discrete way described above, is based on the formulation of the set of the equilibrium equations and the elastic-kinematic equations, which are expressed in matrix form in equation (1):

$$\begin{cases} \mathrm{AX} = \mathrm{F} + \alpha F' \\ \tilde{\mathrm{A}}\mathrm{x} + \mathrm{KX} = \Delta \end{cases} \tag{1}$$

where {F} is the vector of the dead loads acting on the structure, {F'} is the vector of the additional loads that are increased by the incremental load factor α, {X} is the vector of the unknown forces in the links of the joints, [A] is the equilibrium matrix whose coefficients are a function of the block shape and the slope of the joints, [$\tilde{\mathrm{A}}$] is the kinematic matrix, {x} is the vector of the displacements of the block centers of mass and [K] is the matrix whose coefficients are populated by the deformability of mortar, adhesive and FRP links. Vector {Δ}, the "impressed distortion vector", is the key to checking the possible aforementioned failure modes, which are caused by the failure of some joint links.

In the hypothesis that for dead loads F and very low additional forces αF' (i.e.: low values of α) the masonry arch and the reinforcement are undamaged, the solution of equation 1 is provided by the linear elastic solution in equation (2):

$$\mathrm{X}_0 = \mathrm{K}^{-1}\tilde{\mathrm{A}}(\mathrm{A}\mathrm{K}^{-1}\tilde{\mathrm{A}})^{-1} \cdot (\mathrm{F} + \alpha \mathrm{F}') \tag{2}$$

Equation (2) is obtained from equation (1) assuming vector {Δ}={0}.

During the analysis the value of α is increased step by step and, in correspondence to each value, the five limit states, listed in the introduction section, can no longer be verified. These limits are expressed by inequalities in equation (3), that are constraints on equation (1):

$$\begin{cases} \mathrm{M}^{(-)} \le \mathrm{X}_{n,j}^{(m)} \le 0 \text{ (on mortar axial links)} \\ \mathrm{X}_t^{(j)} \le \mathrm{f} \cdot \sum_{j=1}^{4} \mathrm{X}_{n,j}^{(m)} \text{ (on mortar tangent links)} \\ 0 \le \mathrm{X}_j^{(1)} \le \mathrm{G}^{(+)} \text{ (on adhesive axial links)} \\ \left|\mathrm{X}_j^{(2)}\right| \le \mathrm{G} \text{ (on adhesive tangent links)} \\ \mathrm{X}_j^{(3)} \le \mathrm{L} \text{ (on FRP links)} \end{cases} \tag{3}$$

According to the five inequalities in equation (3), to predict the failure modes the algorithm introduces a distortion coefficient [21] in the link whose force, at the current step of the analysis, does not respect the corresponding inequality. The distortion provokes the failure of that link. In detail, the distortions act on tense mortar axial links (provoking tensile cracks), on mortar axial links that are compressed more than their compression strength limit expressed by $M^{(-)}$ in equation (3) (provoking the crushing of masonry), on the mortar tangential links whose shear force exceeds the friction force expressed by the second of equation (3) according to the Coulomb's friction criterion [22,23] (sliding of blocks occurrence), on the "adhesive links" type 1 stretched more than the limit $G^{(+)}$ (delamination occurrence), and on "FRP links" type 3 stretched more than the FRP strength L (FRP rupture occurrence). In this framework, the term "adhesive links" is only symbolic and should be properly considered as "masonry links": in fact, it has been demonstrated that delamination does not occur in the adhesive layer but, rather, inside the masonry, with its consequent ripping [24-28]. Therefore, it is more convenient and correct to consider the rupture limit tension $G^{(+)}$ of such links by referring to the brick rupture limit tension rather than to the adhesive one, which is much higher.

The insertion of a distortion coefficient in a link provokes a modification of the inner stress state and therefore, the static solution vector must be recomputed. The solution vector that also takes into account the effect of the distortion is provided in equation (4):

$$X = X_0 + \left(I - K^{-1}\tilde{A}\left(AK^{-1}\tilde{A}\right)^{-1}A\right) \cdot \Delta \tag{4}$$

At the end of the iterative nonlinear analysis, the load-bearing capacity of the reinforced arch, as well as its failure mode, are assessed, providing information on the intermediate failures in correspondence to each value of the incremental loads. In such a way the numerical procedure provides guidance on the more targeted strengthening intervention of an existing masonry arch as a function of the expected serviceability loads.

4. Numerical and experimental results

To assess the reliability of the numerical procedure proposed above, four models of masonry arches reinforced with carbon FRP composites in scale 1:2 have been performed at the Official Laboratory for Testing of Materials and Structures of the Department of Architecture at the University of Florence (Italy). Experimental results have been compared with the results provided by the numerical procedure.

Such models reproduced a segmental arch with a 150 cm spam, an internal radius corresponding to 86.5 cm, with fixed imposts and inclined by 30° and a square cross section corresponding to 10 x 10 cm.

Figure 2. Construction phases: (a) macro-block; (b) assemblage of macro-blocks on the centering; and (c) curing under a layer of wet paper.

Each arch has been built for subsequent phases. First, some "macro-blocks" of 6 bricks have been bonded, following the intrados curve of the arch to be built (figure 2(a)). After 28 days of curing, they have been positioned on the centering and bonded together so as to obtain the whole arch (figure 2(b)). Finally, the arch has been covered with a layer of wet paper and left to cure for 10 days in order to avoid the occurrence of micro-cracks due to drying and shrinkage (figure 2(c)).

The right and left impost sections have not been laid directly on the testing machine but on bricks adequately shaped and placed in such a way as to simulate the heads of two fixed pillars (figure 3(a)).

The experimental trial was conducted by simulating an increasing point force at the keystone using a hydraulic jack which impressed an increasing vertical displacement. In correspondence to a load of 1980 N the first crack at the keystone appeared and then, later, at the haunches at a load of 2630 N. In correspondence to a load of 4320 N the delamination process began, and, starting from the keystone sections, this process involved little by little, the sections near the left haunch removing part of the masonry (figure 3(b)).

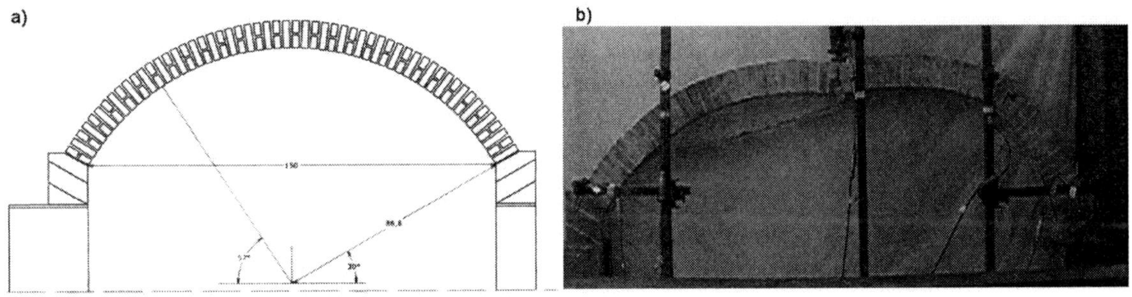

Figure 3. (a) Geometry of arch models performed in a laboratory; and (b) Experimental trial.

In the numerical model the action of the hydraulic jack at the keystone of the arch was simulated inputting two equal point forces at the centre of gravity of both the keystone blocks. During the analysis, these forces were increased little by little.

The numerical model showed (figure 4), step by step, the misshaping delamination process which also occurred experimentally. It highlighted a peak load of the same level as well as the cracking pattern that occurred experimentally, which highlighted both mortar joint failure due to tensile forces and the position of the reinforcement detached in many intrados interfaces.

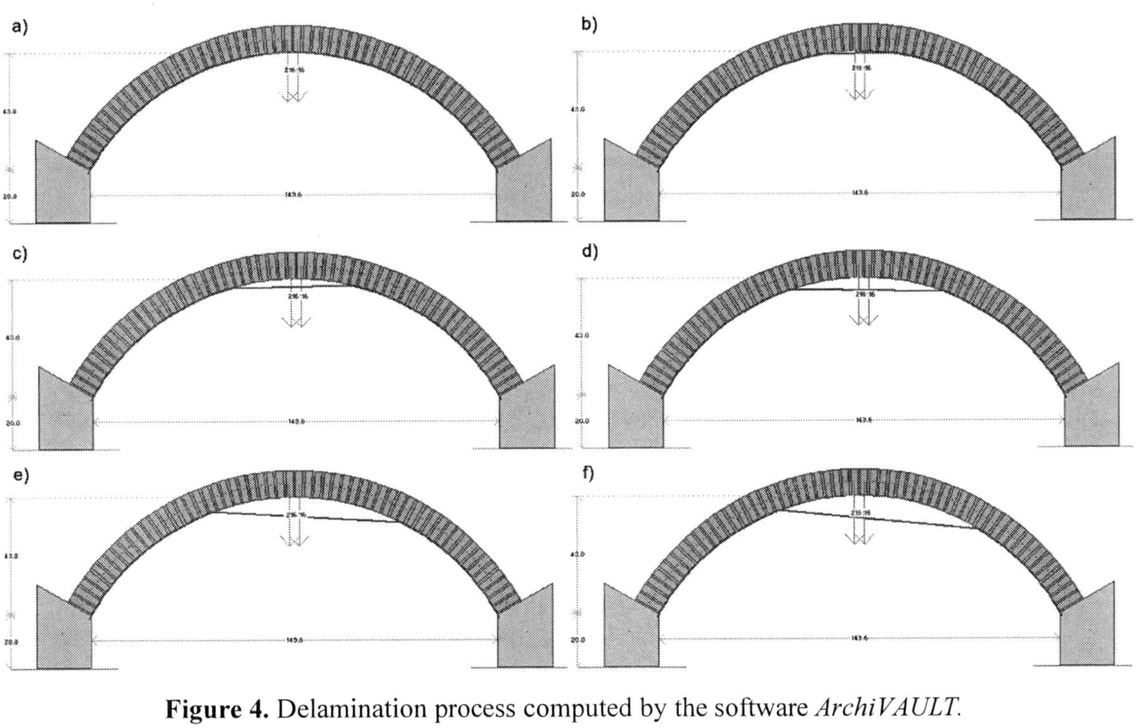

Figure 4. Delamination process computed by the software *ArchiVAULT*.

5. Conclusions

Failure modes of an arch reinforced with FRP materials are different from the typical failure modes of an URM one. For this reason, a nonlinear numerical procedure for the analysis of failure modes, that is an extension of that presented in a previous work, has been described in more detail in this paper and also implemented in the software *ArchiVAULT*, in such a way as to provide a user-friendly design tool for engineers.

In order to provide prediction of failure modes, five limit states have been considered in the proposed procedure. Since the discrete model has been assumed for interpreting the mechanical behaviour of the arch, such limit states are checked in correspondence to the mortar, adhesive and FRP joints.

According to the strength hierarchy criterion, this tool is aimed at providing guidance on a targeted design of the reinforcement system in order to avoid irreversible failure modes, such as masonry crushing, and phenomena that could lead to instant collapse, such as the FRP rupture.

Even if additional investigations should be performed on a higher number of experimental models, the comparison between numerical results and the results of the in laboratory experimental arches seems to confirm the reliability of the procedure proposed.

Acknowledgments

Architect Daniela Sinicropi, that performed the experimental trial in our laboratory, is gratefully acknowledged.

References

[1] Paradiso M, Galassi S, Borri A and Sinicropi D 2013 "Reticolatus": an innovative reinforcement for irregular masonry. A numeric model, *Structures and Architecture: Concepts, Applications and Challenges, Proc. 2nd Int. Conf. on Structure & Architecture ICSA2013,* Paulo J S Cruz (Ed.) (London: CRC Press/Balkema, Taylor and Francis Group), Guimaraes, Portugal

[2] Alecci V, Focacci F, Rovero L, Stipo G and De Stefano M 2017 Intrados strengthening of brick masonry arches with PBO–FRCM composites: experimental and analytical investigations *Compos. Struct.* **176** 898-909

[3] Alecci V, De Stefano M, Focacci F, Luciano R, Rovero L and Stipo G 2017 Strengthening masonry arches with lime-based mortar composite *Buildings* **7** 49

[4] Fagone M and Ranocchiai G 2017 An experimental analysis of the mechanical behaviour of cnchored CFRP-to-masonry reinforcements loaded by out-of-plane actions *Key Eng. Mat.* **747** 204-11

[5] Jurina L 2016 Experimental tests on consolidation of masonry bridges using "ram-reinforced arch method" *Proc. IIX Int. Conf. on Arch Bridges* (5-7 October 2016, Wroclaw, Poland)

[6] Borri A, Castori G and Corradi M 2011 Intrados strengthening of brick masonry arches with composite materials *Compos. Part B-Eng.* **42** 1164-72

[7] Galassi S, Dipasquale L, Ruggieri N and Tempesta G 2018 Andalusian timber roof structure in Chefchaouen, Northern Morocco: construction technique and structural behavior *ASCE's Journal of Architectural Engineering* **24** 10.1061/(ASCE)AE.1943-5568.0000315

[8] Paradiso M, Galassi S and Benedetti S 2014 The peculiar defensive fort of San Fernando de Bocachica, Colombia *Patologia della Costruccion, Tecnologia de la Rehabilitacion y Gestion del Patrimonio, Proc. Latin American Congress on Construction Pathology, Rehabilitation Technology and Heritage Management (Rehabend 2014),* Luis Villegas, Ignacio lombillo, Clara Liano, Haydee Blanco (Eds.), Gráficas Iguña, S.A., Santander, Spain, (in Italian)

[9] Ruggieri N, Galassi S and Tempesta G 2018 Pompeii's Stabian Baths. Mechanical behaviour assessment of selected masonry structures during the 1st century seismic events *Int. J. Archit. Herit.* 10.1080/15583058.2017.1422571

[10] Galassi S, Misseri G, Rovero L and Tempesta G 2017 Equilibrium analysis of masonry domes. On the analytical interpretation of the Eddy-Lévy graphical method *Int. J. Archit. Herit.* **11** 1195-211
[11] Galassi S and Tempesta G 2017 Analysis of masonry block structures with unilateral frictional joints *Proc. of Int. Conf. on Electronic, Control, Automation and Mechanical Engineering (ECAME 2017)* (19-20 November 2017, Sanya, China), *DEStech Transactions on Engineering and Technology Research* 246-51 10.12783/dtetr/ecame2017/18396
[12] Pugi F and Galassi S 2013 Seismic analysis of masonry voussoir arches according to the Italian building code *Ing. Sismica-Ital.* **30** 33-55
[13] Paradiso M, Galassi S and Benedetti S 2014 Stability of masonry arches and vaults: traditional methods and automatic calculation *Patologia della Costruccion, Tecnologia de la Rehabilitacion y Gestion del Patrimonio, Proc. Latin American Congress on Construction Pathology, Rehabilitation Technology and Heritage Management (Rehabend 2014),* Luis Villegas, Ignacio lombillo, Clara Liano, Haydee Blanco (Eds.), Gráficas Iguña, S.A., Santander, Spain (in Italian)
[14] Alecci V, De Stefano M, Galassi S, Lapi M and Orlando M 2018 An assessment of American criterion for detecting plan irregularity *Proc. of 8th European Workshop on the Seismic Behaviour of Irregular and Complex Structures (8EWICS)* (19-20 October 2017, Bucharest, Romania) (in press)
[15] Galassi S, Misseri G, Rovero L and Tempesta G 2018 Failure modes prediction of masonry voussoir arches on moving supports *Eng. Struct.* (in press)
[16] Sinicropi D, Perria E, Galassi S, Paradiso M and Borri A 2014 Artificial aging of mortar prisms reinforced through steel, glass and organic fibers *Key Eng. Mat.*, A Di Tommaso, C Gentilini and G Castellazzi (Eds.) **624** 542-50
[17] Briccoli Bati S, Fagone M and Rotunno T 2013 Lower bound limit analysis of masonry arches with CFRP reinforcements: A numerical method *Journal of Composites for Construction* **17** 10.1061/(asce)cc.1943-5614.0000350
[18] Caporale A, Luciano R and Rosati L 2006 Limit analysis of masonry arches with externally bonded FRP reinforcements *Comput. Method Appl. M.* **196** 247-60
[19] Galassi S 2008 Numerical analysis of masonry vaulted systems reinforced with composite materials (Dept. of Constructions, Florence, Italy)
[20] Galassi S 2018 Analysis of masonry arches reinforced with FRP sheets: experimental results and numerical evaluations *Key Eng. Mat.* (in press)
[21] Colonnetti G 1955 *Scienza delle Costruzioni* (Torino: Edizioni Scientifiche Einaudi)
[22] Coulomb C A 1776 Mémoires de mathématique et de physique présentés à l'Académie Royale des Sciences, par divers Savans, et Ius dans les Assemblées, Paris, Académie Royale des Sciences
[23] Gilbert M and Melbourne C 1994 Rigid-block analysis of masonry structures *Struct. Eng.* **72** 356-60
[24] Yuan H, Teng J C, Seracini R, Wu Z S and Yao J 2004 Full-range behavior of FRP-to-concrete bonded joints *Eng. Struct.* **26** 553-65
[25] Valluzzi M R, *et al* 2012 Round robin test for composite-to-brick shear bond characerization *Mater. Struct.* **45** 1761-91
[26] Rovero L, Focacci F and Stipo G 2013 Structural behavior of arch models strengthened using FRP strips of different lengths *J. Compos. Const.* **17** 249-58
[27] Rotunno T, Rovero L, Tonietti U and Briccoli Bati S 2015 Experimental study of bond behavior of CFRP-to-brick joints *J. Compos. Const* **19** 10.1061/(ASCE)CC.1943-5614.0000528, 04014063
[28] Briccoli Bati S and Rovero L 2008 Towards a methodology for estimating strength and collapse mechanism in masonry arches strengthened with fibre reinforced polymer applied on external surfaces *Mater. Struct.* **41** 1291-306

IOP Conf. Series: Materials Science and Engineering **369** (2018) 012039 doi:10.1088/1757-899X/369/1/012039

Study on rheological properties of CMC/Eu-Tb solutions with different concentrations

Z C Fu[1], J Ye[1] and J Xiong[2,3]

[1]State Key Laboratory of Pulp and Paper Engineering, South China University of Technology, Guangzhou, 510640, China

[2]School of Food Science and Engineering, South China University of Technology, Guangzhou, 510640, China

E-mail: lcjxiong@scut.edu.cn

Abstract. The rheological properties of polymer solution are sensitive to variations in the polymer structure. Carboxymethyl cellulose (CMC) aqueous solution has been used in many fields, such as food, medicine and paper industry. In this paper, the effects of different concentrations (2% - 6%) of CMC/Eu-Tb on their rheological properties were investigeted, including steady-state flow and viscoelastic response. The results show that, the viscosity of CMC/Eu-Tb is lower than that of CMC, at the same concentrations; the products solutions present a nearly Newtonian behavior at the low concentrations (2% - 3%); while at the higher concentrations (4% - 6%), the products solutions present a pseudoplastic behavior; shear-thinning behavior is due to the polymer chains unravel under the action of flow and the molecular chains are oriented in the flow direction. The results also show that the viscosity of the solutions decreases with increasing temperature. Dynamic rheological tests show that CMC/Eu-Tb has viscoelasticity in the concentrations of 2% -6%. At lower concentrations, the elastic modulus G' is slightly higher than the viscous modulus G", and as the concentrations increase, the elastic modulus G' is significantly higher than the viscous modulus G". It means that at the lower solution concentrations, the solutions tend to be less elastic and easier to flow. Most of the energies are lost through the viscous flow. As the solution concentrations increase, the solutions tend to be more elastic, and the system tends to form a gel.

1. Aims

CMC is an important cellulose ether, obtained by carboxymethylation of cellulose. Carboxymethyl cellulose (CMC) has a very wide range of applications [1]. For example, in the food industry, CMC can be used as an emulsifier and stabilizer in ice cream production; in the medical industry, CMC can be used as a tablet binder and film-forming agent, but also a safe and reliable anti-cancer drug carrier; In the paper industry, CMC can be used as coating colors to improve paper properties.

The rheological properties of CMC have been the subject of many studies [2,3]. Mohamed et al. investigated rheological properties of high concentrations of carboxymethyl cellulose solutions [4]. Mamdouh *et al* studied rheological behavior of 1 to 5% carboxymethyl cellulose solutions [5]. Adel Benchabane *et al* investigated the effect of two critical concentrations on the rheological behavior of carboxymethyl cellulose (CMC) solutions [6].

It was confirmed that carboxymethyl cellulose/Tb (III) nanocomposites possess high fluorescence. With their fluorescent properties, the CMC/Tb nanocomposites could be used in such as food packaging, and biomedical devices for potential application [7]. Under ultraviolet light, the green fluorescence is a characteristic emission of Tb^{3+} and the red fluorescence is a characteristic emission of Eu^{3+} [8], so we may change the mixing ratio of Eu, Tb in order to get different colors of complexes. Therefore, CMC/Eu-Tb nano-complexes may have more potential in food packaging, and biomedical devices application.

In this paper, CMC/Eu-Tb nano-complexes were synthesized by reacting Eu^{3+} and Tb^{3+} with carboxymethyl cellulose. The effects of different concentrations (2% - 6%) of CMC/Eu-Tb along with fluorescent properties on their rheological properties were investigeted, including steady-state flow and viscoelastic response. The experimental results of CMC and CMC/Eu-Tb were compared, to investigate the differences of CMC rheological properties reacting with Eu^{3+} and Tb^{3+}.

2. Methods

2.1. The preparation of CMC/Eu-Tb complexes

The CMC has the Degree of Substitution (DS) of 0.81, in food grade, and its 1.0% aqueous solution has a viscosity of 8200 mPa·s at 25℃ [9], (Brookfield viscometer). It was purchased from WEIYI Technology Company in Chang Shu. 1.000 g of CMC was added slowly into a three-necked flask containing 100 mL of deionized water, then the mixture was stirred at room temperature for more than 4 hours until no air bubbles were left. 15ml $TbCl_3$ solution and 15ml $EuCl_3$ solution were taken into a reagent bottle, Both $TbCl_3$ and $EuCl_3$ concentrations were 0.0318 mol/L The mixture was adjusted to pH = 7.0 with NaOH solution and diluted with about 60 mL of deionized water. Then flask containing CMC solution was put on constant temperature (70 ℃) magnetic stirrer. $TbCl_3$ solution and $EuCl_3$ solution were slowly added to the flask for 9h. After that, the suspension was placed in a dialysis bag, repeatedly soaked in dialysis bag with deionized water; the aqueous solution was tested with $AgNO_3$ solution until no white precipitate was obtained. The product was transferred to a watch glass, placed in an oven and dried to constant weight for milling to spare.

2.2. The preparation of CMC/Eu-Tb solutions

Weigh respectively the mass of 0.2, 0.3, 0.4 0.5, 0.6g CMC/Eu-Tb, gradually added to a beaker containing 10 mL of deionized water with stirring. After stirring for 2 hours, the beaker was sealed with plastic wrap, refrigerated and placed in refrigerator overnight.

2.3. Rheological test of CMC/Eu-Tb solutions

Model of rheometer: HAAKE RS600 rheometer

Test fixture: 60 mm parallel plate

Steady-state flow test: the test spacing is 750 μm; the test temperature is from 20℃ to 60℃; maintain for 15 min after pre-shearing; the range of shear rate: 0.01~50 s^{-1}

Frequency scanning test: the test spacing is 500 μm; the test temperature is 25℃; maintain for 5 min after placing samples; the strain is 15%; the range of frequency scanning: 0.01~10 Hz.

3. Results

3.1. Steady-state flow test

In this test, five temperatures of 20, 30, 40, 50, 60 ℃ were selected. At each temperature, the flow behaviors of CMC/Eu-Tb solutions in the concentration range of 2–6% were investigated. The steady-state flow curves of CMC/Eu-Tb measured at various concentrations at different temperatures are shown in figure 1.

As can be seen in figure 1, there is a shear thinning area for each concentration curve. This phenomenon can be explained from the microstructure. When the shear rate is very low, the rate of polymer molecule rewinding is greater than the rate of shear deformation. However, above a certain shear rate, the polymer chains unravel under the action of flow and the molecular chains are oriented in the flow direction, making the rearrangement rate is less than the rate of shear deformation, which leads to shear thinning phenomenon.

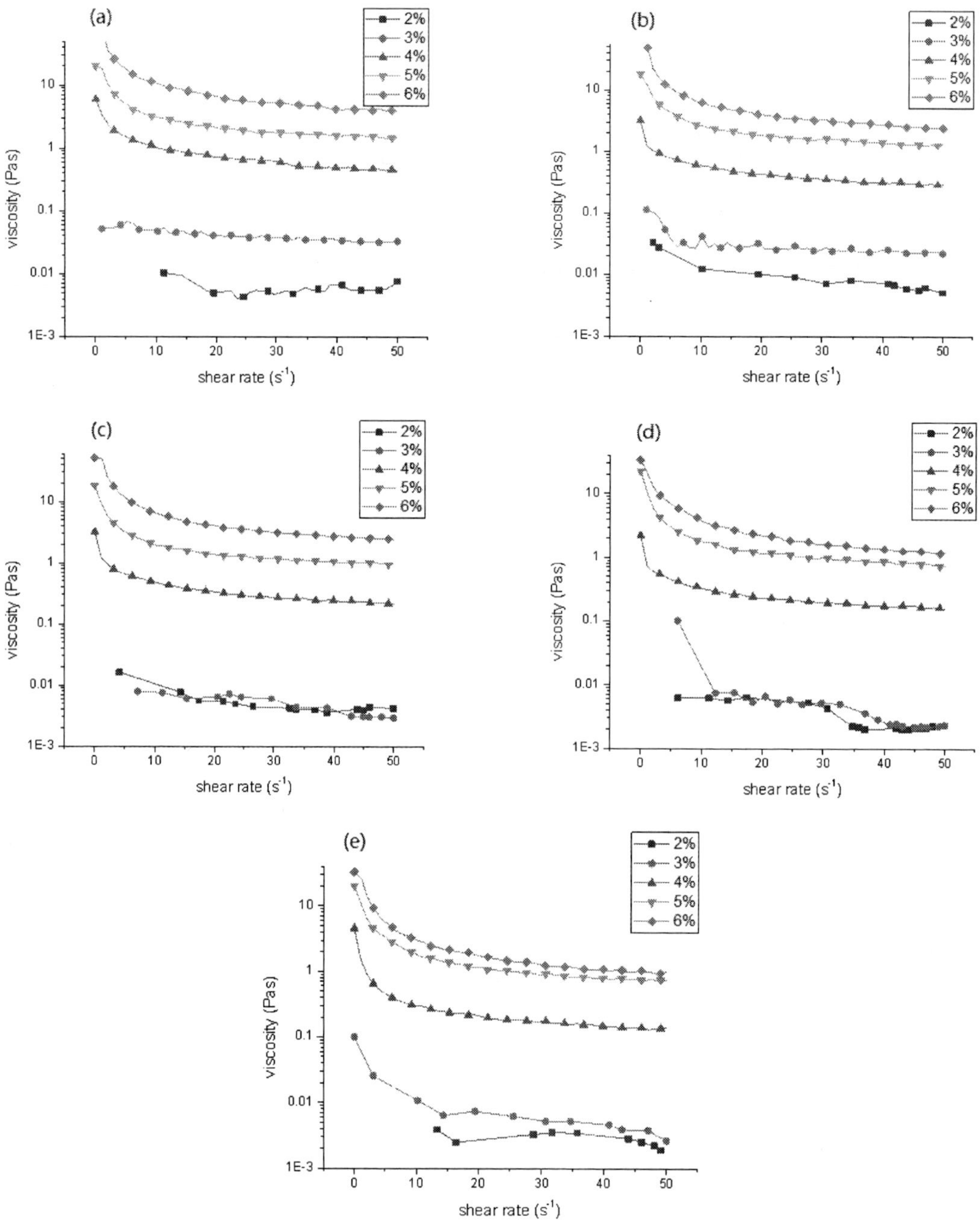

Figure 1. The steady-state flow curves of CMC/Eu-Tb various concentrations at different temperatures. (a)measured at 20℃ (b)measured at 30℃ (c)measured at 40℃ (d)measured at 50℃ (e)measured at 60℃.

The Power-law model fits the relationship between viscosity and shear rate. It can be written as [10]:

IOP Conf. Series: Materials Science and Engineering **369** (2018) 012039 doi:10.1088/1757-899X/369/1/012039

$$\eta_0 = K \dot{\gamma}^{n-1} \quad (1)$$

where η_0 (Pas) is apparent viscosity, $\dot{\gamma}$ (s^{-1}) is shear rate, K(Pas^n) is the consistency index, and n is the flow behavior index. When n=1, it is a Newtonian liquid; when n<1, it is a pseudoplastic fluid; when n>1, it is a swelling fluid.

Equation (1) can be converted to the following:

$$\ln \eta_0 = \ln K + (n-1)\ln \dot{\gamma} \quad (2)$$

Plot using the logarithm of viscosity and shear rate, then n value and K value can be calculated by slope and intercept. The values of n and K of CMC/Eu-Tb at different temperatures are shown in tables 1 and 2.

Table 1. The n value of CMC/Eu-Tb of various concentrations at different temperatures.

T(℃) \ Concentration	n value				
	2%	3%	4%	5%	6%
20	0.8639	0.8042	0.5486	0.5113	0.4715
30	0.7410	0.6441	0.6291	0.5336	0.4773
40	0.7183	0.6693	0.5679	0.5101	0.4301
50	0.8251	0.7897	0.6108	0.4612	0.4194
60	0.6237	0.5010	0.4508	0.4302	0.3659

Table 2. The K value of CMC/Eu-Tb of various concentrations at different temperatures.

T(℃) \ Concentration	K value				
	2%	3%	4%	5%	6%
20	0.0431	0.0728	2.7879	9.7865	33.1254
30	0.0142	0.0842	1.2870	7.8885	20.0554
40	0.0071	0.0104	1.2131	6.3745	23.4530
50	0.0032	0.0051	0.7559	6.2358	12.6962
60	0.0016	0.0182	1.1126	6.6839	12.0155

As can be seen from the tables 1 and 2, 2% and 3% CMC/Eu-Tb solutions present a nearly Newtonian behavior at 20 ℃. As the solutions concentration increases, the flow behavior index decreases sharply from 0.8639 to 0.4715. 4% to 6% CMC/Eu-Tb solutions present a pseudoplastic behavior. At the same temperature, n value decreases with increasing concentration, and K value increases with increasing concentration. However, under the same concentration, n value and K value are not significantly changed with the increase of temperature, but the trend is declining overall.

3.2. Frequency sweep test
Each concentration of CMC/Eu-Tb solutions was doing frequency sweep test. We worked in the linear viscoelastic regime when performing the frequency sweeps. The strain is 15%. The results are shown in figure 2.

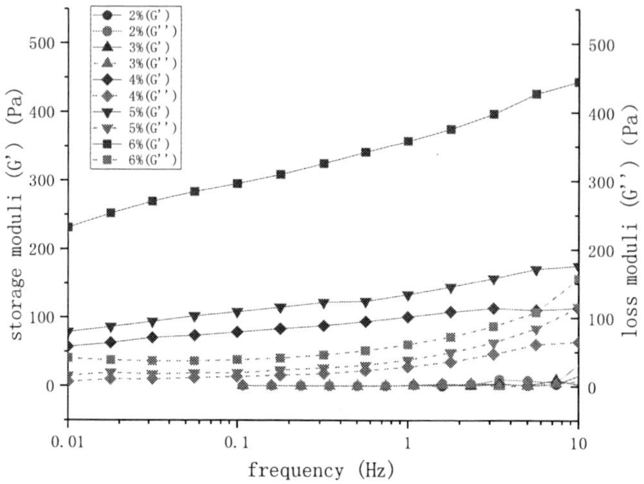

Figure 2. Frequency sweep curves of CMC/Eu-Tb at various concentration.

The results show that in the frequency sweep range, for CMC/Eu-Tb at concentrations of 2% to 3%, the viscous modulus G" is slightly higher than the elastic modulus G'; for CMC/Eu-Tb at concentrations of 4% to 6%, the elastic modulus G' is significantly higher than the viscous modulus G". It means that at the lower solution concentrations, the solutions tend to be less elastic and easier to flow. Most of the energies are lost through the viscous flow. As the solution concentrations increase, the solutions tend to be more elastic, and the system tends to form a gel.

3.3. Comparison of rheological properties of 3% CMC and 3% CMC/Eu-Tb

As can be seen in figure 3, the viscosity of CMC/Eu-Tb is lower than that of CMC, at the same concentrations; In addition, CMC has a lower elastic modulus G' than CMC/Eu-Tb and a higher viscous modulus G" than CMC/Eu-Tb. It is speculated that the degree of polymerization of CMC decreases after reacting with Eu^{3+} and Tb^{3+}, which leads to the phenomenon.

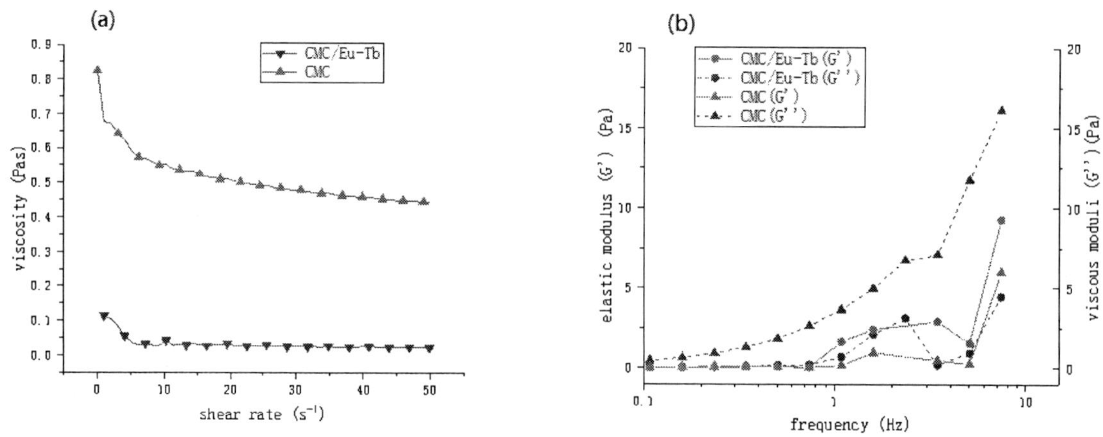

Figure 3. (a) The steady-state flow curves of 3%CMC/Eu-Tb and 3%CMC at 30℃ and (b) Frequency sweep curves of 3% CMC/Eu-Tb and 3% CMC.

4. Conclusions

The viscosity of CMC/Eu-Tb solution decreases with the increase of shear rate. The products solutions present the nearly Newtonian behavior at the low concentrations (2% - 3%); at the higher concentrations (4% - 6%), the products solutions present the pseudoplastic behavior; at the same temperature, flow behavior index decreases with increasing concentration, and consistency index increases with increasing concentration. In the experimental temperature range, the flow activation energy of CMC/Eu-Tb at various concentrations decreases with the increase of shear rate. Compared to CMC, the viscosity of CMC/Eu-Tb decreases.

Acknowledgments

We gratefully acknowledge the financial support provided by State Key Laboratory of Pulp and Paper Engineering (project 2016C12) and the National Natural Foundation of China (project 31270617).

References

[1] Biswal D R and Singh R P 2004 Characterisation of carboxymethyl cellulose and polyacrylamide graft copolymer *Carbohydrate Polymers* **57** 379-87
[2] McCoy S, Zuritz C and Sastry S 1987 *ASAE Paper* No. 87-6536 *ASAE International Winter Meeting* (Chicago, Illinois)
[3] Abdelrahim K, *et al* 1994 Effects of concentration and temperature on carboxymethyl cellulose rheology *Food Science and Technology* **29** 243-53
[4] Mohamed, *et al* 2000 Rheological properties of high concentrations of carboxymethyl Cellulose solutions *Applied Polymer* **79** 1787-801
[5] Mamdouh, *et al* 1997 Rheological properties of carboxymethyl cellulose *Applied Polymer* **64** 289-301
[6] Benchabane A and Bekkour K 2008 Rheological properties of carboxymethyl cellulose (CMC) solutions *Colloid and Polymer Science* **286** 1173-80
[7] Li Q-Y, *et al* 2017 Structures and high fluorescence of novel nanocomposites of sodium carboxymethyl cellulose/Tb(III) prepared at different pHs *Polymer Composites* **38** E498-507
[8] Liu Z L, *et al* 2011 Effect of Eu, Tb codoping on the luminescent properties of Y_2O_3 nanorods *Journal of Luminescence* **131** 12-6
[9] Ye J, Wang B, *et al* 2016 Enhanced fluorescence and structural characteristics of carboxymethyl cellulose/Eu(III) nano-complex: Influence of reaction time *Carbohydrate Polymers* **135** 57-63
[10] Michalis, *et al* 2009 On power-law relationships of the internet topology *ACM SIGCOMM Computer Communication Review* **29** 251-62

Experimental study on the affecting factors of compensating blocks for repairing EDM abrasive electrode

Z B Lin[1], Z N Guo[1], S Y Liang[1], J W Liu[1,4], X Y Yan[2] and H Y Wei[3]

[1]School of Electromechanical Engineering, Guangdong University of Technology, Guangzhou 510006, PR China
[2]State Key Laboratory of Pulp and Paper Engineering, South China University of Technology, Guangzhou 510640, China
[3]Zhongkai University of Agriculture and Engineering, Guangzhou 510225, China

E-mail: fejwliu@126.com

Abstract. Aiming at the problem of electrode wear during EDM, a method of instantaneous repairing EDM abrasive electrode with compensation block was proposed, taking relative mass wear as research index. The results show that diamond abrasive electrode is processed by intermittent spray cooling at 1mm / min feed rate, and the adhesion layer appears on the surface of abrasive electrode. The experimental results show that the relative wear of the electrode with compensation block is lower than that of the electrode without compensation block by comparing the relative wear value of the electrode with different speed, voltage and duty cycle for the 304 stainless steel processed by EDM abrasive electrode 0.7 times. The method of adding compensation block shows that the electrode grinding head can be compensated online to some extent.

1. Introduction

In recent years, as the needs of production and the rapid development of science and technology, the machining accuracy requirements of the work-pieces are also getting higher and higher, especially with the appearance of new materials that are difficult to process. Furthermore, traditional processing methods cannot meet the acquirement of the manufacture of precision, ultra-precision, complicated structures and special shape parts. Thus the special processing method, matches the advantages, gradually becomes the trend of the development of modern processing [1,2]. Among them, Electrical Discharge Machining (EDM) technology, which is one of the most popular processing methods at present and it is in parallel with milling, turning and grinding, has become one of the irreplaceably important processing methods in the modern manufacturing industry because of its unique non-contact electric erosion machining principle and good and stable processing performance. Subsequently, scholars in pursuit of higher EDM machining accuracy and efficiency, put forward the traditional machining methods to EDM process, forming EDM machining, such as EDM grinding [3,4], Ultrasonic EDM [5], Electro-chemical machining [6], EDM milling [7], as well as EDM surface enhancing [8], and so on.

Among them, EDM grinding is developed based on EDM and composite mechanical grinding method. It is a compound processing method, based on impulsive discharge removal material between electrode and work-piece in medium with a certain resistance that the electrode with abrasive grain layer is used as the tool electrode and the grinding machine is used as the main equipment. The rotating electrode in the process of machining is beneficial for the discharge channel to timely eliminate ionization to improve processing efficiency and decentralizes erosion products and discharge points to improve the processing stability.

However, due to the high temperature and high pressure generated during EDM, not only the surface material of the work-piece can be removed, but also the electrode will suffer some wear, which will seriously affect the machining accuracy and shape accuracy [9,10]. Therefore, it is necessary to compensate the electrode wear to reach a certain degree of processing accuracy.

In this paper, in order to research the problem of electrode wear during the time of machining stainless steel by EDM grinding with electroplated diamond grinding electrode, a method of on-line compensation electrode based on aluminum-based compensation block was proposed.

2. Experimental methods and principles

2.1. Experimental principle

The experiment principle diagram is shown in figure 1. Diamond grinding electrode, connected to a pole of the power supply is used as EDM grinding tool electrode to achieve grinding and discharge function. The work piece is fed in normal discharge under the control of the servo controller and returned while abnormal discharge or short circuit. The compensation block is controlled by the motion control loop to achieve feed motion. In the EDM grinding process, the atomized cooling liquid moistens the electrodes and the surface of the compensation block to make the ground debris adhere to the electrode surface. As shown in the compensation area, the debris is adhered to the gap between the abrasive particles of the tool electrode, and with the electrode for rotational movement, the electrode forms metal adhesion layer. When the debris is brought to the processing area, the electrode surface of the debris can compensate for the electrode wear due to spark discharge on the electrode substrate. As the rotation progresses, some of the debris that is not electrically eroded by the spark will fall in the area to be compensated, and then in the compensation area for grinding, the original wear of metal adhesion layer has electrode surface debris compensation again. In order to return after the compensation, wear and then compensate for the processing state.

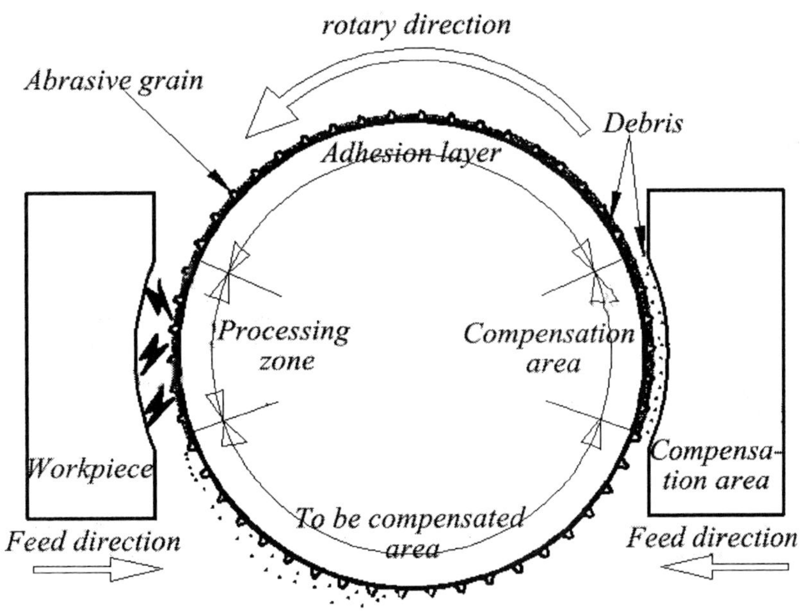

Figure 1. The schematic of experiment processing.

2.2. Experimental equipment and materials

The experiment is carried out on a self-designed test platform, as shown in figure 2. The test uses 6061 Aluminum as a grinding compensation block, which is fixed on the fixture. The fixture, mounted above the screw rail, will have radial movement relative to the abrasive electrode in the machining process through the command of controller. As the figure 3 shows, abrasive electrode whose grinding head is made by 46 mesh galvanized diamond, is mounted on the spindle for rotational movement. As shown in figure 4, the intermittent spraying method using tap water as medium is employed for cooling. Not only can this method cool the heat generated during the grinding process but also make the grinding aluminum scrap effectively adhere to the electrode.

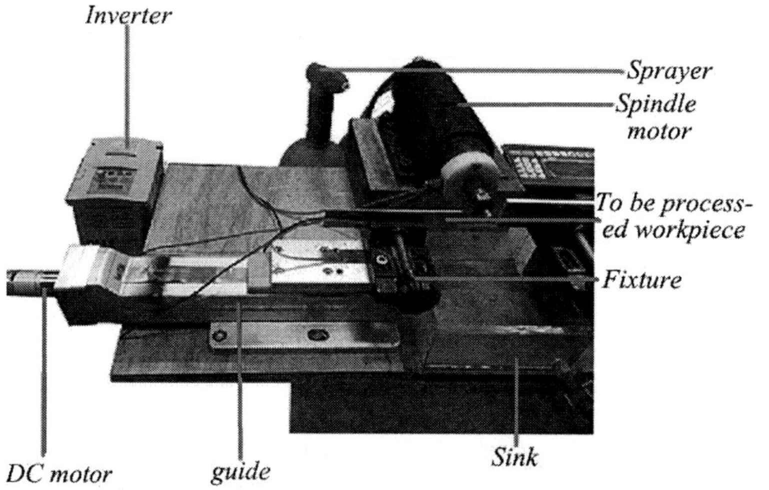

Figure 2. The electric spark grinding platform.

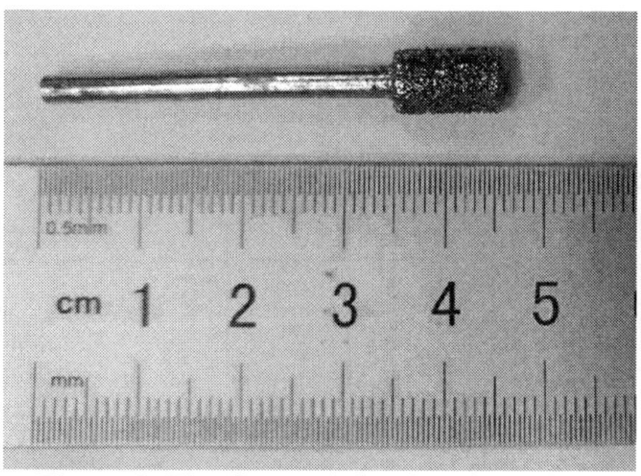

Figure 3. The diamond abrasive electrode.

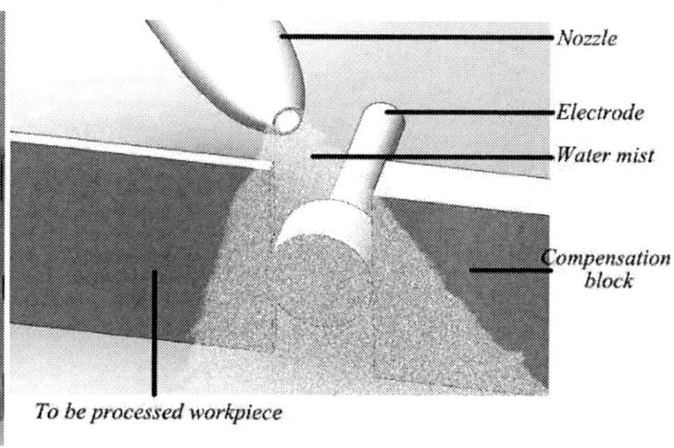

Figure 4. The intermittent spray diagram.

2.3. Tool electrode relative wear ratio

In the process of the EDM, the tool electrode wear rate is an important process index, which includes absolute wear rate v_{Em} and relative wear rate θ_m. v_{Em} refers to the wear rate of tool electrode in unit time and θ_m refers to the percentage of the absolute wear rate(v_{Em}) of the tool electrode and the removal rate(v_w) of the work-piece; the relationship among them can be described as the following equation:

$$v_w = V/t\ (mm^3/min) \tag{1}$$

$$v_{Em} = m/t\ (g/min) \tag{2}$$

$$\theta_m = v_{Em}/v_w \times 100\% \tag{3}$$

Where, V Indicates the volume of the machining work-piece, which can be measured by machining a through hole; m indicates the quality of electrode wear during processing, which can be measured with an electronic analytical balance.

3. Results and discussion

3.1. Effect of spindle speed on relative wear of electrode

Table 1 demonstrates the experimental processing parameters and the experimental results are shown in table 2.

Table 1. The experimental conditions.

Rotating speed/rpm	720	1440	2160
Voltage/V	75		
Pulse width/ms	50		
Duty cycle/%	50		
Peak current/A	3.5		
Feed speed mm/min	56		
Machining time/min	1		
Temperature/℃	Room temperature		
Voltage waveform	Square wave		
Medium	Tap water		
Work-piece	Compensation block (6061Al), work-piece (304 stainless steel)		
electrode/mm	Diamond abrasive electrode (Φ6*10)		

Table 2. The experimental results.

Compensation block	Rotating speed/rpm	Relative wear θ_m				Mean E	Standard deviation
		θ_m1	θ_m2	θ_m3	θ_m4		
	720	0.159	0.173	0.174	0.184	0.172	0.0103
N	1440	0.151	0.161	0.152	0.145	0.152	0.0069
	2160	0.149	0.156	0.155	0.163	0.156	0.0057
	720	0.148	0.152	0.1438	0.132	0.143	0.0085
Y	1440	0.094	0.102	0.097	0.094	0.097	0.0038
	2160	0.106	0.113	0.111	0.115	0.111	0.0042

Figure 5 depicts the relative wear of grinding head electrode with the changes of voltage. It can be seen from the figure that with the increase of rotating speed, the relative wear of the electrode decreases first and then increases no matter whether EDM grinding employs compensating block or not. Along with the increase of the rotation speed, the changes of relative wear value of the electrode without compensation block changes are not particularly obvious. While the electrode with the compensation block has the lowest electrode relative wear value when the rotation speed is 1440 rpm. At this speed, the relative wear value is reduced about 0.6 times when compared with EDM grinding without compensating block. This is mainly because the low speed electrode will beat at a certain degree, making the grinding force from time to time poly increase and the processing unstable. At the same time, the diamond in the electrode is also impacted to varying degrees, resulting in diamond particles off. In addition, when the rotation speed is large, the grinding force increases due to the overall grinding force, so that the relative wear of the electrode will increase slowly. On the one hand, the compensation block limits a side of the electrode reducing comparatively jump, so the relative wear at low speed is also reduced. On the other hand, after the electrode has been filled with the compensating pieces, the electrode substrate originally electrocuted is covered and protected by the crumb layer, thus the relative wear of electrodes is reduced.

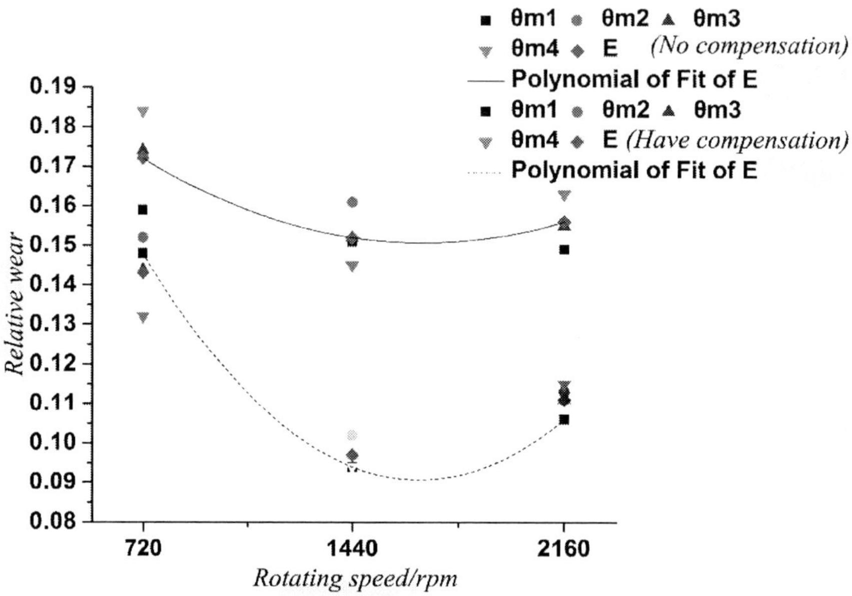

Figure 5. The effect of rotational speed on relative wear of electrodes.

3.2. Effect of voltage on relative wear of electrodes

The rotating speed was set as 1440 rmp, the voltage was set to 75 v and 90 v respectively and the other parameters were the same as those in table 3. The experimental results were shown in table 3.

Table 3. The experimental results.

Compensation block	Voltage/V	Relative wear θm				Mean E	Standard deviation
		θ_m1	θ_m2	θ_m3	θ_m4		
N	75	0.150	0.149	0.152	0.157	0.152	0.0034
	90	0.169	0.162	0.166	0.166	0.166	0.0028
Y	75	0.094	0.101	0.097	0.097	0.097	0.0027
	90	0.096	0.104	0.099	0.098	0.099	0.0036

Figure 6 describes the wear of the electrode relative wear with voltage changes. As the figure illustrates that with the addition of voltage, the relative wear of electrodes increases both in the absence of compensation blocks and with compensation block for EDM grinding. In addition, the wear electrode with the compensation block shows the lowest relative wear of the electrode at the voltage of 75 V, which is about 0.5 times lower than the relative wear of the uncompensated block at this voltage. In addition, the value of electrode wear after adding compensation block is slower than that without compensation block. It is mainly because during the process of EDM grinding with the voltage increases, the ability to remove the material and the electrode wear are improved, resulting in the electrode relative wear increases. Furthermore, after the compensation block is added, the debris layer ground adhered to the electrode surface partially acts as a part of the electrode surface that is removed, so that the relative wear of the electrode increases more slowly. What's more, it can be seen that the relative wear values of the electrodes of 75 V and 90 V are the same.

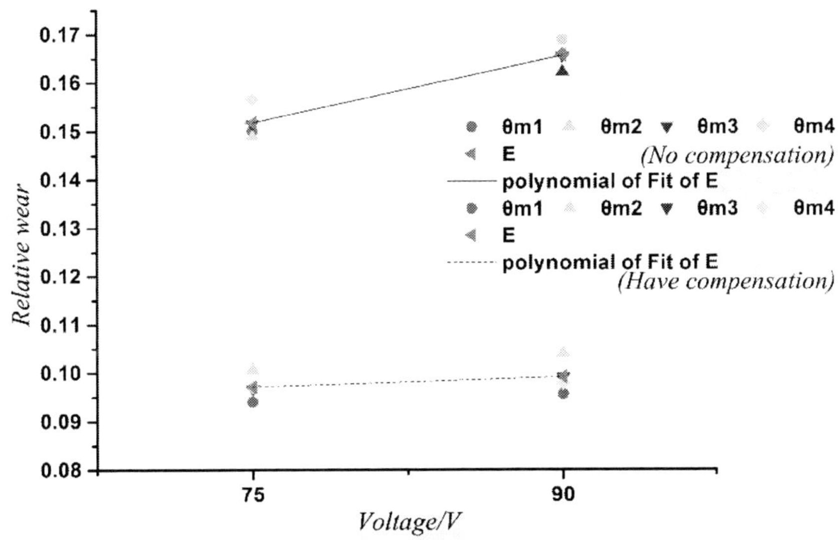

Figure 6. The effect of voltage on relative wear of electrodes.

3.3. The effect of duty cycle in relative wear of electrodes

The voltage was set as 75 v, the rotating speed was set as 1440 rpm, and the duty cycle was set to 20%, 35% and 50% respectively. The data results were shown in table 4.

Table 4. The experimental results.

Compensation block	Duty cycle /%	Relative wear θ_m				Mean E	Standard deviation
		θ_m1	θ_m2	θ_m3	θ_m4		
N	20	0.199	0.210	0.203	0.197	0.202	0.0056
	35	0.165	0.161	0.167	0.176	0.167	0.0061
	50	0.161	0.153	0.156	0.154	0.156	0.0036
Y	20	0.102	0.128	0.107	0.114	0.113	0.0113
	35	0.099	0.092	0.093	0.103	0.097	0.0052
	50	0.100	0.086	0.092	0.090	0.092	0.0062

Figure 7 describes the relationship of the wear of the electrode relative wear and duty cycle. It can be seen from the figure, whether to add compensation block for EDM grinding or not, the relative wear of the cut electrode would tend to decrease with the increase of duty cycle. The wear electrode with compensation block shows a relative wear value of the lowest electrode at a duty cycle of 50%, which is about 0.7 times lower than the relative wear value of the uncompensated block under this duty cycle.

It can be seen that the curve of the relative wear about electrode without employing the compensation block to conduct EDM grinding decline more significantly than that of using the compensation block. In the case of small duty cycle, the effect of relative wear of EDM on the electrode may be greater than that of diamond particle grinding. Therefore, with the decrease of pulse interval, the electrical corrosion between the discharge gaps is increased and more electric erosion products are spattered to the electrode to compensate for the electrode so that the relative wear is reduced.

However, when employing the compensation block for EDM grinding at small duty cycle, the effect of diamond grain grinding on the relative wear of the electrode may be larger than that of EDM, which leads to the larger relative wear value of the electrode. While as the duty cycle increases, the processing energy per unit time increases, which will make the work-piece materials adhere to the electrode surface after electric erosion. Therefore, the relative wear of electrode changes little as the pulse interval decreases.

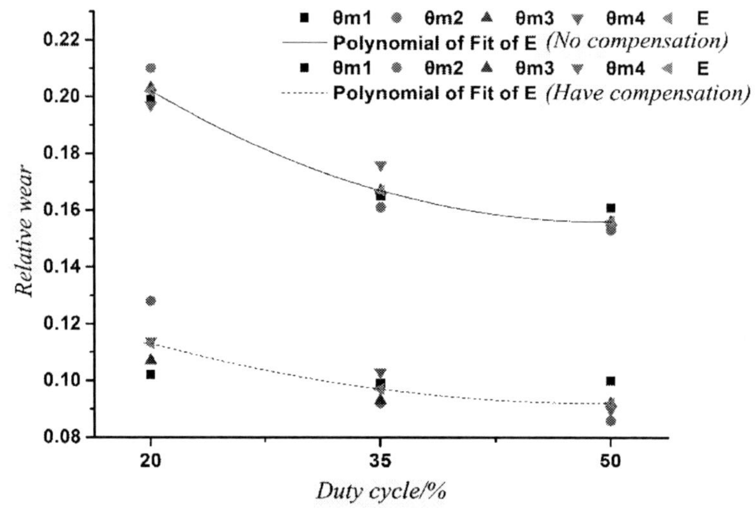

Figure 7. The effect of duty cycle on relative wear of electrode.

4. Conclusion

With the increase of rotational speed, the relative wear of the electrode first decreases and then increases, that is because it is very difficult to form an adhesive layer, whether it is low speed or high speed. The relative wear of the electrode changes slightly with the increase of the voltage, but it shows an increasing trend, as the compensation block is added; the debris layer adhered to the surface of the grinding electrode acts as a part of the removal of the electrode surface, so that the relative wear rate of the electrode increases slowly. As the increase of the duty cycle, the relative wear rate of the electrode shares a downward trend, which because with the duty cycle increases, the energy processed per unit of time increases, causing the removed material to be adhered to the surface of the electrode.

Acknowledgments

This research was supported by The Special Support Plan of the Guangdong province (2014TQ01X542) and the Fundamental Research Funds for the Central Universities (2015ZZ080) and Guangdong Natural Science Foundation of China (2017A030310587) and Science and Technology Planning Project of Guangdong Province 2016A010102017 and the Natural Science Foundation of Guangdong province (2017A0330313330).

References

[1] Pramanik A 2014 Developments in the non-traditional machining of particle reinforced metal matrix composites *Int. J. Mach. Tools Manu.* **86** 44-61
[2] Müller F and Monaghan J 2001 Non-conventional machining of particle reinforced metal matrix composites *J. of Mater. Process Tech.* **118** 278-85
[3] Brian K R, Albert J S, Ronald O S, Jeffery L A, Darryl J G and Marion B G 2002 Wire electrical discharge machining of metal bond diamond wheels for ceramic grinding *Int. J. Mach. Tools Manu.* **42** 1355-62
[4] Balaji P S and Vinod Y 2013 Three dimensional thermal finite element simulation of

electro-discharge diamond surface grinding *Simul. Model Pract. and Th.* **35** 97-117

[5] Kurniawan R, *et al* 2017 Measurement of burr removal rate and analysis of machining parameters in ultrasonic assisted dry EDM (US-EDM) for deburring drilled holes in CFRP composite *Measurement* **110** 98-115

[6] Cheng C P 2010 Efficiency improvement of electrochemical discharge machining by a magnetic field-assisted approach *Non-traditional Machining Society of CMES. Proc. Int. Conf. on Electromachining* (Shanghai: Shanghai Jiao Tong Univ Press) **5** pp 471-5

[7] Segon H 2010 Simulation based off-line tool wear compensation in micro-EDM milling processes *Non-traditional Non-traditional Machining Society of CMES. Proc. Int. Conf. on Electromachining* (Shanghai: Shanghai Jiao Tong Univ Press) **5** pp 611-5

[8] Marashi H, Sarhan A A D and Hamdi M 2015 Employing Ti nano-powder dielectric to enhance surface characteristics in electrical discharge machining of AISI D2 steel *Appl. Surf. Sci.* **357** 892-907

[9] Nguyen V Q, Duong T H and Kim H C 2015 Precision micro EDM based on real-time monitoring and electrode wear compensation *Int. J. Adv. Manuf. Tech.* **79** 1829-38

[10] Wang C, *et al* 2009 Research on Tool Wear Compensation in Micro-EDM *Electromachining & Mould*

Development of 3D woven cellular structures for adaptive composites based on thermoplastic hybrid yarns

C Sennewald[1,3], M Vorhof[1], P Schegner[1], G Hoffmann[1], C Cherif[1], J Boblenz[2], M Sinapius[2] and C Hühne[2]

[1]Institute of Textile machinery and High Performance Material Technology, Technische Universität Dresden, D-01062 Dresden, Germany
[2]Institute of Adaptronics and Function Integration, Technische Universität Braunschweig, Langer Kamp 6, D-38106 Braunschweig, Germany

E-mail: Cornelia.sennewald@tu-dresden.de

Abstract. Flexible cellular 3D structures with structure-inherent compliance made of fiber-reinforced composites have repeatedly aroused the interest of international research groups. Such structures offer the possibility to meet the increasing demand for flexible and adaptive structures. The aim of this paper is the development of cellular 3D structures based on weaving technology. Considering the desired geometry of the 3D structure, algorithms are developed for the formation of geometry through tissue sub-areas. Subsequently, these sub-areas are unwound into the weaving level and appropriate weave patterns are developed. A particular challenge is the realization of compliant mechanisms in the woven fabric. This can be achieved either by combining different materials or, in particular, by implementing large stiffness gradients by means of varying the woven fabrics thickness, whereas differences in wall thickness have to be realized with a factor of 1:10. A manufacturing technology based on the weaving process is developed for the realization of the developed 3D cellular structures. To this end, solutions for the processing of hybrid thermoplastic materials (e.g. tapes), solutions for the integration of inlays in the weaving process (thickening of partial areas), and solutions for tissue retraction, as well as for the fabric pull-off (linear pull-off system) are being developed. In this way, woven cellular 3D structures with woven outer layers and woven joint areas (compliance) can be realized in a single process step and are subsequently characterized.

1. Introduction

Flexible, flexible 3D structures made of fiber-reinforced plastic composites (FRP) with defined stiffness have aroused the interest of international research groups. By directly manufacturing such structures, significant advantages can be achieved in terms of weight reduction, increased efficiency and flexibility. The existing limitations in terms of structure design can be significantly extended by implementing flexible 3D structures with variable shapes. In the aerospace sector in particular, the development goals include increasing agility and aerodynamic efficiency. Despite the high potential of adaptive structures, the advantages of flexible structures are often more than compensated for by additional peripheral or system-related efforts. One approach to avoid the contradiction between structural flexibility on the one hand and stiffness and strength on the other hand within the same structure are pressure-activated cell and honeycomb structures. This also includes pressure-actuated cellular structures (PACS, cf. Figure 1), which have so far only been investigated theoretically [1],[2].

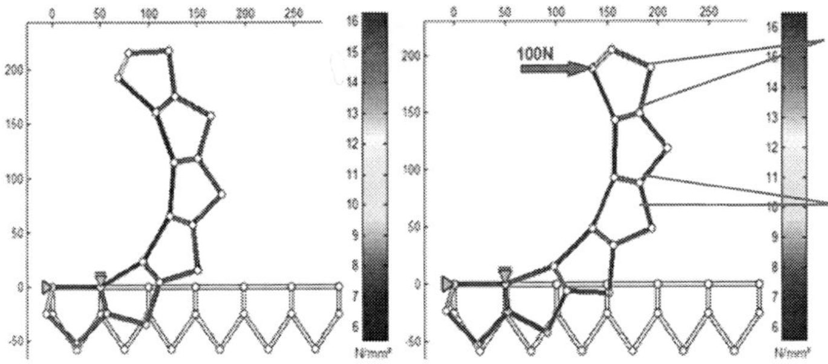

Figure 1. Geometry change of PACS [1].

The aim of this project is to develop theoretical and experimental basics for the development, production and testing of woven PACS. The functionality of the concept is equally guaranteed in the dimensional range between centimeters and meters and can be guaranteed by appropriate geometry and material selection. For the development of the necessary cellular structures as fiber reinforced composite (FRP), textile manufacturing technologies in particular offer numerous solutions for the effective integral production of the necessary 3D semi-finished products [3]-[4]. Since the PACS are characterized by a predominantly biaxial property profile, high stiffness in one direction and low stiffness (flexibility) in the other direction, biaxial structures are particularly advantageous. Weaving is predestined for transverse and longitudinal reinforced biaxial structures and semi-finished products for FRP [5]-[10]. So far, there have been two possible solutions for the integral production of closed-cell woven structures. Several layers of fabric are produced, which are interwoven in line at defined points. The formation of cells between the layers creates a closed-cell honeycomb structure [11],[12]. The main disadvantage of this technology is that no straight yarn layer can be arranged at right angles to the honeycomb direction, which means that the structure has only limited tensile strength and is only suitable to a limited extent for the development of PACS. The second approach uses the woven spacer fabrics developed at the TU Dresden within the framework of Priority Program 639 [13],[14]. This closed-cell structure is integrally woven and possesses biaxial arranged stretched yarn layers. However, the curved structure necessary for PACS is missing. The spacer fabrics have so far been developed for use as lightweight panels that are designed for maximum rigidity and strength throughout the entire structure. For this reason, spacer fabrics were designed homogeneously and with a high thickness, especially in the wall thicknesses. Compliant mechanism integrated in the textile structure, due to very large differences in wall thickness are missing. This does not in any way meet the requirements of PACS. For the development of PACS, integrally produced 3D woven fabrics that combine several characteristics are missing. The production of cellular structures whose wall thicknesses can be adapted to locally variable bending stiffness is not yet known. In addition to the necessary development of the technology and the fabric structures, comprehensive knowledge of theoretical structure development and structure design is also lacking. Closed cellular structures are required for the production of PACS. There are numerous solutions and applications for such structures in which the cells are manufactured and bonded using assembly technology. The almost exclusively used adhesive joints do not meet the requirements of the PACS with regard to strength and reduced bending stiffness, especially at the junctions. Therefore, the focus in the following will be on the consolidation of integrally manufactured cellular semi-finished textile products. Due to their positive matrix properties, thermoplastic FRPs are particularly suitable for the flexible design of the PACS. In addition, the challenge with thermoset FRP is to impregnate the textile semi-finished products in the closed cells. With thermoplastic FRP, the thermoplastic matrix can already be added to the semi-finished product as a thermoplastic fiber in the textile process. These solutions, which were also developed at ITM and are primarily based on the use of hybrid yarns (high-performance and thermoplastic fiber blends), are established. Through the targeted use of hybrid yarns with different blending ratios, which can be adjusted in a defined manner, FRP with graded fiber volume contents and corresponding graded bending properties can also be developed in principle. No investigations have been carried out to date. For the successful implementation of the PACS, solutions are lacking which, above all, can execute the necessary different wall thicknesses with high accuracy and thus ensure a defined bending stiffness. The particular challenge here is to keep the wall thicknesses for bending flexibility to a minimum, especially at the knots. Tried and tested solutions lead to resin accumulation, especially in the knot area.

2. Materials and methods

2.1. Yarn material

As already described, the use of hybrid materials for the weaving of cellular 3D structures is absolutely essential. These hybrid materials consist of a reinforcing component (glass fiber) and a matrix component (thermoplastic fiber). In a thermoforming process, the thermoplastic fiber is melted to form the matrix of the FRP. Only the use of hybrid yarns can ensure even distribution of the matrix material and reliable consolidation of the component in complex structures such as the PACS. In addition, thermoplastic matrices offer the required mechanical properties for the PACS. Two different hybrid yarn materials are used for the development of cellular 3D structures. For a reliable processing of hybrid yarns in the weaving process, it is necessary to either twist or pre-consolidate the hybrid yarns. This is the basis for the selection and development of the yarns. First hybrid yarn was produced with a fineness of 900 tex consisting of glass fiber 2 x 300 tex (GF, 67% content in weight, 46% fiber volume content) and polyamide 1 x 300 tex (PA, 33% content in weight, 54% fiber volume content) with 20 twists per meter. By twisting, the yarn surface is compacted, thus ensuring optimum yarn processing during the weaving process. TecTape with a yarn count of 1800 tex consisting of glass fiber (GF, 67% content in weight, 46% fiber volume content) and polyamide (PA, 33% content in weight, 54% fibre volume content) is used as a further material. The TecTape consists of pre-consolidated spread multifilament yarns, with 2 x Pa with 300 tex fineness on the outer sides and GF Multifilament with 1200 tex in the middle. The TecTape has a tape-like shape with a width of 6-8 mm and a thickness of 0.2 mm and is therefore predestined to produce very thin fabrics.

2.2. Development of algorithms for the geometry formation of cellular 3D structures for PACS

The geometry of the PACS has to be converted into surface elements for the weaving implementation of the cellular 3D structures. This can be done by reshaping 2D layers or by an integral production of the complete 3D geometry in the weaving process. The geometry of the PACS to be achieved is shown in Figure 2.

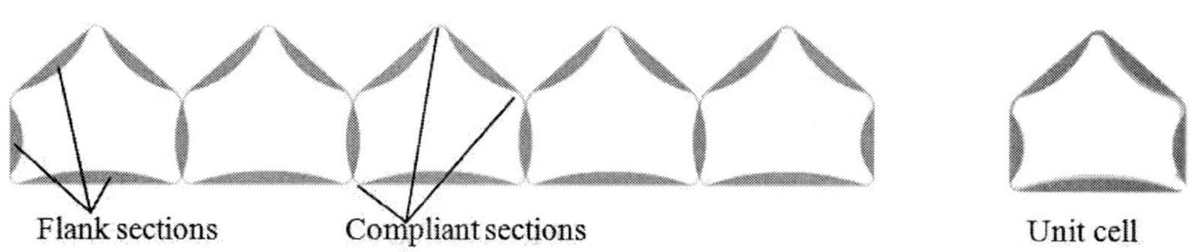

Figure 2. 3D geometry of PACS.

The PACS consist of unit cells which are lined up in the direction of production and which are divided into subareas. In order to generate the overall geometry and for the technical implementation of the weave, the subareas are reassembled periodically. Using combinatorics, the subareas are combined to form unit cells. For the realization of the cellular 3D geometry with a fabric layer, there is exactly one solution. Using 2 layers of fabric, seven different solutions are available. With 3 layers of fabric, 12 different solutions and with 4 layers of fabric 14 solutions can be determined. The solutions for geometry formation are then examined with regard to the fulfilment of the criteria resulting from the application. The criteria are: continuous fabric plies, constant wall thicknesses, continuous fabric plies in the flanks, minimum number of plies in the joint area and the minimum number of plies required. Figure 3 shows a selection of different solutions for composing the geometry of the PACS by fabric layers. The colored lines mark the respective fabric layers. At least at the intersection points of the lines, the subareas are connected to each other in a way that makes them compatible with each other. A variant that meets all the evaluation criteria is, for example, the solution consisting of four partial areas in Figure 3 at the bottom right.

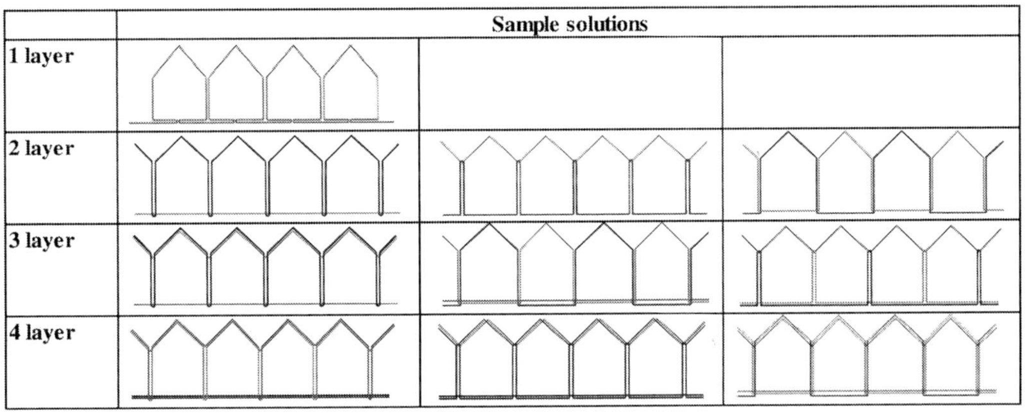

Figure 3. Sample solutions for PACS composed of sub-areas.

3. Results and discussion

3.1. Weave structure development for sections of the PACS

The flanks and compliant sections of the PACS require a separate examination and development. On the one hand, the flanks must be sufficiently rigid (stiffness of the entire PACS) and on the other hand, the compliant sections must be flexible. For this reason, the flanks and the compliant mechanism must show great differences in wall thickness in the individual areas. For this purpose, the technical weaving possibilities (thickening via warp and/or weft threads, inserts) were examined and evaluated. The formation of the compliant sections is achieved by the use of tape yarns and by varying the weave pattern. Without using any trimming the flanks can only be achieved by means of pre-consolidated inserts consisting of several layers used material. The inserts are made by stacking single layers of woven fabric, consolidating them in a thermos press process and cutting into desired size. The processing of inserts has to be taken into account when developing the weaving technology for the whole PACS. The weave pattern design for compliant section and flank section with pre-consolidated inserts is shown in Figure 4.

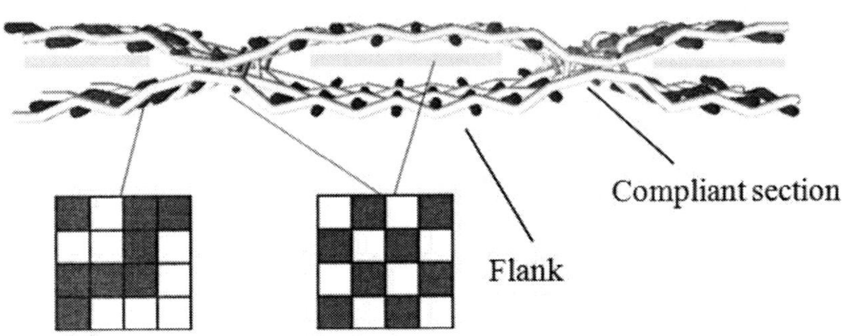

Figure 4. Structure and weave pattern design of compliant and flank section of PACS.

IOP Conf. Series: Materials Science and Engineering **369** (2018) 012041 doi:10.1088/1757-899X/369/1/012041

3.2. Methodology for weaving of PACS composed of partial areas

The technical principles of terry and spacer weaving have to be combined for the technical implementation of PACS by means of weaving. Based on the layer arrangement and the resulting connection points, all parts are transferred to a 2D weave arrangement. The gaps resulting from transfer to the plane are filled with floats (see Figure 5).

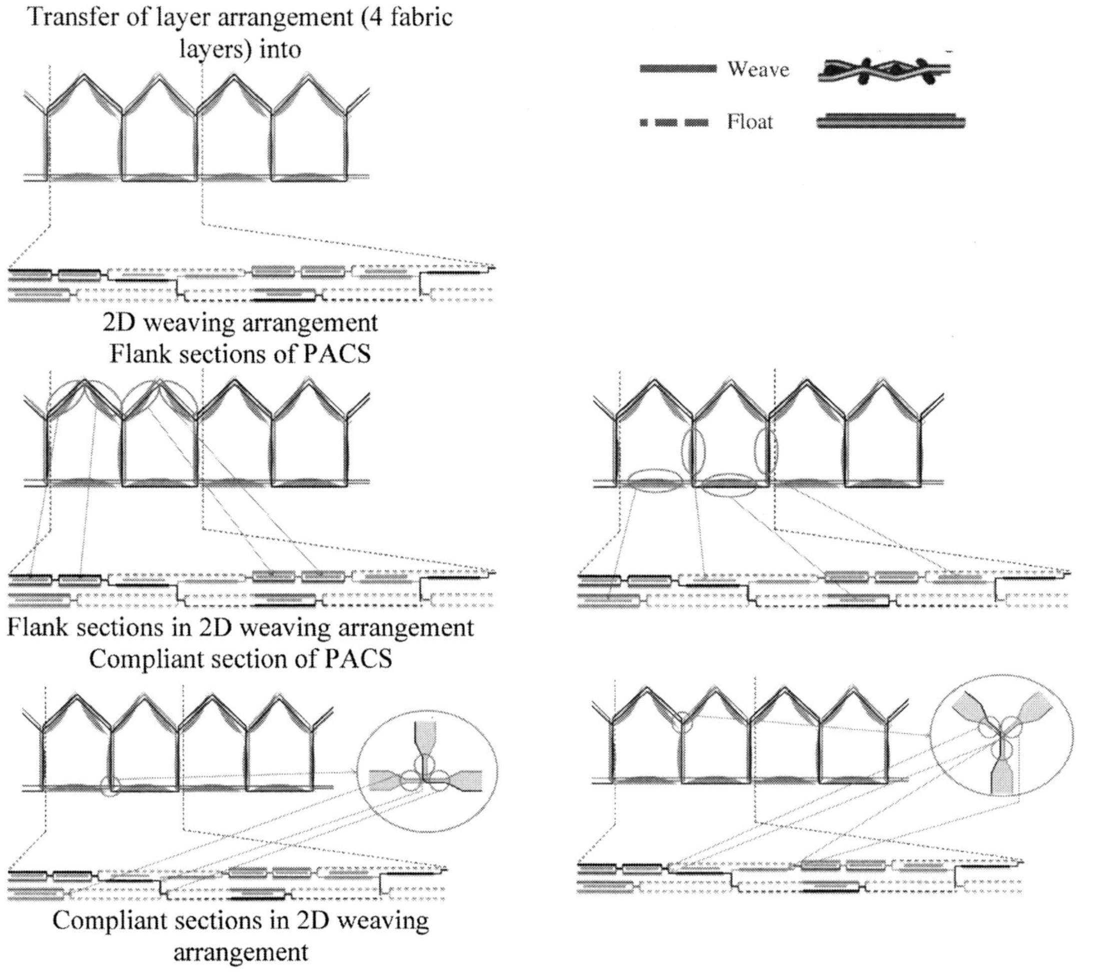

Figure 5. Transfer of layer arrangement into 2D weaving arrangement for weaving of PACS composed of partial sections.

The 2D weaving arrangement again is determined into individual sections, whereas weave patterns for each section must be developed on the basis of the weave patterns developed in 3.1 (Figure 6). In particular, the arrangement of the individual layers on top of each other, their intersection and the position of the floats must be determined. By merging the weave patterns for the individual sections, the total weave pattern for PACS can be developed and the control of the weaving machine can be created.

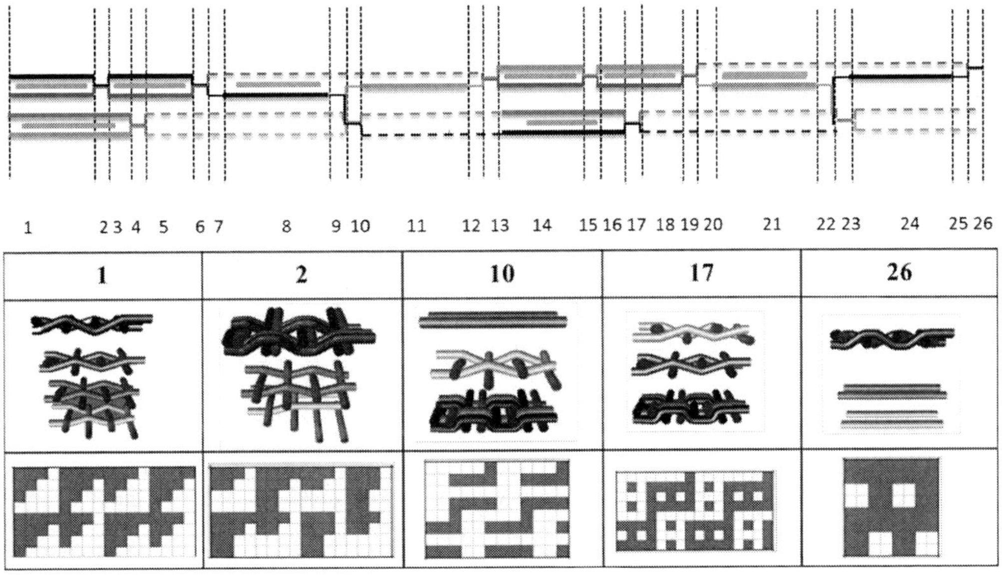

Figure 6. Subareas of 2D weaving arrangement and cross section as well as weave pattern for selected sections.

3.3. Weave trials

The process ability of the hybrid yarns in the weaving process and the realization of the sections of the PACS were carried out during practical weaving and consolidation tests (see Figure 7). It was shown that structures with large gradients of wall thickness can be produced in the weaving process. By using suitable tools, and thermos press processes the woven structures can be consolidated into composite compliant structures.

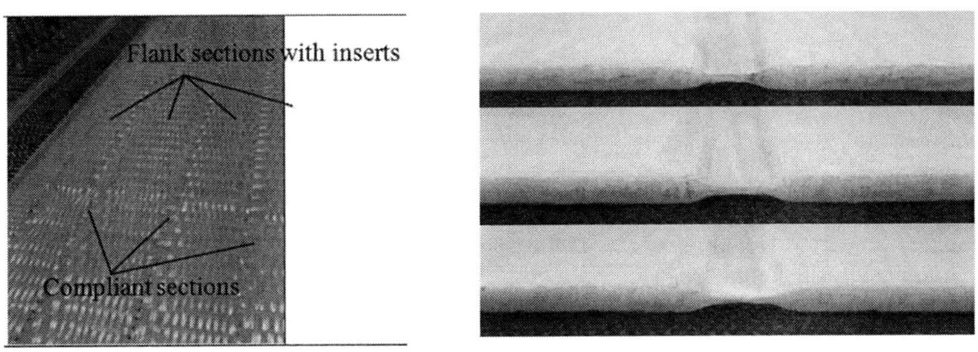

Figure 7. Weaving of compliant and flank sections of PACS using hybrid yarns and consolidated structures with different geometries of compliant section.

4. Summary and outlook

The use of textile reinforced composites in automotive, mechanical and plant engineering sectors is steadily increasing. This is accompanied by an increasing demand for load-adapted, highly flexible and variable structures. In particular, the formation of form-variable structures through selective geometric design in combination with compliant mechanisms shows very great potential for fulfilling the complex requirements. Weaving technology offers outstanding approaches to the implementation of complex 3D geometries. However, there is a lack of methods for the formation of complex 3D geometries such as PACS and for the technical implementation in weaving processes. The first approaches are terry weaving and spacer weaving. However, both methods are not suitable for the technical implementation of PACS. On the one hand the required 3D geometry of the PACS and especially the wall thickness gradients as well as the required different fabric lengths on the outer sides cannot be realized. Therefore, algorithms and methods for the formation of geometry were developed for the realization of PACS by means of weaving. The knowledge gained in the field of geometry formation for the specific arrangement and connection of fabric layers has a generally valid character and can be applied to other 3D geometries. The process chain developed for weaving implementation, i. e. the transfer of layer arrangement into a 2D weaving arrangement and the creation of weave patterns; can also be used for the realization of complex 3D structures in a weaving process. Weave structures have been developed and implemented for the individual sub-areas of PACS. The results obtained serve as a basis for the development of a technology concept and for the constructional and technological work for the development of a weaving machine concept for the realization of the PACS.

Acknowledgments
This article presents selected results of the Collaborative Research Project DFG CH 174/42-1 “Pressure actuated cellular structures (PACS)”. The authors would like to thank the Deutsche Forschungsgemeinschaft (DFG) for the financial support for the project at the Technische Universität Dresden.

References

[1] Gramüller B *et al* 2014 PACS - Realization of an adaptive concept using pressure actuated cellular structures *Smart Materials and Structures* **23.11** 115006
[2] Gramüller B, Hardy K and Christian H 2015 Holistic design and implementation of pressure actuated cellular structures *Smart Materials and Structures* **24** 125027
[3] Cherif C 2015 *Textile Material for Lightweight Constructions* (Springer Berlin)
[4] Bannister M Challenges 2001 for composites into the next millennium-a reinforcement perspective *Composites Part A* **32** doi:10.1016/S1359-835X(01)00008-2
[5] Cherif C *et al* 2012 Development of a process chain for the realization of multilayer weft knitted fabrics showing complex 2D/3D geometries for composite applications *Textile Research Journal* **82** 1195-210
[6] Islam A 2012 3D woven structures and an overview of manufacturing technologies 4th *World Conference on 3D Fabrics and their Applications* (Aachen)
[7] Khokar N 2011 Aligning 3D fabric-forming processes with market requirements *Proceedings of the Third World Conference on 3D Fabrics and Their Applications* (Wuhan, China)
[8] Sharpa K, Bogdanovichb A, Boylec R, Browna J and Mungalov D 2013 Wind blade joints based on non-crimp 3D orthogonal woven Pi shaped preforms *Composites: Part A* **49** 9-17
[9] Khokar N 2014 Development of innovative 2D and 3D fabric-forming processes for manufacturing composite reinforcements *Proceedings of the 8th ADITC* (Dresden, Germany)
[10] Karahan M, *et al* 2012 Low velocity impact characteristic of 3D integrated core sandwich composites *TRJ* **82** doi 10.1177/0040517512436832
[11] Chen X *et al* 2011 An overview on fabrication of three-dimensional woven textile preforms for composites *TRJ* **81.9** 932-44
[12] Chen X *et al* 2008 Design, manufacture, and experimental analysis of 3D honeycomb textile composites Part I: design and manufacture *Textile Research Journal* **78.9** 771-81
[13] Mountasir A *et al* 2015 Competitive manufacturing of 3D thermoplastic composite panels based on multi-layered woven structures for lightweight engineering *Composite Structures* **133** 415-24
[14] Mountasir A *et al* 2016 3D woven near-net-shape preforms for composite structures *Advanced Engineering Materials* **18.3** 391-6

Effect of nitrogen on morphology of ZnO nanorods deposited by CVD method

L N Wang[1,2], J Wang[1], D Zhou[1], N Liu[1], J T Liu[1] and Z H Gao[1]

[1]Key Laboratory of Marine Biophysics of Liaoning Province, Dalian Ocean University, Dalian, China

E-mail: l.n.w@163.com

Abstract. The ZnO nanorods are synthesized by a typical CVD method without any catalyze and substrate. The concentration of nitrogen is changed by a digital gas control system. The influence on the morphology is studied in detail. Under lower amounts of nitrogen the diameter of nanorod is smaller than in the higher one. Meanwhile, the size of the nanorod is bigger in the rich nitrogen amounts. The morphology of the sample is carry out by SEM, and the crystal structure information is giving by XRD. All the measurement told that the nanorods have a great hexagonal wurtzite structure.

1. Introduction

As an excellent semiconductor photoelectric material Zinc oxide (ZnO) can be widely used in many areas, especially in photodetectors [1], light-emitting diode (LED) [2,3], nanogenerators [4,5] and gas sensors [6-8]. With the development of economic society, the photoelectric devices need to be more and smaller, so that the ZnO nanostructures are more useful. After many years of hard work, different morphology of ZnO nanostructure has been made, such as nanowires, nanorods, nanoflowers, nanostars and nanoflakes [9-13]. Among them ZnO nanorods exhibit more advantage, for instance the large surface-to-volume and prefer orientation. During the study process there are also some difficulties need to be overcome, the nanorods have not realized controlling deposition, because it is very sensitive to the experiment condition, such as pressure, gas concentration, deposition temperature, and it may also some other factors which has not been consider. So it is very necessary to discuss the major factor which can affect the morphology of the nanorod. The experiment technique is also a important factor, many methods have been made to deposited ZnO nanostructures, for example aqueous based chemical deposition [14], hydrothermal method [15], thermal evaporation technique [16], electrochemical solution [17], and chemical vapor deposition (CVD) [18]. Among all the methods, CVD technique exhibit is more stable, reproducible and easily to change the experiment parameter.

In this paper, we deposited ZnO nanorods using three different nitrogen concentrations without any substrate, the concentration of nitrogen are 20 sccm, 25 sccm, 30 sccm, respectively. During the whole deposition process, the nitrogen is act as carrier gas. It was found that at different concentration the ZnO exhibit different nanostructures. The mechanism of different morphology was studied in detail.

2. Experiment

The ZnO nanorods were deposited by CVD method without any substrate. The CVD system is composed of three parts: gas controller operation, three heating zone quartz tube furnace and integrated digital control system. The quartz tube furnace was 120cm in length and 8cm in diameter. The melting point of ZnO is about 1975℃, in order to reduce the growth temperature, A mix powder of ZnO 8g and graphite 2g are grinding in agate mortars, the mix powder react readily to produce Zn and CO_2, the melting point of Zn is about 419℃, in the next step the Zn and O_2 react to produce ZnO. Then the mix powder was put into one end of a small quartz tube (30 cm in length, 4 cm in diameter). The small tube is located at the central of the heating zone. The nitrogen is introduce into the system with a constant flow rate of 20 sccm(sample a), after a few minutes the impurity gas was all exhaust, turn on the digital control system to start the deposition, the deposition temperature was 1000℃, when the grow up temperature up to 800℃, introduced the oxygen with constant flow rate of 20 sccm which was act as react gas, the whole grow up process maintained 60 minutes, then the temperature cooled down to room temperature, and the ZnO nanorods can be found at another end of the small tube. Changing the flow rate of nitrogen to 25 sccm (sample b) and 30 sccm (sample c), other parameters remain the same one, the other two samples can also be found.

The morphology of the ZnO nanorods was observed in a hitachi SU1510 scanning electron microscope (SEM) system which operating at 15 KV. The structural properties detail was given by X-ray diffraction (XRD; RigakuD/max3B) which performed in a θ-2θ configuration.

3. Results and discussion

The surface images of ZnO nanorods were shown in figure 1, the sample a,b,c were deposited in nitrogen under a constant flow rate of 20, 25 and 30 sccm severally. In figure 1, a it can be clearly found that the nanorods was short, and the most longest one was about 2.33 μm. There are also some nanotetrapod structure exist in the sample, the nanorods were grow up out of order and there remain some small holes exist. The nanorods have clearly hexagonal wurtzite structure with smooth surface. When the concentration of nitrogen up to 25 sccm, the morphology was clearly changed, the nanorods become more longer with smooth surface, the longest one was about 3.67 μm, meanwhile, the hollows were become bigger which means the density was decrease. In the other hand, there are few nanowires emerged in the picture. The nanorods still have a perfect hexagonal wurtzite structure. With the flow rate continue to grow, the morphology have a great change. As can be seen the nanorods gets more tall and fat, it can up to 10 μm in length, in the same time the structure also have some difference. Two different surface morphologies appeared in the same time, one is the hexagonal wurtzite structure, the other one is the hexagonal pyramids like nanorods, there are also some nanowires can be found around.

Figure 1. SEM images of ZnO nanorods with different amounts of nitrogen (a) 20 sccm, (b) 25 sccm and (c) 30 sccm.

From the SEM image we can see that when the flow rate increased, the nanorods became bigger and fatter. The different structures maybe explained by different mechanism. At the first stage, the carrier gas was not enough, so the process was carry on slowly, ZnO powder in the react mix source are reduced to Zn or its suboxide, due to the lowly flow rate of nitrogen the concentration of oxygen is relative bigger, as a result a lot of ZnO nuclei are formed at the low temperature area which lead to small ZnO nanorods and nanotetrapod structures as sample a. When the flow rate up to 25 sccm, the carrier gas was increase in the same time the concentration of oxygen is decrease which lead to few ZnO nuclei, so the nanorods became a little bigger. However for the case of sample c, it is deposited under rich carrier gas atmosphere, the pressure was change a lot which result in the nanorods get bigger and fatter. It can be deduce that the pyramids-like nanorods also a result of the higher flow rate. When the concentration changed the pressure also changed which lead to a different crystal and size [19].

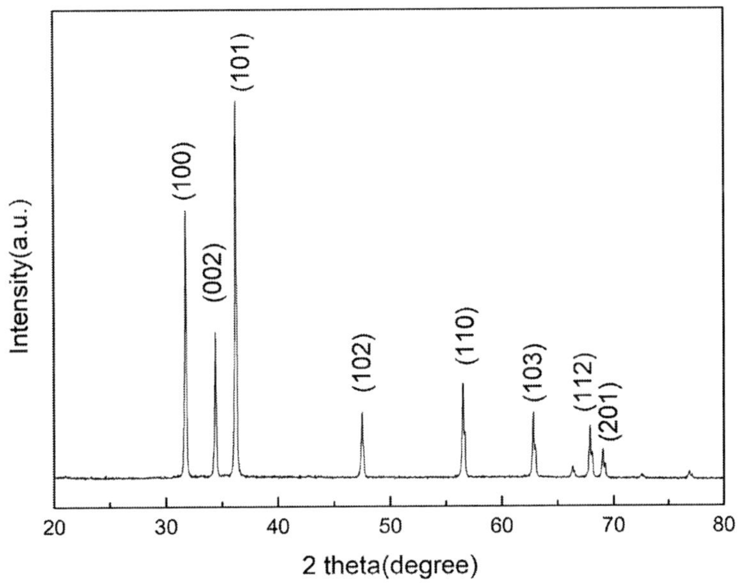

Figure 2. The XRD pattern of ZnO nanorods.

The XRD pattern of ZnO nanorods is shown in figure 2. The XRD test is made in this way; we put the three samples together in a small box and then using θ-2θ scanning. There are eight peaks arrives at 31.74°, 34.42°, 36.25°, 47.43°, 56.45°, 62.80°, 67.95° and 69.03° respectively which corresponding to ZnO(100), (002), (101), (102), (110), (103), (112) and (201) peaks [20]. All these peaks are related to the typical hexagonal wurtzite structure. These peaks all have a great intense especially the (101) peak which means the ZnO nanorods are in a good crystal quality. The lattice constant can be calculated as follows [21]:

$$a = \frac{\lambda}{2\sin\theta}\left(\frac{4}{3}\left(h^2 + hk + k^2\right) + \left(\frac{al}{c}\right)^2\right)^{\frac{1}{2}} \tag{1}$$

$$c = \frac{\lambda}{2\sin\theta}\left(\frac{4}{3}\frac{\left(h^2 + hk + k^2\right)c^2}{a^2} + l^2\right)^{\frac{1}{2}} \tag{2}$$

The (hkl) represent Miller indices, λ=0.15418 nm, θ is the diffraction angle, a and c are the lattice constant, according to equation above, the constant of the most intense (101) peak can be calculated with a=3.265 Å, c=5.214 Å. According to literature (JCPDS NO.36-451), when the ZnO exhibit hexagonal wurtzite structure, the lattice constant are a=3.427 Å, c=5.200 Å. The XRD pattern has a great agreement with the literature which means the nanorods is grown up with hexagonal wurtzite structure.

4. Conclusion

In summary, the ZnO nanorods are synthesized by a simple CVD method without any substrate. By controlling the flow rate of nitrogen, the ZnO nanorods realized controlling synthesis. When the concentration of nitrogen is not enough, the diameter of the nanorods is small, there are also some small hollow exist, when the amounts of the gas carrier increasing, the nanorods become longer and bigger, there are also some nanowires emerged. In a rich nitrogen condition the diameter and length of the nanorods all increase to some extent. It can be deduced that the concentration of nitrogen can affect the pressure of the growing environment, under a rich gas condition the nanorods are long, and otherwise the nanorods are too short. The SEM and XRD all give the information that the nanorods have a great hexagonal wurtzite structure with lattice constant a=3.265 Å, c=5.214 Å.

Acknowledgment

This work was supported by the Nature Science Foundation of Liaoning Province (No. 20170540109 and No. 201601285); and Nature Science Foundation of China (No. 11647104).

References

[1] Liang S, Sheng H, Liu Y, Huo Z, Lu Y and Shen H 2001 *J. Cryst. Growth* **225** 110-3
[2] Tsukazaki A, Ohtomo A, Onuma T, Ohtani M, Makino T, Sumiya M, Ohtani K, Chichibu S, Fuke S, Segawa Y, Ohno H, Koinuma H and Kawasaki M 2004 *Nat. Mater.* **4** 42-6
[3] Tsukazaki A, Kubota M, Ohtomo A, Onuma T, Ohtani K, Ohno H, Chichibu S.F and Kawasaki M 2005 *Japan J. Appl. Phys. Part 2* **44** 643-5
[4] Yin B, Qiu Y, Zhang H, Ji J and Hu L 2014 *Cryst. Eng. Comm* **16** 6831-5
[5] Yin B, Qiu Y, Zhang H, Ji J and Lei J 2015 *J. Mater. Sci: Mater Electron* **26** 742-6
[6] Chen Y, Xu P C, Xu T, Zheng D and Li X X 2017 *Sens. Actuators B* **240** 264-7
[7] Zou A L, Qiu Y, Yu J J, Yin B, Cao G Y, Zhang H Q and Hu L Z 2016 *Sens. Actuators B* **227** 65-72
[8] Zou A L, Hu L Z, Qiu Y, Cao G Y, Yu J J, Wang L N, Zhang H Q, Yin B and Xu L L 2015 *J. Mater. Sci: Mater Electron* **26** 4908-12
[9] Mosquera E, Bernal J, Zarate R, Mendoza F, Katiyar R and Morell G 2013 *Mater. Lett.* **93** 326-9
[10] Zhang S, Yan C L, Zhang H X and Lu G Y 2015 *Mater. Lett.* **148** 1-4
[11] Liu C, Wang B Q, Liu T, Sun P, Gao Y, Liu F M and Lu GY 2016 *Sens. Actuators B* **235** 294-7
[12] Aladpoosh R and Montazer M 2016 *Carbohydr. Polym.* **141** 116-25
[13] Behera B and Chandra S 2016 *Sens. Actuators B* **229** 414-24
[14] Panigrahy B, Aslam M and Bahadur D 2010 *J. Phys. Chem. C* **114** 11758-63
[15] Yoo R, Cho S, Song M J and Lee W 2015 *Sens. Actuators B* **221** 217-23
[16] Fan D H, Zhang R and Li Y 2010 *Solid State Commun.* **150** 1911-4
[17] Liang J K, Su H L, Kuo C L, Kao S P and Cui J W 2014 *Electrochim Acta* **125** 124-32
[18] Ding M, Yao B, Zhao D, Fang F and Shen D 2010 *Thin Solid Films.* **518** 4390-3
[19] Mongia H, Samir F, Frederic S, Farah K, Noureddine J, Bruno V and Arbi M 2014 *J. Alloys Compd.* **615** s472-5
[20] Gayen R N and Paul R 2016 *Thin Solid Films* **605** 248-56
[21] Moulahi A and Sediri F 2016 *Optik* **127** 7586-93

IOP Conf. Series: Materials Science and Engineering **369** (2018) 012043 doi:10.1088/1757-899X/369/1/012043

Highly filled biocomposites based on ethylene-vinyl acetate copolymer and wood flour

P G Shelenkov[1,3], P V Pantyukhov[1,2] and A A Popov[1,2]

[1]Emanuel Institute of Biochemical Physics, Russian Academy of Sciences, 4 Kosygina Street, Moscow, 119334, Russia
[2]Department of Chemistry and Physics, Plekhanov Russian University of Economics, 36 Stremyanny Lane, Moscow, 117997, Russia

E-mail: shell1183@mail.ru

Abstract. Recently, there is a great interest in the world to biodegradable materials based on synthetic polymers in a composition with natural fillers. Highly filled polymer composite materials based on various grades of synthetic block copolymer of ethylene vinyl acetate with wood flour were under investigation. Five grades of ethylene-vinyl acetate copolymer differing in the content of vinyl acetate groups and a melt flow index were used in this work in order to find the best one for highly filled biocomposites. Wood flour content in biocomposites was 50, 60, 70 weight %. The rheological and physico-mechanical characteristics of the resulting biocomposites were studied.

1. Introduction

Superconcentrates are polymer composites with a high filler content. They are intended to be added to pure polymer during the manufacture of finished products in order to impart a certain color or improve the functional characteristics of the product. Today, the most widely uses superconcentrates of dyes, they also called as masterbatches. Superconcentrates of oxo-degrading additives, UV stabilizers, antistatic agents, antioxidants, flavors, etc. are also manufactured. The idea of creating superconcentrates of biodegradable composites based on synthetic polymers with biodegradable fillers has not been studied well.

The most common biodegradable materials, based on synthetic polymer and natural filler, are the composites with wood flour. A number of works are known where composites, based on synthetic polymers filled with wood flour, were studied. Such materials have a wide range of applications in packaging sector (films, bags, containers etc.), in automotive and electronic industry, for furniture, suitcase, grinding discs, safety helmets [1]. Such materials are rather widely described [2-4].

In the previous work it was found, that the particle size of lignocellulosic filler has influence on physical properties and biodegradation of biocomposites based on polyethylene. It was found that as filler particle size decreases, the rate of biodegradation decreases [5]. Sometimes, it is proposed to use not only one polymer, but a mixture of two or more polymers, as a polymer matrix in such biocomposites, for example blend of polyethylene and polypropylene was used as a polymer matrix in the work of E. Mastalygina [6]. In the other work the blend of polyethylene with ethylene with vinyl acetate copolymer was used as a matrix [7]. Superconcentrates contain a small amount of synthetic polymer and after adding superconcentrate to the other pure polymer the composite with two blended polymers and the filler is being obtained.

Currently, a lot of works have been devoted to improving the compatibility, as well as the mechanical and rheological properties of wood-polymer composites, but a little amount of works studied highly-filled biocomposites [8-10].

2. Materials and methods

It was studied the influence of the chemical composition and physical properties of the matrix on the main properties of highly filled composites with wood flour. Five different grades of the ethylene vinyl acetate copolymer (EVA) from LG Chem, supplied by OOO Ecoplastics (Moscow, Russia), differing in the content of vinyl acetate (VA) groups and in the melt flow index (MFI) were taken as matrix (table 1).

Table 1. Characteristics of different trademarks of ethylene vinyl acetate copolymer (EVA). Characteristics were provided by LG Chem.

No	EVA trademark	VA content [weight %]	Melt flow index (MFI), [g/10 min] ASTM D 1238	Tensile strength, [MPa] ASTM D 638	Elongation at break, [%] ASTM D 638	Density, [kg/m^3] ASTM D 1505	Melting temperature, [°C] DSC
1	ES28005	28	5	13.5	800	0.951	72
2	EA28025	28	25	9.5	850	0.951	69
3	EA28150	28	150	4.0	900	0.946	70
4	EA19150	19	150	7.0	800	0.940	80
5	EC15006	15	6	16.0	800	0.936	89

Three EVA trademarks (ES28005, EA28025, EA28150) have the same content VA groups, but different MFI, while two other trademarks (EC15006 and EA19150) differ both by MFI and the content of VA. The effect of these two parameters on the properties of highly filled composites was studied.

As a filler for superconcentrates, wood flour (a mixture of deciduous and coniferous trees) was used because it is wide-spread cellulose-containing filler. Wood flour was supplied by OOO Novotop (Moscow, Russia). Wood flour contains 46% cellulose, 20% lignin, 29% pentosans, 5% polyuronic acid, residual ash content is 10% [11]. Wood flour was filtered by electromagnetic sieve shaker Matest A059-02KIT (Italy) through a sieve with a mesh diameter 0.1 mm.

The main difficulty in processing composites based on thermoplastics and wood flour is the propensity of untreated wood particles to form large agglomerations. Therefore, using traditional methods of mixing, the fibers distribute poorly in the polymer matrix, and it leads to a significant decrease in the reinforcing action of fibers [12]. There are several methods for mixing polymers: extrusion, milling, compounding on a Brabender-type plastograph. In the work [13], various methods of mixing polymer materials were considered, and it was concluded that milling provides the best distribution of the components into the polymer matrix. Therefore, in this work for mixing EVA and wood flour heated mixing rolls UBL6175BL (China) were used. A series of composites was prepared: EVA + wood flour 50%, 60 and 70% with all the used above EVA trademarks. Milling processing mode: mill roll temperatures T1 / T2 = 130/150°C, roll speed 8 rpm.

The rheological properties of the obtained composites were studied by measuring the melt flow index (MFI). MFI was determined according to the standard [14]. The load is 2.16 kgf, the melt temperature is 190°C.

The obtained composites were pressed on a hydraulic press Biolent RP-12 (Russia) at a temperature of 140°C and pressure 40 kgf/cm^2 for 1 minute. As a result, flat sheets of square form with a thickness of 0.3 to 0.6 mm were prepared.

Samples for mechanical tests were cut from the sheets. Measurement of strength and elongation was performed according to ISO 527-4 [15] by the universal testing machine GOTECH AI-7000M

(Taiwan) at a deformation speed of 100 mm/min. Eight specimens were tested for the tensile and flexural properties of each composite, standard deviation was calculated. The density of the composites was studied by the hydrostatic weighing method in accordance with standard [16].

3. Results and discussion

Table 2 shows the measurement data for MFI of composites, with a filling of 50% and 70%. The MFI values are very low even at 50% wood flour content in the composite. Nevertheless, as a result, it was confirmed that with an increase in the MFI of the polymer matrix, the MFI of highly-filled composite increased. With a filling of 70%, the flow was not observed at all.

Table 2. Melt flow index of composites based on EVA and wood flour.

Filler content [weight %]	EVA trademark	Melt flow index (MFI) [g/10min]
0	28005	5
	28025	25
	28150	150
	19150	150
	15006	6
50	28005	0.08
	28025	0.28
	28150	0.65
	19150	0.15
	15006	0.03
70	28005	Not flow
	28025	Not flow
	28150	Not flow
	19150	Not flow
	15006	Not flow

The high viscosity of the composite is due to the fact that wood flour particles are in the form of fibers, i.e. a high L/D ratio (the ratio of the length of the particle to its diameter). It is also possible that part of the filler starts to degrade at the test temperature (190°C) [17]. According to the literature, wood flour starts to undergo thermal degradation (loses 3 wt.%) at 240-250°C in nitrogen atmosphere [18], but this temperature can be lower in the air and under mechanical treatment. Also, with an increase in the content of VA groups in EVA matrix, the MFI of highly filled composite increases. VA groups have good adhesion to the wood flour, that is why biocomposites with the higher VA content have better distribution of wood particles in the matrix and as a consequence less viscosity of biocomposite.

Table 3 shows tensile strength and elongation at break of EVA composites with wood flour. It can be concluded that the viscosity of the polymer matrix has a significant influence on the strength of the composites, the higher the MFI of the matrix, the lower the tensile strength of the composites because MFI is directly related to molecular weight distribution (MWD) of the polymer [19]. Also, the strength of the composites depends on the content of VA groups in the polymer matrix, the lower content of VA groups, the higher the tensile strength. It may be explained, because with an increase in the content of VA groups in EVA, its crystallinity degree decreases and the modulus of elasticity increases [20]. The content of the filler does not significantly affect the strength of the composites up to 60% of the filling, however, for a number of composites, a significant reduction in strength at 70% filling was found. The elongation at break of the composites depends on the MFI of the polymer matrix (the higher the MFI, the lower the elongation), the VA content (the elongation is higher at a high VA content) and the filler content in the composite (the smaller the filler content, the higher the elongation). The density of the composites tends to increase with the increasing filler content. This is due to the higher real density of wood flour (1.54 g/cm^3), compared to the density of pure EVA (0.95 cm^3).

Table 3. Mechanical characteristics of composites on the basis of EVA and wood flour with the filling of 50, 60 and 70 weight %.

Filler content [weight %]	EVA trademark	Tensile strength [MPa]	Elongation at break [%]	Density [kg/m^3]
50	28005	3.3 ± 0.2	57.5 ± 5.8	1.147 ± 0.002
	28025	2.9 ± 0.2	26.4 ± 4.9	1.145 ± 0.001
	28150	2.4 ± 0.1	9.4 ± 5.3	1.141 ± 0.003
	19150	3.1 ± 0.2	4.2 ± 1.3	1.140 ± 0.003
	15006	3.8 ± 0.5	5.5 ± 1.6	1.134 ± 0.002
60	28005	3.3 ± 0.2	9.4 ± 1.6	1.185 ± 0.002
	28025	2.8 ± 0.2	8.3 ± 0.1	1.179 ± 0.003
	28150	2.7 ± 0.2	4.3 ± 0.8	1.184 ± 0.001
	19150	2.8 ± 0.2	1.7 ± 0.1	1.171 ± 0.001
	15006	4.1 ± 0.9	2.0 ± 0.6	1.089 ± 0.011
70	28005	3.4 ± 0.3	4.0 ± 0.8	1.216 ± 0.001
	28025	2.2 ± 0.4	6.7 ± 3.3	1.166 ± 0.004
	28150	2.0 ± 0.3	5.0 ± 1.1	1.189 ± 0.009
	19150	2.1 ± 0.1	2.1 ± 0.7	1.205 ± 0.007
	15006	2.9 ± 0.2	2.1 ± 0.7	1.193 ± 0.009

4. Conclusion

Rheological and mechanical properties of highly filled biocomposites based on ethylene-vinyl acetate copolymer were studied. The main purpose of such materials is using as superconcentrates in the process of manufacturing products from biocomposites. That is why the mechanical characteristics are important only for comparing the composites with each other and estimating the adhesion of the matrix polymer with the filler. It was concluded that the most durable composites are the ones based on EVA 15006, but because of the lowest MFI, they have restrictions on the processing methods. The most optimal combination of characteristics (tensile strength, elongation at break and MFI) are found in the composites based on EVA 28025, which indicates the prospect of using this EVA trademark for manufacturing of superconcentrates.

References

[1] Faruk O, Bledzki A K, Fink H P and Sain M 2014 *Macromol. Mater. Eng.* **299** 9-26
[2] Schneider M H 1994 *Wood an Fiber Science* **26** 142-51
[3] Ayrilmis N, Tasdemir M and Akbulut T 2015 *Polym. Compos.* **38** 863-9
[4] Zykova A, Pantyukhov P and Popov A 2017 *Polym. Eng. Sci.* **57** 756-63
[5] Zykova A K, Pantyukhov P V, Kolesnikova N N, Monakhova T V and Popov A A 2018 *J. Polym. Environ.* **26** 1343-54
[6] Mastalygina E E and Popov A A 2017 *Solid State Phenomena* **265** 221-6
[7] Pantyukhov P, Kolesnikova N and Popov A 2016 *Polym. Compos.* **37** 1461-72
[8] Ou R, Wang Q, Wolcott M P, Sui S and Xie Y 2016 *Polym. Compos.* **37** 553-60

[9] Younes M M, Abdel-Rahman H A, Orabi W and Ismail M R 2008 *Polym. Compos.* **29** 768-72
[10] Pantyukhov P V, Monakhova T V, Kolesnikova N N, Popov A A and Nikolaeva S G 2013 *J. Balk. Tribol. Assoc.* **19** 467-75
[11] Nikitin N I 1962 *Chemistry of Wood and Cellulose* (Leningrad: RAS) (in Russian)
[12] Harutun C K 2003 *Handbook of Polypropylene and Polypropylene Composites* (NY: Marcel Dekker)
[13] Olkhov A A 2001 Ecologically safe self-degradable composite films based on polyethylene and polyhydroxybutyrate (Moscow: Dissertation) (in Russian)
[14] ISO 1133-2:2011 Plastics - Determination of the melt mass-flow rate (MFR) and melt volume-flow rate (MVR) of thermoplastics - Part 2: Method for materials sensitive to time-temperature history and/or moisture 2011 (London: BSI) p 16
[15] ISO 527-4 1997 Plastics-Determination of tensile properties-Part 4: Test conditions for isotropic and orthotropic fibre-reinforced plastic composites 1997 (London: BSI) p 11
[16] ASTM D1505-10 Standard test method for density of plastics by the density-gradient technique 2010 (West Conshohocken, PA: ASTM) p 7
[17] Peacock A 2000 *Handbook of Polyethylene: Structures, Properties, and Applimtions* (NY: Marcel Dekker)
[18] Poletto M 2016 *Revista Arvore* **40** 941-8
[19] Seavey K C, Liu Y A and Khare N P 2003 *Ind. Eng. Chem. Res.* **42** 5354-62
[20] Askadskii A, Popova M, Matseevich T and Afanasyev E 2014 *Advanced Materials Research* **864-867** 640-3

Biodegradation of natural reinforcing fillers for polymer composites

E E Mastalygina[1,2,3], P V Pantyukhov[1,2] and A A Popov[1,2]

[1]Department of Chemistry and Physics, Plekhanov Russian University of Economics, 36 Stremyanny Lane, Moscow, 117997, Russia
[2]Emanuel Institute of Biochemical Physics, Russian Academy of Sciences, 4 Kosygina Street, Moscow, 119334, Russia

E-mail: elena.mastalygina@gmail.com

Abstract. Twelve different natural raw materials were selected as possible fillers for eco-friendly biocomposites. The target was to find the most biodegradable ones. Two mycological tests were held: in the aqueous and agar media. It was found that two tests showed different results. In aqueous media, the fillers with a high content of water-soluble and easy-hydrolysed compounds demostrated the most intensive biofouling. In agar media, the entire filler was exposed to biodigestion by fungi. Therefore, multi-compound fillers with a set of different macro- and microelements were more biodegradable than others.

1. Introduction

Among solid domestic and industrial waste, synthetic polymers, particularly polyolefins, are the most resistant to biodegradation by microorganisms that is due to their chemical nature and high hydrophobicity [1]. In view of this, developed countries are engaged in creating novel biodegradable polymeric materials. Adding natural fillers into synthetic polymers allows developing materials with enhanced biodegradability and partially replacing non-renewable petrochemical resources with renewable ones [2,3]. In such composite materials, the synthetic polymer provides optimal mechanical properties and recycling possibility [4]. In turns, the natural filler content in composites enhances the ability to bioassimilation by microorganisms [5,6]. The composite compound has the greatest effect on its biodegradability that proves the importance of filler kind choice and its investigation. The aim of this work is to determine the most effective natural fillers for developing biodegradable polymeric composite materials. The focus of this study was in determining the degradation and deterioration of a set of natural components used for filled composites due to the action of mold fungi cultures. Results interpretation emphasis is on the correlation between chemical composition and biodegradability of natural fillers.

2. Materials and methods

2.1. Materials

In deciding on a particular set of natural fillers initiating composites biodegradation, the following requirements were taken into account: low cost and affordability; ecological safety of biodegradation products; possibility of grinding on standard grinding equipment; high temperature of thermal destruction.

A set of natural components being waste products of various industries and rarely used for making polymer compositions were studied. The chemical composition of the fillers under investigation was analyzed according to literature data (table 1) [7-12].

Table 1. Chemical composition of natural fillers.

Filler	Composition, %						
	Cellulose	Hemicellulose	Lignin	Starch	Peptides	Lipids	Ash
Oil flax	47÷58	5÷17	10÷21	–	3÷9	2÷4	3÷12
Fibre flax	38÷40	14÷15	23÷24	–	n/a	2÷4	2÷8
Wheat	40÷46	24÷32	6÷8	1÷2	4÷5	1÷2	6÷8
Husk	31÷35	18÷19	27÷29	n/a	3÷5	2÷4	3÷5
Wood	45÷48	10÷13	28÷30	1÷2	n/a	1÷2	1÷2
Leaves	24÷26	3÷5	33÷35	2÷3	n/a	n/a	5÷7
Banana	7÷12	6÷8	6÷10	13	9	13	3÷8
Hay	35÷49	10÷15	23÷27	1	10÷11	2	12÷14

The analyzed fillers include oil flax straw (oil flax) (grade LM98), winter wheat chaff (wheat) (grade Krasnodar-99), fir wood flour (wood), product of keratin hydrolysis of bird's feather (oligopeptides), fibre flax shive (fibre flax), sunflower husk (husk), sodium lignosulfonate (SL), banana skin (banana), birch leaves (leaves), mixed herbs hay (hay).

As reference fillers, powdered cellulose (hereinafter cellulose) (Polycell PCC, TU 5410-029-32957739) and corn starch (starch) (GOST R 51953) having a fixed chemical composition and widely used for development of biodegradable composite materials based on synthetic polymers were studied.

Fillers were preliminary dried for 3 hours at (80 ± 2)°C, crushed in an electric mill with a rotary knife, and screened by means of a sieve analyzer. The filler fraction is not exceed 200 μm, humidity 10 ± 2%.

2.2. Methods

The effectiveness of natural fillers as additives accelerating the biodegradation of polymers was determined by growth intensity of mold fungi. To assess the effect of the input materials on the development of the test cultures, experiments were carried out on aqueous and agar media with filler particles in accordance with GOST 9.049 [13] and ISO EN 846 [14]. In the experiment, test cultures of mold fungi *Aspergillus niger*, *Aspergillus terreus*, *Penicillium chrysogenum* from the collection of Mycology and Algology Department of Lomonosov Moscow State University were used.

For the experiment on aqueous media, the substrate was a solution of bidistilled water containing 10 wt.% of milled filler. For the experiment on agar media, a solution of bidistilled water with 2 wt.% of dry agar was prepared, after sterilization of the solution 10 wt.% of milled filler was added thereto. The substrates were exposed to detention for 1 day and autoclaved at a pressure of 1 atm. at a temperature of 121°C for 30 minutes. After that, the substrate was poured into test tubes and Petri dishes for experiments on aqueous and agar media, respectively. The media were inoculated by spores of pure fungal cultures with a prick method. Bidistillated water and Czapek's medium without saccharose were used as control substrates for experiments in aqueous and agar media, respectively.

Incubation of media contaminated with the test cultures was carried out under conditions of relative humidity over 90% and a temperature of (25±2)°C. The incubation period for agar and aqueous media was 28 and 14 days, respectively, with monitoring after 2, 5, 8, 11, 14, 21 days.

For aqueous media, the biomass of micromycetes and change in the pH-index of the medium were determined. For biomass gain analysing, the grown mycelium of fungi was filtered on a shaker, dried and weighed. By the mass of the grown mycelium, the biodegradability of the water-soluble part of the material was determined.

After the incubation period on the agar media, the degree of mold fungi development was assessed

[14]: linear growth rate of fungal seed colonies, development of the aerial mycelium, and the onset of sporulation.

3. Results and discussion

Under the action of mold fungi, natural fillers degraded to carbon dioxide and water. As the experiment was conducted for 28 days, the total biodegradability was not reached. To assess the fillers' biodegradation, biomass gain and growth intensity of fungal cultures were analysed.

Figure 1 shows total fungal biomass gain for aqueous media with different natural fillers after 14 days from inoculation, recalculated by 1 g of the filler. According to the results of this study, all used fungal cultures had a high ability of growth on natural fillers extracts. The main biodestructor of fillers was *Aspergillus terreus*, which is a universal destructor assimilating the majority of polysaccharides and polypeptides.

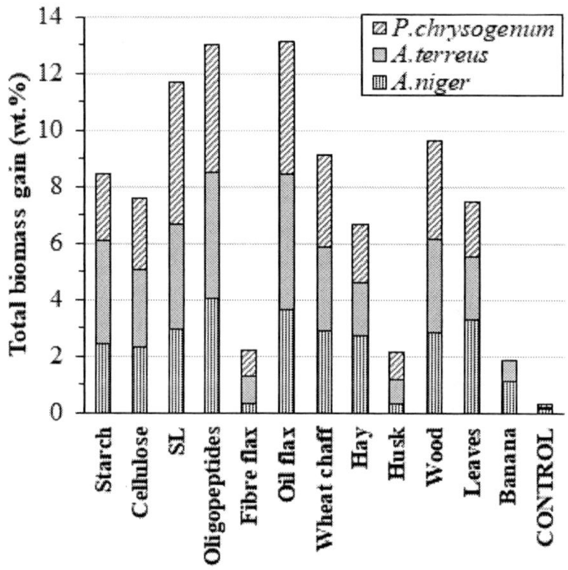

Figure 1. Biomass gain due to fungal growth in aqueous media for different fillers.

The aqueous media with additives of oil flax straw, oligopeptides and lignosulfonate had the maximum total biomass gain (12±0.5 wt.%). Less biomass accumulation was fixed for media with birch leaves, hay, wheat chaff and wood flour (total biomass gain is 8±1 wt.%). The media with fibre flax, banana peel and sunflower husk showed the lowest biomass growth of mold fungi from all the tested fillers, which is probably due to their low water solubility and nature of water-soluble compounds.

It can be seen (figure 2) that the character of biofouling is different for the fillers. The fibers of cellulose swelled up a lot in the water. They were fully covered by fungi, the main mass of fungi was into the liquid media. Oil flax, wood flour, oligopeptides and the majority of other fillers had a «hat» of aerial micellium at the interphase between liquid and air. Starch was degraded by combination of these two mechanisms – it had feculence of fungi into the volume and a small fungal «hat» at the interphase.

The regularities of mold fungi growth on agar medium with different fillers differed from the results of aqueous media test (figure 3). In this test the whole filler was exposed to biodegradation. The media with banana and hay were characterized by a high intensity of fungal development for the majority of mold fungi cultures (figure 4), which is apparently associated with a high content of peptide component and microelements. In fact, mixed herbs hay contains high amount of different components. Banana skin consists of different parts (external, intermediate and internal) with different chemical composition. From the other side, the fillers include one or two main components (starch, cellulose, sodium lignosulfonate, fiber flax, and wheat chaff) were poorly degraded by micromycetes.

Figure 2. Several photographs demonstrating *Penicillium chrysogenum* development on aqueous media with 10 wt.% of fillers after 14 days from inoculation.

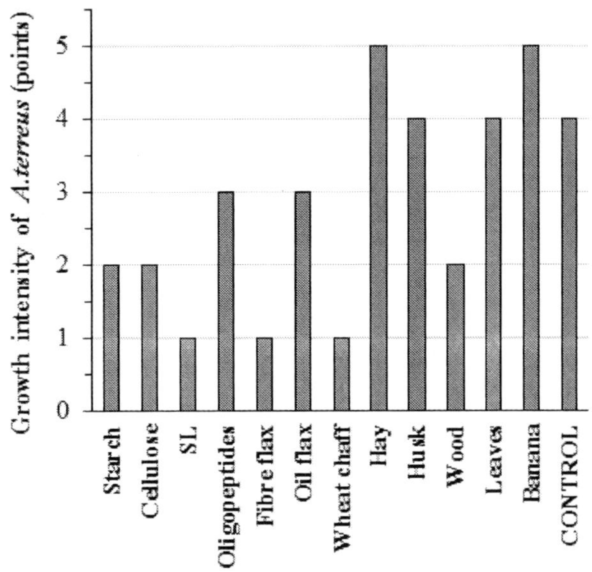

Figure 3. Assessment of growth intensity of *Aspergillus terreus* culture (in points) on agar media with 10 wt.% of fillers after 28 days from inoculation according to ISO 846.

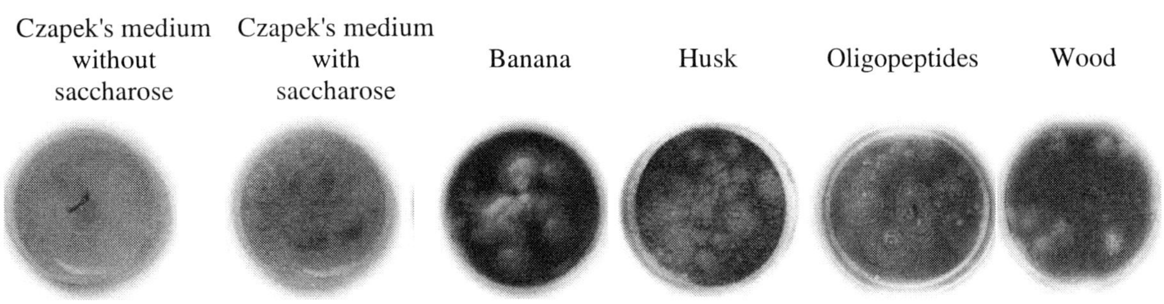

Figure 4. Several photographs demonstrating Aspergillus terreus development on agar media with 10 wt.% of fillers after 28 days from inoculation.

4. Conclusion

Twelve different natural raw materials were selected as the most perspective fillers for biocomposites. Technical parameters of the filler (thermal degradation temperature, length to diameter ratio of the particles, low price, etc.) are important, but in this paper, the main focus was on biodegradation. According to this paper results, at selecting the right filler for initiating biocomposites biogedradation, two parameters should be noted. First, for high biodegradability a composite filler should be multi-compound or it may be a combination of different fillers. Second, a filler should have water soluble and easy-hydrolysed components.

References

[1] Arutchelvi J, Sudhakar M, Arkatkar A, Doble M, Bhaduri1 S and Uppara P V 2008 *Indian Journal of Biotechnology* **7** 9-22
[2] Mohanty A K, Misra M and Drzal L T 2005 *Natural Fibers, Biopolymers, and Biocomposites* (Boca Raton, FL: Taylor & Francis Group)
[3] Faruk O, Bledzki A, Fink H-P and Sain M 2012 *Prog. Polym. Sci.* **37** 1552-96
[4] Zykova A, Pantyukhov P and Popov A 2017 *Polym. Eng. Sci.* **57** 756-63
[5] Mastalygina E E, Kolesnikova N N, Popov A A and Olkhov A A 2015 *AIP Conf. Proc. (Tomsk)* **1683** (Melville, NY: AIP Publishing LLC) 020143
[6] Mastalygina E E, Popov A A and Pantyukhov P V 2017 *IOP Conf. Ser.: Mater. Sci. Eng.* (Guangzhou) **213** (Bristol: IOP Publishing Ltd.) 012011
[7] Mann M E, Cohen R D H, Kernan J A, *et al* 1988 *Anim. Feed Sci. Technol.* **21** 57-66
[8] La Banque de données de l'Alimentation Animale. French feed database. Association Française de Zootechnie 2011 *AFZ, French Feed Database* http://www.feedbase.com/
[9] Rebolé A, Alvira P and González G 1989 *J. Sci. Food Agr.* **48** 141-53
[10] Sobolev M A 1963 *Chemistry of Flax and Bast–Fiber Materials* (Moscow: Gizlegprom) (in Russian)
[11] Conghos M M, Aguirre M E and Santamarıa R M 2003 *Biol. Fertil. Soils.* **38** 282-7
[12] Palma C, Contreras E, Urra J and Martinez M J 2011 *Waste Biomass. Valor.* **2** 77-86
[13] GOST 9.049-91 1991 Unified system of protection against corrosion and aging. Polymer materials and their components. *Laboratory Test Methods for Resistance to Impact Molds of Fungi* (Moscow: Publishing Standards) p 15
[14] ISO EN 846:1997 Plastics *Evaluation of the Action of Microorganisms* 1997 (London: BSI) p 22

Study on improving viscosity of polymer solution based on complex reaction

G Sun[1,3], D Li[1], D Zhang[2]and T H Xu[2]

[1]Exploration and Development Research Institute of Daqing Oilfield Company Ltd., Daqing, Heilongjiang 163712, China
[2]Northeast Petroleum University, Daqing, Heilongjiang 163318, China

E-mail: gangsun@petrochina.com.cn

Abstract. The current status of polymer flooding Technology on high salinity oil reservoir is not ideal. A method for increasing the viscosity of polymer solutions is urgently needed. This paper systematically studied the effect of ions with different mass concentrations on the viscosity of polymer solutions. Based on the theory of complex reaction, a countermeasure of increasing viscosity of polymer solution under conditions of high salinity reservoir was proposed. The results show that Ca^{2+} and Mg^{2+} have greater influence on the solution viscosity than K^+ and Na^+. When the concentration of divalent ions increases from 0 mg/L to 80 mg/L, the viscosity of the polymer solution decreases from 210 mPa·s to 38.6 mPa·s. The viscosity of the polymer solution prepared from the sewage treated with the $Na_2C_2O_4$ increased by 25.3%. Atomic force microscopy test results show that $Na_2C_2O_4$ can effectively shield the divalent metal ions, so that the polymer molecules in the solution stretch more, thereby increasing the solution viscosity. Atomic force microscopy test results show that $Na_2C_2O_4$ can effectively shield the divalent metal ions, so that the polymer molecules in the solution stretch more, thereby increasing the solution viscosity.

1. Introduction

Application of polymer flooding in high salinity oil reservoirs put forward higher requirements for polymer salt resistance[1]. At present, researchers use acrylamide polymers as the basis, through polymers modification or introducing special structural units in the molecular chain, the problem of oil-repelling polymers maintaining its high effective viscosity during the process has been solved to a certain extent[2,3]. However, with the increasing depth of reservoir exploitation, the salinity of reservoir fluid also increases. Therefore, it is urgent to solve the problem of increasing viscosity of high salinity reservoir polymer solution[4-7]. This paper focuses on the effects of salinity and metal ions and their concentration on the viscosity of polymer (HPAM) solution. Based on the theory of complex, $Na_2C_2O_4$ was selected as the metal complex agent to improve the viscosity performance of the polymer solution. With the help of atomic force microscopy tests, the effect of divalent metal ions on the configuration of polymer molecules in solution was qualitatively described, and the mechanism for increasing viscosity of polymer solutions based on complex reactions was clarified.

2. Experimental conditions and methods

2.1. Experimental materials

Polymer: The experimental polymer was partially hydrolyzed polyacrylamide (HPAM) supplied by Daqing Refining & Chemical Company, molecular weight 2500×10^4, solids content 90%, degree of hydrolysis 25%;

Inorganic salts: NaCl, KCl, $CaCl_2$, $MgCl_2$, effective content $\geq$ 96.0%, supplied by Xi'an Chemical Reagent Factory.

2.2. *Experimental apparatus*

Viscosity Measurement Instruments: tThe viscosity of the polymer solution was measured using a Brookfield Model LVDVII+ viscometer (American Boleyfield Engineering Laboratory Co., Ltd.), selected 0# rotor at reservoir temperature to measure the viscosity of the polymer solution with the rotational speed was 6 r/min.

2.3. *Polymer solution preparation*

2.3.1. Simulated water. This paper mainly studied the effect of total salinity and high cation content in formation water on polymer viscosity. The main inorganic salt composition was deionized water of NaCl, $CaCl_2$, and $MgCl_2$.

2.3.2. Polymer solution. Prepared a concentration of 5000 mg/L polymer mother liquor and stir for 3 h at a speed of 100 r/min. The prepared polymer solution with a mass concentration of 5000 mg/L was used as a mother liquor, after being sealed and cured, it could be diluted to a desired solution of any concentration.

3. Effect of sewage metal cation on anti-salt polymers solution viscosity

In the field applications of polymer flooding, sewage with higher ion content polymers are generally used. The degree of salinity and ion composition of the sewage have a great influence on the polymer solution. The composition of metal ions in sewage is complex. The main monovalent metal ions include Na^+ and K^+, the divalent metal ions include Ca^{2+}, Mg^{2+} and Fe^{2+}, and the trivalent metal ions include Fe^{3+}. All of above metal ions have an influence on the viscosity of polymer solution. Thus, the following experiment explores the effect of salinity and various cation concentrations on polymer viscosity.

3.1. *Effects of Na^+ and K^+ on polymer solution viscosity*

In order to verify the effect of Na^+ and K^+ content on the polymer solution viscosity, the influence of other cations on the viscosity was first excluded. Polymer mother liquor with a concentration of 5000 mg/L was prepared using distilled water, which was diluted with simulated water containing different concentrations of Na^+ and K^+ to prepare the target solution. The Na^+ and K^+ contents of the target solution were 0, 200, 400, 600, 800, 1000, 1500, 2000, 2500, 3000, 3500, 4000, 4500, and 5000 mg/L, respectively. The results of target solution viscosity are shown in figure 1.

The NaCl and KCl were added to distilled water to prepare different polymer solutions with different Na^+ and K^+ contents, and their viscosity was measured. The experimental results are shown in figure 1. As shown in the figure, the viscosity of the polymer solution reduced rapidly with the increase of Na^+ and K^+ concentrations. When the Na^+ and K^+ contents were >1000 mg/L, the viscosity decreases gently. Because when Na^+ and K^+ concentrations increase, the negative charges on -COO– of the polymer molecule weaken, and the electrostatic attraction within and between polymers decreased, the friction between molecules decreased, the molecules became curled, therefore viscosity is reduced. The effect of Na^+ and K^+ on the viscosity of the polymer solution was due to the small radius of Na^+ in the case of the same charge, which made it easier to access the polymer chain than K^+, neutralized the negative charge more strongly, and had a greater effect on the viscosity.

IOP Conf. Series: Materials Science and Engineering **369** (2018) 012045 doi:10.1088/1757-899X/369/1/012045

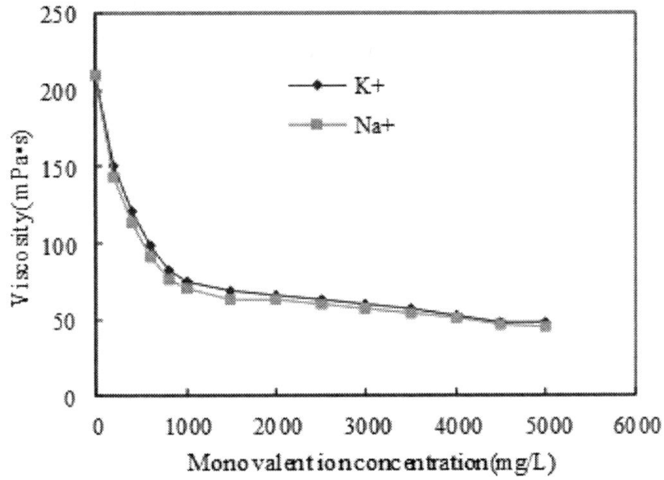

Figure 1. Effects of Na^+ and K^+ on polymer solution viscosity.

3.2. Effects of Mg^{2+} and Ca^{2+} on polymer solution viscosity

In order to verify the effect of Mg^{2+} and Ca^{2+} content on the viscosity of the polymer solution, the influence of other cations on the viscosity was first excluded. Polymer mother liquor with a concentration of 5000 mg/L was prepared using distilled water. The polymer mother liquor was diluted to the target solution with simulated water containing different concentrations of Mg^{2+} and Ca^{2+}. The contents of Mg^{2+} and Ca^{2+} in the target solution were 0, 20, 40, 60, 80, 100, 200, 300, 400 mg/L, respectively. The results of the viscosity measurement of the target solution are shown in figure 2.

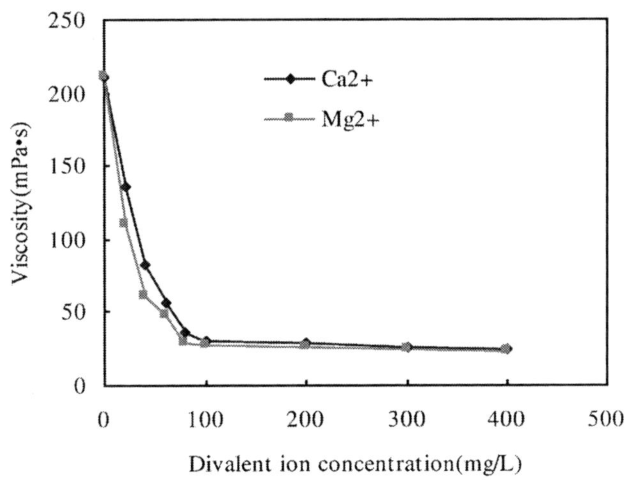

Figure 2. Effects of Ca^{2+} and Mg^{2+} on viscosity of polymer solution.

Different amounts of $MgCl_2$•$6H_2O$ and $CaCl_2$ were added to distilled water to prepare different polymer solutions with different Mg^{2+} and Ca^{2+} contents, and their viscosities were measured. As shown in figure 2, the viscosity of the polymer solution decreased drastically with the increase of Mg^{2+} and Ca^{2+} concentration. When the content of Mg^{2+} and Ca^{2+} increased from 0 mg/L to 80 mg/L, the viscosity of the polymer solution decreased from 210 mPa•s to 38.6 mPa•s, when the divalent potassium ion concentration is greater than 80 mg/L, the decrease in viscosity becomes more gentle.

The higher the content of Mg^{2+}, Ca^{2+}, the greater the loss of viscosity, and the polymer solution will precipitate and fail. This is because Mg^{2+} and Ca^{2+} have a large number of charges, and the ability to shield negative charges on Polymer -COO- is stronger. This results in polycondensation of the polymer, shrinking of the molecular chain, and dehydration. The effect of Mg^{2+} on the viscosity of the polymer solution is greater than that of Ca^{2+} due to the small radius of Mg^{2+} in the case of the same charge. The smaller radius of Mg^{2+} makes it easier to access the polymer chain than Ca^{2+}, shielding the negative charge on the polymer and excluding bound water molecule on the polymer, thus making the polymer curled more in solution, which leads to a greater effect on viscosity.

4. The method of increasing the viscosity of polymer solution

$Na_2C_2O_4$ as a complexing agent can effectively remove the Ca^{2+} and Mg^{2+} in the wastewater from the west block of Changyuan and increase the viscosity of the displacing phase. The addition of $Na_2C_2O_4$ as a complexing agent for the removal of Ca^{2+} and Mg^{2+} eliminates the settling effect of divalent ions on polymer chains and increases the viscosity of the polymer system.

$$Na_2C_2O_4 + M^{2+} = MC_2O_4 \downarrow + 2Na^+ \quad (1)$$

$Na_2C_2O_4$ was selected as the complexing agent of Ca^{2+} and Mg^{2+} and added into HPAM solution with a concentration of 1500 mg/L in different ratios. The effects of addition of 70 mg/L, 140 mg/L, 210 mg/L, 280 mg/L, 350 mg/L $Na_2C_2O_4$ on the viscosity of the HPAM solution were measured at 45°C. The results are summarized in table 1.

Table 1. Viscosity of HPAM solution at different concentrations.

Complexing agent dosage (mg/L)	0	70	140	210	280	350
Viscosity (mPa·s)	53.9	59.6	65.3	71.8	68.5	63.1
Viscosity increase rate (%)	/	10.6	21.1	33.2	27.1	17.0

The variation of the viscosity after adding different concentrations of sodium $Na_2C_2O_4$ to the HPAM solution was plotted as a line graph, as shown in figure 3.

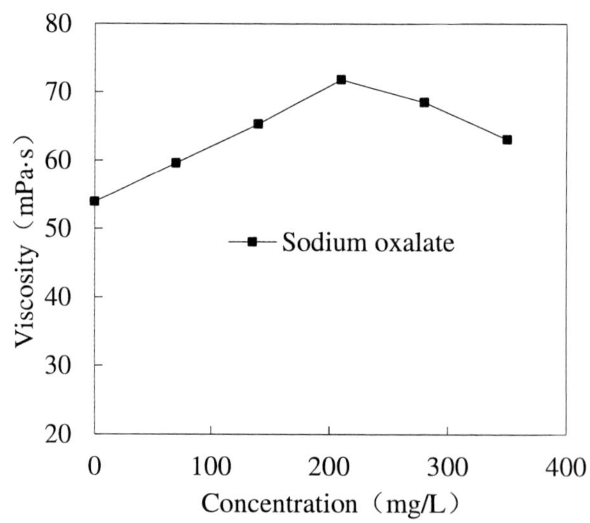

Figure 3. Effect of sodium oxalate concentration on viscosity of HPAM solution.

As can be seen in figure 3, with the increase of $Na_2C_2O_4$ concentration, the viscosity of the partially hydrolyzed polyacrylamide solution increases first and then decreases. $Na_2C_2O_4$ is a kind of metal complex agent for multi-coordination atoms, so it can form stable complexes with high-valence ions such as Ca^{2+} and Mg^{2+} in the polymer solution, and it can effectively improve the solution viscosity when the amount of $Na_2C_2O_4$ is small. When 210 mg/L $Na_2C_2O_4$ was added into the HPAM solution, the viscosity of the polymer solution reached a peak value of 71.8 mPa·s, and the viscosity increase rate reached 33.2%. Further increase of $Na_2C_2O_4$ content resulted in a large increase of Na^+ in the solution and a decrease in the viscosity.

5. Conclusions

As the concentration of Na^+ and K^+ increases, the viscosity of the polymer solution decreases rapidly. However, when the concentration of Na^+ and K^+ is >1000 mg/L, the viscosity decreases gently. The concentration of Mg^{2+} and Ca^{2+} has a great influence on the viscosity of polymer solution. In the low concentration range, a small increase in ion concentration can cause big drop in viscosity of HPAM solution. The selected $Na_2C_2O_4$ as Mg^{2+}, Ca^{2+} complex agent can effectively increase the initial viscosity of HPAM. When 210 mg/L $Na_2C_2O_4$ was added to the HPAM solution, the viscosity of the polymer solution reached a peak value of 71.8 mPa•s, and the viscosity increase rate reached 33.2%.

References

[1] Wang D X 2001 Petroleum Science **4** (Beijing: China University of Petroleum) p 22
[2] Kulicke W M, Kniewske R and Klein J 1982 *Preparation, Progress in Polymer Science* **8** (Pergamon-Elsevier Science) p 373
[3] Zhang J G 2005 *Fault-block Oil & Gas Field* **12** (Zhengzhou: Fault Block Oil and Gas Field Magazine) p 57
[4] Zhao X T ,Wang Z B and Qiu G M 2009 *Chemical Engineering of Oil & Gas* **3** (Beijing Research Institute of Chemical Industry) p 231
[5] Ding Y 2013 *Petrochemical Industry* 42 (Beijing: China Petrochemical Corp) p 973
[6] Chen X R 2009 *Petrochemical Industry* 38 (Beijing: China Petrochemical Corp) p 1132
[7] Wang Y P 2003 *Progress in Chemical Industry* **22** (Beijing: China Association for Science and Technology) p 271

Synthesis of diamond-like hydrocarbon films by atmospheric pressure filamentary dielectric barrier discharge

T Suzuki[1,2] and A Shirakura[1]

[1] Graduate School of Science and Technology, Keio University, 3-14-1 Hiyoshi, Yokohama 223-8522, Japan

E-mail: tsuzuki@mech.keio.ac.jp

Abstract. Diamond-like hydrocarbon (DLHC, a-C:H) films were synthesized under atmospheric pressure filamentary by dielectric discharge (FDBD) to improve their mechanical properties compared to the films synthesized by glow dielectric barrier discharge (GDBD). GDBD is generally used for atmospheric pressure-plasma-enhanced chemical vapour deposition. The discharge form transitioned from GDBD to FDBD by increasing the gap between electrodes from 1 to 4 mm. The hardness of the films increased from 3.7 to 11.9 GPa by using FDBD. The results indicate that hard DLHC films can be synthesized at room temperature and on a large area under atmospheric pressure using FDBD.

1. Introduction

Amorphous carbon films were widely employed in various fields of industrial applications due to their excellent chemical and mechanical properties; high hardness, high abrasion resistance, gas barrier properties [1-2]. Among them, hydrogenated amorphous carbons (a-C:H) are called Diamond-like hydrocarbons (DLHCs), and synthesized by the easily accessible deposition processes such as plasma enhanced chemical vapor deposition (PECVD) at vacuum condition [3]. However, in case of various industrial applications, the vacuum process has several drawbacks, such as high costs and limited coating area associated with vacuum equipment. Generating low temperature plasma under atmospheric pressure (AP) condition is one of the solutions to solve the above problems. The stabilization of atmospheric pressure glow plasma using dielectric barrier discharge (DBD) was reported in 1990 [4-5]. AP-PECVD using DBD has been actively studied for synthesizing low-cost, highly-functional films [6–8]. DBD is classified into glow DBD (GDBD), glow-like DBD, Townsend DBD, and filamentary DBD (FDBD) [9]. The GDBD is conventionally used for the AP-PECVD process. The plasma generated by GDBD uniformly covers the surface and synthesizes smooth films while inhibiting powder formation unlike those generated by other discharges. However, the ion density and the electron density in GDBD are not sufficient to deposit hard DLHC films compared to low-pressure CVD (LPCVD), and only soft DLHC films with higher hydrogen content have been synthesized by GDBD [10-12]. In contrast to GDBD, FDBD consists of many filamentary discharges (streamers) that have significantly higher ion density and electron density. However, the synthesis process using FDBD has several problems; the discharge with high plasma density can damage the substrate, and deposit films with non-uniformity due to a state where the streamers are retained at the same points. In this work, we present the results on synthesis of hard DLHC films at room temperature using FDBD. To generate FDBD with a large number of streamers between the electrodes of DBD equipment, we employed substrates with various levels of surface resistivity and investigated their effects on the film synthesis. The DLHC films were produced using FDBD with various electrodes gaps, and we evaluated the characteristics of the films [13].

2. Materials and Methods

A schematic diagram of the AP-PECVD apparatus used in this study is shown in Figure 1. The plasma was maintained between the parallel copper electrodes: the upper electrode, 100 × 10 mm^2 in size, was connected to a high-frequency pulsed power supply (SPD1PGU1, SK Medical Electronics Co., Ltd., Japan) and overlaid with a alumina dielectric plate (1mm in thickness), and the lower electrode, 200 × 300 mm2 in size, was grounded. Boron-doped single crystal Si (100) wafers (KOUJUNDO CHEMICAL LABORATORY CO., LTD) with a thickness of 380 μm were used as substrates to measure thickness, chemical bonding structure, and hardness. The boron-doped Si substrates were cut into small pieces of 30 mm length × 70 mm width. The surface resistivity of the Si substrates was 47.2 Ω /sq.. as measured by a four-point probe apparatus. To evaluate the effect of the surface conductivity, 50 μm thick polyethylene terephthalate (PET) films coated with transparent conductive polymer using poly (3,4-ethylenedioxythiophene) (PEDOT) were used. The surface resistivity of the un-coated PET film was 109 Ω /sq.., and those of the conductive polymer coated PET films were 150, 200 and 530 Ω /sq..

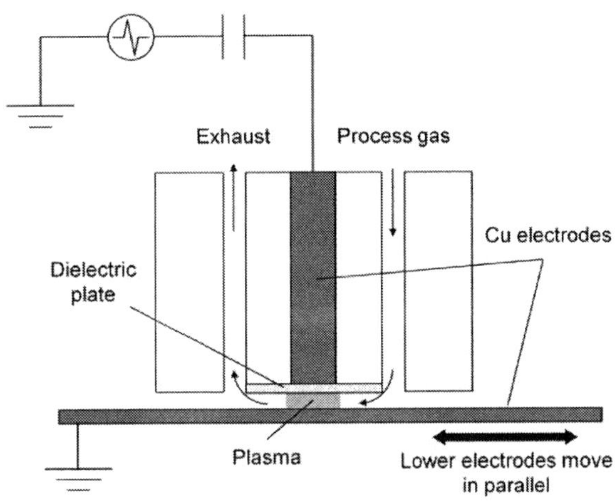

Figure 1. Schematic presentation of the AP-PECVD apparatus.

The substrates were set on the lower electrode moving back and forth for 30 mm range in the direction of plasma length (10 mm) at speeds of 0.1 mm/s. The gap between upper and lower electrodes was set between 1 mm and 4 mm. According to Meek [14], the larger the electrodes gap, the more likely it is to generate streamer discharges because the electron avalanche generated rapidly. The DLHC films were synthesized from CH_4/He/Ar mixture. The gas mixture was introduced through a slotted gas outlet located at the center axis of the upper electrode. He has high ionization energy and decomposes CH_4 efficiently by Penning ionization. Whereas, the diffusion coefficient of Ar is lower than He, hence Ar can generate streamers easily. The flow rates of He, Ar, CH_4 were set at 4 l/min, 1 l/min and 200 ml/min, respectively. The input power was 90 W and the pulse width and the peak negative voltage were set at 5 μs and 7 kV respectively.

The thickness of the DLHC films was measured by a contact-type surface profiler (Dektak 3030, Veeco Instruments Inc., USA). The cross sections of the films were observed by scanning electron microscopy (SEM: S-4700, Hitachi High Technologies Corp., Japan). The hardness of the films was studied using a Nanoindentation Tester (Agilent Nano Indenter G200, Agilent Technologies Inc., USA) with a Berkovich diamond indenter using the Oliver and Pharr method [15]. The structure of the films was studied by Raman spectroscopy (STR300, Seki Technotron Corp., Japan).

IOP Conf. Series: Materials Science and Engineering **369** (2018) 012046 doi:10.1088/1757-899X/369/1/012046

3. Results and Discussion

Figure 2 shows the discharge images viewed from the horizontal direction at various electrodes gaps without substrates. The forms of discharge were transitioned from GDBD to FDBD with an increase of the gap between the electrodes. The glow discharge was clearly confirmed at an electrodes gap of 1 mm, and at 2 mm gap, the discharge form was a mixture of FDBD and GDBD. The discharge form completely shifted to FDBD with a large number of streamers, when the gap was at 3 mm and 4 mm. The results showed that FDBD can be easily generated by increasing the electrodes gap using CH_4/He/Ar mixed gas using the bare lower electrode.

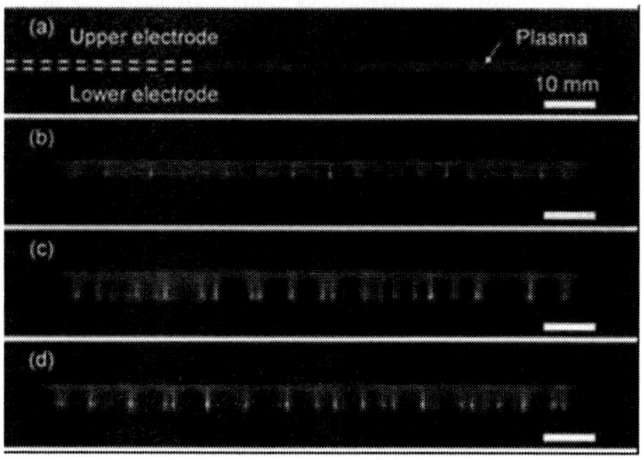

Figure 2. Discharge images without substrates (bare lower electrode). Electrodes gap: (a) 1 mm, (b) 2 mm, (c) 3 mm, (d) 4 mm.

Figure 3 shows the discharge images with the substrates with different surface resistivities. When the surface resistivity increased from 200 to 530 Ω /sq., the discharge form shifted to GDBD due to a decrease in the electric field.

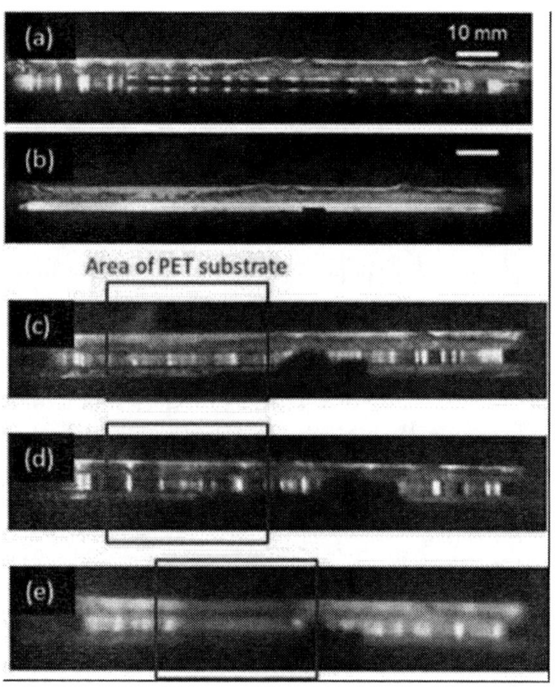

(a) B-doped Si (47.2 Ω/sq.) (FDBD);

(b) PET50 μm (10^9 Ω/sq.) (GDBD)

(c) Conductive polymer-coated PET (150 Ω/sq.) (FDBD);

(d) Conductive polymer-coated PET (200 Ω/sq.) (FDBD);

(e) Conductive polymer-coated PET (530 Ω/sq.) (GDBD).

Figure 3. Discharge images at the electrodes gaps of 4 mm with substrates.

The deposition rate of the DLHC films as a function of the electrodes gap is shown in Figure 4. As the electrodes gap increased from 1 mm to 4 mm, the deposition rate increased from 1.1 nm/s to 1.9 nm/s.

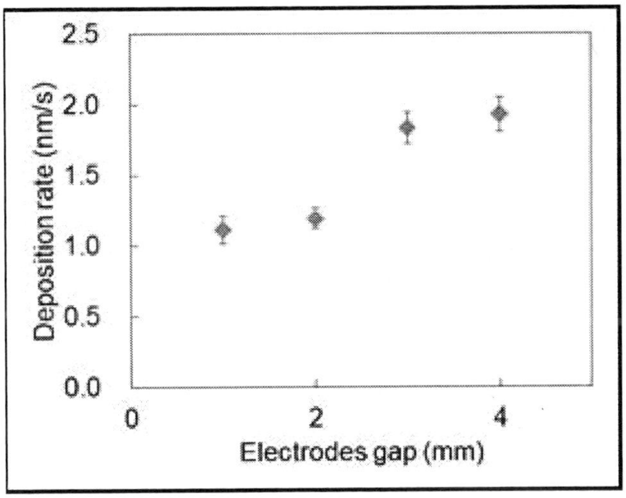

Figure 4. Relationship between deposition rate and electrodes gap of DLHC films.

Figure 5 shows SEM micrographs of the cross-section of the DLHC films synthesized with 1 mm and 4 mm of electrode gaps. The thicknesses of the DLHC films were coordinated to ca. 500 nm by adjusting the time of deposition. A smooth surface without the powder formation was confirmed on the DLHC films synthesized using both GDBD and FDBD. The thickness of the film synthesized using FDBD was about 540 nm ± 20 nm at all points, indicating DLHC films were uniformity produced. This is because the streamers shift from a higher resistivity point to a lower resistivity point on the surface in accord with the deposition of insulating DLHC.

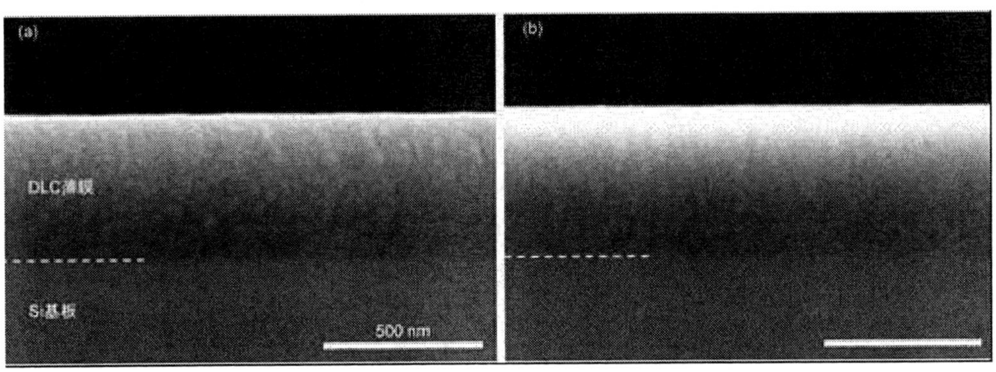

Figure 5. SEM images of cross-section of DLHC films.

Electrodes gap: (a) 1 mm; (b) 4 mm.

The hardness of the DLHC films as a function of the electrodes gaps is shown in Figure 6. When the electrodes gap increased from 1 mm to 3 mm the hardness of the films increased from 3.7 GPa to 11.9 GPa. The hardness became constant when the electrodes gap was increased to 4 mm. These results indicate that the DLHC films deposited by FDBD are harder than those deposited by GDBD.

The hardness of DLHC films using GDBD is reported to be achieved at 3 GPa[16-17]. The Raman spectra (not shown) of the DLHC films with various electrodes gaps showed that higher background intensity of the photoluminescence due to the polymeric structure with high content of hydrogen for the electrodes gaps of 1mm and 2mm. In contrast, when the electrodes gap increased to 3 or 4mm, the background intensity was greatly reduced that indicated high energy of FDBD make hydrogen abstraction from C-H bond.

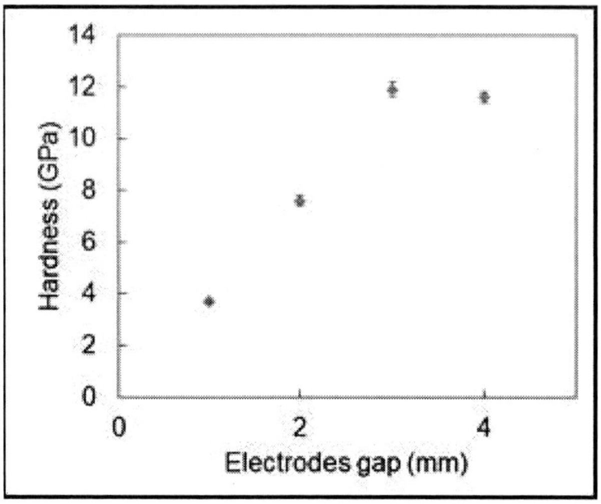

Figure 6. Relationship between hardness of the DLHC films and electrodes gap.

4. Conclusions

Hard DLHC films were synthesized at room temperature using FDBD with various electrodes gaps under atmospheric pressure. The form of discharge was transitioned from GDBD to FDBD with an increase of the electrodes gap to 3 mm or over when the surface resistivity of substrates was lower than 530 Ω/sq.. and mixture of He and Ar were used as a dilution gas. The hardness of the DLHC films increased from 3.7 GPa to 11.9 GPa with increasing the electrodes gap from 1 mm to 3 mm.

References

[1] Lifshitz Y 1999 *Diamond Relat. Mater.* **8** 1659-1676
[2] Abbas G A, Roy S S, Papakonstantinou P and McLaughlin J A 2005 *Carbon* **43** 303-9
[3] Wu Y, Sugimura H, Inoue Y and Takai O 2003 *Thin Solid Films* **435** 161-4
[4] Yokoyama T, Kogoma M, Kanazawa S, Moriwaki T and Okazaki S 1990 *J. Phys.* **D 23** 374-7
[5] Yokoyama T, Kogoma M, Moriwaki T and Okazaki S 1990 *J. Phys.* **D 23** 1125-8
[6] Trunec D, Navratil Z, Stahel P, Zajickova L, Bursikova V and Cech J 2004 *J. Phys.* **D 37** 2112-20
[7] Massines F, Gherardi N, Fornelli A and Martin S 2005 *Surf. Coatings Technol.* **200** 1851-55
[8] Sawada Y, Ogawa S and Kogoma M 1995 *J. Phys.* **D 28** 1661-9
[9] Massines F, Sarra-Bournet C, Fanelli F, Naude N and Gherardi N 2012 *Plasma Process. Polym.* **9** 1041-73
[10] Kugimiya T, Kannaka M, Yokomizo M, Nakaue A and Takamatsu H 2006 *J. Electrochem. Soc.* **153** C282-8
[11] Fanelli F, Fracassi F and d'Agostino R 2005 *Plasma Process. Polym.* **2** 688-94
[12] Noborisaka M, Horikoshi R, Nagashima S, Shirakura A and Suzuki T 2013 *Thin Solid Films* **527** 114-99

[13] Mori T, Futagami Y, Kishimoto E, Shirakura A and Suzuki T 2015 *J. Vac. Sci. Technol.* **A 33** 060607 1-4
[14] Meek J M and Craggs J D 1953 *Electrical Breakdown of Gases* (Clarendon Press, Oxford)
[15] Oliver W C and Pharr G M 1992 *J. Mater. Res.* **7** 1564-83
[16] Liu D, Yu S, Ma T, Song Z and Yang X 2000 Jpn. *J. Appl. Phys.* Part 1 **39** 3359-60
[17] Sakata T, Kodama H, Hayashi H, Shimo T and Suzuki T 2010 *Surf. Coat. Technol.* **205** S414-7

Institute of Physics
Dirac House, Temple Back
Bristol BS1 6BE UK

ISSN: 1757-8981
ISBN 978-1-5108-6484-9

CURRAN ASSOCIATES INC.
proceedings.com